THE STRUCTURE OF APPLIED GENERAL EQUILIBRIUM MODELS

THE STRUCTURE OF APPLIED GENERAL EQUILIBRIUM MODELS

Victor Ginsburgh
and
Michiel Keyzer

The MIT Press
Cambridge, Massachusetts
London, England

First MIT Press paperback edition, 2002

This book was set in Times Roman by Doyle Graphics, Tullamore, Ireland.

Printed and bound in the United States of America.

Library of Congress Cataloging-in-Publication Data

Ginsburgh, Victor.
The structure of applied general equilibrium models / Victor Ginsburgh and Michiel Keyzer.
p. cm.
Includes bibliographical references and index.
ISBN 0-262-07179-7 (hardcover : alk. paper), 0-262-57157-9 (pb)
1. Equilibrium (Economics)—Mathematical models. 2. Competition—Mathematical models. I. Keyzer, Michiel. II. Title.
HB145.G563 1997
330′.01′5195—dc21 96-47835
CIP

Contents

Introduction xiii
Acknowledgments xvii
Notation xix

1 **Competitive Equilibrium** 1

1.1 Basic Concepts 1
1.1.1 Commodities and Agents; Demands and Supplies 2
1.1.2 The Behavior of Producers and Consumers 3
1.1.3 General Competitive Equilibrium 4
1.1.4 Characteristics of a General Competitive Equilibrium 4
1.2 Excess Demand Equilibrium 4
1.2.1 Assumptions on Excess Demand 4
1.2.2 An Existence Proof 5
1.2.3 Multiplicity of Equilibrium 10
1.2.4 Lack of General Properties of Excess Demand 13
1.2.5 Computation of Equilibria 14
1.3 Competitive Equilibrium 15
1.4 Efficiency and Equity 18
1.4.1 Pareto-Efficiency 18
1.4.2 The First Welfare Theorem 19
1.4.3 The Welfare Optimum 20
1.4.4 The Second Welfare Theorem 24
1.4.5 The Single-Consumer Case 25
1.4.6 Welfare Consequences of Reforms 26
1.5 Possible Extensions of the Competitive Model 29

2 **Optimal Production and Consumption Allocations** 35

2.1 Producer Behavior 35
2.1.1 The Producer 35
2.1.2 Technology of the Firm 36
2.1.3 Profit Maximization 41
2.1.4 Producer Behavior in Applied Models 55
2.2 Consumer Behavior 59
2.2.1 The Consumer 60
2.2.2 Assumptions on Consumer Behavior 61
2.2.3 Utility Maximization 65
2.2.4 Recovering Consumer Behavior via Integrability or via Revealed Preference 71
2.2.5 Consumer Behavior in Applied Models 74
2.3 Supply and Demand under the Standard Assumptions 81

2.4 Welfare 84
2.4.1 The Welfare Optimum 84
2.4.2 Welfare Measures 87

3 **Applied General Equilibrium** 91
3.1 The Negishi Format 92
3.1.1 Existence of Equilibrium: Negishi's Theorem 92
3.1.2 Multiplicity of Equilibrium 96
3.2 The Full-Format 97
3.3 The Open Economy Format 101
3.3.1 Existence of Equilibrium 101
3.3.2 Extensions of the Open Economy Model 106
3.4 The CGE Format 107
3.4.1 The Basic CGE Model and the CGE Format 108
3.4.2 Extensions of the Basic CGE Model 110
3.5 Comparing the Formats 112
3.6 Model Implementation 115
3.6.1 Empirical Elaboration 116
3.6.2 The Treatment of Missing Commodities and Markets 122
3.6.3 Simulations 127
3.6.4 A GAMS Application of the Basic CGE Model 129

4 **International Trade** 137
4.1 Reinterpreting the Competitive Equilibrium Model as a World Model 137
4.2 International Trade in the General Equilibrium Model 138
4.2.1 The Multi-country Case 138
4.2.2 The Two-Country Case and the Small-Country Assumption 142
4.3 Welfare Implications 143
4.3.1 The International Competitive Equilibrium 143
4.3.2 The Competitive Equilibrium under the Small-Country Assumption 143
4.4 Practices in Applied Modeling 144
4.4.1 International Trade in the CGE Model 145
4.4.2 Issues in Model Specification 147
4.4.3 Empirical Elaboration: The SAM for a Country Model with International Trade 153
4.4.4 Simulations 155

5 **Taxes, Tariffs, and Quotas** 157
5.1 Government and Taxation 157
5.1.1 Government 157

5.1.2 A Short Typology of Taxes 157
5.1.3 Taxes May Affect Allocations 161
5.2 Taxes and Quotas in General Equilibrium 162
5.2.1 Indirect Taxes on Consumption 163
5.2.2 Indirect Taxes on Production and Intermediate Demand 168
5.2.3 Tariffs and Quotas in International Trade 170
5.2.4 Government Activity: A Summary 173
5.3 Welfare Implications 173
5.3.1 Indirect Taxes on Consumption in the Closed Economy 174
5.3.2 Indirect Taxes on Production and Intermediate Demand in the Closed Economy 178
5.3.3 Tariffs and Quotas in the International Trade Model 178
5.3.4 Indirect Taxes and Tariffs in the Small Open Economy 179
5.3.5 Welfare Analysis in the Absence of Compensation 181
5.3.6 Consequences of the Propositions for Policy Design 182
5.4 Practices in Applied Modeling 183
5.4.1 Taxes and Tariffs in the CGE Model 183
5.4.2 Issues in Model Implementation 185
5.4.3 Empirical Elaboration: The SAM with Taxes and Tariffs 187
5.4.4 Simulations 188

6 Price Rigidities 193
6.1 Price Rigidities and Supporting Mechanisms 193
6.2 Price Rigidities in General Equilibrium 194
6.2.1 Centrally Held Stocks 195
6.2.2 Shifting the Adjustment to Consumers 198
6.2.3 Drèze Rationing 202
6.2.4 Benassy Rationing: Fix-Price Exchange Equilibria 208
6.2.5 Virtual Tax Rationing 211
6.2.6 Soft Rationing 212
6.2.7 Shifting the Adjustments Also to Producers 213
6.3 Welfare Implications 215
6.3.1 Centrally Held Stocks 215
6.3.2 Consumer (and Producer) Rationing 216
6.4 Practices in Applied Modeling 218
6.4.1 Fixed Prices in the CGE Model 218
6.4.2 Issues in Model Implementation 223
6.4.3 Empirical Elaboration 225
6.4.4 Simulations 226

7 Finite-Horizon Dynamics 229

7.1 Elements of Dynamics 229

7.2 Finite-Horizon Dynamics 230

7.2.1 T-Period Competitive Equilibrium 230

7.2.2 Temporary Equilibrium 240

7.2.3 Single-Period Equilibrium 247

7.2.4 Dynamics in the Presence of International Trade, Taxes, and Price Rigidities 249

7.3 Welfare Implications 252

7.3.1 Intertemporal Pareto-Efficiency 252

7.3.2 Static Pareto-Efficiency 253

7.4 Practices in Applied Modeling 253

7.4.1 Dynamics in the CGE Model 255

7.4.2 Issues in Model Implementation 257

7.4.3 Empirical Implementation 263

7.4.4 Simulations 266

8 Infinite-Horizon Dynamics 267

8.1 Dynastic Models 267

8.1.1 Specification of the Dynastic Negishi Program 267

8.1.2 Negishi Equilibrium: Fixed Discount Factor β 270

8.1.3 Existence and Efficiency of Equilibrium with Recursive Utility Functions 272

8.1.4 Multiplicity and Indeterminacy of Equilibrium 278

8.1.5 Dynamics and Steady States 279

8.1.6 Finite-Horizon Approximations 283

8.1.7 Real Business Cycles 285

8.2 Overlapping Generations Models of Pure Exchange 291

8.2.1 Alternative OLG Models 291

8.2.2 Specification of the Model with Financial Claims and Existence of Equilibrium 294

8.2.3 Inefficiency of Equilibrium 304

8.2.4 Indeterminacy of Equilibrium 307

8.2.5 Dynamics and Steady States 307

8.2.6 Finite-Horizon Approximations 309

8.3 Overlapping Generations Models with Production 310

8.3.1 Specification of the Model 310

8.3.2 Existence of Equilibrium 312

8.3.3 Inefficiency of Equilibrium 316

8.3.4 Indeterminacy of Equilibrium 317

8.3.5 Dynamics and Steady States 317

8.3.6 Finite-Horizon Approximations 318

8.4 Practices in Applied Modeling 319
8.4.1 Issues in Model Implementation 319
8.4.2 Simulations 322

9 Externalities 325
9.1 Positive Externalities 325
9.1.1 Nonrival Commodities in Consumption 325
9.1.2 Interdependent Utilities: Empathy 332
9.1.3 Interdependent Consumptions 336
9.1.4 Interdependence in Production 339
9.1.5 Endowments Dependent on Consumption: Efficiency Wages 341
9.2 Pollution and Negative Externalities 345
9.2.1 Pollution in the Competitive Model 345
9.2.2 The Treatment of Externalities 347
9.3 Perspectives for Implementation in Applied Models 347

10 Nonconvexities 351
10.1 Within-Firm Nonconvexities 352
10.2 Indivisibilities at Firm Level 355
10.2.1 Decentralization of a Welfare Optimum: The General Case 357
10.2.2 Setup Costs: Decentralization of a Welfare Optimum 359
10.2.3 Setup Costs: Equilibrium with Given δ_j Zero Transfers among Consumers 361
10.2.4 Increasing Returns to Scale: Decentralization of a Welfare Optimum 363
10.2.5 Increasing Returns: Equilibrium with Rules to Determine q_j and Zero Transfers among Consumers 365
10.2.6 Increasing Returns: Equilibrium for Concave-Convex Cost Functions 370
10.2.7 Solving Model with Indivisibilities 371
10.3 Nonconvexities at Above-Firm Level 372
10.3.1 Nonrival Goods Produced Intentionally: Research and Development 372
10.3.2 Nonrival Goods Produced as By-products: Economies of Scope and Learning by Doing 374
10.4 Nonconvexities in Consumption 376
10.4.1 Nonconvexities in the Utility Function and Indivisible Commodities 376
10.4.2 Nonconvexity and Efficiency Wages 379
10.5 CGE Applications 384

11 Imperfect Competition 387

11.1 The Main Issues in Partial Equilibrium 388

11.1.1 An Example: Cournot Duopoly 388

11.1.2 Main Issues 390

11.1.3 The Stages of Anticipation and Market Clearing 391

11.2 The Main Issues in General Equilibrium 393

11.2.1 An Example with Two Strategic Consumers 393

11.2.2 A Representation That Avoids Inverse Demand Functions 396

11.2.3 A First Sketch of Alternative Specifications for the Anticipation Stage 397

11.2.4 Introducing Producers 398

11.2.5 Producers as Noncompetitive Agents 400

11.2.6 The Normalization Issue 402

11.2.7 Free Entry 403

11.2.8 Pareto-Efficiency 403

11.3 Alternative Specifications for the Anticipation and the Market-Clearing Stages 404

11.3.1 The Anticipation Stage 404

11.3.2 The Market-Clearing Stage 406

11.4 Simplified Representations of the Anticipation Stage 408

11.4.1 Negishi's Subjective Demand Approach 408

11.4.2 Generalizing Negishi's Approximation 409

11.4.3 A Markup Representation 411

11.4.4 Objective Markups 414

11.5 Practices in Applied Modeling 416

11.5.1 Issues 417

11.5.2 Applications 420

11.5.3 Possible Extensions and Directions for Future Research 422

12 Money and Incomplete Asset Markets 427

12.1 Cash-in-Advance 427

12.1.1 Flexible Supply of Money 428

12.1.2 Scarcity of the Medium of Exchange: General Equilibrium with Cash-in-Advance 430

12.1.3 Welfare Properties of the Equilibrium 435

12.2 Incomplete Asset Markets 436

12.2.1 An Example of Restricted Access: The Case of Commodity Markets 437

12.2.2 Concepts and Assumptions 438
12.2.3 Issues and Approaches Related to the Existence Proof 441
12.2.4 Approximate Equilibrium with Financial Assets: An Existence Proof 444
12.2.5 Properties of the Equilibrium: Pricing of Assets, Nonarbitrage and Efficiency 452
12.2.6 Introducing Firms 455
12.2.7 Returns in State *s* Depend on Prices in State *s* Only 457
12.3 Practices in Applied Modeling 458
12.3.1 Determination of the Absolute Price Level: Nonhomogeneities 459
12.3.2 Exchange Rates 460
12.3.3 Transactions Demand for Money 460
12.3.4 Financial Assets 461
12.3.5 Models with Nonhomogeneities, Transactions Demand for Money, and Financial Assets 462

A **Mathematical Background** 465
A.1 Sets and Functions 465
A.2 The Convex Program 470
A.3 The Parametric Convex Program 472
A.4 The Fixed-Point Mapping 476
A.5 The Nonconvex Program 478
A.6 Infinite Dimensions 479
A.7 Multiplicity of Solutions of $F(x) = 0$ 479
A.8 Solving a System of Equations $F(x) = 0$ 483
A.9 Solving a Nonlinear Complementarity Problem 488

B **A GAMS Application** 493
B.1 Defining the Model in GAMS 493
B.2 Computation of Equilibrium 509
B.3 Accounting 510
B.4 Report Writing 514
B.5 Running the GAMS Job 519
B.6 Listing of Results 521

C **Assumptions** 529

References 535
Index 551

Introduction

This book describes the structure of general equilibrium models. It is written for the researcher who intends to construct or study applied general equilibrium (AGE) models and has a special interest in their theoretical background. Both general equilibrium theory and AGE modeling continue to be active fields of research, but the styles of presentation differ greatly. Whereas the applied model builder often finds the style of theoretical papers inaccessible, the theoretician can hardly recognize the concepts he is used to in the list of equations of applied models. The main purpose of the book is to present the theoretical models in a unified way and to indicate how the main concepts can find their way into applications.

To make the models more accessible and their structure more transparent, we unify the presentation in four ways. First, we use standardized, though possibly not the weakest, assumptions on basic model components like utility functions and production sets (chapters 1 and 2), and we only deviate from the standard when the topic requires it. Second, in chapter 3 we define five basic formats to represent and to analyze the same model. For every specific model to be discussed in chapters 4 to 12, we apply the format that proves most convenient. Third, beside fixed point theorems (appendix A.4), we almost exclusively use theorems from the theory of convex programming as mathematical background (appendixes A.1 to A.3). Finally, every model in chapters 4 to 12 analyzes a topic within a common basic scheme, as follows:

1. The chapter starts introducing the topic (e.g., taxes or finite horizon dynamics) and then proceeds with the discussion of the issues related to the topic itself as well as its incorporation in a general equilibrium model. This may require nonstandard assumptions (e.g., relaxing convexity).

2. Existence proofs are given, using the most convenient format defined in chapter 3. The proofs are not meant as a theoretical contribution and mainly serve to highlight the roles of the various assumptions and to indicate how the fixed point mapping can be set up for computing a solution via fixed point algorithms. Issues in computation are briefly dealt with in appendixes A.7 to A.9.

3. Various properties of the equilibrium solution of the model are analyzed, with a focus on efficiency. We derive conditions under which the inefficiencies due to specific imperfections (taxes, price rigidities, external effects) can be reduced through Pareto-improving reforms. However, these reforms only represent an idealized situation, since they require losers to be compensated and all imperfections to be reduced simultaneously.

4. Policy reforms can rarely eliminate or even reduce all imperfections at the same time. Therefore their consequences cannot be predicted from theory alone, and numerical simulation is called for. This requires construct-

ing applied models that can be used to run alternative scenarios. Every chapter ends with a section that describes how applied models that incorporate the theoretical concepts can be built and briefly surveys existing applications.

The following limitations should be mentioned:

1. For clarity of exposition, the topics will often be treated in isolation. For example, we limit the presentation of finite and infinite horizon dynamics to the case of the closed economy. Occasionally some issues need to be treated together (e.g., taxes, tariffs, and foreign trade in chapter 5).
2. In chapter 2 we compare several analytical forms (CES, translog, etc.) for the functions that should be estimated econometrically, but in the sections on applications, we do not enter into the details of these specifications.
3. We sketch the framework for national and social accounting that is needed as a database (the social accounting matrix, SAM) to calibrate a model, but we hardly discuss the elaborate process that leads from statistical publications to such a database.
4. Computation of equilibria has received much attention in the early years of AGE modeling. Advances in computing capacity and speed as well as the availability of user friendly packages, such as GAMS, have almost eliminated computational concerns for most applications. Therefore computational issues are only treated briefly in appendixes.
5. The numerical solutions of AGE models should be made easily interpretable for the policy analyst who will be reluctant to decipher the standard printings from software packages. The presentation should thus be customized, and the results should be cast in the form of tables the analyst is familiar with. Deriving such tables for every model is hardly meaningful without a numerical illustration, and giving this in every chapter would cause excessive duplication.

We overcome some of these limitations and avoid repetition by describing, in appendix B, a complete numerical application in GAMS language, incorporating taxes, trade, price rigidities, buffer stocks, transportation costs, and simple dynamics. The application uses simple functional forms and yields social and national accounts in report quality form.[1]

To summarize, chapters 1 to 3 and appendix A set the framework for the topics covered in chapters 4 to 12. Chapter 1 provides an elementary but almost comprehensive treatment of the competitive model. It ends, in section 1.5, with an overview of the properties that could be relaxed so as to make the model more useful for policy analysis and introduces the subjects covered in chapters 4 to 12. Chapter 2 deals with relatively standard elements of the theory of producer and consumer behavior, most of which can be found in textbooks such as Varian's *Microeconomic Analysis.* The reader who is familiar with this material can proceed to section 2.3, which summarizes the microeconomic assumptions that are used in later chapters. Section 2.4 presents results on welfare analysis that are needed in proofs. Chapter 3 describes the basic formats and introduces the methods of proof used in the book. Its last section discusses the main steps in constructing a numerical application and gives a full GAMS application for the simplest possible model. The reader may choose freely the order in which he wants to study the topics covered in chapters 4 to 12. Occasionally he may have to consult material from earlier chapters and from appendix A, and this will be pointed out explicitly when needed.

A user's guide and library of GAMS-models (Keyzer 1997) accompany this volume. Both are freely accessible at the MIT Press Internet site (http://mitpress.mit.edu). The guide extends the material covered in appendix B. The library contains a set of computer programs with illustrative applications for most of the models covered in chapters 3 to 12 of this book, except for the infinite horizon models of chapter 8.

Acknowledgments

A book that covers so many subjects could not have been written without the support of many colleagues at the Center for Operations Research and Econometrics (CORE), in Louvain-la-Neuve, the Centre for World Food Studies (SOW-VU) in Amsterdam, and the Centre d'Économie Mathématique et d'Économétrie (CEME) in Brussels. We also want to express our gratitude for their careful reading to Manjira Datta (chapter 8), Renato Flores (chapters 1 to 6 and 11), Chris Elbers (chapter 3 and appendix A), Jean Gabszewicz (chapter 11), Marjorie Gassner (appendix A), Jan Willem Gunning (sections on applications), Gerard van der Laan (chapter 6 and section 12.2), Geert Overbosch (all chapters), Herakles Polemarchakis (section 12.2), Aldo Rustichini (chapter 8), T. N. Srinivasan (chapters 1 to 3), and Wim van Veen (chapters 1 to 3). Parts of the book were taught in graduate courses at Université d'Aix-Marseille, Université Libre de Bruxelles, Université de Paris II, and Vrije Universiteit Amsterdam, and the comments by many students led to several clarifications. We also thank three referees for their constructive comments.

Writing this book took several years and diverted us during many weeks from our other duties. We want to acknowledge the understanding of our colleagues, especially Wouter Tims, who strongly promoted this activity. Special thanks are also due to Jean Waelbroeck who, several years ago, introduced the first author to the field of general equilibrium and supervised the second author's dissertation on a general equilibrium model of international trade. Finally, we are grateful to Terry Vaughn from MIT Press for his patience and confidence.

Financial support was obtained from the Services Fédéraux des Affaires Scientifiques, Techniques et Culturelles under contract PAI no. 26 and the Fonds National de la Recherche Scientifique (FNRS) in Belgium, and the Centre for World Food Studies of the Vrije Universiteit Amsterdam (SOW-VU) in the Netherlands.

Notation

We denote the n-dimensional real space by R^n, the nonnegative orthant by R^n_+, and the positive orthant by R^n_{++}. Let x and y be two vectors in R^n. The following notations are equivalent:

Enumeration	all $h \Leftrightarrow h = 1, \ldots, n$
Summation	$\Sigma_h x_h \equiv \sum_{h=1}^{n} x_h$
If and only if	iff or $\Leftrightarrow$
Complementary slackness	$(x \geqslant 0 \perp y \geqslant 0) \Leftrightarrow (x \geqslant 0,\ xy = 0,\ y \geqslant 0)$
Inner product[1]	$xy \equiv x \cdot y \equiv \Sigma_h x_h y_h$
Inequality	$x \geqslant y \Leftrightarrow x_h \geqslant y_h$ for all h, and similarly for $\leqslant$; $x > y \Leftrightarrow x_h > y_h$ for all h, and similarly for $<$.
Partial derivative (Jacobian)	$F'(x) \equiv \partial F(x)/\partial x$
Fixed point or equilibrium	superscript *
Optimal value of a variable[2]	superscript $^\circ$
Vector norm[3]	$\|x\|_p \equiv (\Sigma_{h=1}^n \|x_h\|^p)^{1/p}$ for given $1 \leqslant p < \infty$; $\|x\|_\infty \equiv \max_{1 \leqslant h \leqslant n} \|x_h\|$, for the limiting case; $\|x\| \equiv \|x\|_p$ for some p to be specified.
Matrix norm	$\|A\| \equiv \sup_{\|x\|=1} \|Ax\|$

Notation in the Mathematical Program

The notation will be introduced by means of an example. Define the function $u: X \subset R^n \to R$ and the sets $Y_j \subset R^n$, $j = 1, \ldots, J$, and consider the mathematical program:

$\max u(x),$

$x \geqslant 0,\ y_j$, all j,

subject to

$x \leqslant \Sigma_j y_j \qquad (p),$

$y_j \in Y_j.$

For simplicity we write "max" instead of "sup" because we only consider programs for which the maximum will be attained. The choice variables x and y_j are placed under the maximand. Here x is constrained to be

1. When no confusion is possible, we avoid using the transposition sign for inner products and for premultiplication of a matrix by a vector.

2. When no distinction with equilibrium values is needed, the superscript * is also used.

3. More general definitions can be used. Here we restrict ourselves to l_p-norms.

nonnegative while y_j is unconstrained. We write "all j" as equivalent for "$j = 1, \ldots, J$"; "all j" is assumed to apply for all constraints that use the subscript without summation sign (here it applies to the second but not to the first constraint). The brackets (p) denote a vector of Lagrange multipliers associated with the constraint $x \leqslant \Sigma_j y_j$ (the vector p is a member of the set of multipliers).

We also use a more compact notation; for example, for the utility maximization of the consumer, we write

$$V(p, h) = \max_{x \geqslant 0}\{u(x) \mid px \leqslant h\}$$

or

$$x(p, h) = \operatorname{argmax}_{x \geqslant 0}\{u(x) \mid px \leqslant h\},$$

where p and h are the given price and income, respectively, $V(p, h)$ is the value function, and $x(p, h)$ is the unique optimal choice. If the solution is not unique, we consider one optimal choice x° and write

$$x^\circ \in \operatorname{argmax}_{x \geqslant 0}\{u(x) \mid px \leqslant h\}.$$

1 Competitive Equilibrium

In this chapter we describe an economy, say, of a village, a town, or a country, that does not entertain trading relations with an outside world. We call this a closed economy.

After introducing basic concepts in section 1.1, we prove existence of equilibrium in the simplest possible aggregate framework, without reference to individuals' behavior, and we discuss computation of equilibrium and properties like multiplicity and stability in section 1.2. In section 1.3 we impose a first set of assumptions on the behavior of producers and consumers and show that under these assumptions, a competitive equilibrium exists. Section 1.4 is devoted to the welfare properties of equilibrium allocations. Finally, section 1.5 points to the need for extensions that will yield an applied general equilibrium model with a more realistic structure.

1.1 Basic Concepts

1.1.1 Commodities and Agents; Demands and Supplies

Let us consider an economy with r commodities indexed by $k = 1, 2, \ldots, r$. The commodity space is thus an r-dimensional space, denoted by R^r, and all vectors belong to that space.

An orange available today in New York is more or less the same as an orange in Marseilles tomorrow, but each will be defined as a different commodity in a general equilibrium model, which distinguishes commodities by location and date of delivery. By definition, each commodity k will be traded at a single price p_k. Agents are characterized by preferences over commodities and by capabilities to satisfy these preferences through actions like production, purchases, sales, storage, and consumption, now or in the future. This situation may be formalized by representing each agent as maximizing his utility subject to technological and trading constraints. In the simplest case the budget constraint, which requires expenditure not to exceed revenue, is the only restriction on exchanges. Under these conditions and some specific assumptions to be discussed later on, we can decompose the decisions made by each agent into two subproblems:[1] profit maximization subject to technological constraints, and utility maximization subject to a budget constraint.

These subproblems enable us to consider two types of agents who make decisions: producers (or firms) and consumers. There will be n producers, indexed by $j = 1, 2, \ldots, n$ who will produce (and sell) commodities using (and buying) some other commodities, like labor, steel, and machines. Let $y_j(p)$ be the production plan of producer j, where p denotes the price vector; outputs will carry a positive sign and inputs a negative sign. There will be

1. See Koopmans (1957).

m consumers, indexed by $i = 1, 2, \ldots, m$. Every consumer offers for sale his commodity endowment ω_i and expresses the wish to buy a commodity bundle $x_i(p)$ at given prices p.

We define the excess demand vector $z(p)$ as

$$z(p) = \Sigma_i x_i(p) - \Sigma_j y_j(p) - \Sigma_i \omega_i.$$

The typical component $z_k(p)$ of this vector will represent the excess of demand over supply (which includes the initial endowment) of commodity k.

A natural solution concept is to require that no commodity be in excess demand; otherwise, some agents would not be able to carry out their demands. We accept, however, that there can be excess supply for some commodities, which can be disposed of freely (the free disposal assumption). An excess demand equilibrium is then defined as follows:

DEFINITION **1.1** (Excess demand equilibrium) The price $p^* \geqslant 0$, $p^* \neq 0$, and the excess demand $z(p^*)$ define an excess demand equilibrium if $z(p^*) \leqslant 0$.

1.1.2 The Behavior of Producers and Consumers

Our description of the agents' behavior starts with the assumption that prices exist for all commodities and that all agents take these prices as given, none of them being sufficiently "large" or "important" to think that he can influence a price, even less set a price. Prices are thus considered as signals on the basis of which agents compute their plans. Such an institutional setting is commonly referred to as perfect competition.

Production Plans

Each producer j is endowed with a technology, represented by a set Y_j, which belongs to R^r and is the set of feasible production plans. A producer formulates a production plan y_j that must be feasible: This is expressed as $y_j \in Y_j$. Obviously further assumptions that characterize the mathematical properties of the set Y_j are needed. The competitive model assumes that from the set of feasible plans y_j, the producer chooses those that maximize his profit, defined as $\Sigma_k p_k y_{jk}$ or py_j.

The problem of producer j can thus be stated as follows: Given the price vector p, and the technological set Y_j, producer j chooses y_j so as to maximize profits py_j subject to a feasibility constraint $y_j \in Y_j$, or

$$\Pi_j(p) = \max_{y_j}\{py_j \mid y_j \in Y_j\}, \qquad (1.1)$$

where $\Pi_j(p)$ is the resulting maximal profit.

Consumption Plans

The choice made by consumer i is restricted in two ways. First, his consumption plans need to be feasible: He cannot consume negative quantities of any commodity: $x_i \in R^r_+$. Second, he is faced with a budget constraint: He cannot spend more than his income h_i. Since at given prices p, a consumption plan x_i costs px_i, his budget constraint can be written as

$$px_i \leqslant h_i.$$

The income h_i of the consumer consists of two parts: The proceeds $p\omega_i$ of selling the endowment ω_i and distributed profits. The latter are defined as follows: It is assumed that consumer i owns a nonnegative share θ_{ij} in firm j and that he receives dividends $\theta_{ij}\Pi_j(p)$ from this firm. All profits are distributed so that $\Sigma_i\theta_{ij} = 1$ for every j. Consumer i's income is now

$$h_i = p\omega_i + \Sigma_j\theta_{ij}\Pi_j(p).$$

Consumer i is also characterized by a utility function $u_i(x_i)$, which associates to every consumption plan x_i a utility level $u_i(x_i)$; under further assumptions on $u_i(x_i)$, this makes it possible to consistently rank alternative consumption bundles.

The problem of consumer i can now be stated as follows: Given the price vector p and the revenue h_i, consumer i chooses x_i so as to maximize his utility $u_i(x_i)$ subject to a feasibility constraint $x_i \geqslant 0$ and to his budget constraint $px_i \leqslant h_i$, or

$$\max_{x_i \geqslant 0}\{u_i(x_i) \mid px_i \leqslant h_i\}. \tag{1.2}$$

1.1.3 General Competitive Equilibrium

The excess demand equilibrium of definition 1.1 is a general equilibrium because it covers all agents and all commodities of the economy. We can now define a general competitive equilibrium as an excess demand equilibrium, in which producers and consumers behave according to (1.1) and (1.2), respectively.

DEFINITION 1.2 (General competitive equilibrium) The allocation y_j^*, all j, x_i^*, all i, supported by the price vector $p^* \geqslant 0$, $p^* \neq 0$ is a general competitive equilibrium if the following conditions are satisfied:

1. For every producer j, y_j^* solves $\max_{y_j}\{p^*y_j \mid y_j \in Y_j\}$.
2. For every consumer i, x_i^* solves $\max_{x_i \geqslant 0}\{u_i(x_i) \mid p^*x_i \leqslant h_i^*\}$, where $h_i^* = p^*\omega_i + \Sigma_j\theta_{ij}p^*y_j^*$.
3. All markets are in equilibrium, $\Sigma_i x_i^* - \Sigma_j y_j^* - \Sigma_i\omega_i \leqslant 0$.

In definition 1.2 there are four components. First, agents have behavioral rules that they follow to compute their optimal decisions. Second, in doing so, they take into account signals, without trying to affect these: Producers react on prices only, consumers take into account prices and their income. Third, there is a price for every commodity, so a competitive market exists for every commodity. Fourth, there are conditions on excess demands, which agents do not take into account when making their decisions and which are satisfied in equilibrium.

1.1.4 Characteristics of a General Competitive Equilibrium

So far, endowments ω_i and shares in profits θ_{ij} are the only explicit parameters. In applied models other parameters like tax rates or institutional rigidities will appear, and the model will be used to compute solutions when variations are imposed on some of these parameters. When analyzing the response of the model to such changes, several issues have to be addressed. To introduce these, we consider the response of the system $z(p, \omega) \leqslant 0$ to variations in ω only, where $\omega = (\omega_1, \omega_2, \ldots, \omega_m)$ is the vector of endowments of all consumers.

Various questions now come up: (1) What are natural properties for $z(p, \omega)$ (assumptions)? (2) Does the system $z(p, \omega) = 0$ have a solution with nonnegative prices (existence)? (3) How many such solutions are there (multiplicity)? (4) Do we know anything on the direction of change of prices and allocations when ω changes (general properties of $z(p, \omega)$)? (5) How to compute solutions given a numerical specification of $z(p, \omega)$? (6) Are some policy changes "better" than others (welfare analysis)? Finally, when building an applied general equilibrium model, the main issue is (7) how to specify numerically the model (empirical implementation). We will discuss these questions by turns.

1.2 Excess Demand Equilibrium

1.2.1 Assumptions on the Excess Demands

We now formulate assumptions on the excess demand function $z(p)$, along the lines of Arrow and Hahn (1971, ch. 2). In section 1.3 we will specify assumptions on individual agents and the properties of the excess demand function will follow from there:

ASSUMPTION **Z1** (Single-valuedness and continuity) $z(p)$ is a single-valued and continuous function, which is defined for $p \geqslant 0$, $p \neq 0$.[2]

2. We use the following notation: $p \geqslant 0$ means that $p_k \geqslant 0$ for all k, $p \geqslant 0$ and $p \neq 0$ mean that $p_k > 0$ for some k, $p > 0$ means that $p_k > 0$ for all k.

ASSUMPTION **Z2** (Continuous differentiability) $z(p)$ is continuously differentiable for $p > 0$.

ASSUMPTION **Z3** (Homogeneity) $z(p)$ is homogeneous of degree zero in p.

ASSUMPTION **Z4** (Walras's law) $pz(p) = 0$.

ASSUMPTION **Z5** (Desirability) $p_k = 0$ implies that $z_k(p) > 0$ for $k = 1, 2, \ldots, r$.

In assumption Z1 single-valuedness is fairly restrictive and will not be satisfied in many practical applications (e.g., when production takes place under constant returns to scale).[3] The assumption is made here in order to keep the first proof of existence of an equilibrium simple, and it will be relaxed later on. The continuity assumption is also restrictive, particularly when $p_k = 0$ for some k, because at that price we may expect excess demand to rise to infinity and hence to be discontinuous.

Assumption Z2 is restrictive, but it will only be used when we discuss multiplicity of equilibria and the response to changes in parameters. Assumption Z3 implies that for every scalar $\lambda > 0$, $z(\lambda p) = z(p)$. We can thus multiply each nonzero, nonnegative price vector by some positive number, without changing the value of the excess demand function $z(p)$: The absolute level of prices does not affect outcomes. Without loss of generality, and since we require some prices to be positive anyway, we can assume that $\Sigma_k p_k = 1$. Prices now belong to a set $S^r = \{p \mid \Sigma p_k = 1, p_k \geqslant 0\}$, called the price simplex. This scaling is known as price normalization. Other normalizations are possible. For example, we may choose commodity 1 as numéraire, that is, set $p_1 = 1$. But this can only be done if p_1 is positive in equilibrium.

Assumption Z4 (Walras's law) plays an important role in general equilibrium models. Replacing $z(p)$ by its components yields

$$p\Sigma_i x_i(p) = p\Sigma_j y_j(p) + p\Sigma_i \omega_i.$$

The assumption requires that for every nonzero, nonnegative price, the value of aggregate demand must be equal to the value of aggregate supply.

Finally, assumption Z5 states that demand for a commodity will be larger than supply whenever its price is zero. This assumption is not essential. It merely ensures that all prices are positive in equilibrium.

1.2.2 An Existence Proof

Existence proofs play an essential role in economic theory. This is because a model that does not possess any solution is inconsistent and therefore meaningless, but also because the existence proof itself highlights the role of

3. In that case $y_j \in Y_j$ implies that $\lambda y_j \in Y_j$ for every positive scalar λ. We will later see that this implies nonuniqueness of optimal y_j.

the assumptions made and, by that, facilitates the search for weaker assumptions. In doing so, it enlarges the field to which the theory applies.

The applied modeler may think that he can dispense of existence proofs because a reasonable model can always be calibrated so as to possess a solution. However, such a calibration will not help when the modeler seeks to compute a new solution after having changed the value of some parameter. Then he needs to know the range of parametric variations for which a solution will exist at all. The theoretical assumptions limit the range of variations of these parameters, and the existence proof makes clear why these restrictions are needed. Finally, the existence proof is useful because it will in general construct a fixed point mapping that can serve as the basis of the fixed point algorithms that solve the model numerically.

We will now prove that an equilibrium exists in this economy, using the assumptions made in section 1.2.1. The proof of this basic result rests on a fixed point theorem, due to Brouwer, that states that a continuous function G that maps a compact convex set A into itself ($G: A \to A$) has a fixed point x^*, that is, a point such that $x^* = G(x^*)$.[4]

Hence Brouwer's theorem imposes three requirements: (1) the function $G(x)$ should be continuous, (2) it should map from a compact, convex domain A, (3) the mapping should be "into itself": The range A should be the same as the domain. We illustrate the role of each of these requirements. In figure 1.1a we draw a function that meets all the requirements with the unit interval as the set A. There are three fixed points (intersections with the 45° line). Note first that unless the function is tangent to the 45° line, the number of fixed points will always be finite, and there will be at least one intersection. Then, unless the function starts at $(0, 0)$ and ends at $(1, 1)$, the number of fixed points must be odd. In figure 1.1b the mapping is not single valued (it is not a function but a correspondence); nevertheless, it maps from A into A and is convex valued (it is obviously so when single valued, but it also is at x° where it is set valued, since the interval BB' is a convex set). In this case a fixed point will exist, but now by virtue of the Kakutani theorem[5] instead of the Brouwer theorem. In figure 1.1c the correspondence is compact and maps into A, but it is not convex valued (and not continuous) at x° and there is no fixed point. Figure 1.1d shows that the three requirements are sufficient but not necessary: A fixed point exists, though the function is not continuous, not defined everywhere, not compact valued (it goes to infinity), and does not map into itself (C does not lie in A). In figure 1.1e the domain of the function is compact, and the function is continuous, so its range is compact. But the range is not contained in the

4. See theorem A.4.1 in the mathematical appendix A.

5. See theorem A.4.2.

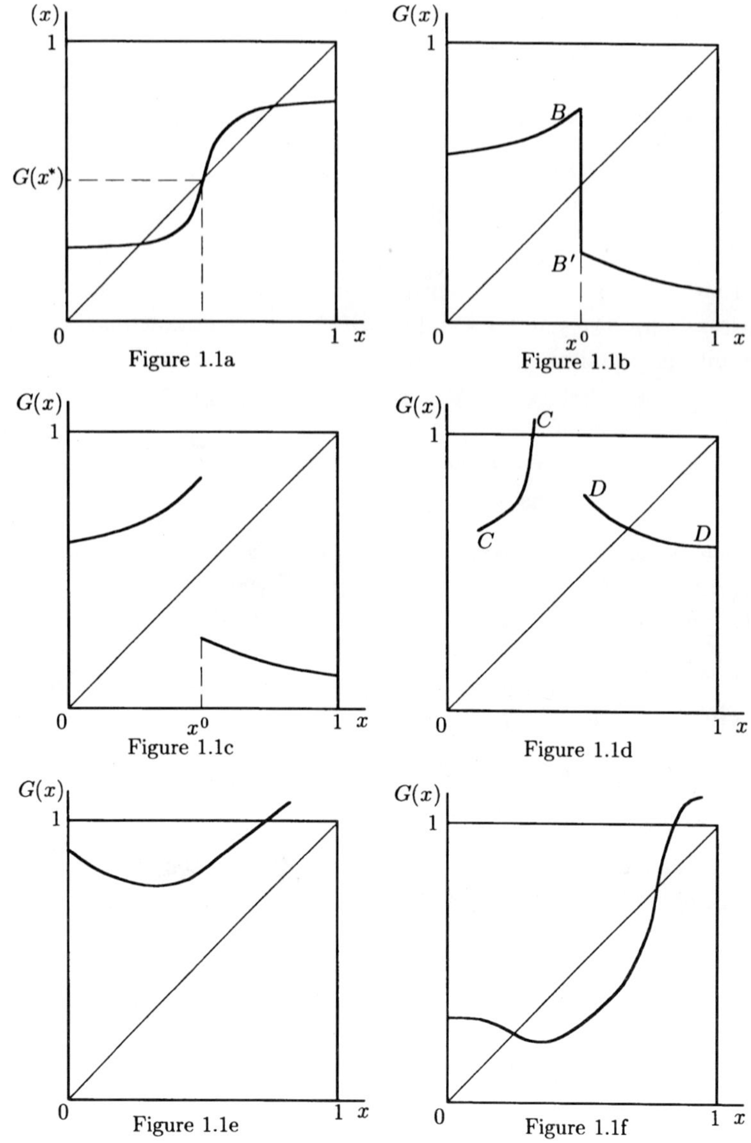

Figure 1.1
Illustrations of Brouwer's theorem in S^1.

domain, and no fixed point exists. Finally, figure 1.1f illustrates that the requirement (3) of "mapping into itself" is not necessary.

Though the proof of existence of equilibrium is standard, we give it because it clarifies the role played by the various assumptions.

PROPOSITION 1.1 (Existence of an excess demand equilibrium) If assumptions Z1, Z3, and Z4 hold, then there exists a price vector $p^* \in S^r$ such that $z(p^*) \leqslant 0$.

Proof To apply the Brouwer theorem, we first define a continuous function G that maps the simplex S^r into itself. Second, we show that at such a fixed point, the equilibrium conditions are satisfied.

1. Definition of the function. Let

$$G_k(p) = \frac{p_k + \max[0, z_k(p)]}{\Sigma_j p_j + \Sigma_j \max[0, z_j(p)]}.$$

Then, since $p \in S^r$ and $z(p)$ is continuous, $G(p)$ maps the simplex into a compact set. Moreover $\max[0, z_k(p)] \geqslant 0$, $\Sigma_j \max[0, z_j(p)] \geqslant 0$, and $\Sigma_j p_j = 1$. Therefore the denominator is strictly positive; hence the function $G(p)$ is continuous and maps S^r into itself. At this point we can invoke Brouwer's theorem. There exists a p^* such that $p_k^* = G_k(p^*)$. By the definition of $G_k(p)$, in the fixed point, we have

$$p_k^* = \frac{p_k^* + \max[0, z_k(p^*)]}{1 + \Sigma_j \max[0, z_j(p^*)]}. \tag{1.3}$$

2. We have still to show that $z_k(p^*) \leqslant 0$ in the fixed point. Indeed, multiplying the two sides of (1.3) by the denominator leads to

$$p_k^* \Sigma_j \max[0, z_j(p^*)] = \max[0, z_k(p^*)]. \tag{1.4}$$

Multiplying each term of (1.4) by $z_k(p^*)$ and summing over all k yields

$$\Sigma_k p_k^* z_k(p^*)\, \Sigma_j \max[0, z_j(p^*)] = \Sigma_k z_k(p^*) \max[0, z_k(p^*)]. \tag{1.5}$$

On the left-hand side of (1.5), there is a term $\Sigma_k p_k^* z_k(p^*)$ that is equal to zero, by assumption Z4, and hence

$$\Sigma_k z_k(p^*) \max[0, z_k(p^*)] = 0.$$

Each term in this sum is equal to 0 if $z_k(p^*) \leqslant 0$ and to $[z_k(p^*)]^2$ if $z_k(p^*) > 0$. The zero terms do not contribute to this sum. All others are positive, but then the expression on the left-hand side cannot be equal to zero. Therefore none of the $z_k(p^*)$ can be positive. ■

Proposition 1.1 guarantees that there exists at least one equilibrium. But, as suggested by figure 1.1a, there may exist many equilibria. We return to this in section 1.2.3.

We note that by the homogeneity assumption Z3 and because we disregard the case with $p = 0$, it was possible to normalize prices to the simplex S^r, the compact, convex domain of the mapping. Assumption Z4 (Walras's law) was used to prove that the fixed point is an equilibrium. Hence assumptions Z1, Z3, and Z4 are sufficient for existence of an equilibrium. However, models that fail to satisfy some of these may possess solutions, as was illustrated in figures 1.1d and f.

The intuition behind the artificial function $G(p)$ in the proof is to increase the price p_k of a commodity that is in excess demand ($z_k(p) > 0$). However, such a mapping should not be interpreted as a theory of the dynamics of price adjustment. Producers and consumers carry out their plans only in equilibrium; the model has no clear "out-of-equilibrium" interpretation; in particular, no exchange is specified to take place out of equilibrium. In section 1.2.5 we further discuss the issue when we consider computation of equilibrium.

Since $z_k(p^*) \leqslant 0$ in equilibrium, we do not exclude strict inequality and hence allow for excess supply of some commodity. But then, as stated in proposition 1.2, the associated price p_k^* will be zero.

PROPOSITION 1.2 (Free goods) Under assumptions Z1, Z3, and Z4, $z_k(p^*) < 0$ implies that $p_k^* = 0$.

Proof In equilibrium $z(p^*) \leqslant 0$ and $p^* \geqslant 0$, so $p_k^* z_k(p^*) \leqslant 0$ for every k. Now, if, for some k, $z_k(p^*) < 0$ and $p_k^* > 0$, then $p_k^* z_k(p^*) < 0$. However, for Walras's law to hold, it must be that $p_h^* z_h(p^*) > 0$ for at least one commodity. This cannot happen, since $p_k^* z_k(p^*) \leqslant 0$ for every k, a contradiction. ■

Proposition 1.2 shows that all goods in excess supply have a zero price. Of course in the real world economy, they may have a positive price. For example, on the labor market, unemployment may prevail at a positive wage rate. We will return to this in chapter 6.

In the existence proof we do not use assumption Z5. However, when excess demand is obtained from aggregation of consumers' and producers' plans, it is difficult to maintain continuity of $z(p)$ if the price of some commodity is zero. We will make assumptions on consumers' utility functions and endowments that make prices positive in equilibrium. We will be interested only in goods that carry positive prices and will discard from our model goods we know to be available freely.

PROPOSITION 1.3 (Positive prices) Under assumptions Z1 and Z3 to Z5, equilibrium prices p^* are positive.

Proof Evidently $p_k^* = 0$ implies that $z_k(p^*) > 0$ by assumption Z5 and this cannot be an equilibrium. ■

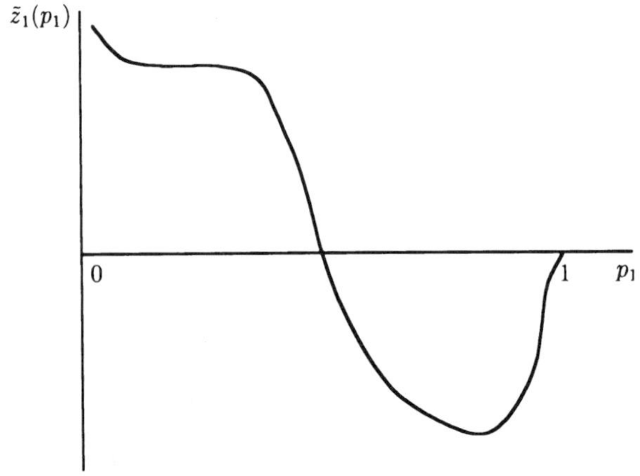

Figure 1.2
The excess demand function $\tilde{z}_1(p_1)$ in the two-commodity case.

When assumption Z5 holds, and if there are only two commodities, an easy geometrical argument for existence of equilibrium can be given. Under assumption Z5 no price can be zero in equilibrium. Since $p_2 = 1 - p_1$, we can draw the curve $\tilde{z}_1(p_1) = z_1(p_1, 1 - p_1)$, shown in figure 1.2. At $p_1 = 0$, we have $\tilde{z}_1(p_1) > 0$, by assumption Z5; by continuity, $\tilde{z}_1(p_1) > 0$ for p_1 close to zero. This makes it possible to start the curve above the p_1-axis. On the other hand, for $p_1 = 1$, Walras's law implies that $\tilde{z}_1(p_1) = 0$. However, for p_1 close to 1, we have p_2 close to zero, and therefore $\tilde{z}_2(p_1) > 0$, $\tilde{z}_1(p_1) < 0$, by Walras's law, where $\tilde{z}_2(p_1) = z_2(p_1, 1 - p_1)$. Hence $\tilde{z}_1(p_1)$ must lie under the p_1-axis for p_1 close to 1. Since $\tilde{z}_1(p_1)$ is a continuous function, it must cross the p_1-axis at least once, and there exists a value p_1 (and p_2), say, $p_1^* \in (0, 1)$ such that $\tilde{z}_1(p_1^*) = 0$ (and $\tilde{z}_2(p_1^*) = 0$), again by Walras's law.

1.2.3 Multiplicity of Equilibrium

The multiplicity issue is important in applied modeling because we need to know whether a change in parameter will lead to a unique new solution; otherwise, the impact of the change is ambiguous. We also want this solution to vary continuously under the change. We discuss multiplicity under the additional assumption of continuous differentiability (assumption Z2) and of desirability (assumption Z5). Clearly, for positive prices, continuous differentiability implies single-valuedness and continuity.

Because of homogeneity (assumption Z3), we can normalize prices on the simplex and define the price of commodity r residually. We define the excess demand function for the $r-1$ commodities as

$$\tilde{z}_k(\tilde{p}) = z_k(p_1, p_2, \ldots, 1 - \Sigma_{k \neq r} p_k),$$

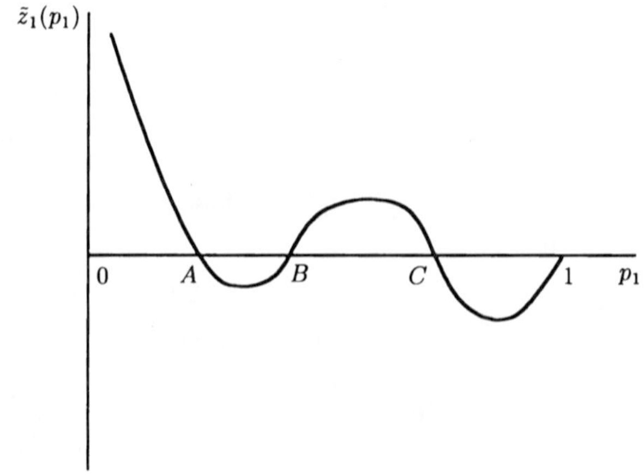

Figure 1.3
The excess demand function $\tilde{z}_1(p_1)$ in the two-commodity case with three equilibria.

where this function is only defined for $\Sigma_{k \neq r} p_k \leqslant 1$. We are thus only concerned with $r - 1$ excess demands, and, due to Walras's law, we can always recover net demand for commodity r as $z_r = -\Sigma_{k=1}^{r-1} p_k z_k/(1 - \Sigma_{k \neq r} p_k)$. In this section we only want to give the geometric intuition and therefore return to a two-commodity economy[6] represented in figure 1.3.

In this figure we are concerned only with intersections on the open interval (0, 1) on which there are three equilibria, denoted by A, B, and C. Note that $\partial \tilde{z}_1/\partial p_1$ is zero at some points. Here, this happens at three points only, but if there were flat horizontal parts in the curve, there would be an infinity of prices where this would happen. If, moreover, a flat were tangent to the horizontal axis, there would be an infinite number of equilibria. This leads to the following geometric characterization of the number of equilibria:

1. If $\partial \tilde{z}_1/\partial p_1$ does not change sign (here $\partial \tilde{z}_1/\partial p_1$ must then be negative), there is only one equilibrium, and if the derivative changes sign, several equilibria may exist.
2. There is only a finite number of $\tilde{z}_1$-values for which $\partial \tilde{z}_1/\partial p_1 = 0$. Hence it would be a "coincidence" if such a value of $\tilde{z}_1$ happened to be zero. Even if this were the case, a small upward (or downward) shift of the function (resulting from a small change in the parameters of the model) will either eliminate this equilibrium, or generate two distinct solutions. In this sense equilibria for which $\partial \tilde{z}_1/\partial p_1 = 0$ are "rare," but if this happens, the p_1-values can form a continuum and hence be infi-

6. A more general treatment is given in section A.7 of the mathematical appendix.

nite in number. An equilibrium at which $\partial \tilde{z}_1/\partial p_1 \neq 0$ is called a regular equilibrium.

3. Assumption Z5 (desirability) ensures that equilibrium prices must lie within the $(0, 1)$ interval. Then, by assumption Z2 (differentiability), the derivative cannot change sign an infinite number of times on this finite interval.[7] Therefore the number of regular equilibria is finite, and regular equilibria are locally unique.

4. Regular equilibria for which the derivative $\partial \tilde{z}_1/\partial p_1$ is positive alternate with equilibria for which this derivative is negative.

5. The number of regular equilibria can also be shown to be odd. Suppose, indeed, that in figure 1.3, A and B are intersections with the p_1-axis but that there is no intersection C. Now, since for $p_1 = 1$, $\tilde{z}_2(p_1)$ is positive, this is also true for $p_1 = 1 - \varepsilon$ by assumption Z2. Therefore $\tilde{z}_1(p_1)$ must, by Walras's law, be negative at this point, and there must be a crossing right of B. Hence the number of regular equilibria must be odd.

The geometric intuition behind the discussion of the two-good economy extends to the r-good economy and can be applied to all the models covered in this book, except to those of chapter 8 where the time horizon is infinite.[8]

Assumptions on the behavior of individual consumers and producers will in general not be sufficient to prove nonsingularity of $\partial \tilde{z}_1/\partial p_1$, the Jacobian of the excess demand function. Therefore, theory cannot exclude the possibility of multiple equilibria. However, continuous differentiability in equilibrium and desirability, two properties that can be derived from individual behavior, are sufficient to ensure that equilibria will in general be regular, finite in number, and thus, locally unique. Then, if $\tilde{z}_1(p_1; \omega)$ is differentiable with respect to ω, we can compute the derivative of equilibrium prices with respect to changes in the endowments $\omega = (\omega_1, \omega_2, \ldots, \omega_m)$ as follows. First, we take the total derivative of $\tilde{z}_1(p_1; \omega) = 0$:

$$\left(\frac{\partial \tilde{z}_1}{\partial p_1}\right) dp_1 + \left(\frac{\partial \tilde{z}_1}{\partial \omega}\right) d\omega = 0.$$

Second, at a regular equilibrium, we have $\partial \tilde{z}_1/\partial p_1 \neq 0$, so, by the implicit function theorem,[9] we can compute

$$\frac{dp_1}{d\omega} = -\left(\frac{\partial \tilde{z}_1}{\partial p_1}\right)^{-1} \left(\frac{\partial \tilde{z}_1}{\partial \omega}\right). \tag{1.6}$$

7. Note that if the function is continuous on the interval, and continuously differentiable except at a finite number of points, the number of regular equilibria is also finite.

8. This is discussed in sections A.7 to A.9 of the mathematical appendix. In particular, theorems A.7.1 and A.7.3 (uniqueness), A.7.6 (finite number of regular equilibria), and A.8.3 (odd number of regular equilibria) give formal characterizations for the r commodity case.

9. See theorem A.7.7 for a discussion of the r commodity case.

This result ensures (also when $r > 2$) that at regular equilibria, prices p_1 will vary continuously under a small change in parameters.

1.2.4 Lack of General Properties of Excess Demand

The total differential (1.6) gives the basic information on the response of the model to variations in parameters ω, since once the new prices are known, we can compute demands and supplies of consumers and producers for the new parameter values. We need to know the properties of $dp_1/d\omega$, and in particular, its sign, since this would make it possible to anticipate the effect of parameter changes. By (1.6), knowledge of the sign of $dp_1/d\omega$ requires knowledge of the sign of $(\partial \tilde{z}_1/\partial p_1)$ and of $(\partial \tilde{z}_1/\partial \omega)$. As is obvious from figure 1.3, even in the two-commodity case, this sign depends on the equilibrium one happens to consider.[10] In the case with more than two commodities, the situation is worse. In fact Debreu (1974), Mantel (1974), and Sonnenschein (1972, 1973) have shown that in general, at positive prices, excess demand functions have no other property than those of assumptions Z1, Z3, and Z4. The result goes a step further. It proves that for any excess demand function that satisfies these three assumptions, one can construct well-behaved producer and consumer problems (1.1)–(1.2) that generate it. In section 2.2.3 we will see that this difficulty is caused by the consumer demand functions, when there is more than one consumer.

The Debreu-Mantel-Sonnenschein result has far-reaching consequences. It shows that even under severe restrictions on production sets and on utility functions (to be introduced in section 1.3), the excess demand function may turn out to lose most of its properties in the process of aggregation over consumers. In particular, its Jacobian may become singular at some positive price, and even if it is regular, the signs of the elements of its inverse cannot be predicted.

However, the result does not exclude the possibility of the Jacobian having stronger properties for given numerical values of parameters in an applied model. For example, if the Jacobian matrix $\partial \tilde{z}/\partial p$ has a specific structure, it may be possible to determine the signs of the elements of $(\partial \tilde{z}/\partial p)^{-1}$. For instance, if the matrix $-\partial \tilde{z}/\partial p$ has the property of gross substitutability,[11] its inverse is nonnegative and has a positive diagonal. Then, if ω is increased (supply shock) and if $\partial \tilde{z}/\partial \omega \leqslant 0$ (i.e., if the supply is larger than its impact on demand via income), then $d\tilde{p}/d\omega \leqslant 0$, so prices cannot increase. Gross substitutability is, however, a very restrictive as-

10. One may wish to restrict attention to equilibria for which the determinant of the Jacobian has a particular sign (theorem A.7.4), but for $r > 2$ this will not determine the sign pattern of its inverse unless the Jacobian has a very particular structure.

11. See definition A.7.5 and theorem A.7.4.

sumption, and as soon as it is relaxed (even if one retains the assumption that it is a P-matrix), very little structure remains. The P-matrix assumption ensures that $-(\partial\tilde{z}/\partial p)^{-1}$ will have a positive diagonal and therefore that the effect of a demand shock in commodity k on its own price will be positive; however, the consequences of changes in ω are unclear.

Since theory does not lead to clear predictions on the impact of parameter changes, there is a need to implement the model numerically. This is an important reason for building applied general equilibrium models.

1.2.5 Computation of Equilibria

Computation of equilibrium is relatively easy in the two-commodity case of figure 1.3. It is possible to scan the p_1-axis, evaluating $\tilde{z}_1(p_1)$ at regular intervals until one finds the equilibria A, B, and C. Obviously this simple method is not operational when there are more than two commodities, since the number of points at which the function has to be evaluated increases with the power of $r - 1$.

When prices are positive in equilibrium (assumption Z5), the excess demand problem to be solved is simply $\tilde{z}(\tilde{p}) = 0$, a system of $(r - 1)$ nonlinear equations in $(r - 1)$ unknowns.[12] This is a standard problem in numerical analysis. The design of algorithms usually proceeds in two stages. In the first stage, we define a differential equation $d\tilde{p}/dt = G(\tilde{p}(t))$, $p(0) = p_0$ given, that converges to an equilibrium solution $G(\tilde{p}^*) = 0$ such that $\tilde{z}(\tilde{p}^*) = 0$. In the second stage, the differential equation is approximated through discretization, for example by a difference equation (or by a pivoting algorithm). Appendixes A.8 and A.9 treat this problem in more detail. Here we only show that because of the lack of general properties of the excess demand function, we cannot expect easily interpretable differential equations to converge to the solution in all cases.

Consider, for example, the differential equation:

$$\frac{d\tilde{p}}{dt} = \tilde{z}(\tilde{p}(t)) \qquad \text{with } p(0) = p_0. \tag{1.7}$$

If the Jacobian $\partial\tilde{z}/\partial\tilde{p}$ of the excess demand function along the path starting at p_0 has eigenvalues with negative real parts, this differential equation will converge.[13] This will be the case if $-\partial\tilde{z}/\partial\tilde{p}$ is diagonally dominant or has gross substitutability, but such a property cannot be ensured because of the lack of general properties of $\tilde{z}(\tilde{p})$.

12. Since all prices are positive in equilibrium, $z(p^*) = 0$, and by Walras's law, one balance constraint can be dropped and one price chosen as numéraire. One is left with a system of $r - 1$ equations in $r - 1$ unknowns.

13. To avoid prices becoming negative, assumption Z5 has to be strengthened so that some excess demand will go to infinity whenever some price goes to zero.

We can also think of using algorithms that do not require such strong properties but only need the Jacobian to be nonsingular (like the Newton method).[14] However, such an algorithm may not converge either because the nonsingularity condition may be violated somewhere along the path, again due to lack of general properties.

Thus theory suggests that more complex algorithms are needed,[15] and applied general equilibrium has gained popularity particularly since Scarf (1967, 1973) has designed and applied globally convergent algorithms. Moreover experience shows that simple algorithms that do not have the property of global convergence also work, in practice.

The differential equation (1.7) describes a process of price adjustments that is commonly referred to as tatonnement. It has the well-known interpretation of an auctioneer who adjusts the price of commodity k in proportion to its excess demand. If such a process were convergent, we could conceive of an economy in which this process takes place at the beginning of the period, before goods are exchanged: Price adjustment could then be looked at as being part of the model itself. Unfortunately, tatonnement algorithms may fail to converge, and the processes described by globally convergent algorithms lack such an elegant economic interpretation.[16]

Therefore the equilibrium conditions can, in general, not be thought of as resulting from a dynamic process. In definition 1.2 they are merely a condition, not a description of reality. The definition only says that if markets are in equilibrium, and if producers and consumers behave in the way specified, the solution is said to be a competitive equilibrium. When taken literally, this principle does not allow us to calculate the effects of parameter changes because after such a change some of the equilibrium conditions could be violated or agents could pay different prices for the same commodity. Yet such an equilibrium calculation is precisely what is of interest in policy analysis. To perform this analysis, all the constraints of the model should have a descriptive status, and in the case of applied general equilibrium, this is the interpretation that we adopt.

1.3 Competitive Equilibrium

In this section we introduce assumptions on the production sets of producers and on the utility functions and the endowments of consumers. These will enable us to derive assumptions Z1 and Z3 to Z5 and thus to prove

14. See also theorem A.8.1.

15. See theorem A.8.3.

16. Note that some triangulation algorithms (van der Laan and Talman 1985) incorporate tatonnement rules, but they overrule these when there is danger of cycling. There is no pure tatonnement interpretation for this case either.

existence of the competitive equilibrium specified in definition 1.2.[17] As a first approach we impose restrictive, but straightforward assumptions, that will be relaxed from chapter 2 onward:

ASSUMPTION P1 (Production sets) The production set Y_j of every producer j has possibility of inaction ($0 \in Y_j$); it is compact and strictly convex.

ASSUMPTION C1 (Utility functions) The utility function $u_i: R^r_+ \to R_+, u_i(x_i)$ of every consumer i is continuous, strictly concave, increasing, and, without loss of generality, $u_i(0) = 0$.

ASSUMPTION C2 (Endowments) The endowment ω_i of every consumer i is strictly positive.

In assumption P1 the possibility of inaction is imposed to ensure that the set Y_j is nonempty and that profits py_j are nonnegative in (1.1). Compactness guarantees that profits are bounded, and strict convexity that the optimal y_j is unique (theorem A.1.1). From this follows that $y_j(p)$ is a continuous function (maximum theorem A.3.1, assertion 4).

For consumption, positive endowments (and nonnegative profits) ensure that income will be positive for all consumers at all prices on the simplex. In chapter 2 we will see that increasingness of utility functions avoids zero prices in equilibrium. Strict concavity of $u_i(x_i)$ makes demand functions continuous when $p > 0$ (maximum theorem A.3.1, assertion 3). However, if some price is zero, the set of attainable consumptions is unbounded, and utility may become infinite, so demand functions can be discontinuous.

At positive prices, assumptions Z1 and Z3 to Z5 can be derived immediately as follows:

ASSUMPTION Z1 (Single-valuedness and continuity) Since every supply and demand function has this property, it also holds for aggregate excess demand.

ASSUMPTION Z3 (Homogeneity) In every producer and consumer optimization problem, we can replace p by λp, for given $\lambda > 0$. For the producer, this amounts to a scaling of the parameters of his objective function; this shifts the value of his profit but does not change his decisions. For the consumer, the scaling changes both sides of the linear budget constraint and therefore does not affect the set of feasible options. Since preferences are not changed, the optimal decision is not affected.

ASSUMPTION Z4 (Walras's law) Walras's law can be obtained by summation over consumers i of expenditures px_i and revenues $p\omega_i + \Sigma_j \theta_{ij} py_j$. Since income is positive and $u_i(x_i)$ is increasing, consumer i will spend all

17. Assumption Z2 is not needed here.

his income.[18] Then we obtain

$$\Sigma_i px_i = \Sigma_i p\omega_i + \Sigma_i\Sigma_j\theta_{ij}py_j.$$

Since $\Sigma_i\theta_{ij} = 1$ all j, this reduces to

$$p(\Sigma_i x_i - \Sigma_i\omega_i - \Sigma_j y_j) = 0,$$

which is Walras's law.

ASSUMPTION Z5 (Desirability) For a consumer with income larger than a given positive ε, increasingness of $u_i(x_i)$ implies that this consumer's demand will exceed total supply when a price drops toward zero.[19]

The above properties also hold when some price p_k is zero, except that $x_i(p)$ will then be discontinuous. This discontinuity prevents us from invoking proposition 1.1 to prove existence of a competitive equilibrium, but we will see that the difficulty is easily overcome.

PROPOSITION 1.4 (Existence of a general competitive equilibrium) Let assumptions P1, C1, and C2 hold. Then there exists a general competitive equilibrium as specified in definition 1.2.

Proof Since Y_j is bounded for all j, there exists a finite and positive consumption level $\bar{x}$ such that $\bar{x} > \Sigma_j y_j + \Sigma_i\omega_i$ for all $y_j \in Y_j$. We then construct an excess demand function by using supply functions $y_j(p) = \operatorname{argmax}\{py_j \mid y_j \in Y_j\}$ and consumer demand functions $x_i(p, h_i, \bar{x}) = \operatorname{argmax}\{u_i(x_i) \mid px_i \leqslant h_i,\ 0 \leqslant x_i \leqslant \bar{x}\}$. Since $\bar{x} > 0$, the constraint set is nonempty and compact. Then strict concavity ensures that $x_i(p, h_i, \bar{x})$ is continuous (maximum theorem A.3.1, assertion 3). The function is also homogeneous of degree zero in (p, h_i). Since $p\bar{x}$ exceeds $p(\Sigma_j y_j + \Sigma_i\omega_i)$, it also exceeds $h_i = p\omega_i + \Sigma_j\theta_{ij}py_j$. Hence, with $u_i(x_i)$ increasing, consumer i will always be able to spend all his income, and Walras's law holds as well. Therefore all the conditions of proposition 1.1 are fulfilled, and an excess demand equilibrium will exist. In this equilibrium, $\bar{x}$ cannot be binding for any consumer; consumer i actually solves (1.2), which shows that the excess demand equilibrium is a competitive equilibrium. ■

Proposition 1.4 shows that the equilibrium price vector p^* provides sufficient information. Given this price, every agent makes his best decision on his own and will get exactly what he plans. Decisions of all agents can be decentralized and are compatible.

18. Since utility is increasing, a solution where not all income is spent would not be optimal for the consumer.

19. In chapter 2, proposition 2.11, this will, under further assumptions, be proved to hold as a property.

Decentralization is only one property of the competitive equilibrium. The interest in the general equilibrium model largely stems from a further property, of a normative nature: the Pareto-efficiency of equilibrium allocations.

1.4 Efficiency and Equity

1.4.1 Pareto-Efficiency

Until now we have described equilibrium as the outcome of price-taking behavior of individual agents under the constraint that demand should not exceed supply. We have made no assessment of the quality of the resulting allocations. Now we will show that some normative properties can be established, in particular, the Pareto-efficiency of allocations, which is defined as follows:

DEFINITION 1.3 (Pareto-efficient allocation) An allocation $x_i \geqslant 0$, all i, $y_j \in Y_j$, all j, is Pareto-efficient if it is feasible, that is, if

$$\Sigma_i x_i \leqslant \Sigma_j y_j + \Sigma_i \omega_i,$$

and if there exists no other feasible allocation x'_i, all i, y'_j, all j, such that

$$u_i(x'_i) \geqslant u_i(x_i) \qquad \text{all } i,$$

while, for at least one consumer, say, s,

$$u_s(x'_s) > u_s(x_s).$$

In other words, an allocation is Pareto-efficient if it is impossible to find another feasible allocation (which is said to be Pareto-superior) that makes at least one consumer better off in terms of his utility without making anyone worse off.

Recall that assumptions P1, C1, and C2 are sufficient but not necessary to prove the existence of equilibrium and we will encounter models that violate some of these assumptions, for example, convexity. Therefore we now use assumptions that are as mild as possible. In particular, increasingness of utility functions is replaced by nonsatiation[20] and strict concavity by concavity (occasionally even by mere continuity). Strict convexity of production sets is replaced by convexity (sometimes even convexity will be dropped), but we will maintain the assumptions of possibility of inaction

20. Nonsatiation means that the utility function is everywhere increasing in at least one commodity (see definition A.1.12). This is weaker than increasingness in assumption C1.

and compactness of Y_j. As a consequence all the propositions that are proved in this section will also hold under assumptions P1, C1, and C2, that is, in the model for which existence of equilibrium has already been proved.

1.4.2 The First Welfare Theorem

The first welfare theorem shows that (if it exists) a competitive equilibrium is Pareto-efficient under assumptions that are milder than those in proposition 1.2.

PROPOSITION 1.5 (First welfare theorem) Let the production set of producer j be nonempty and compact, and let consumer i's utility function be continuous and nonsatiated. Then, if it exists, the competitive equilibrium allocation x_i^*, all i, y_j^*, all j, with price vector p^* is Pareto-efficient.

Proof Consider the consumer problem at prices p^*, with income $h_i^* = p^*\omega_i + \Sigma_j \theta_{ij} \Pi_j(p^*)$

$$x_i^* = \operatorname{argmax}\{u_i(x_i) \mid p^* x_i \leqslant h_i^*, x_i \geqslant 0\}. \tag{1.8}$$

We first show that for any nonnegative consumption x_i^0, the following two implications hold:

1. $u_i(x_i^0) > u_i(x_i^*)$ implies that $p^*x_i^0 > p^*x_i^*$.
2. $u_i(x_i^0) \geqslant u_i(x_i^*)$ implies that $p^*x_i^0 \geqslant p^*x_i^*$.

To verify (1), assume that the second inequality does not hold. This would mean that x_i^0 is feasible in (1.8) and therefore would contradict optimality of x_i^*. To verify (2), we distinguish two cases. If the first inequality is strict, and the second violated, we are back in case (1); if $u_i(x_i^0) = u_i(x_i^*)$ and $p^*x_i^0 < p^*x_i^*$, then, by nonsatiation of utility, there exists a vector x_i'' such that $u_i(x_i'') > u_i(x_i^*)$ and $p^*x_i'' = p^*x_i^*$, and this again contradicts optimality of x_i^*.

We are now ready to prove that the equilibrium allocation x_i^*, all i, y_j^*, all j is Pareto-efficient. Assume that there exists an alternative feasible allocation x_i', all i, y_j', all j, possibly not a competitive equilibrium, that is Pareto-superior to the equilibrium allocation such that

$$u_i(x_i') \geqslant u_i(x_i^*) \qquad \text{for all } i,$$

while, for at least one consumer, say, s,

$$u_s(x_s') > u_s(x_s^*).$$

By (1) and (2) above,

$$p^*x_i' \geqslant p^*x_i^* = p^*\omega_i + \Sigma_j\theta_{ij}p^*y_j^*$$

for all i except s, and

$$p^*x_s' > p^*x_s^* = p^*\omega_s + \Sigma_j\theta_{sj}p^*y_j^*.$$

Summation of these m inequalities yields

$$p^*\Sigma_i x_i' > p^*\Sigma_i\omega_i + \Sigma_i\Sigma_j\theta_{ij}p^*y_j^* = p^*\Sigma_i\omega_i + p^*\Sigma_j y_j^*, \tag{1.9}$$

Since the allocation x_i', all i, y_j', all j, is feasible, it satisfies

$$\Sigma_i x_i' \leqslant \Sigma_i\omega_i + \Sigma_j y_j'.$$

Replacing $\Sigma_i x_i'$ by $\Sigma_i\omega_i + \Sigma_j y_j'$ in (1.9) cannot reverse the inequality, which now reads as

$$p^*\Sigma_i\omega_i + p^*\Sigma_j y_j' > p^*\Sigma_i\omega_i + p^*\Sigma_j y_j^*,$$

or

$$p^*\Sigma_j y_j' > p^*\Sigma_j y_j^*. \tag{1.10}$$

Since Y_j is nonempty and compact, the optimal choices y_j^* of individual producers also solve the collective profit maximization problem $\max\{\Sigma_j p^*y_j \mid y_j \in Y_j, \text{ all } j\}$. Therefore inequality (1.10) contradicts collective profit maximization. Hence the assertion that the allocation x_i^*, all i, y_j^*, all j, is not Pareto-efficient is false. ■

The first welfare theorem shows that there can be no allocation that is Pareto-superior to the one prevailing at the equilibrium p^*. In particular, there cannot exist another equilibrium $p^{**} \neq p^*$ that is Pareto-superior to it. This leads to the following proposition.

PROPOSITION 1.6 (Equilibra cannot dominate each other) Let the assumptions of proposition 1.5 hold, and suppose that x_i^* and x_i^{**}, all i, are consumption bundles belonging to two equilibria. Then there must exist at least one pair of consumers, say, the pair s and t, for which $u_s(x_s^*) > u_s(x_s^{**})$ and $u_t(x_t^*) < u_t(x_t^{**})$.

Proof The proof is obvious ■

1.4.3 The Welfare Optimum

In the rest of the book we will often make use of the following welfare program:

$$W(\alpha) = \max \Sigma_i\alpha_i u_i(x_i)$$

$$x_i \geqslant 0, \text{ all } i, \; y_j, \text{ all } j$$

subject to (1.11)

$\Sigma_i x_i - \Sigma_j y_j \leqslant \Sigma_i \omega_i$,

$y_j \in Y_j$,

for given $\alpha \in S^m = \{\alpha \mid \alpha_i \geqslant 0, \Sigma_i \alpha_i = 1\}$.

An allocation that is optimal in (1.11) will be referred to as a welfare optimum. A welfare optimum can be thought of as a central plan that allocates goods over agents. Such a program has properties described in the following two propositions.

PROPOSITION 1.7 (Pareto-efficiency of a welfare optimum) If for positive welfare weights, (1.11) has a bounded optimal solution, then this solution is Pareto-efficient.

Proof If $\alpha_i > 0$ for all i, a Pareto-superior alternative allocation would lead to a higher value of the objective function, and therefore contradict optimality of the original solution. ■

PROPOSITION 1.8 (Representation of a Pareto-efficient allocation by a welfare optimum) Assume that the production set Y_j has possibility of inaction, is compact and convex.[21] Assume further that the utility function of consumer i is continuous, concave, nonsatiated, and that $u_i(0) = 0$. Finally, assume that $\Sigma_i \omega_i > 0$. Then any Pareto-efficient allocation with positive utilities $u_1^0, u_2^0, \ldots, u_m^0$ is a welfare optimum with positive welfare weights.

Proof We consider the following convex program:

$$
\begin{aligned}
&\max \rho, \\
&x_i \geqslant 0,\ \text{all } i,\ y_j,\ \text{all } j,\ \rho, \\
&\text{subject to} \\
&u_i(x_i) \geqslant \rho u_i^0 \qquad (\alpha_i), \\
&\Sigma_i x_i - \Sigma_j y_j \leqslant \Sigma_i \omega_i \qquad (p), \\
&y_j \in Y_j.
\end{aligned}
\tag{1.12}
$$

Since the u_i^0 are Pareto-efficient, they are feasible in (1.12) and hence allow an optimal value $\rho \geqslant 1$. Pareto-efficiency also implies that no single consumer can gain without reducing the utility of another consumer. Therefore consumers cannot gain jointly so that $\rho = 1$ is the maximum value, which implies that the Pareto-efficient allocation is also optimal in (1.12).

21. This assumption is weaker than strict convexity in assumption P1.

By possibility of inaction, and since $\Sigma_i \omega_i > 0$, while ρ can be set to zero, Slater's constraint qualification holds.[22] Hence nonnegative Lagrange multipliers α_i and p exist. Suppose that $\alpha_s = 0$ for some s; this means that adding a small positive perturbation ε_s to the right-hand side of the associated constraint (such a perturbation does not violate constraint qualification because of nonsatiation and because $u_s^0 > 0$) would not reduce the value of the objective ρ. Hence we can increase the utility of consumer s without reducing the utility of any other consumer, which is a contradiction of Pareto-efficiency of the allocation. Therefore the value function of the program must be decreasing in ε_s, so, by theorem A.3.4 and definition A.1.16, α_s cannot be zero.

The Lagrangian of (1.12) is

$$L(\rho, x, y, \alpha, p) = \rho + \Sigma_i \alpha_i (u_i(x_i) - \rho u_i^0) - p(\Sigma_i x_i - \Sigma_j y_j - \Sigma_i \omega_i).$$

For arbitrary but feasible x_i's and y_j's and optimal ρ (equal to unity), the associated Lagrange inequality is[23]

$$L(\rho, x, y, \alpha^*, p^*) \leqslant L(\rho^*, x^*, y^*, \alpha^*, p^*),$$

where the superscript * denotes optimal values. This inequality coincides with the inequality of a welfare program, with the Lagrangian

$$L(x, y, \alpha, p) = \Sigma_i \alpha_i u_i(x_i) - p(\Sigma_i x_i - \Sigma_j y_j - \Sigma_i \omega_i). \tag{1.13}$$

Since all α_i are positive, they can as in (1.11) be normalized on S^r. Therefore the Pareto-efficient allocation is a welfare optimum. ■

We note that consumers with zero utility u_i^0 would receive a zero quantity of any commodity useful to another consumer, since $u_i(0) = 0$ and $u_i(x_i)$ is nonsatiated. When faced with a Pareto-efficient allocation where some consumers have zero utility, we can set the consumption of these consumers to zero and restrict attention to those with positive utility.

Proposition 1.4 gives sufficient conditions for existence of a competitive equilibrium. Proposition 1.5 shows that a competitive equilibrium is Pareto-efficient. Finally, proposition 1.8 establishes that a Pareto-efficient allocation can be represented through a welfare optimum with welfare weights on the simplex. Since the assumptions needed to prove propositions 1.5 and 1.8 are weaker than those needed in proposition 1.4, it follows that under the latter set of assumptions, a competitive equilibrium can be represented through a welfare optimum with nonzero welfare weights. This is known as Negishi's theorem, and it will play a crucial role in the existence proofs of subsequent chapters. In chapter 3 we will prove this result in a direct way.

22. See theorem A.2.1.

23. See theorem A.2.1, assertion 7.

So far we have seen that a competitive equilibrium is a welfare optimum. We now study the converse problem of whether a welfare optimum can be obtained in a decentralized way through a competitive equilibrium. This is indeed the case if one allows for transfers among consumers. For this we define

DEFINITION 1.4 (Competitive equilibrium with transfers) The allocation y_j^*, all j, x_i^*, all i, supported by the price vector $p^* \geqslant 0$, $p^* \neq 0$ is a competitive equilibrium with transfers T_i^* if the following conditions are satisfied:

1. For every producer j, y_j^* solves $\max_{y_j}\{p^* y_j \mid y_j \in Y_j\}$.
2. For every consumer i, x_i^* solves $\max_{x_i \geqslant 0}\{u_i(x_i) \mid p^* x_i \leqslant h_i^*\}$, where $h_i^* = p^*\omega_i + \Sigma_j \theta_{ij} p^* y_j^* + T_i^*$.
3. All markets are in equilibrium, $\Sigma_i x_i^* - \Sigma_j y_j^* - \Sigma_i \omega_i \leqslant 0$.
4. Transfers add up to zero: $\Sigma_i T_i^* = 0$.

We now state and prove:

PROPOSITION 1.9 (A welfare optimum is a competitive equilibrium with transfers) Assume that the production set Y_j has possibility of inaction, is compact and convex. Assume further that the utility function of consumer i is continuous, concave, nonsatiated and that $u_i(0) = 0$. Finally, assume that $\Sigma_i \omega_i > 0$. Then a welfare optimum of (1.11) with $\alpha > 0$ is a competitive equilibrium with transfers as in definition 1.4.

Proof We must prove that in an optimum of (1.11), all the conditions of definition 1.4 are satisfied and that the Lagrange multipliers p^* can be interpreted as equilibrium prices (up to a scaling factor).

Consider the welfare program (1.11). Given the possibility of inaction, the program is feasible; given the continuity of utility functions and compactness of Y_j, it is compact valued. Since it is also convex, and Slater's constraint qualification holds (set $y_j = 0$ for all j and $x_i > 0$ such that $\Sigma_i x_i < \Sigma_i \omega_i$), Lagrange multipliers p exist (see theorem A.2.1). Since $\alpha_i > 0$ for all i, and since utility functions are nonsatiated, it follows that $p^* \neq 0$. We may therefore consider the Lagrangian (1.13) and the associated inequality:

$$\Sigma_i \alpha_i u_i(x_i) - p^*(\Sigma_i x_i - \Sigma_j y_j - \Sigma_i \omega_i) \leqslant \Sigma_i \alpha_i u_i(x_i^*) - p^*(\Sigma_i x_i^* - \Sigma_j y_j^* - \Sigma_i \omega_i), \tag{1.14}$$

for all $x_i \geqslant 0$ and all $y_j \in Y_j$, and where the superscript * denotes optimal values.

We now prove condition (1) of definition 1.4. In (1.14) we set $x_i = x_i^*$ for all i and $y_j = y_j^*$ for all j, except for $j = s$; then (1.14) implies that

$$p^*y_s \leqslant p^*y_s^* \qquad \text{for all } y_s \in Y_s,$$

which shows that every producer s maximizes profits so that $p^*y_s^* = \Pi_s(p^*)$.

Turning to condition (2), in (1.14) we set $x_i = x_i^*$ for all i, except for $i = s$, and $y_j = y_j^*$ for all j. Now (1.14) implies that

$$\alpha_s u_s(x_s) - p^*x_s \leqslant \alpha_s u_s(x_s^*) - p^*x_s^* \qquad \text{for all } x_s \geqslant 0, \tag{1.15}$$

which, if one multiplies both sides by $\lambda_s = 1/\alpha_s$, yields

$$u_s(x_s) - \lambda_s p^*x_s \leqslant u_s(x_s^*) - \lambda_s p^*x_s^* \qquad \text{for all } x_s \geqslant 0. \tag{1.16}$$

This is the Lagrangian inequality that corresponds to the consumer s problem (1.2), with income $h_s^* = p^*x_s^*$; this proves assertion (2).

Condition (3) is obviously satisfied, and complementary slackness ensures that commodities in excess supply will have zero price.

To verify condition (4), we use the fact that in the optimum, $p^*(\Sigma_i x_i^* - \Sigma_j y_j^* - \Sigma_i \omega_i) = 0$ (theorem A.2.1, assertion 7) which, since $p^* \Sigma_j y_j^* = \Sigma_i \Sigma_j \theta_{ij} \Pi_j(p^*)$, ensures that the transfers $T_i^* = p^*x_i^* - (p^*\omega_i + \Sigma_j \theta_{ij} \Pi_j(p^*))$ add up to zero.

Finally, since all four conditions hold, the Lagrange multipliers p^* can be interpreted as equilibrium prices (up to a scaling factor which would map these multipliers on the simplex). ■

Proposition 1.9 shows that the welfare optimum generates allocations that are optimal for producers and consumers at prices p^* as well as compatible. However, it may also generate incompatible allocations. When such prices p^* and the implied transfers T_i^* are communicated to the agents, they can make decisions on their own that do not satisfy the commodity balances. The reason is that choices made by agents are not necessarily unique[24] (the production set of producer j is only required to be convex rather than strictly convex,[25] and the utility function of consumer i is concave rather than strictly concave[26]). In chapter 3 we will see that incompatibility can be avoided whenever the utility functions are strictly concave. Finally, we note that Lagrange inequalities (1.15) and (1.16) indicate that at prices p^* the multiplier λ_i associated with the budget

24. This is shown in the following example. Assume that there is only one producer and one consumer in our economy and that in equilibrium the price for commodity 1 is $p_1^* = 1$. At that price suppose that the producer's optimal supply of commodity 1 is any $y_1 \in [2, 4]$, while the consumer's plan is any $x_1 \in [4, 6]$. The allocation $y_1 = 2$, $x_1 = 6$ is evidently not an equilibrium, though there exists an equilibrium allocation $y_1 = x_1 = 4$.

25. See theorem A.1.1.

26. See theorem A.1.3.

constraint in (1.2) is i's marginal utility of income and is equal to the inverse of the welfare weight.

1.4.4 The Second Welfare Theorem

Proposition 1.8 establishes that any Pareto-efficient allocation can be represented by a welfare program with positive weights α_i for consumers with positive utility u_i^0 and zero consumption for the others. Proposition 1.9 shows that a welfare program with positive welfare weights is a competitive equilibrium with transfers. Therefore we have proved the second welfare theorem which can be stated as follows:

PROPOSITION 1.10 (Second welfare theorem) Assume that the production set Y_j has possibility of inaction, is compact and convex; assume further that the utility function of consumer i is continuous, concave, and nonsatiated with $u_i(0) = 0$ and that $\Sigma_i \omega_i > 0$. Then any Pareto-efficient allocation with positive utilities $u_1^0\, u_2^0, \ldots, u_m^0$ is a competitive equilibrium with transfers as in definition 1.4.

Proof The result follows from propositions 1.8 and 1.9 ■

The first welfare theorem shows that for given ownership of endowments and shares in firms, a competitive equilibrium will be Pareto-efficient. However, the resulting allocation may be considered to be unacceptable from an equity perspective. The second welfare theorem shows that if transfers can be mobilized, one can achieve distributional objectives while the resulting equilibrium will also be Pareto-efficient, though obviously not Pareto-superior to the first one. However, the second welfare theorem assumes that some exogenous force sets the desirable welfare weights, without taking into account the fact that consumers were assumed to be selfish; while of course being happy to receive transfers, they have no desire to give any. If willingness to give transfers is represented as part of the consumers' preferences (empathy), then a competitive equilibrium with voluntary transfers will account for equity considerations and be Pareto-efficient at the same time. To this, we will return in chapter 9.

The transfers are also referred to as lump-sum subsidies (when positive) or lump-sum taxes (when negative), and they do not impair on efficiency. In chapter 5 we will see that only very few taxes qualify as lump-sum taxes, and that those taxes that are easy to raise (e.g., indirect taxes) in fact do reduce efficiency.

1.4.5 The Single-Consumer Case

Let the assumptions of proposition 1.9 hold. Then, if there is only one consumer, the welfare program reduces to

$$\max u(x),$$
$$x \geqslant 0,\ y_j \text{ all } j,$$
subject to
$$x - \Sigma_j y_j \leqslant \omega \qquad (p), \tag{1.17}$$
$$y_j \in Y_j.$$

As can easily be seen, satisfaction of the single budget constraint follows from satisfaction of the commodity balances, so the optimum of this convex program is a competitive equilibrium.

In this case, proving existence and computing an equilibrium reduce to analyzing and solving a convex program, and there is no need for fixed point arguments or algorithms. This illustrates that the main complication in the structure of a general equilibrium model results not from the number of commodities or firms but from the fact that there are several heterogeneous consumers; this became already apparent in section 1.2.4 in connection with the general properties of the excess demand function.

The welfare program (1.11) can also be viewed as a single-consumer model whose utility function is $\Sigma_i \alpha_i u_i(x_i)$, with budget constraint $p\Sigma_i x_i \leqslant p\Sigma_i \omega_i + \Sigma_j \Pi_j(p)$. To exploit results from the theory of convex programming, we will often use this construction.

1.4.6 Welfare Consequences of Reforms

The welfare program (1.11) may be interpreted as the planning model of a government. In this simple setting, welfare weights are the only policy objectives that the planner can change. The resulting plan can be implemented either through central quantity allocations or through prices and lump-sum transfers. Though this setting is highly stylized, it has all the ingredients necessary for analysis of the policy reforms in the models of subsequent chapters.

We call policy reform a change of welfare weights, and we will see, particularly in connection with taxes (chapter 5), prices rigidities (chapter 6), and noncompetitive pricing (chapter 11), that many reforms can be interpreted in this way. Indeed in these models we show that the distortion caused by taxes, rigidities, and noncompetitive pricing can be represented through an additional consumer whose utility has a positive weight in the social welfare function. A reform that reduces such distortions in a proportional manner can be looked at as reducing the welfare weight of this additional consumer and will show the impact of the change in this welfare weight on the utility of the group of all other consumers. This justifies the emphasis on the two-consumer case with consumers 1 to m in the first group and the "additional consumer" as the second group.

In general, any policy driven change of parameters can be interpreted as a reform, but the literature is usually concerned with reforms that have some normative content. Weakly Pareto-superior reforms (reforms such that no consumer is worse off after the reform) and efficiency-improving reforms (which are closer to Pareto-efficiency after the reform) have been given particular attention. The first type combines efficiency and equity considerations and will require compensating transfers; the second only looks for efficiency, without providing compensation to losers.

Reform with Compensation

Consider program (1.11), which we constrain further in order to maintain weak Pareto-superiority. For this, we impose pre-reform utility levels u_i^0 (optimal in welfare program 1.11) as lower bounds; we assume that all pre-reform welfare weights are positive:

$$
\begin{aligned}
&\max \Sigma_i \alpha_i u_i(x_i),\\
&x_i \geqslant 0, \text{ all } i,\ y_j, \text{ all } j,\\
&\text{subject to}\\
&u_i(x_i) \geqslant u_i^0,\\
&\Sigma_i x_i - \Sigma_j y_j \leqslant \Sigma_i \omega_i,\\
&y_j \in Y_j.
\end{aligned}
\tag{1.18}
$$

Clearly the solution of program (1.18) will, at pre-reform welfare weights, coincide with the solution of welfare program (1.11). Changing a welfare weight will not affect the allocation because, by proposition 1.7, the earlier allocation was Pareto-efficient. Thus the reform has no effect.

Now consider the case of a reduction of the welfare weight of consumer s, without compensating him, that is, without imposing a lower bound for him in

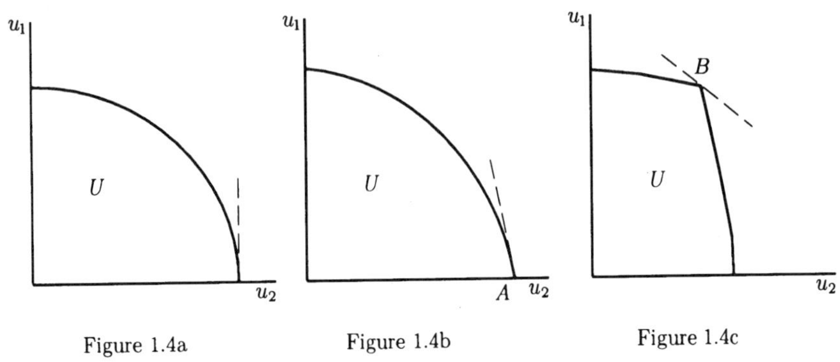

Figure 1.4
The utility possibility set U.

program (1.18). Then consumer s may lose and other consumers may gain, but they can certainly not lose because of the lower pre-reform bound on their utility.

Reform without Compensation

Intuition suggests that if in the welfare program (1.11), that is, in the program without lower bounds on utility, the weight of consumer s is reduced, then, as in (1.18), his utility cannot increase. In chapter 2 (proposition 2.14, assertion 4), we will show that this is indeed so in the m consumer case. Here we restrict the discussion to $m = 2$.

We first rewrite program (1.11) as

$$\begin{aligned} &\max \Sigma_i \alpha_i u_i, \\ &u_i, \text{ all } i, \\ &\text{subject to} \\ &(u_1, u_2, \ldots, u_m) \in U, \end{aligned} \tag{1.19}$$

where U is the utility possibility set, defined as

$$U = \{(v_1, v_2, \ldots, v_m) \in R^m_+ \mid v_i \leqslant u_i(x_i),\ x_i \geqslant 0, \text{ all } i,\ y_j \in Y_j, \\ \text{all } j,\ \Sigma_i x_i - \Sigma_j y_j - \Sigma_i \omega_i \leqslant 0\}.$$

Under the assumptions of proposition 1.8, the utility possibility set U is nonempty, compact, and convex. It is represented in figure 1.4 for the case of two consumers. In each of the panels we represent program (1.19) by taking a tangent to the frontier of U, the slope of which is $-\alpha_2/\alpha_1$. In figure 1.4a the utility possibility frontier at u_1 or $u_2 = 0$ is perpendicular to one of the axes. As a result, for any positive (α_1, α_2) both utilities will be positive. By contrast, in figure 1.4b the slope is nonzero in point A. Therefore, though both α_1, α_2 are positive, u_2 is zero. In figure 1.4c, at point B, utilities of both consumers remain unaffected under a small change in welfare weights. Note also that in figures 1.4a and b, a reduction in α_2/α_1 will always lead to a reduction of u_2 and an increase of u_1.

These remarks lead to the following proposition.

PROPOSITION 1.11 (Changes of welfare weights in the two-consumer case) Let the assumptions of proposition 1.9 hold, and consider the welfare program (1.19) for the two-consumer case, with $\alpha_2 \geqslant 0$ and $\alpha_1 > 0$. After a reduction of α_1,

1. if there is compensation for consumer 1 (or both consumers), the allocation is unchanged and no consumer is affected;

otherwise,

2. consumer 1 cannot gain and consumer 2 cannot lose.

Moreover

3. if the boundary of U is smooth (figures 1.4a and b), consumer 1 will lose and consumer 2 will gain;

4. if convexity of U is strict (figures 1.4a, b, and c) and consumer 1 loses, then consumer 2 will gain;

5. if convexity of U is strict, then bringing α_1 to zero will lead to a gain for consumer 2.

Proof The proof is obvious by inspection of figure 1.4. ■

Note that the proposition relies on the compact valuedness of the set U, but several of the five properties above can be seen to hold in the absence of convexity.

Other Reforms

Reforms following from changes in welfare weights can be implemented in a decentralized way through the prices and the transfers generated by the welfare program provided that the agents' choices are unique (otherwise the incompatibility problem may arise). However, the informational requirement of the welfare program is very demanding, and in practice, reforms are often carried out by redistributing endowments or shares in firms.[27] If the generated equilibrium model is accepted as a description of reality, it can be used to compute solutions before and after the policy change and to compare outcomes, in particular, utility levels. Some consumers will gain, others will lose; indeed, it is not possible for all consumers to gain, since the pre-reform solution was Pareto-efficient.[28]

The Pareto-criterion only checks whether a consumer gains or loses in terms of his utility. However, changes in utilities do not give a cardinal indication of the size of welfare gains and losses for any given consumer (and a fortiori they are not comparable across individuals). Therefore, in applications, there is a need for other welfare criteria. These will be discussed in section 2.4.2.

1.5 Possible Extensions of the Competitive Model

The competitive model discussed so far is highly stylized and may appear to be unrealistic; it has often been criticized on these grounds. For example, one may argue that firms set prices rather than taking these as given and that consumers do not know their income with certainty when they decide on consumption. However, the basic model has played a role of increasing

27. In reality many other parameters, like tax rates, can be changed by the government.

28. Obviously, if the pre-reform solution were not Pareto-efficient, as is usually the case in applied models, the reform would lead to Pareto-superior (or Pareto-inferior) results.

significance because the two welfare theorems provide far-reaching guidance for policy prescriptions. The first theorem shows that, under rather weak assumptions, Pareto-efficiency is implemented by a competitive equilibrium (possibly mimicked by central planning), while the second theorem states that distributional objectives should be implemented by lump-sum transfers (or possibly by other means that lead to the same outcome).

The theorems have the following policy consequences:

1. *Existence of markets.* Markets should exist for all commodities in all periods. Transaction costs should be avoided, since they reduce welfare.

2. *Distortions.* Price-distorting taxes or tariffs, price rigidities, and quantitative restrictions cannot lead to Pareto-efficient outcomes, by comparison to a situation without such taxes and rigidities.

3. *Transfers.* Though income transfers to consumers may be desirable socially, Pareto-efficiency can be maintained without them; if given, transfers should be lump sum.

4. *Producer support.* There is no need to support (or tax) producers through subsidies or transfers. Those who would make a loss at the going prices should not produce, and those who can make a positive profit should be allowed to produce as much as they like. Producers should maximize profits; any other rule of conduct that leads to different outcomes is Pareto-inefficient.

5. *Competitiveness.* Markets should be competitive: All agents should take prices as given.

These are strong and clear guidelines that essentially advocate free competition. Any possibly more realistic model that does not generate a competitive equilibrium will lack the important property of Pareto-efficiency. The guidelines therefore say that it is misleading to call the competitive model unrealistic. They tell governments to transform their economies into a competitive general equilibrium, that is, to "make" the model realistic. Government should protect property rights and ensure that the five competitive conditions above hold. However, several arguments have been advanced against these guidelines because they rest on particular assumptions concerning social institutions, technologies, and preferences that may be violated by empirical evidence, in particular:

6. *Rivalry.* A consumer only benefits from the commodities he buys, and a commodity allocated to him cannot at the same time benefit another consumer; the same holds for producers.

7. *Selfishness.* Consumers only derive utility from their own consumption, not from utility of other consumers.

8. *Fixed endowments.* Endowments are given parameters and hence are not affected by allocations.

9. *Convexity of production sets.* Production sets are convex, and all commodities are perfectly divisible.

Relaxation of the assumptions of the purely competitive model will necessarily lead us to redefine the role of institutions and to widen their scope. Though in applied models many of these assumptions will be relaxed at the same time, we address them by turns as follows.

1. *Commodity balances for all commodities.*[29] Access to some markets will be restricted and involve (possibly prohibitive) costs. Transactions may also be impossible without a means of exchange (coupons or money). Such extensions of the competitive model will be introduced in section 3.6.2 for commodity markets, and in chapter 12 for coupons and money.

2. *Prices as the only parameters individual consumers and producers take as given.* We will introduce other parameters, like taxes and rations (to implement price rigidities), in chapters 5 and 6 and show that they are distortionary. In those chapters a government will be introduced that levies taxes and redistributes the proceeds to agents, but that also appears as an additional consumer with a utility function and a budget constraint of his own.

3. *Distributional objectives implemented through lump-sum transfers.* It is difficult to design lump-sum schemes. In practice, many such schemes work like taxes (chapter 5).

4. *Producers maximizing profits at competitive prices.* As long as production sets allow for inaction and are convex, this principle will be maintained. In chapter 5 we show how producer taxes and subsidies affect allocations and in chapter 10 we relax the assumptions on technology.

5. *Agents taking prices as given.* When agents, particularly producers, become relatively large, it may become relevant for them to anticipate the consequences on markets of their decisions. Chapter 11 will address problems of imperfect competition.

6. *Rivalry.* Nonrival goods, which affect the agents either positively (public services, like broadcasting) or negatively (pollution), will be a topic of chapter 9. Government consumption will then be decided upon by agents themselves, and unlike the model in chapter 5, there will be no separate utility function for the government.

29. This requirement is not only demanding theoretically but also impossible to meet in applied models. Therefore shortcuts are unavoidable, even if they go against theoretical elegance.

7. *Selfishness.* Altruism will also be introduced in chapter 9, and the associated equilibrium with voluntary transfers will make it possible to achieve equity objectives in a competitive setting.

8. *Fixed endowments.* Endowments will usually include labor, but labor can only be "produced" by workers who consume at least a minimal basket of commodities. This relation between labor supply and consumption is known as the efficiency-wage relation (covered in chapters 9 and 10).

9. *Convex technology sets with all commodities perfectly divisible.* Setup costs and increasing returns, both of which violate the convexity assumption, will be looked at as resulting from indivisibilities and considered more fully in chapter 10. We will be able to prove existence of equilibrium only in very few cases with nonconvexity.

These extensions will lead to various versions of a static model, which may be viewed as a closed-country model or as a world model of international trade. In such a world model, countries can be represented by subsets of producers and consumers who formulate supplies and demands that, through summation, generate a net "national" demand for all commodities. For each national model we can derive a balance of payments, national accounts, and an account of revenues and expenditures. The opening-up of the closed-economy model to international trade will be dealt with in chapters 4 (under free trade) and 5 (when tariffs and quota are imposed).

The static model can also be given a finite-horizon dynamic interpretation if one distinguishes commodities by date of delivery (and possibly agents by date of birth and death). Such an approach will be pursued in chapter 7, but it assumes that agents have perfect foresight with respect to prices. We will also deal with cases where this assumption is not satisfied (temporary equilibrium and recursive dynamics). Finally, the finite-horizon assumes that the world ends after T periods, and this truncation generates distortions basically due to inappropriate valuation of terminal stocks. It is therefore natural to consider (in chapter 8) infinite-horizon models that have indeed received much attention in the literature.

There are obviously also many isues that we do not address. For example, when we discuss decentralization and efficiency, we only look for Pareto-efficient schemes and do not study how property rights can be enforced and how appropriate (price) signals can be generated to ensure incentive compatibility.

More specifically, the coverage of the subjects is restricted in several respects. Imperfect competition is dealt with either in an ad hoc way (through markup pricing) or in a more rigorous way, but then without existence proofs. In chapter 12 the treatment of uncertainty is limited to the analysis of the model with incomplete asset markets. Infinite numbers of commodities and agents appear in the infinite-horizon model but otherwise receive scant attention.

To incorporate all the extensions that were mentioned, some further preparation and standardization of assumptions is needed. These are the subject of chapters 2 and 3 which, together with the mathematical appendix A, provide the basic tools for a unified treatment of the subject.

2 Optimal Production and Consumption Allocations

In chapter 1 the behavior of producers and consumers was described through optimization. Assumptions P1, C1, and C2 were used to prove the existence of equilibrium, and some weaker assumptions, to analyze welfare properties. In this chapter we relax assumptions P1, C1, and C2, in order to obtain a version that will serve as our standard throughout the book.

The chapter does not deal with general equilibrium per se. Sections 2.1 and 2.2 deal with producers and consumers, respectively, using the producer and the consumer programs (1.1) and (1.2) to derive demand and supply choices under different assumptions. We provide some microeconomic justification for these, and discuss limitations and ways to test their validity. This should also clarify their implications for applied model building. Both sections contain reference material for the remainder of the book, which can be found in standard microeconomic textbooks. It is only included here for the convenience of the reader, who may readily proceed to section 2.3 which summarizes the discussion via standard assumptions and basic propositions that are needed in the sequel. Finally, section 2.4 is devoted to a further analysis of the welfare program and to welfare measures.

If one is interested in the optimal choices y_j of producer j and x_i of consumer i, it is natural to restrict attention to commodities that enter the production set of the producer as inputs or outputs, or the utility function of the consumer. In general equilibrium, one needs to determine aggregate supply and demand ($\Sigma_j y_j$ and $\Sigma_i x_i$) and all these vectors must be defined in the same full commodity space of dimension r. In the presentation we do not distinguish the subset of commodities that is of interest to a particular agent. Unless specified otherwise, all agents are assumed to face the same commodity space. The propositions of this chapter easily extend to the case where an agent does not consume or produce all commodities, by defining the production sets and the utility functions on his particular commodity space and by setting to zero his supply and demand of the remaining commodities.

2.1 Producer Behavior

In the competitive equilibrium model, producer j is assumed to maximize his profit at given prices p on his production set Y_j. Formally, the producer has to compute a plan y_j which solves

$$\Pi_j(p) = \max_{y_j}\{py_j \text{ for } y_j \in Y_j\}. \tag{2.1}$$

2.1.1 The Producer

Before turning to the assumptions on Y_j and the properties of this program, it may be useful to digress on the interpretation of the index j. In other words, who is j? So far we have assumed that every producer is represented separately, and we have been loose in our definition of producers and firms: Any agent who behaves as in (2.1) is a producer.

In a general equilibrium model, the concept of a producer is far less clear than that of a consumer. The consumer is typically an individual; his existence does not have to be explained by economic theory. All one may argue about is whether the consumption decision is made by an individual, by a family, or by an even larger unit. For production, one may assume that there exists a physical technology that describes the set of activities that agents could engage in. This may be represented by an aggregate production set Y.[1] However, the general equilibrium model also assumes that there are n firms, each with a given technology set Y_j and $Y = \Sigma_j Y_j$.[2] In reality the organization into firms is not only a technological datum: Specialization of firms depends on market conditions (e.g., on transaction costs). For aggregate production $\Sigma_j Y_j$, it does not matter whether one considers the technologies to belong to one or several firms. As long as assumption P1 holds, problems (2.1) solved for $j = 1, 2, \dots, n$, will lead to the same optimal outcome as the problem of collective profit maximization:

$$\max p\Sigma_j y_j,$$

$$y_j, \text{ all } j,$$

subject to

$$y_j \in Y_j.$$

However, consumers are assumed to own shares in firms and not in production processes, so the way of organizing the economy into firms may affect their income. Clearly, if profits are zero (as will be the case under constant returns to scale), the distribution of shares does not matter.

Note that since by assumption P1 producers may be inactive, we can consider the list 1 to n to contain all potential producers, and those who do not produce in equilibrium are not absent but simply inactive.

2.1.2 Technology of the Firm

Assumption P1 consists of three parts: Possibility of inaction, compactness, and strict convexity. Before turning to the discussion of these assumptions, we give some graphical illustrations of production sets in a two-commodity economy.

1. From this set the aggregate consumption set X can be derived as $X = \{x \geqslant 0 \mid x \leqslant \Sigma_i \omega_i + y, y \in Y\}$, where X has to be nonempty and compact.

2. $\Sigma_j Y_j$ denotes the union of the sets.

Possibility of Inaction

Possibility of inaction gives to each producer the freedom not to produce. This ensures nonnegativity of profits. In figures 2.1a and b, we have depicted compact and strictly convex production sets (the shaded areas). In figure 2.1a the firm can produce commodity 2 using commodity 1 as an input, but it cannot remain inactive, since $0 \notin Y_j$. In figures 2.1b and e, the firm can choose to remain inactive.

Compactness

In figure 2.1 all production sets are compact. Compactness may be decomposed into closedness and boundedness. Intuitively closedness means that the boundary of the set belongs to the set. To see that this is not overly

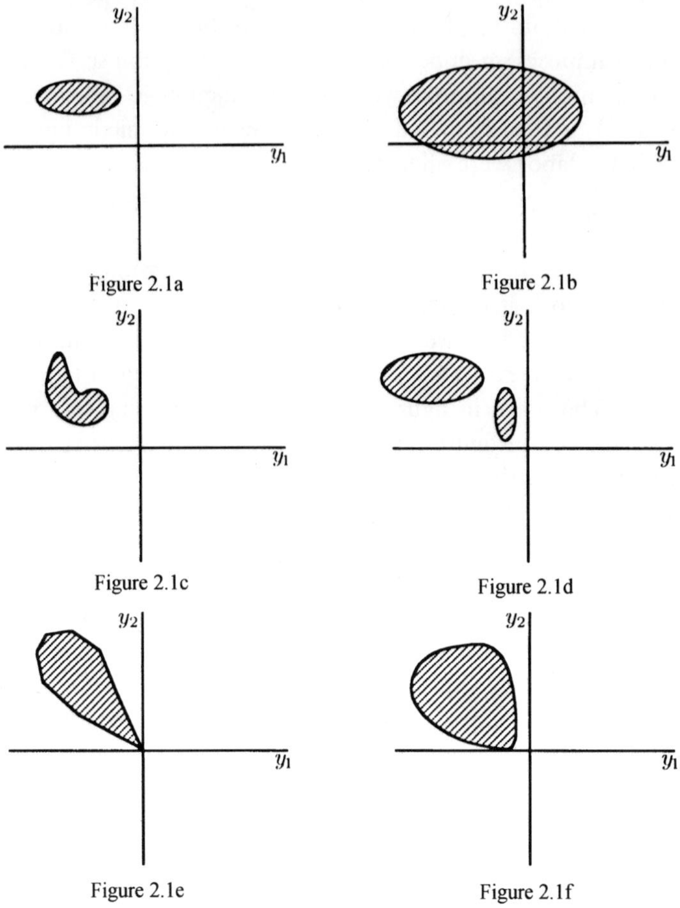

Figure 2.1
Representation of production sets.

restrictive, consider the following example in which a technology needs liquid water under room conditions. At 0°C the water freezes, and at 100°C it boils. Suppose that the intensity of the production process, say, pumping, determines the temperature, with freezing at low and boiling at high intensity. Since the technology needs liquid water, ice and gas are not allowed, and therefore the production set is open. However, for all practical purposes, it is natural to introduce a safety interval [0.1°C, 99.9°C] within which the technology must operate, and this is a closed (and bounded) set. Such phenomena often occur in production processes, where breakdown sets the limits on technology.

Boundedness means that no vector $y_j \in Y_j$ can have an infinite element. Over a finite time horizon, this assumption seems reasonable. Although one may argue that bounds should only be imposed on the set Y (i.e., on the production side of the economy as a whole), these aggregate bounds necessarily constrain the supply of each firm. For instance, the total labor endowment imposes an upper bound on the production set Y_j of each firm, which needs additional labor whenever production is expanded. The production set Y_j is then bounded (from above) by the maximum labor input if all available labor were allocated to it.

Strict Convexity

Strict convexity is the most restrictive condition of assumption P1. Convexity (and, a fortiori, strict convexity) of Y_j is violated in figures 2.1c and d (and also in figure 2.1a if we assume that $0 \in Y_j$). If there is no possibility of inaction, strict convexity (but not convexity) is violated in figure 2.1f along the y_1-axis. The set Y_j in figure 2.1c exhibits "increasing returns to scale." The nonconvexity in figure 2.1d can be avoided by splitting the firm into two firms with strictly convex sets each. The set Y_j of figure 2.1b satisfies assumption P1. Finally, note for later reference that in figures 2.1a and b the boundary of the Y_j set is smooth, so there will be a unique tangent at every point; this is not the case in figures 2.1c, e, and f, where kinks appear.

If strict convexity is assumed, it is not possible to represent constant returns to scale and other linearities in technologies that are often present in applied models. In most chapters of the book, we relax the strict convexity condition of assumption P1, and assume instead that:

ASSUMPTION **P1′** (Production sets) The production set Y_j of every producer j has possibility of inaction ($0 \in Y_j$); it is compact and convex.

Other Possible Assumptions on the Production Set

Production theory often makes assumptions that are thought to add realism to the description of technology but are not needed for the derivation of well-behaved supply functions (or correspondences). We mention three such assumptions: (1) free disposal, (2) no free lunch, and (3) irreversibility of production processes.

1. *Free disposal.* All surpluses can be disposed of at no cost. Formally, this is incorporated via the weak inequality sign in the commodity balance and does not need a separate assumption on Y_j.

2. *No free lunch.* Positive production needs inputs, or formally,

$Y_j \cap R^r_{++} = \varnothing$.

In assumption P1, and in figure 2.1b, commodities can be produced without apparent use of inputs. In a model where all commodities (goods and services) are represented and can be recorded in physical terms, it is clear that no firm should have the possibility to produce without inputs. The no-free-lunch assumption formalizes this requirement. However, in applied modeling, many inputs (climate, fresh water, etc.) that can be recorded will nevertheless be left out of the model for convenience; other inputs (management skills, creativity, etc.) cannot even be recorded, let alone accurately observed, and will have to be left out. At any rate, no applied model will include all inputs (or outputs), and we may therefore accept strictly positive points in Y_j, as those in figure 2.1b.

3. *Irreversibility of production processes.* Production processes work only one way, or formally, if $y_j \in Y_j$, then $-y_j \notin Y_j$.

The idea behind irreversibility is that, even in cases where it is possible to undo a physical transformation, in every production process inputs will be required in positive amounts, and it is not possible to combine production processes in such a way that overall there is no need for inputs. For example, in transportation, one can return a commodity to its origin but not without using energy; also if one produces hot water with electricity and electricity with hot water, one needs to supply either electricity or hot water from outside. Again, if we do not represent all inputs, reversibility cannot be excluded in our representation of production sets. Therefore the production set represented in figure 2.1b violates irreversibility.

Linear Activity Analysis as a Special Case

In many applications, modelers use what is known as linear activity analysis, which represents producer behavior via linear programming. We will characterize the constraint set of such a program in terms of the above assumptions on production sets.

Define the production set Y_j as

$$Y_j = \{y_j \mid y_j = B_j q_j;\ A_j q_j \leqslant b_j \text{ and } q_j \geqslant \underline{q}_j\}. \qquad (2.2)$$

In (2.2), B_j is an $r \times s$ matrix, A_j is a $t \times s$ matrix, b_j is a vector of firm specific given resources, y_j represents supplies (positive for outputs, negative for inputs),[3] q_j is a vector of activities, and $\underline{q}_j$ (possibly zero) a vector of lower bounds on activity levels.

To illustrate such a set, consider j to represent a baker who produces two types of bread (commodities 1 and 2), using two types of flour (commodities 3 and 4); assume that he cannot change his production capacity within the time horizon of the model. Dropping the producer index j, the constraints $y = Bq$ can be specified as

$$\begin{bmatrix} y_1 \\ y_2 \\ y_3 \\ y_4 \\ y_5 \end{bmatrix} = \begin{bmatrix} 1 & 0 & 0.6 & 0 & 0 \\ 0 & 1 & 0.6 & 0 & 0 \\ 0 & 0 & 0 & -1 & 0 \\ 0 & 0 & 0 & 0 & -1 \\ 0 & 0 & 0 & 0 & 0 \end{bmatrix} \begin{bmatrix} q_1 \\ q_2 \\ q_3 \\ q_4 \\ q_5 \end{bmatrix},$$

while the $Aq \leqslant b$ constraint is

$$\begin{bmatrix} 1.1 & 1.05 & 1.15 & -0.9 & -0.85 \\ 0.2 & 0.1 & 0.25 & 0.15 & 0.19 \end{bmatrix} \begin{bmatrix} q_1 \\ q_2 \\ q_3 \\ q_4 \\ q_5 \end{bmatrix} \leqslant \begin{bmatrix} 0 \\ 20 \end{bmatrix}.$$

In the example, q_1, q_2, and q_3 are the levels of the activities that produce (and sell) breads of two types; activities q_4 and q_5 purchase the inputs and produce the dough. The third activity jointly produces the two types of bread. The first inequality in the $Aq \leqslant b$ constraints is the "commodity balance" for dough, which states that demand for dough should not be larger than supply and reflects substitutability of dough among its uses; the zero on the right-hand side shows that there is no initial stock of dough. Note also that dough is an "internal" commodity not appearing in the y vector. The last inequality is the balance for capacity, the right-hand side of which is fixed available capacity, since no new production capacities can be created within the time span represented in the model.

It is easy to verify that the above assumptions on production sets are satisfied. Indeed, if $\underline{q}_j = 0$, then $0 \in Y_j$. Closedness of Y_j follows from the weak inequalities, while convexity (not strict convexity) follows from linearity of the constraints. We still have to check boundedness. Suppose that the

3. In our setting, outputs have positive coefficients, and inputs negative coefficients in B_j. In linear programming, the opposite is more common.

assumption on irreversibility does not hold; this would imply that B_j and A_j contain two columns with the same numerical values, but of opposite sign; as a result the associated activity levels could become unbounded, since the baker could then produce bread from dough and dough from bread, without loss. So irreversibility is necessary for boundedness of Y_j. The no-free-lunch assumption requires that no activity has only nonnegative entries in B_j and nonpositive entries in A_j; clearly what matters for boundedness is that each activity directly or indirectly requires at least one input that cannot be purchased. In our example this is the fixed production capacity. Observe that neither irreversibility, nor no-free-lunch ensure this. Thus boundedness has to be imposed explicitly.

Figure 2.2 represents compact production sets which are linear, as in our example. In figure 2.2a the set satisfies all the assumptions. In figure 2.2b irreversibility is violated; in 2.2c there is free lunch. Note that in all three cases compactness is maintained.

We have so far required the production sets to be bounded. Inspection of figures 2.1 and 2.2 suggest that, for a profit-maximizing producer, it is only boundedness from above that should matter, meaning that profits should never become infinite. We will confirm this when turning to producers' behavior.

2.1.3 Profit Maximization

Profit-maximizing behavior has been questioned in two respects: Do producers have access to all the information that they need to optimize? Is profit maximization a rational rule of conduct for the owners of the firm? With respect to the first issue, in reality producers are obviously faced with uncertainty about their external environment (prices, solvency of their customers, etc.) and about their own production set (productivity of workers, reliability of equipment, etc.) Moreover, in some markets, it is uncertainty itself that makes profits possible. Although uncertainty can be incorporated into theoretical models, this often leads to formulations that

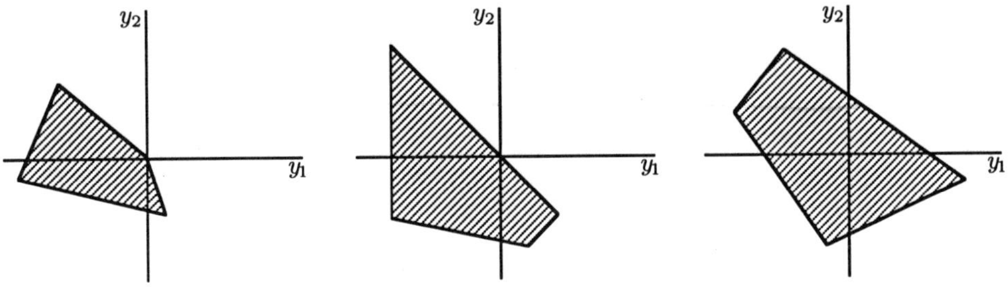

Figure 2.2a Figure 2.2b Figure 2.2c

Figure 2.2
Compact linear production sets.

are too intractable for applied work. However, disregarding uncertainty completely may lead to extreme supply reactions and overspecialization. Therefore, in applications, one maintains profit maximization but accounts for uncertainty by restricting the production set through additional constraints. For example, one excludes highly unreliable techniques or limits the options for substitution.

With respect to the rationale for profit maximization, it may be argued that the producer should be characterized by utility-maximizing behavior and satisfy his own preferences rather than maximizing the profit of the firm. This can be represented by treating the producer like a consumer who maximizes utility. Using proposition 1.9, one can show that profit maximization is then also the appropriate rule of behavior for this producer.

Deciding to what extent profit maximization is realistic or not is, eventually, an empirical matter which can be tested by checking whether data violate this principle. We return to this in section 2.2.4.

It is also important to recall that the two welfare theorems show that profit maximization is socially desirable. Even if individual producers were to become "more clever," more knowledgeable about their environment, and sufficiently powerful to affect prices, collectively, they would not be able to improve upon the Pareto-optimal outcome.

The profit-maximizing behavior of the producer on his production set is illustrated in figure 2.3a, where the arrow points into the direction of profit maximization. The tangent to Y_j, which is shown at the optimal point, has slope $-p_1/p_2$ and intercept Π_j/p_2. The area southwest of the dotted lines is there to show that Y_j only needs to be bounded from above preventing Π_j from becoming infinite.

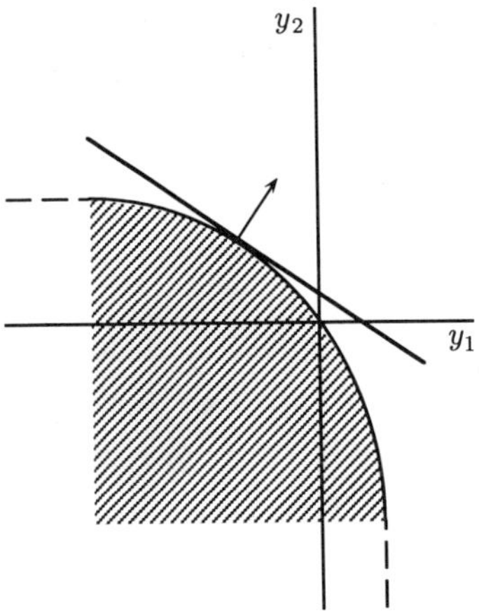

Figure 2.3a

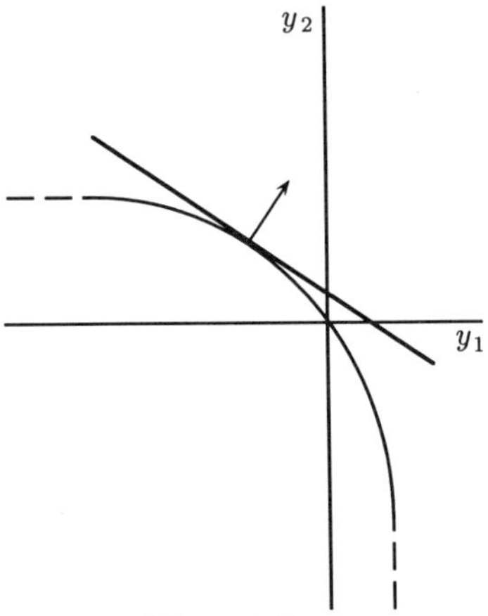

Figure 2.3b

Figure 2.3
Profit maximization on Y_j.

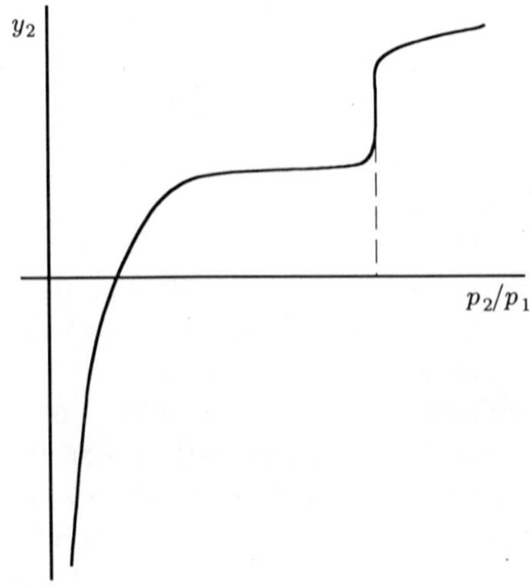

Figure 2.4
The supply correspondence.

As long as prices are positive, the only part of Y_j that matters is the curved part on the upper right, as shown in figure 2.3b. This curve is the transformation function; it maps out the set of efficient points, that is, the maximal y_{2j} subject to $(y_{1j}, y_{2j}) \in Y_j$ and $y_{1j} \geqslant y_{1j}^0$, where y_{1j}^0 is feasible and given. At any other point there is waste of input, or output, or both.

Varying prices in (2.1) generates the supply function, which is shown in figure 2.4 for the two-commodity case. This figure points out that the supply function is neither concave nor convex; it has horizontal segments, which correspond to kinks in the transformation function, and vertical segments, which correspond to linear segments of the transformation function (this is only possible if Y_j is not strictly convex). Positive values of y_{2j} represent outputs; when p_2/p_1 becomes small enough, producer j starts using y_{2j} as an input, most likely, to supply y_{1j}.

Note, however, that, as is illustrated in figure 2.3b, we do not require compactness of Y_j to obtain bounded optimal solutions y_j in (2.1). It suffices that the set $\{y_j \mid y_j \in Y_j, py_j \geqslant 0\}$ is nonempty, compact.[4]

Figure 2.3 also illustrates that assumption P1 is sufficient to ensure continuity of the supply function. When Y_j is not strictly convex, supply is a correspondence as in figure 2.4; if Y_j is represented by linear constraints, the supply function exclusively consists of horizontal and vertical segments.

4. See theorem A.2.1 and A.3.1.

The Transformation and the Production Function

The production set is an abstract construct. One may want to have an explicit representation of this set, as in (2.2), with several constraints. Sometimes one finds it convenient to work with a single constraint; this can be achieved by considering the transformation function (and in special cases, the production function). To simplify notation and when no confusion is possible, we drop the producer index j; for the individual producer set, we use Y_0. We first derive the transformation function from the production set Y_0, and then from a set of constraints that represent Y_0.

In figure 2.3 we have represented the transformation function of a producer. For values (y_1, y_2) that do not belong to the production set Y_0, that is, lying northeast of the curve in figure 2.3b, this function is not defined. Nevertheless, to obtain a transformation function defined for all $y \in R^r$, we assign the value $-\infty$ whenever this occurs. This can be obtained more formally by defining the set

$$\Psi(y_1^0, y_2^0, \ldots, y_{r-1}^0) = \{(y_1, y_2, \ldots, y_r) \in Y_0, y_k \geqslant y_k^0, k = 1, 2, \ldots, r-1\};$$

the production function $f: R^{r-1} \to R$ is then defined as

$$f(y_1^0, y_2^0, \ldots, y_{r-1}^0) = \begin{cases} \max\{y_r \mid (y_1, y_2, \ldots, y_r) \in \Psi(y^0)\} & \text{if } \Psi(y^0) \text{ is nonempty,} \\ -\infty, & \text{otherwise.} \end{cases}$$

The function $f(.)$ will be concave (by the perturbation theorem A.3.2), nonincreasing (if free disposal is assumed), and $f(0) \geqslant 0$ (by possibility of inaction).[5]

Finally, the transformation function is obtained as $F(y_1^0, y_2^0, \ldots, y_r^0) = y_r^0 - f(y_1^0, y_2^0, \ldots, y_{r-1}^0)$. Another way of obtaining this function is to represent the production set by constraints $g_h(y) \leqslant 0$, $h = 1, 2, \ldots, H$, as was done in (2.2) for the linear case. Each function $g_h(y)$ is continuous and convex, nondecreasing, and satisfies $g_h(0) \leqslant 0$. The transformation function (in implicit form) can now be obtained as

$$F(y) = \max_h g_h(y);$$

the transformation function replaces the H constraints by a single one.[6] This approach amounts to restricting attention to the constraint that is most

5. In this construction we have used the rth commodity as maximand; it is clear that any other choice would do as well. For the two-commodity case represented in figure 2.3, this amounts to maximizing y_2 for given levels of y_1. This is a representation of the northeast boundary of the production set, as shown in figure 2.3b.

6. Note that $F(y)$ will be convex in y, since convexity is maintained under the $\max_h$ operation. It will be nondecreasing in y (if free disposal is assumed) and satisfy $F(0) \leqslant 0$ (by possibility of inaction).

binding at given y. It leads to the profit-maximizing program with a single constraint:

$$\max_{y} py,$$

subject to

$$F(y) \leqslant 0.$$

In the production function one has to distinguish outputs from inputs. To do so, we adopt the notation q for outputs and v for inputs. We also use p_q and p_v for output and input prices, respectively. The vector q (and p_q) may have a dimension different from v (and p_v). The transformation constraint can now be written

$$F(q, v) \leqslant 0. \tag{2.3}$$

Under the further separability assumption that $F(q, v) = F_q(q) - F_v(v)$, this becomes

$$F_q(q) - F_v(v) \leqslant 0. \tag{2.4}$$

Moreover, if there is a single output, (2.4) reduces to the usual production function constraint

$$q \leqslant f(v). \tag{2.5}$$

Usually q and v will have to be nonnegative, but when one does not impose the restriction on v, (2.5) can be looked at as another specification of the transformation function.

The Profit Function and Net Supply

We now characterize the profit function[7]

$$\Pi(p) = \max_y \{py \mid y \in Y_0\} \tag{2.6}$$

and the associated net supply. This is the subject of the following propositions:

PROPOSITION 2.1 (The profit function and net supply under assumption P1) Consider the production set Y_0 satisfying assumption P1 and the associated profit function (2.6).

7. See also McFadden (1978a, b).

For positive p, $\Pi(p)$ is

1. convex in p;
2. nonnegative, homogeneous of degree one, nondecreasing in output prices, nonincreasing in input prices;
3. continuously differentiable.

Moreover

4. Hotelling's lemma holds, $y(p) = \partial\Pi(p)/\partial p$;
5. $y(p)$ is a continuous function;
6. $y(p)$ is homogeneous of degree zero in p.

Proof The constraint set Y_0 is nonempty and compact, so the program has a bounded optimal solution.[8] Property 1 follows from perturbation theorem A.3.2. Uniqueness of $y(p)$ follows from strict convexity of Y_0 (theorem A.1.1). Continuous differentiability of the objective py is obvious, so the envelope theorem A.3.3 can be invoked to establish properties 3 and 4. For property 2, nonnegativity follows from possibility of inaction, and homogeneity of degree one of $\Pi(p)$ follows from the linearity of the objective. The nondecreasingness in outputs and nonincreasingness in inputs is a consequence of property 4. Property 5 follows from properties 3 and 4. Property 6 follows from the homogeneity of degree one of $\Pi(p)$ and Hotelling's lemma. ■

PROPOSITION 2.2 (The profit function and net supply under assumption P1′) Consider the production set Y_0 satisfying assumption P1′ and the associated profit function (2.6).

For positive p, $\Pi(p)$

1. is continuous and convex in p;
2. is nonnegative, homogeneous of degree one, nondecreasing in output prices, nonincreasing in input prices;
3. it has a subdifferential $Y(p)$.[9]

Moreover

4. Hotelling's lemma holds, $y \in Y(p)$;
5. $Y(p)$ is uppersemicontinuous in p and convex valued;
6. rescaling prices does not affect outputs and inputs, $Y(\lambda p) = Y(p)$ for all given $\lambda > 0$.

8. Note that existence of a bounded optimal solution does not require Y_0 to be bounded. See also theorem A.2.1.

9. For the definition of a subdifferential, see definition A.1.16.

Proof The constraint set Y_0 is nonempty and compact, so the program has a bounded optimal solution. Property 1 again follows from perturbation theorem A.3.2. For property 3, since $\Pi(p)$ is convex, it has a subdifferential. To prove property 4, we modify the steps of the proof of the envelope theorem A.3.3. The objective function is continuously differentiable with respect to p. Therefore in the optimum the total differential of $\pi = py^*$ is $d\pi = \partial(py^*)/\partial y \cdot dy + \partial(py^*)/\partial p \cdot dp$ for given optimal y^*. Optimality of y^* implies that $d\pi = 0$ whenever $dp = 0$; hence $\partial(py^*)/\partial y \cdot dy = 0$ for all feasible variations dy. This implies that $y^* \in Y(p)$, the subdifferential of $\Pi(p)$. For property 2, while homogeneity of degree one of $\Pi(p)$ follows from the linearity of the objective, the nondecreasingness in output prices and nonincreasingness in input prices is, by definition A.1.17, a trivial consequence of property 4. Property 5 follows from the maximum theorem A.3.1. Property 6 follows from properties 2 and 4. ∎

The proposition that follows establishes sufficient conditions for the differentiability of the net supplies $y(p)$ obtained in proposition 2.1, using the production function constraint (2.5), without constraining the sign of inputs. We treat the first commodity as the only output q and assume that all commodities, including the first, can be inputs so that the vector of inputs is of dimension r. Of course commodities not entering the production process may be dispensed of, and the result of proposition 2.3 will hold if the dimensions of the vectors are adjusted appropriately.

PROPOSITION 2.3 (Differentiability of the net supply function) Consider the production function $f: R^r \to R$, $f(v)$ and the profit function $\tilde{\Pi}(\tilde{p}) = \max_v \{f(v) - \tilde{p}v\}$, where $\tilde{p} = p/p_1$ is the vector of positive normalized prices of inputs. Assume that $\tilde{\Pi}(\tilde{p})$ is compact valued for all positive $\tilde{p}$ and that $f(v)$ is "locally" strongly concave, that is, strongly concave for every optimal input demand $v(\tilde{p})$.[10] Then, the normalized profit function $\tilde{\Pi}(\tilde{p})$ is strongly convex, and the net input demand $v(\tilde{p})$ as well as the optimal output $q(\tilde{p})$ are continuously differentiable.

Proof The first-order condition of the problem includes $\tilde{p} = \partial f(v)/\partial v$, and its total differential is $d\tilde{p} = G dv$, where G is the Hessian of $f(v)$ evaluated in the optimum. Since $f(v)$ is strongly concave in the optimum, the matrix G is negative definite, and hence nonsingular. Therefore we may write $\partial v/\partial \tilde{p} = G^{-1}$. Since G^{-1} is a continuous function of $v(\tilde{p})$, input demand is continuously differentiable and so is $q(\tilde{p})$. Moreover G^{-1} is negative definite and, by Hotelling's lemma, $\partial v/\partial \tilde{p} = -\partial^2 \tilde{\Pi}(\tilde{p})/\partial \tilde{p}^2$. Therefore $\tilde{\Pi}(\tilde{p})$ is strongly convex. ∎

10. See definition A.1.15 (v).

Since the normalized profit function is twice continuously differentiable, its Hessian is symmetric, and $\partial v/\partial \tilde{p}$ is symmetric ($\partial v_k/\partial \tilde{p}_h = \partial v_h/\partial \tilde{p}_k$). Moreover strong convexity implies that the diagonal terms of this Hessian are positive, so $\partial v_k/\partial \tilde{p}_k$ is negative. Finally, the nonnormalized profit function $\Pi(p) = p_1\tilde{\Pi}(p/p_1)$ is twice differentiable, homogeneous of degree one, and hence convex but not strongly convex.

The discussion shows that there are various ways of representing producer behavior. One could choose to maximize profits subject to a production set or to the transformation function that represents this set. This is known as the primal approach. Alternatively, one can use the profit function and its derivatives, the supply and input demand functions, and this is the dual approach.

The Restricted Profit Function

Consider again the profit maximization problem with outputs q and inputs v. Assume that some elements of q or v are fixed exogenously. Let $q = (q_0, q_{00})$ and $v = (v_0, v_{00})$, where, say, q_0 and v_0 are the fixed elements. The resulting profit function $\Pi(p_q, p_v; q_0, v_0)$ is called a restricted profit function. Its properties are the same as those of the profit function (as long as the fixed levels remain feasible) and propositions 2.1 to 2.3 hold.

In some cases inputs are fixed, and outputs variable. Alternatively, outputs may be kept fixed, while inputs are variable: Profit maximization then reduces to cost minimization. In particular, when there is a single output, one obtains the (negative of the) cost function.

The Cost Function and Input Demand

The cost function is defined as[11]

$$c(p_v, q) = \min_{v \geqslant 0}\{p_v v \mid f(v) \geqslant q\}. \tag{2.7}$$

We describe its properties as well as those of the associated input demand.

PROPOSITION 2.4 (The cost function and input demand) Consider the continuous, strictly quasiconcave production function $f: R^r_+ \to R$, $f(v)$ and the associated cost function $c(p_v, q) = \min_{v \geqslant 0}\{p_v v \mid f(v) \geqslant q\}$ for given positive prices p_v and given output level $q \in f(R^r_+)$.

Then $c(p_v, q)$ is

1. continuous in (p_v, q), concave in p_v and convex in q;
2. homogeneous of degree one in p_v, nondecreasing in p_v and nondecreasing in q;

11. See also McFadden (1978a, b).

3. continuously differentiable with respect to p_v;
4. if $f(v)$ is nonsatiated, then $c(p_v, q)$ is increasing in q.

Moreover

5. Shephard's lemma holds, $v(p_v, q) = \partial c(p_v, q)/\partial p_v$;
6. $v(p_v, q)$ is homogeneous of degree zero in p_v.

Proof The constraint set $X = \{v \geqslant 0 \mid f(v) \geqslant q\}$ is nonempty because $q \in f(R^n_+)$. Since $v \geqslant 0$, it is bounded from below; therefore the program has a bounded optimal solution. Property 1 follows from the perturbation theorem A.3.2. For properties 3 and 5, we consider the program $c(p_v, q) = \min_{v \geqslant 0}\{p_v v \mid v \in X\}$. Uniqueness of $v(p_v, q)$ follows from strict quasiconcavity of $f(v)$. Continuous differentiability of the objective $p_v v$ is obvious, so that the envelope theorem A.3.3 can be invoked to establish properties 3 and 5. When proving property 2, we use nonnegativity of v and property 5 to ensure that $c(p_v, q)$ is nondecreasing in p_v. Homogeneity follows from the linearity of the objective function of the program. The cost function is nondecreasing in q because, if q and q' are two feasible output levels with $q < q'$, the optimal input demand for q' can always produce q, due to the weak inequality (free disposal), thus proving that $c(p_v, q)$ is nondecreasing in q. Input demand $v(p_v, q)$ is homogeneous of degree zero in p_v, since it is obtained as the derivative of a function that is homogeneous of degree one in p_v, proving property 6. To prove property 4, nonsatiation of $f(v)$ ensures that any Lagrange multiplier associated with the constraint $f(v) \geqslant q$ will be positive, so the cost function is increasing in q. ■

Note that our assumption on production functions is less restrictive than the one on the production sets, since we require strict quasiconcavity instead of strict concavity. This is in order to cover the case of constant returns to scale (and, in chapter 10, increasing returns to scale); a strictly quasiconcave function that also is homogeneous of degree one will be concave.

PROPOSITION 2.5 (Differentiability of the input demand function) Consider the continuous production function $f: R^r_+ \to R$, $f(v)$ and the associated cost function $c(p_v, q) = \min_{v \geqslant 0}\{p_v v \mid f(v) \geqslant q\}$, for given positive prices p_v and given output level $q \in f(R^r_+)$. Assume that $f(v)$ is nonsatiated and "locally" strongly concave, that is, strongly concave in the optimum and that optimal input demand $v(p_v, q)$ is positive.

Then

1. $v(p_v, q)$ is continuously differentiable and $c(p_v, q)$ is twice differentiable;
2. the matrix $\partial v/\partial p_v$ is negative semidefinite.

Proof To prove property 1, note that since $v(p_v, q)$ is positive, the first-order condition $p_v = \rho\partial f(v)/\partial v$ holds with equality. Hence, since v is unique, $f(v)$ is differentiable and p_v is positive, the multiplier ρ is unique and, by nonsatiation of $f(v)$, it is positive. Hence $f(v) \geqslant q$ holds with equality. The total differentials are

$$dp_v = \rho G dv + g d\rho \tag{2.8}$$

and

$$g^T dv = dq, \tag{2.9}$$

where g and G are, respectively, the gradient $\partial f(v)/\partial v$ and the Hessian of $f(v)$ evaluated at v (and g^T is the transpose of the column vector g). Since the Hessian is negative definite, it is nonsingular, so we may premultiply (2.8) by $g^T G^{-1}$, which yields, after substitution of (2.9),

$$g^T G^{-1} dp_v = \rho dq + (g^T G^{-1} g) d\rho.$$

Hence $d\rho/dq = -\rho(g^T G^{-1} g)$ is positive, since ρ and g are positive, and G^{-1} is negative definite. Also $d\rho/dp_v = 1/(g^T G^{-1} g) G^{-1} g$ so that

$$\frac{\partial v}{\partial p_v} = \left(\frac{1}{\rho}\right)\left[G^{-1} - \frac{1}{(g^T G^{-1} g) G^{-1} g g^T G^{-1}}\right] \tag{2.10}$$

and

$$\frac{\partial v}{\partial q} = \frac{1}{(g^T G^{-1} g) G^{-1} g}. \tag{2.11}$$

Since G^{-1} and g are continuous functions of the continuous function $v(p_v, q)$, the derivatives in (2.10) and (2.11) are continuous in (p_v, q) also, so $v(p_v, q)$ is continuously differentiable. By Shephard's lemma, $\partial v/\partial p_v = \partial^2 c/\partial p_v^2$, $\partial v/\partial q = \partial^2 c/\partial p_v \partial q$. Finally, since ρ is equal to $\partial c/\partial q$, it follows that $\partial\rho/\partial q = \partial^2 c/\partial q^2$, and all the second-order derivatives are well defined.

To prove property 2, note that $\partial v/\partial p_v$ is the principal submatrix of the Hessian of the cost function, with derivatives w.r.t. p_v; hence it must be negative semidefinite (and symmetric). It cannot be negative definite, since $c(p_v, q)$ is homogeneous of degree one in p_v, and the derivatives are homogeneous of degree zero. ■

For the case of strong quasi-concavity, which is relevant when there are constant returns to scale, see Barten and Böhm (1982).

Since the cost function is twice continuously differentiable, its Hessian is symmetric and $\partial v/\partial p_v$ is symmetric. Moreover the diagonal terms $\partial v_k/\partial p_{vk}$ of this Hessian are nonpositive. Since the cost function is homogeneous of degree one in p_v, it cannot be strongly concave, so the properties of its Hessian are weaker than for the profit function in proposition 2.3.

The Profit Function with Outputs and Inputs Represented Separately

When as in (2.4) outputs and inputs are represented separately, and if the transformation functions $T_q(q)$ and $T_v(v)$ are concave and also have strictly convex isoquants, profit maximization can be written

$$\max_{q \geqslant 0, v \geqslant 0} \{p_q q - p_v v \mid T_q(q) - T_v(v) \leqslant 0\},$$

or

$$\max_{q \geqslant 0, v \geqslant 0} \{p_q q - p_v v \mid T_q(q) \leqslant \zeta \text{ and } T_v(v) \geqslant \zeta\},$$

where ζ is a scalar. Obviously the problem may be decomposed as follows:

$$R(p_q, \zeta) = \max_{q \geqslant 0} \{p_q q \mid T_q(q) = \zeta\}, \tag{2.12}$$

where $R(p_q, \zeta)$ is the (restricted) revenue function and

$$c(p_v, \zeta) = \min_{v \geqslant 0} \{p_v v \mid T_v(v) = \zeta\}, \tag{2.13}$$

where $c(p_v, \zeta)$ is the cost function. Both problems are coordinated by

$$\Pi(p_q, p_v) = \max_{\zeta \geqslant 0} \{R(p_q, \zeta) - c(p_v, \zeta)\}. \tag{2.14}$$

We see that ζ is determined in a primal way by solving (2.14); then, for optimal ζ, supplies and input demand functions can be obtained as derivatives of (2.12) and (2.13), as in propositions 2.1 and 2.4.

Constant Returns to Scale

Under constant returns to scale, the production set Y_0 has the property that for every $\lambda \in [0, \infty)$ and every $y \in Y_0$, one also has $\lambda y \in Y_0$, which implies that the set Y_0 cannot be compact. If it is closed, it cannot be strictly convex because the property above also has to hold for points on the boundary.

Profit maximization under constant returns is shown in figure 2.5, where we represent a linear production set with two activities; one output q and one input v. The set evidently fails to satisfy compactness and strict convexity. In this setting, profit maximization has three possible outcomes: (1) zero production and zero profit, (2) indeterminate nonnegative production and zero profit and (3) infinite production and infinite profit.

In equilibrium, only outcomes 1 and 2 can prevail; in situation 2 the scale of production is undetermined. This expresses that prices alone do not provide sufficient information for the producer to set up a unique production plan. As a consequence the profit function will be of little help in determining supplies and input demand; further restrictions must be added,

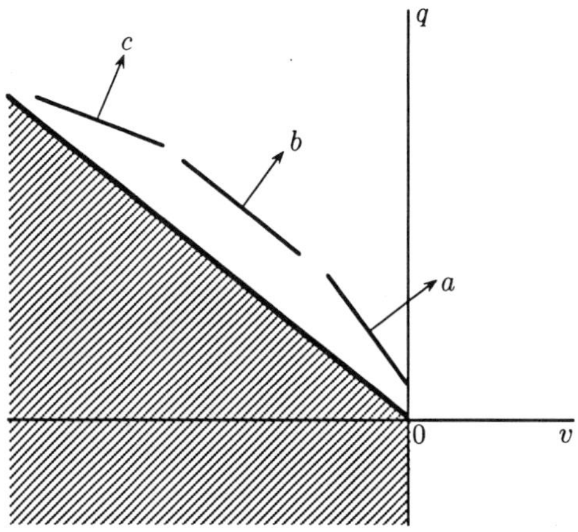

Figure 2.5
Profit maximization under constant returns to scale on an unbounded production set.

and this is done by exogeneously setting some inputs or outputs as in the restricted profit and in the cost function.

Constant returns have the implication that the cost function is homogeneous of degree one in $q: c(p_v, q) = c(p_v, 1)q$, where $c(p_v, 1)$ is referred to as the unit cost function. Obviously all the properties of the cost function hold for the unit cost function. In particular, using Shephard's lemma, we obtain

$$v(p_v, 1) = \frac{\partial c(p_v, 1)}{\partial p_v},$$

where $v(p_v, 1)$ is the vector of input coefficients that minimize unit costs at prices p_v. A similar relation holds when the transformation function is separable; in that case we also have

$$q(p_q, 1) = \frac{\partial R(p_q, 1)}{\partial p_q}$$

as a column in the output matrix. Then the unit profit may be defined as the following restricted profit function:

$$\Pi(p_q, p_v; 1) = R(p_q, 1) - c(p_v, 1).$$

Decreasing Returns to Scale

In figure 2.3 the production set is strictly convex from above, and the input requirement per unit of output rises with scale.[12] This illustrates the

12. Fixed setup costs are also often introduced in microeconomics textbooks; then $c(p_v, q) = \alpha(p_v, q) + \gamma(p_v)$.

property of decreasing returns to scale, as implied by assumption P1. For a (single output) production function, this means that $f(\lambda v) < \lambda f(v)$ for all $\lambda > 1$.

Even if the underlying technology has constant returns to scale, keeping some inputs (or outputs) fixed, as in the restricted profit function, will lead to a decreasing returns to scale specification. If the inputs kept fixed are not represented explicitly, decreasing returns to scale has to be allowed for. Profit will then be the revenue (or rent) accruing to these fixed factors, as is illustrated by the following example. Consider a production process with a unique output q and three inputs v_1 v_2, and v_3; the third input is fixed. The technology is described by a Cobb-Douglas production function $f(v_1, v_2, v_3) = \beta(v_1)^{\alpha_1}(v_2)^{\alpha_2}(v_3)^{\alpha_3}$ with $\Sigma_k \alpha_k = 1$. For fixed v_3 this function becomes $\gamma(v_1)^{\alpha_1}(v_2)^{\alpha_2}$ with $\gamma = \beta(v_3)^{\alpha_3}$ and $\alpha_1 + \alpha_2 < 1$ showing decreasing returns to scale.

If decreasing returns prevail, the profit maximization problem can be written

$$\Pi(p_q, p_v) = \max_{q \geqslant 0}\{p_q q - c(p_v, q)\},$$

so the profit function can always be recovered from the cost function.

Increasing Returns to Scale

This subject will be extensively covered in chapter 10. Here we only note that a firm that faces constant returns to scale in most of its activities may still incur costs that do not rise continuously or even fall with the scale of operations. Examples are a fixed setup cost, as in figures 2.1a and f, and a nonconvexity, as in figure 2.1c. None of these features needs to make the input requirements set $V(q) = \{v \mid f(v) \geqslant q\}$ nonconvex, and as long as convexity holds, the problem remains tractable, at least in the single output case. If, for example, the production function $f(v)$ is strictly quasi-concave, ensuring convexity of $V(q)$,[13] one could use cost functions, and most of the properties of the case where $f(v)$ is strictly concave will carry over.

Total, Average, and Marginal Cost Schedules

Consider the cost function. As in standard microeconomics textbooks (and in figures 2.1a, c, d, and f), it includes a setup cost $\gamma(p_v)$:

$$c(p_v, q) = \alpha(p_v)q^\beta + \gamma(p_v).$$

If $0 < \beta < 1$, there are increasing returns to scale (irs); for $\beta = 1$, there are constant returns (crs) and when $\beta > 1$ returns are decreasing (drs). Figure 2.6

13. If $f(v)$ is quasi-concave, $V(q) = \{v \mid f(v) \geqslant q\}$ is a convex set, by definition of quasi-concavity. $V(q)$ is precisely the input requirements set.

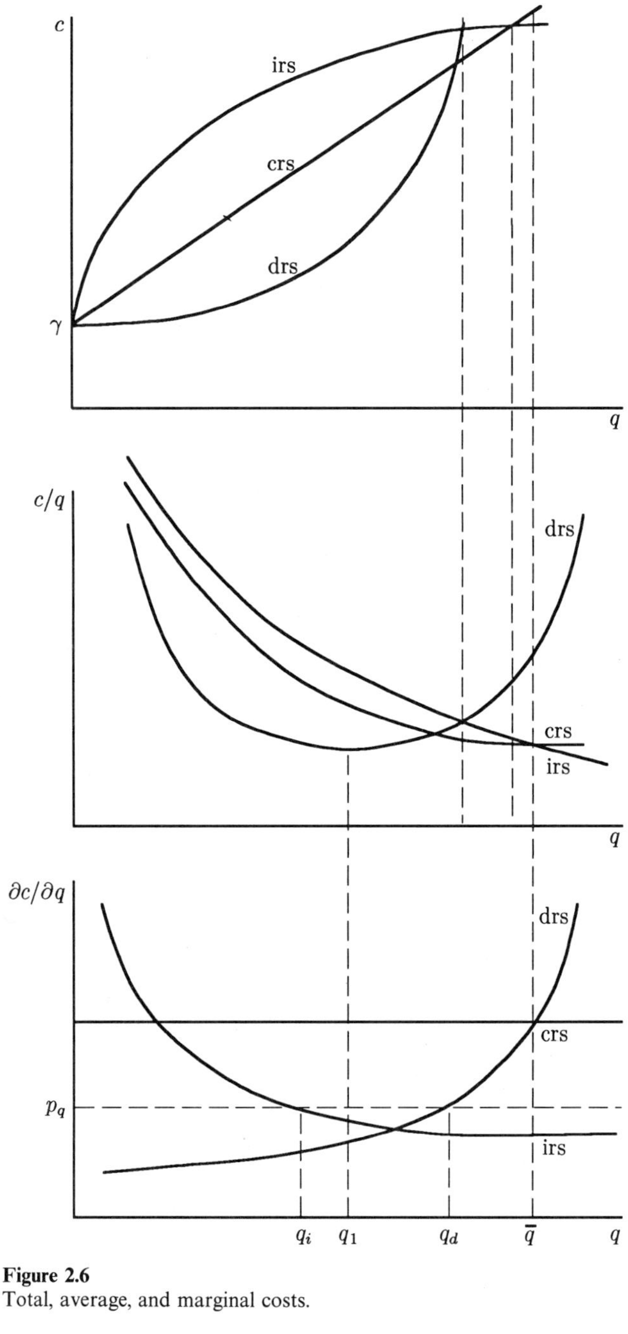

Figure 2.6
Total, average, and marginal costs.

gives the three associated schedules. At given output price p_q, q_d will be optimal under decreasing returns; $\bar{q}$, the upper bound on production, will be optimal under constant and increasing returns. The point q_i, which equates price and marginal cost under increasing returns, corresponds to minimum rather than maximum profit. The point q_1 corresponds to minimum average cost of the decreasing returns schedule and is sometimes referred to as the long-run scale of production.

2.1.4 Producer Behavior in Applied Models

Choosing between the Primal and the Dual Approach

Since the primal representation (which uses production sets) and the dual representation (where net supply and input demand are explicit functions of prices) are formally equivalent, one may wonder why the dual approach is of any interest at all. Production sets describe technical relations, possibly supplemented by behavioral constraints. Strictly convex production sets generate continuous supply and input demand functions. Then, if all the parameters of this set are known, the problem is fully specified, and one might argue that there is no need for the dual representation. However, such representation makes it possible to replace the maximization problem by continuous functions, and this may be expedient for computation and parameter estimation.

Nonetheless, the dual approach sacrifices the explicit incorporation of physical balances such as the balance for dough in our example of activity analysis representation of production behavior. Moreover, to make the duality approach work, one must choose functional specifications. These specifications can only have a finite number of parameters because the number of statistical observations is finite. Since only a few specifications are known that have the properties imposed or derived in propositions 2.1 to 2.5 for all prices and production levels, it is unrealistic to presume that a dual function will incorporate all a priori information available on the technology. This is why there may be a case for using mixed approaches, to which we return later.

Functional Forms

Following the primal approach, we represent the production set Y by a functional form with a finite number of parameters, as in (2.2):

$$Y = \{y \mid y = Bq;\ Aq \leqslant b \text{ and } q \geqslant \underline{q}\}. \tag{2.15}$$

The parameters B, A, b, and $\underline{q}$ have to be chosen so that the set Y satisfies assumption P1′. We can also specify functional forms for the transformation function $F(y)$ or for the production function $f(v)$. Recall that in order to satisfy assumption P1 or P1′, $F(y)$ has to be continuous, (strictly) quasiconvex, nondecreasing, and such that $F(0) \leqslant 0$, whereas $f(v)$ has to be (strictly) concave, nondecreasing, and $f(0) \geqslant 0$. For this, we could choose analytical forms and then face the following dilemma: Either choose (nonflexible) forms that are restrictive but possess the desired properties for all feasible levels of q, v, and y or opt for so-called flexible forms, which will, in general, possess these properties only in a limited domain. Flexible forms usually have more parameters and are obtained by a second-order Taylor approximation of some nonlinear form.[14]

Nonflexible forms are often derived as special cases or generalizations of the constant elasticity of substitution (CES) form:

$$q = [(a_1 v_1)^{-\rho} + (a_2 v_2)^{-\rho}]^{-1/\rho}. \tag{2.16}$$

The inputs v_1 and v_2 could be outputs of CES production functions themselves, and so on, such that a multilevel CES production function is obtained.[15] The elasticities of substitution $\sigma = -1/(1+\rho)$ can obviously be given different values in the various levels.[16] When $\sigma = 0$, we have a Leontief function

$$q = \min\{a_1 v_1, a_2 v_2\}, \tag{2.17}$$

where a_1 and a_2 are the inverse of the input coefficients for v_1 and v_2, respectively. When $\sigma = -1$, we have the Cobb-Douglas function

$$q = \gamma_0 v_1^{\gamma_1} v_2^{\gamma_2}, \tag{2.18}$$

with $\gamma_1 + \gamma_2 = 1$, since the CES has constant returns.[17]

For flexible forms, consider the quadratic form $q = a \cdot z + \frac{1}{2} z \cdot Bz$, which is a second-order approximation of an arbitrary nonlinear function. If B is negative definite, the form will be concave. It will be decreasing for appropriate z values (but not necessarily for all) and have possibility of inaction ($z = 0$); it will also generate negative values for q as z becomes large. This does not conform with the theoretical requirements, though in practice, the problem may be of little importance. To impose some proper-

14. There are also other methods of obtaining flexible forms, see Barnett, Geweke, and Wolfe (1991). See also Fuss et al. (1978), McFadden (1978a, b), and Lau (1986) on functional forms in general.

15. Recently techniques have been developed through which such multilevel functions can be parameterized so as to capture locally any regular configuration of input demands and second-order curvature conditions. See Perroni and Rutherford (1995).

16. One can also use the CES for more than two inputs, but then all pairwise substitution elasticities will be equal.

17. The extension to the case with more than two inputs is straightforward. The Cobb-Douglas function is also used to represent decreasing ($\gamma_1 + \gamma_2 < 1$) and increasing ($\gamma_1 + \gamma_2 > 1$) returns to scale.

ties on the flexible forms,[18] we apply monotonic transformations to the variables v and q, like logarithms[19] or square roots, to obtain

$$z_0(q) = a \cdot z_1(v) + \tfrac{1}{2} z_2(v) \cdot B z_2(v), \tag{2.19}$$

where the estimate of B has to be negative definite.[20] For the translog, the relations are $z_0(q) = \ln q$, $z_1(v) = \ln v$, and $z_2(v) = \ln v$;[21] for the generalized Leontief function,[22] they are $z_0(q) = q$, $z_1(v) = v$, and $z_2(v) = v^{1/2}$.

Turning to the producer behavior, we note that profit maximization under the linear activity set (2.15) yields a linear program, which has to be solved (in primal or dual form). It generates (uppersemicontinuous) supply and input demand correspondences, as in proposition 2.1, and does not lead to analytical supply and input demand functions. When the numerical values of all the parameters are known (e.g., from engineering data), there is no need for explicit analytical forms, and the only difficulty is that due to the uppersemicontinuity, it may be hard to solve the general equilibrium model with this linear technology.

Nonlinear representations of production sets will not have this shortcoming if the $g_h(y)$ functions make the $F(y)$ function strictly convex. However, for such forms, it is more difficult to obtain engineering data. Then we must estimate either these functions or the associated supply and input demand functions that result from profit maximization (if returns are decreasing) or from cost minimization (if there is a single output).

Direct estimation is usually problematic, since all (nonfixed) outputs and inputs are endogenous decision variables of the producer model. Therefore additional exogenous variables are needed for parameter estimation. When the model of a producer is looked at as a separate component,[23] prices are given, and this makes them the natural candidate exogenous variables. Under the hypothesis of profit maximization (or cost minimization) the relations between these exogenous prices and supply and input demand are as described in propositions 2.1 to 2.5. In the dual approach we estimate the system of output and input demand functions, and since these are obtained

18. See, for example, Diewert and Wales (1987), Fuss, McFadden, and Mundlak (1978), Jorgenson (1986), Nadiri (1982), and Varian (1992).

19. Note that as soon as z is not v (e.g., $z = \ln v$), (2.19) is no longer a Taylor approximation.

20. This imposes restrictions on the matrix B, which can be implemented by considering the Cholesky decomposition $B = L'DL$, where L is a lower triangular matrix (with unit diagonal) and D the diagonal matrix of the (negative) eigenvalues. The parameters to be estimated are then the matrices L and D. See Barten and Geyskens (1975).

21. Here $\ln v$ denotes the vector of logarithms of v, and $v^{1/2}$ denotes the vector of square roots.

22. For the generalized Leontief, the nonnegativity of q for nonnegative v may be violated; this cannot happen for the translog.

23. Of course, in the full general equilibrium model, prices are not exogenous, but here one disregards this feature.

as derivatives of the same profit (or cost) function, many crossequation restrictions will have to be satisfied.

The CES family of production functions is of special interest, since it yields convenient analytical expressions. For example,[24] if the production function is a CES of the form (2.16), the cost function is also of the CES type:[25]

$$c(p_{v_1}, p_{v_2}, q) = q\left[\left(\frac{p_{v_1}}{a_1}\right)^{-r} + \left(\frac{p_{v_2}}{a_2}\right)^{-r}\right]^{-1/r},$$

where $r = \rho/(1-\rho)$, and the input demand functions are

$$v_1(p_{v_1}, p_{v_2}, q) = q\left(\frac{p_{v_1}}{a_1}\right)^{-1/(1+\rho)}\left[\left(\frac{p_{v_1}}{a_1}\right)^{-r} + \left(\frac{p_{v_2}}{a_2}\right)^{-r}\right]^{-(1/r+1)},$$

$$v_2(p_{v_1}, p_{v_2}, q) = q\left(\frac{p_{v_2}}{a_2}\right)^{-1/(1+\rho)}\left[\left(\frac{p_{v_1}}{a_1}\right)^{-r} + \left(\frac{p_{v_2}}{a_2}\right)^{-r}\right]^{-(1/r+1)}.$$

However, such analytical expressions are available only for few forms outside the CES class,[26] and flexible forms, like the translog, do not, under cost minimization, lead to an explicit form for the cost function (let alone to a translog form). Therefore, to derive input demand in the non-CES case, we have to start with a flexible form for the cost function itself and obtain input demand from Shephard's lemma. In addition to the requirements of concavity, nondecreasingness, and nonnegativity, the cost function has to be homogeneous of degree one, and obviously a pure second-order Taylor approximation will not do. This is why $z(v)$ transforms are always introduced in the flexible form of a cost function. For example, if the cost function is of the translog form, we have

$$\ln c(p_v, q) = \alpha_0 + \Sigma_k \alpha_k \ln p_{v_k} + \tfrac{1}{2}\Sigma_k \Sigma_h \beta_{kh} \ln p_{v_k} \ln p_{v_h} + \ln q.$$

To impose the properties of propositions 2.4 and 2.5, the parameters will have to satisfy adding up of cost shares, homogeneity of degree one in prices ($\Sigma_k \alpha_k = 1, \Sigma_h \beta_{kh} = 0$) as well as concavity (a sufficient condition[27] is negative semidefiniteness of B). Note that Shephard's lemma $v(p_v, q) = \partial c(p_v, q)/\partial p_v$ implies that

$$w_k(p_v, q) = \frac{\partial \ln c(p_v, q)}{\partial \ln p_{v_k}}, \tag{2.20}$$

24. See, for example, Varian (1992, 55).

25. We only discuss the cost function; as we show in section 2.1.3, the profit function can be recovered from it.

26. See Sato (1976).

27. This condition is not easily implemented in econometric estimation. See Diewert and Wales (1987).

where $w_k(p_v, q)$ is the share of input k in total cost and is given by

$$w_k(p_v, q) = \alpha_k + \Sigma_h \beta_{hk} \ln p_{v_h}.$$

Finally, flexible forms can also be used as profit functions for the multiple variable output case, but with convexity rather than concavity assumptions.

Functional Separability

In applied work, one often estimates profit or cost functions for commodity groups separately. Indeed, the number of commodities could be too large for simultaneous estimation or it could lead to implausible crosselasticities between commodities in different groups. Also data for subgroups may not cover the same time period or firms, and then one would impose separability as follows:

$$f(v^1, v^2, \ldots, v^s) = h[g^1(v^1), g^2(v^2), \ldots, g^s(v^s)],$$

where v^k denotes the vector of commodities that enter $g^k(.)$;[28] $h[.]$ is assumed to be continuous, strictly concave, increasing, and the $g^k(.)$ are continuous, strictly concave, nondecreasing. We may now obtain the associated cost function as

$$c(p_v, q) = \min_{v^k \geq 0,\, \text{all } k}\{\Sigma_k p_v^k v^k \mid h[g^1(v^1), g^2(v^2), \ldots, g^s(v^s)] \geq q\},$$

where p_v^k is the price vector associated to inputs v^k. This can be rewritten as

$$c(p_v, q) = \min_{g_0^k \geq 0,\, \text{all } k}\{\Sigma_k c^k(p_v^k, g_0^k) \mid h[g_0^1, g_0^2, \ldots, g_0^s] \geq q\}, \tag{2.21}$$

where

$$c^k(p_v^k, g_0^k) = \min_{v^k \geq 0}\{p_v^k v^k \mid g^k(v^k) \geq g_0^k\}, \qquad k = 1, 2, \ldots, s.$$

The expression on the right-hand side of (2.21) is known as a mixed primal-dual formulation: The cost functions are replacing the primal production functions $g^k(v^k)$ but the primal production function $h[.]$ is maintained. One can then use engineering data on the function $h[.]$ while the input demand functions are being estimated.

2.2 Consumer Behavior

In the competitive equilibrium, consumer i is assumed to maximize his utility under his budget constraint at given prices p and income h_i. Formally the consumer has to compute a plan x_i that solves

$$\max_{x_i \geq 0}\{u_i(x_i) \mid p x_i \leq h_i\}. \tag{2.22}$$

28. Usually each commodity will appear only once as argument of the function g^k.

2.2.1 The Consumer

Like for the producer we can ask the question "who is consumer i?" So far we have assumed that every consumer is represented separately and that anyone who maximizes utility under his own budget constraint is a consumer; some are individuals, some are households or even extended families. If there is more than a single individual, as in a family, utilities may be member specific, but there is only one budget constraint. What characterizes the family as a consumer is that it decides on its total consumption x_i by maximizing an "aggregate family" utility function rather than adopting a fixed rule for distributing income among its members and letting each member solve his consumer problem. Alternatively, if there is such a sharing rule, every member becomes a single consumer.

However, if one accepts the principle that consumers can be identical, they can be represented by a single one; that this is possible can be verified as follows: Let $\tilde{x}_i$ and $\tilde{h}_i$ denote consumption and income of individual consumer i, $u_i(\tilde{x}_i)$ is his utility function and n_i is the number of identical consumers who will make identical optimal plans. Here $\Sigma_i \tilde{h}_i = h_i$ or $n_i \tilde{h}_i = h_i$. It is easy to check that the following four programs lead to the same optimal aggregate consumption:[29]

1. Maximization of individual utility subject to an individual budget constraint $\max\{u_i(\tilde{x}_i)$ subject to $p\tilde{x}_i \leqslant \tilde{h}_i, \tilde{x}_i \geqslant 0\}$.

2. Maximization of utility per caput subject to a per caput budget constraint $\max\{u_i(x_i/n_i)$ subject to $px_i/n_i \leqslant h_i/n_i$, $x_i \geqslant 0\}$ for n_i given.

3. Maximization of utility per caput subject to a group budget constraint $\max\{u_i(x_i/n_i)$ subject to $px_i \leqslant h_i, x_i \geqslant 0\}$ for n_i given.

4. Maximization of group utility subject to a group budget constraint $\max\{n_i u_i(x_i/n_i)$ subject to $px_i \leqslant h_i, x_i \geqslant 0\}$ for n_i given.

Here the group can also be a household and $\tilde{h}_i$ is the income of a household member. Since n_i, a demographic datum, is fixed, the list (1 to m) of consumers excludes any endogenous demographic adjustment. In section 7.4, n_i will be allowed to vary endogenously over time.

We now review the assumptions made so far on the utility function of consumer i, on his budget set, and on the behavior itself. For the producer the essential feature is the production set, which is easy to describe without reference to behavior. This is more difficult for the consumer, and we cannot separate the two here.

29. Note that in program 4 below we sum over individual utilities. This is not a problem here because we assume that all consumers receive the same bundle of goods; in general, it is not a harmless assumption. In section 2.4 welfare criteria will be discussed and the implications of various weighting schemes for obtaining aggregate utility will be studied.

2.2.2 Assumptions on Consumer Behavior

We repeat the assumptions C1 and C2 made in chapter 1.

ASSUMPTION **C1** (Utility functions) The utility function $u_i: R^r_+ \to R_+, u_i(x_i)$ of every consumer i is continuous, strictly concave, increasing, and, without loss of generality, $u_i(0) = 0$.

ASSUMPTION **C2** (Endowments) The endowment ω_i of every consumer i is strictly positive.

Endowments

Assumption C2 is needed to ensure that income $h_i = p\omega_i + \Sigma_j \theta_{ij} p y_j$ is positive in equilibrium (where, as we have seen $p > 0$ and $py_j \geqslant 0$). It is obviously unrealistic to require that every consumer has positive endowments in all commodities. It is, however, natural to assume that they all possess a positive endowment of some commodity (each consumer owns his labor endowment at least). For one consumer, say, the first one, we have to maintain the more restrictive assumption of positive endowment, but since this endowment may be arbitrarily small, the assumption is relatively innocuous. This leads to assumption C2′, which is standard in subsequent chapters:

ASSUMPTION **C2′** (Endowments) $\omega_i \geqslant 0$, $\omega_i \neq 0$ for all i and $\omega_i > 0$ for $i = 1$.

The assumption is thus stronger for consumer 1; this is used in existence proofs.

The Consumption Set

The nonnegativity of consumption bundles has two implications: There is no minimal requirement for survival, and all nonnegative consumption bundles are feasible (commodities are perfectly divisible).

In general equilibrium theory, the restriction $x_i \in X_i$ (where the consumption set X_i is closed, bounded from below and convex) is sometimes used instead of $x_i \in R^r_+$, where R^r_+ is the nonnegative orthant. This set X_i may be viewed as a household technology set or as a restriction of R^r_+ to avoid consumptions below subsistence level. In applications, one has to represent this set explicitly by equations. For example, minimum requirements can be imposed by introducing a utility independent consumption vector $\bar{x}_i(p, h_i)$, the committed or subsistence demand. This leads to the following utility maximization model:

$$\max_{x_i \geqslant 0}\{u_i[x_i - \bar{x}_i(p, h_i)] \mid px_i \leqslant h_i, x_i \geqslant \bar{x}_i(p, h_i)\}.$$

As long as $\bar{x}_i(p, h_i)$ is continuous and homogeneous of degree zero in (p, h_i) for $p > 0$, $h_i > 0$, and satisfies $p\bar{x}_i(p, h_i) \leqslant h_i$, such a specification can be accommodated. Of course this formulation is too general to identify separately the parameters of the utility function and of the "subsistence function"; it may, however, be useful to characterize a social minimum $\bar{x}(p, h)$ that is common to all consumers.

Indivisibilities in consumption will be considered in section 10.4.1, and a more elaborate treatment of subsistence demand is given in sections 9.1.5 and 10.4.2, where we study the relation between consumption and the supply of labor (the efficiency wage relation).

The Utility Function

The increasingness property of utility in assumption C1 can be relaxed to nonsatiation for all but one consumer (e.g., the first consumer):

ASSUMPTION **C1′** (Utility functions) The utility function $u_i: R^r_+ \to R_+, u_i(x_i)$ is continuous, strictly concave,[30] nonsatiated and satisfies $u_i(0) = 0$ for all i; for $i = 1$ it is increasing with respect to all commodities.

Sometimes we relax strict concavity and replace it by strict quasiconcavity (e.g., when utility is homogeneous of degree one) or by concavity (when we approximate utility by linear segments). Consumer 1 receives a special treatment to ensure that he has positive income ($\omega_1 > 0$ in assumption C2′). Since his utility function is increasing in all commodities (assumption C1′), he will, at positive income, be able to use up all surpluses on the market, and this ensures positive equilibrium prices.

We discuss the properties of the utility function in the following order: continuity, $u_i(0) = 0$, increasingness and nonsatiation, and, finally, concavity.

Continuity is a natural requirement[31,32,33] for a function that is supposed to measure well-being. One could argue that such a function cannot be continuous everywhere, since it should approach $-\infty$ in the neighborhood of unacceptable commodity bundles. However, when decisions are decentralized, this going to minus infinity is of little help: Utility does not need to

30. It is sufficient to assume that there exists a monotonic transformation that makes the utility function strictly concave.

31. Existence and continuity of a utility function can be derived from four basic axioms on preferences: reflexivity, transitivity, completeness, and continuity. See, for example, Debreu (1959).

32. A (strictly) concave function is continuous on the (relative) interior of its effective domain. Here the domain is R^r_+, which is a closed set. Therefore concavity does not ensure continuity of the function at the boundary of the set, and this is why continuity has to be imposed explicitly.

33. Note that the utility function $\Sigma\gamma_k \ln x_k$ is not continuous at $x_k = 0$ but that the associated demand functions are well-defined at positive prices.

go to $-\infty$ for a consumer to avoid a particular consumption bundle, and if he cannot avoid it (e.g., because he is too poor), attributing $-\infty$ to all attainable bundles only incapacitates him in making any choice.[34]

The homogeneity property, $u_i(0) = 0$, is introduced for notational convenience. Recall that if $u_i(x_i)$ is a utility function, any monotonic transformation (e.g., a shift in $u_i(0)$) will yield the same ranking of alternative commodity bundles. Hence we may shift $u_i(0)$ at will, unless we wish to compare cardinal utilities across agents. The homogeneity assumption is made to maintain symmetry with the treatment of the production function and will be used when we allow the number of consumers to vary (e.g., in chapter 7).

The properties that will be discussed now (increasingness, concavity, etc.) may seem much less natural, and they cannot be derived from the basic axioms of choice (reflexivity, transitivity, completeness, and continuity of preferences); they are imposed directly on the utility function. We will see that they are less restrictive than may seem at first.

Consider the set $X_i(v_i) = \{x_i \geqslant 0 \mid u_i(x_i) \geqslant v_i\}$. This is the set of all consumption bundles that give to consumer i a utility not less than v_i. Assume that this set is like the one represented in figure 2.7a, enclosed by the line BB'.[35] The consumption points that lie in the area between the AA' and the BB' lines do not belong to this set. Assume also that the consumer characterized by such a set has to choose a utility maximizing bundle, given his budget constraint, and that v_i is this maximal utility. As long as prices remain positive, the only points that matter are those on the southwestern (fat) part of BB'. Evidently the same demand would have been recovered had we enlarged the set to the one above the line AA' in figure 2.7b (or in figure 2.7a).

Figure 2.7 will help us in our discussion of the remaining elements of assumption C1′. We also need the definitions of increasingness and nonsatiation for differentiable utility functions.[36] The utility function $u(x)$ with derivative $u'(x)$ is increasing if $u'(x) > 0$, nondecreasing if $u'(x) \geqslant 0$, and monotonic if $u'(x) \geqslant 0$ and $u'(x) \neq 0$. Moreover it is nonsatiated if $\partial u(x)/\partial x_k > 0$ for some commodity k and nonstationary if $u'(x) \neq 0$. All these properties are said to hold locally if they hold in the optimum.[37]

Monotonicity does not allow points belonging to the area between lines AA' and BB' of figure 2.7a to give less utility than v_i, thus excluding

34. At centralized level, for a Pareto-efficient allocation supported by an equilibrium with transfers, there is more reason to represent unacceptable consumption bundles explicitly.

35. Note that the set enclosed by CC' is a subset of BB' and that on CC', utility is higher than on BB'.

36. The definitions of increasingness and nonsatiation are given in definition A.1.17 for concave, possibly nondifferentiable functions.

37. Note that local nonsatiation excludes so-called thick indifference curves.

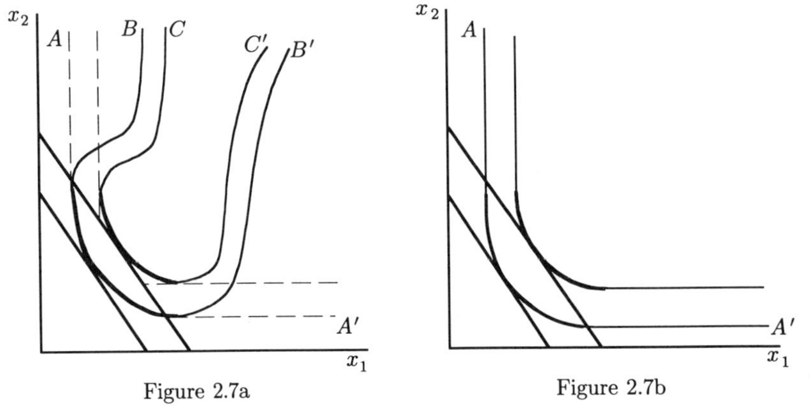

Figure 2.7
A representation of consumer choices.

positively sloped segments of an indifference curve. Such points will never be chosen by a budget-conscious consumer anyway, as can also be seen from the first-order conditions of the consumer problem:

$$\frac{\partial u_i(x_i)}{\partial x_{ik}} \leqslant \lambda_i p_k \qquad \perp x_{ik} \geqslant 0,$$
$$px_i \leqslant h_i \qquad \perp \lambda_i \geqslant 0. \tag{2.23}$$

Under (local) nonsatiation, λ_i will be positive, since, at any price and income, there will be at least one commodity that increases utility. If $p_k > 0$, then either $x_{ik} = 0$ or $\partial u_i(x_i)/\partial x_{ik} > 0$. Hence, replacing the requirement of increasingness by nonsatiation allows us to deal with commodities that are not consumed, but increasingness will still hold in the optimum for all commodities with a positive price that are consumed in positive amounts.

Increasingness can be criticized on empirical grounds, since it implies that increased consumption of a given commodity, say, wine, keeps on raising utility. Nonsatiation is much less restrictive in this respect. However, since prices are nonnegative, it will be impossible to have (2.23) generate[38] negative slopes of $\partial u_i(x_i)/\partial x_{ik}$ at positive levels of x_{ik}, as long as the assumption of local nonsatiation is maintained. Only when a consumer is given commodities freely, and asked whether he would like to have more, can one expect to find evidence of negative marginal utility.[39] It would

38. In some sense this procedure generates only "efficient" points, like the production function, with utility as sole output.

39. This is similar to generating the whole production set in the theory of the producer, including inefficient points.

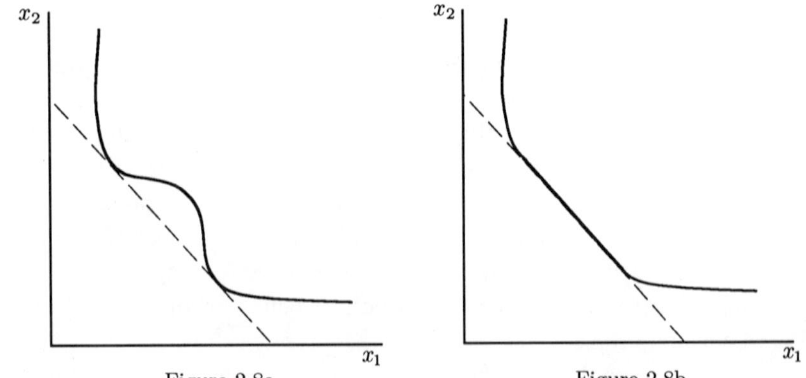

Figure 2.8a

Figure 2.8b

Figure 2.8
Absence of strict convexity in the $X_i(v_i)$ set.

require such experiments to generate a utility function with indifference curves like the one in figure 2.7a.

Let us finally consider the concavity of $u_i(x_i)$. Recall that for a concave function, the set $X_i(v_i) = \{x_i \geqslant 0 \mid u_i(x_i) \geqslant v_i\}$ is convex. In figure 2.7a the set $X_i(v_i)$, enclosed by the BB' line, is not convex; we can argue again that a budget-conscious consumer will never be observed choosing optimal points that generate utility v_i on the "non-fat" part of BB'. It is thus not restrictive to exclude such points, by taking $X_i(v_i)$ to be a convex set. Two difficulties remain: The "fat" part may exhibit nonconvexities as in figure 2.8a or flats as in figure 2.8b. Both situations may lead to nonunique optima for the consumer, but again, empirically, one will never observe optimal choices on the nonconvex part of the indifference curve. Thus, for the same reasons as for nonsatiation, it is not restrictive to assume convexity of $X_i(v_i)$. The solution will be unique if convexity is strict.

So far we have only used properties of the $X_i(v_i)$ set that make the indifference curves (strictly) convex, but any (strictly) quasiconcave function will also generate such sets. However, to maintain desirable properties for the welfare function $\Sigma_i \alpha_i u_i(x_i)$, we will use strict concavity of utility as our standard assumption rather than strict quasiconcavity.[40]

2.2.3 Utility Maximization

The utility function can be arrived at from two different angles: the four axioms[41] of choice or, as we will see in section 2.2.4, the revealed preference

40. See section 2.4.1 for a further discussion.

41. Transitivity of choices has been rejected in several empirical studies; this implies rejection of the utility function.

axiom, which only looks at the southwestern part of the indifference curve in figure 2.7. The axioms of choice lead to a continuous utility function; maximization of this utility function under the budget constraint leads to points that can be thought of as generated by a continuous, concave, and monotonic (and hence nonsatiated) function. This is, as we will see, also the property arrived at via a reasoning based on revealed preference.

Nonetheless, it remains questionable whether this is a meaningful representation of consumer behavior. Some objections against the model are similar to those mentioned for producer behavior; others are more specific to consumption.

First, one could question whether the consumer, like the producer, has all the information he needs to compute his optimal choice. He could face uncertainty with respect to prices, income, quality of goods, and utility they provide, and so on. Although Arrow (1970) already provides an extension of the axioms of choice that lead to the theory of expected utility maximization, applied models usually represent uncertainty only through imperfect substitution.

Second, one could ask whether individual utility maximization is a realistic rule of conduct. There is the obvious fact that only consumption of physical goods and services enters the utility function, and this implies that the consumer is purely materialistic. He does not care about other consumers' utility; he is also isolated, since he is not affected directly by the consumption levels of his fellow consumers. There is little one can do to overcome the limitation of materialism: Economic statistics hardly measure other variables,[42] but altruism and social interaction can be introduced by extending the list of arguments in the utility function; this will be examined in chapter 9.

Third, the specification of the choice set as the intersection of the budget constraint with the nonnegative orthant may be a simplification, since it excludes activities like buying and selling or housekeeping and disregards possible technological constraints. One could defend the approach by arguing that (1) all nontechnological constraints (social, psychological, moral) are incorporated in the utility function and (2) the household model with production activities can be decentralized (e.g., the welfare program in proposition 1.9) into a consumer problem, a producer problem, and market-clearing conditions for intrahousehold goods.[43]

42. Lancaster (1966, 1971) has suggested that commodities carry multiple characteristics: The consumer does not buy a car; he acquires speed, comfort, reliability, etc. It is the set of characteristics that appear in the utility function. This will be dealt with in section 3.6.2.

43. This implies that the household model must be treated as an equilibrium model in which the prices of intrahousehold commodities are cleared on the intrahousehold market. See Koopmans (1957) for the decentralization aspect, Becker (1993) for the theory, and de Janvry and Sadoulet (1990) for applications of applied general equilibrium methods to the modeling of individual households. See also Hirschleifer (1985) for a broader survey of the limitations and potential of the utility maximizing paradigm.

Fourth, the consumer is confronted with his budget constraint as the only limitation from his environment; he is thus "free to choose" any commodity bundle that lies within his reach. Still the consumer is not truly free, since he is the slave of his preferences. In microeconomics, freedom of choice is merely the absence of impediments on choice. By itself, it does not bring any utility and has no social value.[44]

From Utility Maximization to Expenditure Minimization

In general equilibrium models, one uses the Marshallian demand functions $x(p, h)$, which follow directly from utility maximization. However, the properties of these functions are more easily derived from the Hicksian demand functions $\xi(p, v)$ which solve the consumer cost minimization problem $\min_{x \geqslant 0}\{px \mid u(x) \geqslant v\}$ because this enables us to readily apply proposition 2.4 for cost functions and to derive other properties.

We have seen in the proof of the first welfare theorem (proposition 1.5) that under fairly weak assumptions, the utility maximization problem (2.22) with optimal choice x_i^* implies that at positive prices p^* and income h_i^*, the expenditure $p^*x_i^*$ which achieves the utility level $u_i(x_i^*)$ is being minimized, since for any $x_i^0 \geqslant 0$,

1. $u_i(x_i^0) > u_i(x_i^*)$ implies that $p^*x_i^0 > p^*x_i^*$,
2. $u_i(x_i^0) \geqslant u_i(x_i^*)$ implies that $p^*x_i^0 \geqslant p^*x_i^*$.

The Expenditure Function and Hicksian Demand

To simplify notation, we drop the subscript i and define the expenditure function

$$e(p, v) = \min_{x \geqslant 0}\{px \mid u(x) \geqslant v\}, \tag{2.24}$$

where v is the given level of utility to be achieved at minimal cost. We discuss its main properties and those of the associated demand functions.[45]

PROPOSITION 2.6 (The expenditure function and Hicksian demand) Consider the continuous, strictly (quasi)concave, nonsatiated utility function $u: R_+^r \to R, u(x)$, and the associated expenditure function $e(p, v) = \min_{x \geqslant 0}\{px \mid u(x) \geqslant v\}$ for given positive prices p and given utility $v \in u(R_+^r)$.

Then $e(p, v)$ is

1. continuous; it is concave in p and convex in v;
2. homogeneous of degree one in p, nondecreasing in p, and nondecreasing in v;

44. Elaborating on this important issue leads us into deep philosophical debates, which we have to avoid here.

45. See also Barten and Böhm (1982) and Deaton and Muellbauer (1980, ch. 2).

3. continuously differentiable with respect to p;
4. increasing in v.

Moreover, for Hicksian demand $\xi(p, v)$,

5. Shephard's lemma holds, $\xi(p, v) = \partial e(p, v)/\partial p$;
6. $\xi(p, v)$ is homogeneous of degree zero in p.

Proof The proof is the same as for proposition 2.4. ■

PROPOSITION **2.7** (Differentiability of the Hicksian demand function) Consider the utility function $u: R^r \to R, u(x)$ and the associated expenditure function $e(p, v) = \min_{x \geqslant 0}\{px \mid u(x) \geqslant v\}$ for given positive prices p and given $v \in u(R^r_+)$. Assume that $u(x)$ is nonsatiated and locally strongly concave (i.e., strongly concave at the optimum). Then

1. Hicksian demand $\xi(p, v)$ is continuously differentiable and $e(p, v)$ is twice differentiable if it is positive;
2. the matrix $\partial\xi/\partial p$ is negative semidefinite.

Proof The proof is as for proposition 2.5. ■

Like the cost function, the expenditure function is twice continuously differentiable; its Hessian is therefore symmetric, and $\partial\xi/\partial p$ is symmetric; moreover, the diagonal terms $\partial\xi_k/\partial p_k$ of this Hessian are nonpositive. The expenditure function is also homogeneous of degree one in p, so that it cannot be strongly concave.[46]

The Indirect Utility Function and Marshallian Demand

When we derive the Marshallian demand, we restrict attention to positive prices and income; we also make use of the increasingness property in assumption C1′ and consider the utility level v such that the associated expenditure $e(p, v)$ exhausts all income $h: e(p, v) = h$. Then, since $e(p, v)$ is homogeneous of degree one in p and increasing in v, we may define normalized prices $\psi = p/h$ and obtain the so-called indirect utility function in implicit form $e(\psi, v) = 1$ or, in explicit form, $v = \tilde{v}(\psi)$.[47]

Since all income is spent, the corresponding utility level will be optimal in (2.22) and

$$\tilde{v}(\psi) = \max_{x \geqslant 0}\{u(x) \mid \psi x \leqslant 1\}. \tag{2.25}$$

46. If the utility function is locally strongly quasi-concave, like the Cobb-Douglas utility function, the bordered Hessian is nonsingular. See definition A.1.15 (v).

47. The explicit form exists because $e(p, v)$ is increasing in v (by proposition 2.6, assertion 4).

The properties of the indirect utility function and of the associated Marshallian demand function are the subject of the following two propositions:[48]

PROPOSITION 2.8 (The indirect utility function and Marshallian demand functions) Assume that $u(x)$ is strictly concave, nonsatiated, and that prices p and income h are positive. Consider (2.25). Then

1. $\tilde{v}(\psi)$ is continuous, quasiconvex, and the Marshallian demand functions $x(p, h)$ are continuous, homogeneous of degree zero in (p, h) so that one may write $\zeta(p/h) = x(p, h)$;
2. $x(p, h) = \partial e(p, v)/\partial p$ for $v = \tilde{v}(p/h)$.

Moreover, if $u(x)$ is differentiable in the optimum with respect to some commodity k consumed in positive amount (or equivalently, if $e(p, v)$ is continuously differentiable with respect to v), then

3. $\tilde{v}(\psi)$ is continuously differentiable;
4. "Roy's identity" holds, $\zeta_k(\psi) = (\partial\tilde{v}(\psi)/\partial\psi_k)/(\Sigma_j \psi_j \partial\tilde{v}(\psi)/\partial\psi_j)$.

Finally, if $u(x)$ is locally strongly concave and $\zeta(\psi)$ is positive, then

5. $\zeta(\psi)$ is continuously differentiable;
6. Slutsky's equation holds, $\partial x_k(p, h)/\partial p_j = \partial^2 e/\partial p_k \partial p_j - x_j(p, h)\partial x_k(p, h)/\partial h$.

Proof In property 1, continuity of $\tilde{v}(\psi)$ follows from the maximum theorem A.3.1. For quasiconvexity, see Varian (1992, p. 103).[49] By nonsatiation, the budget constraint holds with equality so that $h = e(p, v)$. Hence the expenditure function can be used, and property 2 is an immediate consequence of assertion 5 in proposition 2.6; since $\partial e/\partial p$ and $\tilde{v}(\psi)$ are continuous, it follows that $x(p, h)$ is continuous. Homogeneity of degree zero in (p, h) follows from homogeneity of degree zero of $\tilde{v}(\psi)$.

For property 3, note that $\partial e(\psi, v)/\partial v$ is well-defined and positive. Hence we can apply total differentiation to $e(\psi, v) = 1$, and this yields

$$\frac{\partial\tilde{v}}{\partial\psi} = -\frac{\partial e/\partial\psi}{\partial e/\partial v}. \tag{2.26}$$

Since the right-hand side is continuous, this proves property 3. To prove Roy's identity (property 4), we simply note that, by Shephard's lemma, $\partial e/\partial\psi = x$, and we compute marginal utility of income as $\lambda = 1/(\partial e/\partial v)$. This

48. See also Barten and Böhm (1982) and Deaton and Muellbauer (1980. ch. 2).

49. Varian proves this under the weaker assumption of strict quasi-concavity of the direct utility function. However, under our assumptions no stronger result can be obtained.

reduces (2.26) to

$$\frac{\partial \tilde{v}}{\partial \psi_k} = -\lambda x_k. \tag{2.27}$$

Premultiplying both sides of (2.27) by ψ_k, and then summing over k, yields $\lambda = -\Sigma_j \psi_j \partial \tilde{v}(\psi)/\partial \psi_j$, since $\Sigma_k \psi_k x_k = 1$, so Roy's identity follows.

Property 5 can be proved as follows: By property 2, the Marshallian demand function satisfies

$$\zeta(\psi) = \xi(\psi, \tilde{v}(\psi)). \tag{2.28}$$

We know from proposition 2.7 that the expenditure function will be twice differentiable and that the Hicksian demand function is continuously differentiable. By property 3, the indirect utility function is continuously differentiable. Therefore the Marshallian demand function is continuously differentiable. Finally, to prove property 6, note that the derivative of $\zeta(\psi)$ can be computed from (2.28) as

$$\frac{\partial \zeta_k}{\partial \psi_j} = \frac{\partial \xi_k}{\partial \psi_j} + \left(\frac{\partial \xi_k}{\partial v}\right)\left(\frac{\partial \tilde{v}}{\partial \psi_j}\right). \tag{2.29}$$

By Shephard's lemma, and since $\partial \xi_k/\partial \psi_j = \partial^2 e/\partial p_k \partial p_j$, (2.29) may be written as

$$\frac{\partial \zeta_k}{\partial \psi_j} = \frac{\partial^2 e}{\partial p_k \partial p_j} - \left(\frac{\partial \xi_k/\partial v}{\partial e/\partial v}\right) \cdot x_j.$$

Applying the chain rule to $\xi(p, \tilde{v}(p/h))$ shows that the term between brackets is equal to $(\partial x_k/\partial h)$, proving the Slutsky equation. ■

The Marshallian demand function has very few properties beyond those listed in proposition 2.8. For example, even when one restricts the nonsatiation assumption to increasingness of the utility function, then for some commodities, it still may happen that demand is zero or that it falls with rising income.

The Jacobian of the Excess Demand Function

We now account for the endogeneity of income in the general equilibrium framework. For this we consider the excess demand $z(p) = \Sigma_i x_i(p, h_i(p)) - \Sigma_j y_j(p) - \Sigma_i \omega_i$, where $h_i(p) = p\omega_i + \Sigma_j \theta_{ij} \Pi_j(p)$ and where it is assumed that both $x_i(p, h_i)$ and $y_j(p)$ are continuously differentiable, and we compute the Jacobian of $z(p)$. For consumer i the element (k, s) of this Jacobian is

$$\frac{\partial x_{ik}(p, h_i(p))}{\partial p_s} = \frac{\partial x_{ik}(p, h_i)}{\partial p_s} + \frac{\partial x_{ik}(p, h_i)}{\partial h_i} \cdot \frac{\partial h_i}{\partial p_s},$$

and from the Slutsky condition 6 in proposition 2.8, it follows that

$$\frac{\partial x_{ik}(p, h_i(p))}{\partial p_s} = \frac{\partial^2 e_i}{\partial p_k \partial p_s} - \frac{\partial x_{ik}(p, h_i)}{\partial h_i}(x_{is}(p, h_i) - \omega_{is} - \Sigma_j \theta_{ij} y_{js}).$$

We write this equation in matrix form as

$$M_i = H_i - g_i z_i^T,$$

where $M_i = \partial x_i(p, h_i)/\partial p$ is the Jacobian of the Marshallian demand, $H_i = \partial \zeta_i(p, v_i)/\partial p$ is the Jacobian of the Hicksian demand, $g_i = \partial x_i(p, h_i)/\partial h_i$ and z_i^T is the transpose of $(x_i(p, h_i) - \omega_i - \Sigma_j \theta_{ij} y_j)$. The matrix M_i is thus obtained as the difference between a negative semidefinite matrix H_i and an asymmetric matrix $g_i z_i^T$ of rank one. As a consequence M_i will be asymmetric; the real parts of its eigenvalues can have any sign and several can be zero, so it may have rank less than $(r - 1)$. Moreover

$$\Sigma_i M_i = \Sigma_i H_i - \Sigma_i g_i z_i^T,$$

or, in matrix terms,

$$M = H - GZ^T, \tag{2.30}$$

where G is the matrix with rows g_i and Z, the matrix with rows z_i. By homogeneity, M is singular and so are H and G. Given a matrix M derived from an arbitrary excess demand function that satisfies homogeneity and Walras's law, given also an arbitrary negative semidefinite matrix H and an arbitrary (but of rank $(r - 1)$) matrix Z, we can compute a unique matrix G by solving the system (2.30), dropping one row and one column. Debreu (1974), Mantel (1974), and Sonnenschein (1972, 1973) have shown that this matrix G can be derived from an expenditure function with Hessian H if the distribution of endowments is chosen appropriately. In other words, for an arbitrary aggregate demand function satisfying homogeneity and Walras's law, we can construct a system of $(r - 1)$ utility-maximizing consumers. As M is otherwise arbitrary, this shows that if there are $(r - 1)$ well-behaved consumers, the aggregate consumer demand may nonetheless lack any other properties than continuity, homogeneity, and Walras's law.[50]

Note that if there is a single consumer, in equilibrium, the second term in (2.30) vanishes, and the Jacobian is negative semidefinite.

2.2.4 Recovering Consumer Behavior via Integrability or via Revealed Preference

Instead of characterizing consumer behavior as being derived from utility maximization under a budget constraint, we now turn the question around and try to recover such a behavior from given demand functions or from

50. The importance of the lack of general properties has already been stressed in section 1.2.4.

observations on quantities and prices. Neither utility nor utility maximization is observable, but quantities, prices, and income are. Data on these can be used to estimate functional demand relationships. Is it possible to infer (or reject) consumer's utility-maximizing behavior from such functions, thus validating the construct?[51] The literature distinguishes two approaches for validation: integrability and revealed preference.

Integrability of the Demand Functions

The integrability approach considers any continuously differentiable system of r functions $x(p, h)$, where demand depends on prices and income that satisfy adding up (i.e., $\Sigma_k p_k x_k(p, h) = h$) and homogeneity of degree zero. Assume that we write $x_k(p, h) = \xi_k(p, v(p/h))$, where $v(p/h)$ denotes some scalar function. Given these demand functions, it is possible to use the Slutsky equation to calculate the Slutsky matrix H in (2.30), at given prices and income. If it is symmetric,[52] there exists a twice continuously differentiable (expenditure) function with the Slutsky matrix as its Hessian. The demand functions are then said to be integrable. If the Slutsky matrix is also negative semidefinite, then the expenditure function will be concave in p.[53]

Revealed Preference in Relation to Observations of Prices and Demand

In the revealed preference approach, we can recover the utility function and the utility-maximizing behavior from a given finite number of T observations ($t = 1, 2, \ldots, T$) on prices p^t and quantities q^t. For this we need to define:

DEFINITION 2.1 (Revealed preference) Given an observation (p^t, x^t) and a consumption bundle x, we say that, x^t is

1. directly revealed preferred (R^0) to x if $p^t x^t \geqslant p^t x$;
2. strictly (directly) revealed preferred (P^0) to x if $p^t x^t > p^t x$;
3. revealed preferred (R) to x if $p^t x^t \geqslant p^t x^s$, $p^s x^s \geqslant p^s x^v, \ldots, p^w x^w \geqslant p^w x$.

The generalized axiom of revealed preference (GARP) reads as:

ASSUMPTION GARP If $x^t R x^s$, then it cannot be that $x^s P^0 x^t$.

51. The techniques that will be explained for utility functions can also be used to recover production functions (a single output) from input demand functions; the same can be done for the transformation curve (with several outputs and inputs).

52. Assume that the function $g: R^n \rightarrow R^n$ is continuously differentiable and has a symmetric Jacobian. Then there exists a function $f: R^n \rightarrow R$ such that $\partial f/\partial x = g(x)$. This is known as the symmetry principle or integrability condition.

53. See, for example, Varian (1992, 126) and Deaton and Muellbauer (1980).

The rationale for GARP is that a bundle x^s that is not chosen, even though it is not more expensive, cannot be preferred to the chosen bundle x^t.[54]

PROPOSITION 2.9 (Afriat's theorem) For a given finite set of observations (p^t, x^t), the following conditions are equivalent:

1. All pairs of observations satisfy GARP.
2. There exist numbers $u^t > 0$, $\lambda^t > 0$, $t = 1, 2, \ldots, T$ such that $u^t \leqslant u^s + \lambda^s p^s(x^t - x^s)$ for all t, s.
3. There exists a utility function $u(x)$ satisfying monotonicity and concavity that can be constructed as $u(x) = \min_t(u^t + \lambda^t p^t(x^t - x))$ and that is compatible with expenditure minimization in the sense that whenever $p^t x^t \geqslant p^t x$ for any x, then $u(x^t) \geqslant u(x)$.

Proof See Afriat (1967, 1973). ■

This proposition is important because it shows that for any given, finite set of observations, we can test whether GARP holds; if so, a monotonic concave utility function can be constructed which, when maximized under the budget constraint, generates all the observations of the set. Hence GARP can replace the basic axioms of choice as well as the monotonicity and concavity assumptions.

The result also implies that quasiconcavity of utility is not a testable generalization on a finite data set and that concavity is not more restrictive. Observe, however, that the utility function in condition 3 of Proposition 2.9 is piecewise linear, so it is not strictly concave. To obtain single-valuedness of the demand function, the GARP assumption needs to be replaced by a stronger one.[55]

Implications for Testing Producer Behavior

Recall that the producer cost minimization problem in the single output case has the same format as the consumer expenditure minimization problem. To test integrability, the same analysis can be performed if one replaces $v(p/h)$ by q, the scale of production; $\xi_k(p, q)$ is then the demand for input k.

Testing revealed preference in production amounts to testing the validity of profit maximization or cost minimization, on the basis of observations on prices and net supply or input demand. However, while utility is not

54. This can be related to the fact that for utility functions, utility maximization implies expenditure minimization. But here no utility function is specified.

55. See Varian (1982a).

observable, the scale of production is; therefore it is possible to distinguish between concavity and quasiconcavity of the production function.[56]

2.2.5 Consumer Behavior in Applied Models

Choosing between the Primal and the Dual Approach

Since little a priori information is available on consumer preferences, one could be led to conclude that for consumption, the dual approach is preferable. However, such a statement needs some qualification.

First, in the primal approach we may have clear views on separability, for example, among individuals in a household or among groups of commodities, over time and on some variables, like public goods, that are exogenous to the household. However, this does not in general make the dual formulation separable,[57] and it is the dual separability that is convenient in estimation. Second, in household modeling the dual formulation is made difficult because there are no data available for prices of intrahousehold commodities. Third, the welfare program of chapter 1 which is often used to represent a competitive equilibrium[58] also follows a primal formulation. It is used in applications because its mathematical programming format has the advantage that for given welfare weights, the model can be solved by standard programming algorithms.

The primal and the dual approach are equivalent if utility functions are of the CES type:[59] In estimating demand functions that belong to a CES expenditure function, we can recover the primal utility function in CES form. However, since among the frequently used forms, only those of the CES-family are self-dual, for other forms such as the flexible forms (translog, generalized Leontief, etc.), we have to make a choice between the primal and the dual formulations. Obviously, if we specify numerical values for the parameters of the utility function, we can use a primal formulation. However, as soon as we estimate demand functions for which there is no analytical primal form, we cannot use a mathematical programming format for the general equilibrium model.[60] In applied general equilibrium m-odeling the choice between primal and dual will then mainly depend on considerations of computational expediency: In the primal approach we would solve one or several mathematical programs (with some iterative

56. See also Varian (1982b).

57. See Blackorby, Primont, and Russell (1978) on the nonequivalence between primal and dual separability.

58. See section 1.4.3, and the discussion after proposition 1.8.

59. See section 2.1.4.

60. Such a format will be introduced in chapter 3, section 3.1.

procedure on parameters of these programs to find a fixed point); in the dual approach we would solve a system of nonlinear equations.[61]

Functional Forms

Most of the discussion on production and cost functions applies clearly to utility and expenditure functions, and we do not repeat the various functional forms treated in section 2.1.4. However, in applications, the requirement of strict concavity of the utility function is often relaxed to strict quasiconcavity, and this leads to more general functional forms for the expenditure function, such as[62]

$$\ln e(p, v) = b(p) + a(p)K(v). \tag{2.31}$$

This yields as a special form

$$\ln e(p, v) = \alpha_0 + \Sigma_k \alpha_k \ln p_k + \tfrac{1}{2}\Sigma_k \Sigma_h \beta_{kh} \ln p_k \ln p_h + \Sigma_k \gamma_k \ln p_k \ln v,$$

which is known as the "almost ideal demand system." The parameters have to satisfy the properties of adding-up of expenditure shares and homogeneity of degree one in prices ($\Sigma_k \alpha_k = 1$, $\Sigma_h \beta_{kh} = 0$, $\Sigma_k \gamma_k = 0$, $\beta_{hk} = \beta_{kh}$). For this form it is not possible to impose strict quasiconcavity for all prices.

Shephard's lemma leads to the following expression for the budget shares

$$w_k(p, v) = \alpha_k + \Sigma_h \beta_{hk} \ln p_h + \gamma_k \ln v. \tag{2.32}$$

Since v cannot be observed, it has to be inferred by setting $e(p, v) = h$. Hence we have to estimate the coefficient α_0 in the expenditure function together with the coefficients of the demand system.[63]

Demand systems can be estimated from budget surveys or from aggregate time series, or even better, from panel data obtained by following a sample of households over time. Estimation via budget surveys can be used to measure the α_k and γ_k coefficients, but unless prices differ substantially among households in the sample, household data will not generate a reliable estimate of the β_{hk} coefficients, except if one has access to panel data. Otherwise, one has to rely on aggregate time series for expenditure.

Functional Separability

Functional separability has special relevance in consumer theory, though most of what will be said here also applies to production theory. We briefly

61. See, for example, Deaton (1986) for a detailed treatment of what follows.

62. See Deaton and Muellbauer (1980).

63. For further details, see Deaton (1986). In practice, one often approximates v by deflated income, but this approximation conflicts with the specification of the expenditure function.

discuss separability of the expenditure function, before turning to separability of the utility function.

Consider the following specification for the expenditure function:

$$e(p, v) = b(p) + P(p)K(v), \tag{2.33}$$

which is called the Gorman polar form. Equation (2.33) differs from (2.31), since there is no logarithm on the left-hand side. In (2.33), $P(p)$ is an aggregate price index, and $b(p)$ is the cost of committed expenditure; $K(v)$ is a nonnegative, increasing function of utility. Both $P(p)$ and $b(p)$ have to be homogeneous of degree one, nondecreasing and concave. Shephard's lemma will now lead to

$$\xi_k(p, v) = \bar{x}_k(p) + \frac{\partial P}{\partial p_k} K(v), \tag{2.34}$$

where $\bar{x}_k(p) = \partial b/\partial p_k$ is committed demand on commodity k, while $b(p) = \Sigma_k p_k \bar{x}_k(p) = \Sigma_k p_k(\partial b/\partial p_k)$ is total committed expenditure. In this case the utility function $u(z)$ is homothetic; that is, $F(z) = K(u(z))$ is homogeneous of degree one in z, where z satisfies $z = (x - \bar{x}(p))$. To obtain the Marshallian demand functions, we derive the indirect utility function from (2.33) as

$$K(v) = \frac{h - b(p)}{P(p)}.$$

Substituting this into (2.34) gives the Marshallian expenditure system:

$$p_k x_k(p, h) = \frac{\partial \ln P}{\partial \ln p_k}(h - \Sigma_k p_k \bar{x}_k(p)) + p_k \bar{x}_k(p). \tag{2.35}$$

Note that if $P(p)$ is Cobb-Douglas and $\bar{x}$ is independent of p (i.e., $b(p)$ is linear), then the well-known linear expenditure system (LES) follows. The CES functions that are often used in AGE models also fit within the Gorman polar form, which is of special relevance in relation to aggregation over consumers, as will be seen below.

Separability of the Utility Function

DEFINITION 2.2 (Separable utility function) If the utility function $u(x)$ can be written

$$u(x^1, x^2, \ldots, x^s) = W[u^1(x^1), u^2(x^2), \ldots, u^k(x^k), \ldots, u^s(x^s)],$$

where W is continuous, strictly quasi-concave, increasing, and u^k is continuous, strictly concave, increasing in its arguments, then $u(x)$ is separable.[64]

64. This assumption refers to two-level separability. Separability can be extended to an arbitrary number of levels provided that the functions not at the highest level are all monotonic strictly concave so that $u(x)$ remains strictly quasi-concave, nondecreasing. See theorem A.1.4.

Three interpretations are of particular relevance: k can refer to a subconsumer within a group of consumers for which a so-called social welfare function $u(x)$ is defined; it can denote time in case $u(x)$ is an intertemporal utility function, or it can also refer to commodity groups. The three cases will be discussed by turns.

When the grouping refers to subconsumers, the maximization of $u(x)$ subject to the collective budget constraint allocates collective income to individuals, which in fact amounts to the specification of a tax or transfer function that is consistent with welfare maximization. This may be seen as follows: Consider the social welfare maximization problem:

$$\max W[u^1(x^1), u^2(x^2), \ldots, u^k(x^k), \ldots, u^s(x^s)],$$

x^k, all k,

subject to

$$\Sigma_k p x^k \leqslant \Sigma_k h^k,$$

where h^k is the (pre-transfer) income of consumer k. This may be rewritten as a two-level problem:

Level 1

Consumer k's utility maximization,

$$v^k\left(\frac{p}{e^k}\right) = \max\{u^k(x^k) \mid px^k \leqslant e^k, x^k \geqslant 0\} \qquad \text{for } k = 1, 2, \ldots, s,$$

where $v^k(p/e^k)$ is consumer k's indirect utility function and his expenditure e^k is obtained from level 2 maximization.

Level 2

Optimal expenditure allocation,

$$\max W\left[v^1\left(\frac{p}{e^1}\right), v^2\left(\frac{p}{e^2}\right), \ldots, v^k\left(\frac{p}{e^k}\right), \ldots, v^s\left(\frac{p}{e^s}\right)\right],$$

$e^k \geqslant 0$, all k,

subject to

$$\Sigma_k e^k \leqslant \Sigma_k h^k.$$

As long as indirect utility is concave in e^k, the level 2 optimization problem follows the standard setup of utility maximization under a budget constraint, with prices equal to 1. This makes it possible to obtain expenditure allocation functions e^k, from which one can derive the so-called direct tax functions $T^k = h^k - e^k$.[65]

Note that in specifying such a social welfare function W, we assume that utility levels can be compared between consumers. Clearly a special case of this welfare function is

$$u(x^1, x^2, \ldots, x^m) = \Sigma_k \alpha_k u^k(x^k)$$

which is the objective function of the welfare program of chapter 1.

When the grouping refers to time, the superscript s can be interpreted as the lifetime of the consumer, while k denotes his age. The consumer model is then referred to as the life-cycle model: The consumer maximizes his lifetime utility subject to a lifetime wealth constraint, which implies that he can borrow and lend but cannot leave any debt behind. Such a model does not only generate consumption decisions but also saving and debt over the lifetime. In chapters 7 and 8 we will see that it is useful to consider the following type of recursive separability:

$$u(x) = W(x^1, W(x^2, W(\ldots, W(x^{s-1}, u^s(x^s) \ldots)))).$$

This allows to focus on the recursive single-period utility function:

$$u^k = W(x^k, u^{k+1}).$$

In particular, when W is linear in u^{k+1}, we have $u^k = f(x^k) + \beta u^{k+1}$. Thus we obtain

$$u(x) = \Sigma_k (\beta)^k f(x^k),$$

the well-known additively separable intertemporal utility function, where $0 < \beta < 1$ is the discount factor.[66]

Separability will be used later to represent more elaborate tax systems, intertemporal producer, and consumer behavior as well as public goods and voluntary transfers.

Finally, when applied to commodity groups, separability allows nesting in the utility function. Here a particular commodity belongs to one group only, and $u^k(.)$, assumed to be homogeneous of degree one, can be interpreted as the utility derived from commodity group k but also as the "aggregate" quantity of a commodity. The consumer problem is then

$$\max W[u^1(x^1), u^2(x^2), \ldots, u^s(x^s)], \qquad x^k \geqslant 0, \text{ all } k,$$

65. This will make it possible to estimate a direct tax system in a way similar to a demand system. See also section 5.1.2.

66. Here $(\beta)^k$ denotes β to the power k.

subject to

$\Sigma_k p^k x^k \leqslant h.$

This problem is solved in two levels:

Level 2

Aggregate optimal expenditure allocation,

$\max W[u^1, u^2, \ldots, u^s],$

$u^k \geqslant 0$, all k,

subject to

$\Sigma_k P^k u^k \leqslant h,$

where P^k is an aggregate price index: P^k is a function of p^k.

Level 1

Given the optimal u^k's, we can compute $h^k = P^k u^k$ and solve for each k separately:

$\max u^k(x^k),$

$x^k \geqslant 0,$

subject to

$p^k x^k \leqslant h^k$

Since $u^k(x^k)$ is homogeneous of degree one, the price index P^k for group k is the price index of a Gorman form with zero commitments: $e^k = P^k(p^k)u^k(x^k)$. In practice, a functional form is specified for $P^k(p^k)$, in order to solve level 2 allocations. The resulting optimal u^k is used to solve level 1 as $x^k = u^k \partial P^k / \partial p^k$.

Aggregation over Consumers

Consider the expenditure system (2.35) for consumers $i = 1, 2, \ldots, m$:

$$p_k x_{ik}(p, h_i) = \frac{\partial \ln P}{\partial \ln p_k}(h_i - \Sigma_k p_k \bar{x}_{ik}(p)) + p_k \bar{x}_{ik}(p), \qquad \text{all } k.$$

In aggregating these expenditures over consumers, we obtain

$$p_k \Sigma_i x_{ik}(p, h_i) = \frac{\partial \ln P}{\partial \ln p_k}(\Sigma_i h_i - \Sigma_k p_k \Sigma_i \bar{x}_{ik}(p)) + p_k \Sigma_i \bar{x}_{ik}(p), \qquad \text{all } k.$$

Clearly, for $h = \Sigma_i h_i$, $\bar{x}_k(p) = \Sigma_i \bar{x}_{ik}(p)$ and $x_k(p, h_i) = \Sigma_i x_{ik}(p, h_i)$, the aggregate expenditure system is

$$p_k x_k(p, h) = \frac{\partial \ln P}{\partial \ln p_k}(h - \Sigma_k p_k \bar{x}_k(p)) + p_k \bar{x}_k(p), \text{ all } k;$$

hence, one can conceive of an aggregate consumer with income h and committed demand $\bar{x}(p)$, whose expenditure function is (2.33) and whose behavior is therefore consistent with utility maximization subject to an aggregate budget constraint with income h.

The essential properties are that (1) the function $\partial \ln P/\partial \ln p_k$, which is the marginal budget share for commodity k, does not vary among consumers (common expenditure function for uncommitted income $P(p)$) and (2) the marginal budget shares $\partial \ln P/\partial \ln p$ should not depend on income h (homotheticity of the utility functions). The CES and Cobb-Douglas forms that are often used in AGE models ensure that these aggregation conditions hold.

Econometricians have devised less restrictive forms than the Gorman polar form to obtain specifications of aggregate demand that result from utility maximization by a representative consumer. These forms are generally based on (2.31) instead of (2.33). For consumer i the associated budget shares are[67]

$$\frac{\partial \ln e_i(p, v)}{\partial \ln p} = \frac{\partial b(p)}{\partial \ln p} + \left(\frac{\partial a(p)}{\partial \ln p}\right) K(v_i),$$

where by (2.31),

$$K(v_i) = \frac{\ln h_i - b(p)}{a(p)}.$$

This leads to the expenditure system

$$w_{ik}(p, h_i) = \frac{1}{a(p)} \frac{\partial a(p)}{\partial \ln p_k}(\ln h_i - b(p)) + \frac{\partial b(p)}{\partial \ln p}, \qquad \text{all } k.$$

Aggregating these shares over consumers and grouping the constant terms leads to

$$\frac{\Sigma_i h_i w_{ik}(p, h_i)}{\Sigma_i h_i} = \frac{1}{a(p)} \frac{\partial a(p)}{\partial \ln p_k}\left(\frac{\Sigma_i h_i \ln h_i}{\Sigma_i h_i} - b(p)\right) + \frac{\partial b(p)}{\partial \ln p}, \qquad \text{all } k.$$

Clearly, for $\ln \bar{h} = \Sigma_i h_i \ln h_i/\Sigma_i h_i$, and $w_k(p, h_i) = \Sigma_i h_i w_{ik}(p, h_i)/\Sigma_i h_i$, the aggre-

67. Note that here one assumes that both $a(p)$ and $b(p)$ are the same for all consumers.

gate expenditure system is

$$w_k(p, \bar{h}) = \frac{1}{a(p)} \frac{\partial a(p)}{\partial \ln p_k} (\ln \bar{h} - b(p)) + \frac{\partial b(p)}{\partial \ln p}, \qquad \text{all } k. \tag{2.36}$$

There again is a consumer with income $\bar{h}$, the representative consumer, whose budget shares are determined according to (2.36) and whose behavior is consistent with utility maximization.

However, there is an important difference with respect to the Gorman form: The system (2.36) allows for differences in budget shares among consumers (and has equation (2.32) as a special case), but the income $\bar{h}$ is not equal to total income $\Sigma_i h_i$.[68] If distribution of income, say, within some social group, is known and remains constant around the mean (e.g., if all consumers in this group own endowments and shares in profits in the same proportions), then $\bar{h}$ can readily be computed, and the general equilibrium model can work with a single representative consumer per class.

However, if the only data available are aggregate time series (for a class or for the whole population), one cannot obtain $\bar{h}$ and has to work with aggregate income h. This may introduce an important aggregation error, unless the Gorman form holds for every consumer in the class.

2.3 Supply and Demand under the Standard Assumptions

The discussion of sections 2.1 and 2.2 has covered many topics that are useful in applications. For convenience we repeat now the main assumptions that are important in theoretical general equilibrium analysis. These assumptions will serve as standard assumptions in chapters 3, 4, and 5. They will be modified afterward to account for additional characteristics of the models.

ASSUMPTION S (Standard assumptions) The standard assumptions are assumptions P1′, C1′, and C2′.

Recall that this implies the following:

Production sets. The production set Y_j of every producer j has possibility of inaction ($0 \in Y_j$); it is compact and convex (assumption P1′).

Utility functions. The utility function $u_i: R^r_+ \rightarrow R_+$, $u_i(x_i)$ is continuous, strictly concave, nonsatiated, and satisfies $u_i(0) = 0$ for all i; for $i = 1$ it is increasing with respect to all commodities (assumption C1′).

Endowments. $\omega_i \geqslant 0$, $\omega_i \neq 0$ for all i; for $i = 1$, $\omega_i > 0$ (assumption C2′).

68. It is rather the geometric mean of the individual incomes, with exponents equal to the income shares of the consumers.

For producers we have assumed that production sets are convex (and not strictly convex) to allow for constant returns to scale and linear activity analysis formulations that are often used in applied general equilibrium models. A single value of production will follow by adjusting supply to a single-valued demand (a function of prices and income), where single-valuedness is obtained through strict concavity of utility functions in C1′.[69] We do not make differentiability assumptions because these are not needed for proofs of existence and efficiency of equilibrium and, at any rate, are not sufficient to ensure differentiability of net supply and consumer demand.

We also summarize the properties of net supply and consumer demand that are used in chapters 3 to 12.

PROPOSITION 2.10 (Properties of net supply and consumer demand) Under assumption S:

1. the profit function $\Pi_j(p)$ is nonnegative, continuous, convex, homogeneous of degree one in prices;

2. the supply correspondence $Y_j(p)$ is uppersemicontinuous in p and convex-valued; rescaling prices does not affect outputs and inputs, $Y_j(\lambda p) = Y_j(p)$ for all given $\lambda > 0$;

3. the Marshallian demand function $x_i(p, h_i)$ is continuous at positive prices p and income h_i; it is also homogeneous of degree zero in (p, h_i) and satisfies $\Sigma_k p_k x_{ik}(p, h_i) = h_i$.

Proof See propositions 2.2 and 2.8 ∎

Consumer $i = 1$ has a utility function that is increasing in all commodities and a strictly positive endowment. We need this assumption for the proofs of existence of equilibrium in subsequent chapters, since this makes equilibrium prices positive. The property is specified in the following proposition.[70]

PROPOSITION 2.11 (Desirability when $u(x)$ is increasing in all commodities) Consider the consumer problem (2.22) where $u(x)$ is continuous, strictly concave, increasing, and assume positive (p^0, h). Then, for any given supply y^0, there exists a value $\varepsilon > 0$ such that the demand for commodity k, $x_k(p^0, h)$ exceeds y_k^0 whenever the price p_k^0 falls below the value ε.

69. This assumption may look too restrictive for applications as it seems to exclude well-known homothetic forms like the CES. But whenever we use assumption C1′ in subsequent chapters, "strict concavity" of a function may be replaced by the requirement that there exists a monotonic transformation that makes it concave with strictly convex isoquants. A simple example of such a nonconcave function is the Cobb-Douglas for which the sum of exponents exceeds unity. This function is also strictly quasi-concave, since its isoquants are strictly convex. Unless we use welfare optima, we can drop the requirement that a monotonic transformation should exist and merely assume strict quasi-concavity.

70. To lighten notation, we drop the consumer index $i = 1$.

Proof Choose a value $\varepsilon > 0$, and define the rationed consumer problem:

$$\max_{x \geqslant 0}\{u(x) \mid px \leqslant h, x \leqslant \bar{x}\}, \tag{2.37}$$

where $\bar{x}$ is a ration. Since the constraint set is convex, nonempty, and compact, and $u(x)$ is strictly concave, the optimal solution $\tilde{x}(p, h)$ of (2.37) is unique. Hence, by the maximum theorem A.3.1, it is continuous for all nonnegative p. We start at prices $p_k = p_k^0 \geqslant \varepsilon$, with ε chosen so that $p_k^0 = \varepsilon$ for at least one k, and set the ration such that $\bar{x}_k > h/\varepsilon$ for all k (meaning that if a price p_k is not below ε, optimal consumption will be less than $\bar{x}_k$), and $\bar{x}_k > y_k^0$. We now reduce p to zero for all commodities k for which $p_k^0 = \varepsilon$. Let K be the set of such commodities, and let p^1 be the new price vector. By increasingness of utility, we have $\tilde{x}_k(p^1, h) = \bar{x}_k$ (which exceeds y_k^0) for all $k \in K$, and therefore $\eta(p^1) = \min_{k \in K}(\tilde{x}_k(p^1, h) - y_k^0) > 0$. Second, for all commodities $k \in K$, we raise the price back to p^0. Since $\tilde{x}(p, h)$ is continuous in p, it follows that for p_k^0 sufficiently close to zero (i.e., for ε sufficiently small), $\eta(p)$ will be positive at p^0 also, since it is continuous. Now, by construction, $\tilde{x}_k(p^0, h) < \bar{x}_k$, so it solves the unrationed consumer problem. Hence $\tilde{x}_k(p^0, h)$ is equal to the optimal solution of the unrationed consumer problem $x_k(p^0, h)$. Therefore, the unrationed demand $x_k(p^0, h)$ exceeds y_k^0 for all $k \in K$. ■

This proposition will be used to show that $p_k > \varepsilon$ in equilibrium; note that in the proposition we have taken income h to be constant. In general equilibrium models income will depend on prices. Therefore, to ensure that demand exceeds any upper bound, we must guarantee that income exceeds a lower bound that does not depend on ε. For example, if prices are on the simplex and the consumer owns a positive endowment, we have $h \geqslant \min_k \omega_k$ as a lower bound, which does not depend on ε. This explains why, in assumption C2′, consumer $i = 1$ is required to possess a positive endowment in all commodities.

In many theoretical models, particularly in the literature on general equilibrium that takes a "differentiable viewpoint," net supply and demand must be continuously differentiable.[71] We have already used such an assumption in sections 1.2.3 and 1.2.5 when discussing multiplicity and computation of equilibrium. Recall from proposition 2.3, that for the producer, net supply will be continuously differentiable if the production set is locally strongly convex, that is, if assumption P1″ is satisfied:

ASSUMPTION P1″ (Local strong convexity of the boundary of Y_j) The production set Y_j of every producer j has possibility of inaction ($0 \in Y_j$); it is compact and has a boundary that is locally strongly convex in the optimal points chosen by the producer.

71. See, for example, Mas-Collel (1985).

As far as the consumer is concerned local strong concavity of the utility function is also sufficient in propositions 2.7 and 2.8, but only if consumption is positive. Therefore, to guarantee differentiability, one has to impose an additional assumption that makes consumption positive:

ASSUMPTION C1″ (Local strong concavity of $u_i(x_i)$ and infinite marginal utilities) The utility function $u_i : R^r_+ \to R_+, u_i(x_i)$ is continuous, strictly concave, nonsatiated, and satisfies $u_i(0) = 0$ for all i; for $i = 1$ it is increasing with respect to all commodities. Moreover it is locally strongly concave in the optimal points chosen by the consumer and satisfies the boundary condition $\lim_{x \to x^0} \|\partial u_i(x_i)/\partial x_i\| = \infty$ whenever $x^0_k = 0$.

This assumption means that whenever $x_{ik} = 0$ for some k, the utility function has an infinite slope in some direction. At that point the marginal utility of some commodity is infinite, and the consumer with positive income will never choose this point.

It is now easy to prove:

PROPOSITION 2.12 (Positive consumption) Let assumption C1″ hold. Then, at positive prices and income, consumption is positive for every consumer who solves (2.22).

Proof Since prices and income are positive, the constraint set is compact and satisfies constraint qualification.[72] Therefore λ_i, the Lagrange multiplier associated with the budget constraint, exists and is bounded, and the first-order conditions are

$$\frac{\partial u_i}{\partial x_i} \leqslant \lambda_i p \qquad \perp x_i \geqslant 0.$$

Since $\partial u_i/\partial x_i$ is bounded, x_i must be positive. ■

2.4 Welfare

In section 2.4.1 we apply our standard assumption to establish properties of the welfare optimum which will be used frequently. In section 2.4.2 we return to the discussion of reform scenarios and the welfare measures that are used for comparing outcomes between scenarios.

2.4.1 The Welfare Optimum

Proposition 1.9 describes the important property that a welfare optimum can be decentralized through a competitive equilibrium with transfers. Here

72. See theorem A.2.1.

we describe other, more technical properties of the welfare optimum that will be needed later on. Recall that the welfare optimum may be written

$W(\alpha, \omega) = \max \Sigma_i \alpha_i u_i(x_i),$

$x_i \geqslant 0$, all i, y_j, all j,

subject to

$\Sigma_i x_i - \Sigma_j y_j \leqslant \omega \qquad (p),$

$y_j \in Y_j,$

for given $\alpha \in S^m = \{\alpha \mid \alpha_i \geqslant 0, \Sigma_i \alpha_i = 1\}$ and where $\omega = \Sigma_i \omega_i$. Let x_i° be an optimal consumption and u_i° the associated optimal utility.

Before characterizing the optimal allocations of this program, we state a rather elementary property of the welfare function itself.

PROPOSITION 2.13 (Concavity of the welfare function) If utility functions are strictly concave (as in assumption C1′), then $W(\alpha, \omega)$ is concave if $\alpha \geqslant 0$ and strictly concave if $\alpha > 0$.

Proof In terms of theorem A.1.4, assertion 1, we can choose $h(f, g)$ linear where f and g are two utility functions; therefore the sum of any two utility functions is concave. Repeating this, we obtain a concave welfare function. For strict concavity we proceed in the same way, using assertion 4, where $\alpha_i u_i(x_i)$ is strictly concave since $\alpha_i > 0$, and therefore $\Sigma_i \alpha_i u_i(x_i)$ also is strictly concave. Note, however, that if $\alpha_i = 0$ for some i, then $\alpha_i u_i(x_i)$ is not strictly concave. Therefore $\Sigma_i \alpha_i u_i(x_i)$ is no longer strictly concave in $(x_1, x_2, \ldots, x_m)$. ■

PROPOSITION 2.14 (Properties of the welfare optimum) Let assumption S be satisfied. Then the following properties hold:

1. $\alpha_i u_i(x_i^\circ) \geqslant p x_i^\circ$ for all $p \in M^\circ(\alpha, \omega)$, where $M^\circ(\alpha, \omega)$ is the set of Lagrange multipliers;[73] in particular, if $\alpha_i = 0$, then $p x_i^\circ = 0$.
2. $x_i^\circ(\alpha, \omega)$ is a continuous function if $\alpha_i > 0$ or if $\alpha_1 > 0$.
3. $u_i^\circ(\alpha, \omega) = \partial W / \partial \alpha_i$ for $\alpha_i > 0$ or $\alpha_1 > 0$; $\alpha_i u_i^\circ(\alpha, \omega)$ is continuous for all $\alpha \in S^m$.
4. $W(\alpha, \omega)$ is convex nondecreasing in α and concave nondecreasing in ω.
5. If the utility possibility set is (locally) strongly convex, then $u_i^\circ(\alpha, \omega)$ is increasing in α_i.
6. $W(\alpha, \omega)$ is homogeneous of degree one in α; $u_i^\circ(\alpha, \omega)$ and $x_i^\circ(\alpha, \omega)$ are

73. See the maximum theorem A.3.1.

homogeneous of degree zero in α for $\alpha_i > 0$ or $\alpha_1 > 0$, and the correspondence $M^\circ(\alpha, \omega)$ of Lagrange multipliers is homogeneous of degree one in α.[74]

7. $0 \notin M^\circ(\alpha, \omega)$.

8. If $\alpha_1 > 0$, then commodity balances will hold with equality.

9. If $\alpha_1 > 0$, then $p > 0$ for all $p \in M^\circ(\alpha, \omega)$.

10. If $\alpha_1 > 0$, and $u_1(x_1)$ is continuously differentiable on R^r_{++} and satisfies the boundary condition $\lim_{x \to x^\circ} \|\partial u_1(x_1)/\partial x_1\| = \infty$ whenever $x_k^\circ = 0$, then $M^\circ(\alpha, \omega)$ is single-valued and continuous.

Proof The objective function is continuous and concave. By assumption P1′, $y_j = 0$ is feasible, since $0 \in Y_j$, and since $x_i = 0$ is feasible as well, the constraint set is nonempty. It is compact, since Y_j is compact, and $x_i \geqslant 0$. So a bounded primal optimal solution exists. Moreover, since $\omega > 0$, by assumption C2′, Slater's constraint qualification holds, and bounded Lagrange multipliers exist. We now prove assertions 1 to 10.

1. To prove this assertion, we use a Lagrange inequality as in the proof of proposition 1.9:

$$\Sigma_i \alpha_i u_i(x_i) - p(\Sigma_i x_i - \Sigma_j y_j - \omega) \leqslant \Sigma_i \alpha_i u_i(x_i^\circ) - p(\Sigma_i x_i^\circ - \Sigma_j y_j^\circ - \omega) \tag{2.38}$$

for all $x_i \geqslant 0$ and all $y_j \in Y_j$ and $p \in M^\circ(\alpha, \omega)$.

We set $y_j = y_j^\circ$ for all j and $x_i = x_i^\circ$ for all i, except for $i = s$. Then (2.38) implies that $\alpha_s u_s(x_s^\circ) - p x_s^\circ \geqslant \alpha_s u_s(x_s) - p x_s$ for all $x_s \geqslant 0$. This inequality must hold for all $x_s \geqslant 0$, including $x_s = 0$. Since $u_s(0) = 0$, it follows that $\alpha_s u_s(x_s^\circ) - p x_s^\circ \geqslant 0$. Now $\alpha_s = 0$ implies that $p x_s^\circ = 0$.

2. For $\alpha_i > 0$, this follows from the strict concavity of $u_i(x_i)$ and proposition 2.13. Therefore the maximum theorem A.3.1, assertion 3, shows continuity of $x_i^\circ(\alpha, \omega)$. For $\alpha_i = 0$ and $\alpha_1 > 0$, consumer 1, whose utility function is increasing in all commodities, will leave no consumption to consumer i. Therefore $x_i^\circ = 0$, and consumption is unique for $\alpha_i = 0$ as well and hence continuous, again by the maximum theorem.

3. The first part follows from the envelope theorem A.3.3; continuity of $\alpha_i u_i^\circ(\alpha, \omega)$ follows from property 2 above if $\alpha_1 > 0$, and since $\lim_{\alpha_i \to 0} \alpha_i u_i^\circ(\alpha, \omega) = 0$, it is continuous for $\alpha_i = 0$ also.

4. Convexity and nondecreasingness in α follows from the perturbation theorem A.3.2, assertion 2; concavity in ω follows from the same theorem A.3.2, assertion 1. Nondecreasingness in ω is obvious.

74. The correspondence $P(\alpha, \omega)$ is homogeneous of degree one in α if $\lambda p \in P(\lambda\alpha, \omega)$ whenever $p \in P(\alpha, \omega)$.

5. (Local) strong convexity of the utility possibility set implies that the Hessian of $W(\alpha, \omega)$ has positive diagonal, and by property 3 above, this Hessian $\partial^2 W/\partial\alpha^2$ is equal to $\partial u^\circ(\alpha, \omega)/\partial\alpha$.

6. Multiplying welfare weights by a common factor $\lambda > 0$ does not change the allocations. Therefore $W(.)$ is homogeneous of degree one in α, and $u_i^\circ(\alpha, \omega)$ and $x_i^\circ(\alpha, \omega)$ are homogeneous of degree zero. That the correspondence $M^\circ(\alpha, \omega)$ of Lagrange multipliers is homogeneous of degree one follows from the fact that Lagrange multipliers are subgradients of $W(\alpha, \omega)$ with respect to ω (see definition A.1.16 and theorem A.3.4). Let the subgradient of W with respect to ω be defined as

$$\partial_\omega W(\alpha, \omega) = \{p \in R^r \mid W(\alpha, \omega) \geqslant W(\alpha, \omega') - p(\omega' - \omega) \text{ for all } \omega' \in R^r_+\}; \tag{2.39}$$

then, since $W(\lambda\alpha, \omega) = \lambda W(\alpha, \omega)$ and $W(\lambda\alpha, \omega') = \lambda W(\alpha, \omega')$, it follows that if $p \in \partial_\omega W(\alpha, \omega)$, then $\lambda p \in \partial_\omega W(\lambda\alpha, \omega)$.

7. This follows from increasingness in some commodity of the utility of every consumer and from the fact that $\alpha \in S^m$ (see theorem A.2.1, assertion 6).

8. Since $\alpha_1 u_1(x_1)$ is increasing in all commodities, an allocation with strict inequality in the balance of any commodity cannot be optimal.

9. For a given commodity h, consider the subdifferential (2.39), and choose $\omega'_k = \omega_k$ for all $k \neq h$ and $\omega'_h > \omega_h$. Since $\alpha_1 u_1(x_1)$ is increasing in all commodities, $W(\alpha, \omega') > W(\alpha, \omega)$. But then, for p to be a subgradient, p_h must be positive in (2.39).

10. Under the continuous differentiability and the boundary conditions, x_1 will be positive. Then, from the first-order condition with respect to x_1, we have $\alpha_1 \partial u_1(x_1)/\partial x_1 = p$ in any optimum. Since $x_1(\alpha, \omega)$ and $\alpha_1 \partial u_1(x_1)/\partial x_1$ are single-valued and continuous, so is $P(\alpha, \omega)$. ■

We note that if α_i and α_1 are zero, optimal consumption x_i° will not be single-valued if consumer i derives utility from commodities that no one with positive welfare weight can use. It may also be stressed that if $u_i(x_i)$ were only required to be strictly quasiconcave, $\Sigma_i \alpha_i u_i(x_i)$ would not necessarily have this property, and the resulting program would not be convex. It is possible to maintain strict quasiconvexity of the objective by specifying it as $\min_i \alpha_i u_i(x_i)$, and all the analysis that uses the welfare optimum would carry through. The formulation with the linear welfare function is, however, more common and more transparent.

2.4.2 Welfare Measures

To compare outcomes in two situations (parameter settings), say, a reference situation 0 and an alternative 1, we need performance criteria. These may be direct measures of consumption like food consumption expressed in calories or expenditure in constant prices. They may also be utility-based measures.

An obvious and valid measure is utility itself. This has, however, the following disadvantages: (1) There is no unique cardinal scale; utility can be measured as $u(x)$ or as $\ln u(x)$ without changing demand. Though if $u_1 > u_0$, we also have $\ln u_1 > \ln u_0$, so the numerical values (e.g., percentage increases) will look different. (2) In policy analysis, we are often not interested in numerical outcomes of the utility of every consumer, and then aggregation is needed that gives outcomes for well-identified groups of consumers (rural, urban, etc.). For this it is convenient to express welfare changes in terms of income, such as compensated variations (CV) or equivalent variations (EV).

Both measures compute differences in expenditures for each consumer separately, and since they are expressed in value terms, they can be summed over consumers to obtain an aggregate measure. The measures are defined as

$$\mathrm{CV} = e(p_1, u_1) - e(p_1, u_0),$$

$$\mathrm{EV} = e(p_0, u_1) - e(p_0, u_0).$$

CV expresses the minimal change of income necessary to reach u_1 at prices p_1, and for EV, this is at prices p_0. After summation over individuals, the two measures may lead to different assessments, with gains according to one aggregate measure and losses according to the other.

Moreover these measures only calculate actual expenditure. By contrast, partial equilibrium analysis uses the concept of producer and consumer surplus as its criterion. These are the welfare triangles obtained through integration of the demand and the supply functions. For the producer, the integration leads to profit, which can be computed directly. For consumers, the situation is more complex. It may be seen from the Slutsky equation (section 2.2.3) that since direct integration over prices of the Marshallian demand function has no theoretical interpretation, it is difficult to interpret such integration in a general equilibrium model.

In the welfare program, total welfare $\Sigma_i \alpha_i u_i(x_i)$ can also be decomposed into consumer and producer surplus. Producer surplus includes profits $\Sigma_j p y_j$ as well as proceeds from sales of initial endowments $\Sigma_i p\omega_i$. Aggregate consumer surplus is thus

$$\Sigma_i \alpha_i u_i(x_i) - \Sigma_j p y_j - \Sigma_i p\omega_i, \tag{2.40}$$

and we can define the surplus of consumer i as

$$\mathrm{CS}_i = \alpha_i u_i(x_i) - h_i, \tag{2.41}$$

where $px_i = h_i$, $h_i = p\omega_i + \Sigma_j \theta_{ij} p y_j + T_i$ and T_i is a lump-sum transfer. By proposition 1.9 the welfare program decentralizes into a competitive equilibrium with transfers in which $\alpha_i = 1/\lambda_i$, where λ_i is the marginal utility of income.[75] The term $\alpha_i u_i(x_i)$ is thus, like income, expressed in monetary terms and referred to as "full expenditure" (expenditure plus consumer surplus).

An alternative way of obtaining CS_i is to define the extended utility function

$$\tilde{u}_i(x_i, n_i) = n_i u_i\left(\frac{x_i}{n_i}\right)$$

and add the constraint $n_i = 1$ to the consumer problem, which can be written

$$\max \tilde{u}_i(x_i, n_i),$$

$$x_i \geqslant 0,\ n_i \geqslant 0,$$

$$\text{subject to} \tag{2.42}$$

$$px_i \leqslant h_i \qquad (\lambda_i),$$

$$n_i = 1 \qquad (\zeta_i).$$

The function $\tilde{u}_i(x_i, n_i)$ is now homogeneous of degree one in (x_i, n_i) for positive n_i and, in particular, at $n_i = 1$. Assuming differentiability of $u_i(x_i)$, we can apply Euler's rule[76] to obtain

$$\tilde{u}_i(x_i, n_i) = \frac{\partial \tilde{u}_i(x_i, n_i)}{\partial x_i} x_i + \frac{\partial \tilde{u}_i(x_i, n_i)}{\partial n_i} n_i.$$

Since $n_i = 1$, and at $n_i = 1$, $\partial \tilde{u}_i(x_i, n_i)/\partial x_i = \partial u_i(x_i)/\partial x_i$, this implies that

$$\tilde{u}_i(x_i, n_i) = \frac{\partial u_i(x_i)}{\partial x_i} x_i + \frac{\partial \tilde{u}_i(x_i, n_i)}{\partial n_i} \tag{2.43}$$

It follows from the first-order conditions of the consumer problem that

$$\frac{\partial u_i(x_i)}{\partial x_i} x_i = \lambda_i p x_i,$$

where λ_i, the marginal utility of income, is positive. The consumer surplus

75. See equations (1.15)–(1.16) in chapter 1.

76. See theorem A.1.5, assertion 4.

can therefore be obtained from (2.43) as

$$\frac{1}{\lambda_i}\frac{\partial \tilde{u}_i(x_i, n_i)}{\partial n_i} = \frac{1}{\lambda_i}\tilde{u}_i(x_i, n_i) - px_i, \tag{2.44}$$

which may be rewritten

$$\mathrm{CS}_i = \frac{1}{\lambda_i} u_i(x_i) - h_i,$$

leading back to (2.41). This shows that consumer surplus can be interpreted as the shadow price ζ_i that accrues to n_i in (2.42), expressed in monetary terms after multiplication by $1/\lambda_i$. This approach will be used in models with infinite horizons (chapter 8, section 8.1).[77]

Likewise, for the CV and EV measures, we can compute $\alpha_i u_i$ for each consumer in the reference and in the alternative solution. To do this, we must use common welfare weights in the two situations. Hence we can define "compensated full expenditure" (CFE) and "equivalent full expenditure" (EFV) as

$$\mathrm{CFV}_i = \alpha_{i1}(u_{i1} - u_{i0}),$$

$$\mathrm{EFV}_i = \alpha_{i0}(u_{i1} - u_{i0}),$$

where subscripts 0 and 1 refer to the reference and the alternative solutions of the general equilibrium model, respectively. These measures can also be summed over consumers to yield the associated social welfare measures, which, however, can have different signs.

Obviously it is also possible to use any linear or nonlinear welfare function $W(u_1(x_1), u_2(x_2), \ldots, u_m(x_m))$,[78] but all these measures are based on ex post computations and do not affect allocations. A planner can also impose a welfare function on the model and maximize it subject to commodity balances and technology constraints, in two different situations (e.g., good and bad harvests due to weather conditions). This will, like in the second welfare theorem, lead to allocations that can be decentralized through an equilibrium, which cannot be implemented without transfers. Clearly in this situation the objective of the planner's program is the welfare measure, which applies to both situations, so there can be no contradiction.

77. In models with infinitely lived consumers, expenditure will contain an infinite number of terms. Then it will prove convenient to use (2.44) to compute expenditure as the difference between full expenditure and consumer surplus.

78. These have to assume various degrees of interpersonal comparisons of utility; see Roberts (1980). An advantage of using the general equilibrium welfare weights is that one obtains "effective" welfare weights from the solution of the model and the problem of interpersonal comparison is avoided, at least in the reference solution.

3 Applied General Equilibrium

In the literature, general equilibrium models are presented in very different ways. Some are written in terms of excess demand, as in section 1.3, others in terms of welfare programs, as in section 1.4. Yet applied models often use other representations. In this chapter we will formally specify the various representations used in theory as well as in applications and will refer to these as formats.

It is important to study the various formats in some detail because this facilitates further analysis of the competitive model and naturally leads to extensions of this model. When the structure of the model is changed, say, to introduce international trade, taxes, or external effects, or when some assumptions are relaxed, such as allowing for nonconvexities, one format is usually best suited for proofs of existence of equilibrium; another should possibly be used for analysis of efficiency, and yet others for applied work and for computation.

However, it should be stressed that all formats describe the same model and lead to the same equilibrium solutions. Actually we will prove that under our standard assumption S of section 2.3, the model written in each of the formats has the following general competitive equilibrium as its solution:

DEFINITION 3.1 (General competitive equilibrium)[1] The allocation y_j^*, all j, x_i^*, all i, supported by the price vector $p^* \geqslant 0$, $p^* \neq 0$ is a general competitive equilibrium if the following conditions are satisfied:

1. For every producer j, y_j^* solves $\max_{y_j}\{p^* y_j \mid y_j \in Y_j\}$.
2. For every consumer i, x_i^* solves $\max_{x_i \geqslant 0}\{u_i(x_i) \mid p^* x_i \leqslant h_i^*\}$, where $h_i^* = p^*\omega_i + \Sigma_j \theta_{ij} p^* y_j^*$.
3. All markets are in equilibrium, $\Sigma_i x_i^* - \Sigma_j y_j^* - \Sigma_i \omega_i \leqslant 0$.

This model is usually analyzed by defining the excess demand correspondence $Z(p)$ and showing that there exists an equilibrium value p^* such that $z^* \leqslant 0$ for some $z^* \in Z(p^*)$. This is the format used by Arrow, Debreu, and McKenzie to prove existence of equilibrium, and we will refer to it as the excess demand format.

In chapter 1 we used this format under the assumption of strictly concave utility functions and strictly convex production sets, so excess demand is single-valued. In the classical Arrow-Debreu model[2] these assumptions are relaxed to convex production sets and quasiconcave utility functions, leading to excess demand correspondences.

In chapter 2 we have introduced and provided justification for a set of standard assumptions, that is somewhat more restrictive. In particular,

1. This is definition 1.2; it is repeated and renumbered here for convenience.

2. See Debreu (1959).

though we relax strict convexity of production sets, we still require strict concavity of utility functions. In this chapter we will prove existence of a competitive equilibrium under these more general assumptions by means of formats other than the excess demand format.

Sections 3.1 to 3.3 present formats that use mathematical programming representations and have proved useful in theoretical as well as applied work. Each of these formats lends itself to the formulation of an applied general equilibrium (AGE) model: Functional forms can be estimated by econometrics and calibration, and the model can be solved numerically. However, there is an additional format that is widely used in applications: the computable general equilibrium (CGE) format which uses assumptions that are somewhat more restrictive than our standard assumptions. In section 3.4 we specify these assumptions which lead to a basic version of the CGE model.[3] Section 3.5 compares the various formats, stressing their relative merits.

Construction of an applied general equilibrium obviously needs more than choosing a format. Issues ranging from development of the database to estimation of parameters and computation of equilibrium solutions have to be addressed. Section 3.6 is devoted to such aspects of model implementation; it also includes, in GAMS language, a full computer implementation of the basic CGE model of section 3.4. A more elaborate example that combines topics covered in chapters 4 to 7 can be found in appendix B.

While in chapter 1 the assumptions made to prove existence were more restrictive than assumption S, the welfare properties (propositions 1.5 to 1.11) were analyzed under assumptions that were less restrictive. Therefore the general competitive equilibrium of definition 3.1 also possesses these properties and no further discussion on efficiency is needed in this chapter.

In summary, the main function of this chapter is to make the final preparations for the later chapters on specific topics in applied general equilibrium modeling. After a general discussion of the topic, each chapter considers:

1. the representation of the topic in a general equilibrium framework, followed by a proof of existence of equilibrium using a particular format;

2. the efficiency properties using the Negishi format;

3. practices in applied modeling: CGE formulation and issues in model implementation.

3. Because of these specific assumptions, we may refer to CGE as a model rather than as a format.

3.1 The Negishi Format

3.1.1 Existence of Equilibrium: Negishi's Theorem

In section 1.4.3 we showed that a competitive equilibrium can be represented through a welfare optimum with nonzero welfare weights α_i, which are such that all consumers satisfy their budget constraints. This is known as the Negishi theorem (1960), which we now prove in a direct way under our standard assumption S.

DEFINITION 3.2 (Negishi format) The Negishi format is defined as the welfare optimum:

$W(\alpha) = \max \Sigma_i \alpha_i u_i(x_i)$,

$x_i \geqslant 0$, all i, y_j all j,

subject to

$\Sigma_i x_i - \Sigma_j y_j \leqslant \Sigma_i \omega_i \qquad (p)$,

$y_j \in Y_j$,

with welfare weights $\alpha \in S^m$, such that the budget constraints

$px_i = p\omega_i + \Sigma_j \theta_{ij} \Pi_j(p)$

hold for every consumer i. Here p is a vector of Lagrange multipliers associated to the balance constraints, x_i is optimal in the program, and $\Pi_j(p) = \max\{py_j \mid y_j \in Y_j\}$ is the profit function of producer j.

We now state and prove Negishi's theorem.

PROPOSITION 3.1 (Negishi's theorem) Under assumption S, there exists $\alpha^* \in S^m$ in the program of definition 3.2, such that the resulting allocation is a competitive equilibrium as specified in definition 3.1.

Proof We consider the program of definition 3.2. The constraint set of the program is bounded and satisfies Slater's constraint qualification. From proposition 2.14, assertions 1–3 and 6–9, it follows that at every optimum (here denoted by a superscript $^\circ$, rather than $*$, to distinguish optimal from equilibrium values) a number of conditions hold.

1. $\alpha_i u_i(x_i^\circ) \geqslant px_i^\circ$ for all $p \in M^\circ(\alpha)$; in particular, if $\alpha_i = 0$, then $px_i^\circ = 0$.
2. $x_i^\circ(\alpha)$ is a continuous function if $\alpha_i > 0$ or if $\alpha_1 > 0$ (and may be set-valued otherwise).
3. $\alpha_i u_i^\circ(\alpha)$ is continuous for all $\alpha \in S^m$.
4. $W(\alpha)$ is homogeneous of degree one in α; $u_i^\circ(\alpha)$ and $x_i^\circ(\alpha)$ are homogene-

ous of degree zero in α (for $\alpha_i > 0$ or $\alpha_i = 0$ and $\alpha_1 > 0$) and the correspondence $M^\circ(\alpha)$ of Lagrange multipliers is homogeneous of degree one.

5. $0 \notin M^\circ(\alpha)$.
6. If $\alpha_1 > 0$, then commodity balances will hold with equality.
7. If $\alpha_1 > 0$, then $p > 0$ for all $p \in M^\circ(\alpha)$.

We now proceed in two steps. We first construct a fixed point correspondence and show that a fixed point exists; then we prove that in the fixed point, all the conditions of a competitive equilibrium (definition 3.1) are satisfied.

Step 1 Construction of the Fixed Point Correspondence

We construct a fixed point mapping by defining a correspondence and a function. The correspondence generates Lagrange multipliers, and the function updates welfare weights.

Consider the correspondence $M^\circ: A \Rightarrow \bar{M}$, $M^\circ(\alpha)$, where $A\ (=S^m)$ is the set of welfare weights and $\bar{M}$ is the convex hull of $M^\circ(A) = \{p \mid p \in M^\circ(\alpha), \alpha \in A\}$, the set of Lagrange multipliers of the program. By condition 4, constraining α to lie on the simplex, does not affect allocations; by condition 5, $0 \notin M^\circ(\alpha)$. By the maximum theorem A.3.1, assertions 5a–5b, this correspondence is uppersemicontinuous in A and compact, convex-valued.

The continuous function $h: A \times \bar{M} \to A$, $h(\alpha, p)$ adjusts welfare weights as follows: To keep expenditure of consumer i continuous for every $\alpha \in A$, we first define the function $e_i(\alpha, p) = \min\{px_i^\circ(\alpha), \alpha_i u_i(x_i^\circ(\alpha))\}$, which by conditions 2 and 3 above is continuous for all α. Next we store the current value α_i as $\hat{\alpha}_i = \alpha_i$ and obtain the new value of α_i, increasing it when there is a positive budget surplus $s_i(\alpha, p) = \max\{p\omega_i + \Sigma_j \theta_{ij} \Pi_j(p) - e_i(\alpha, p), 0\}$:

$$\alpha_i = \mu(\hat{\alpha}_i + \delta s_i), \tag{3.1}$$

where δ is a positive constant and μ is a scalar such that $\Sigma_i \alpha_i = 1$. Since at least one of the $\hat{\alpha}_i$'s is positive, such a scaling is always possible. By proposition 2.1, assertion 3, $\Pi_j(p)$ is continuous, and since $e_i(\alpha, p)$ is continuous, $s_i(\alpha, p)$ is continuous, and (3.1) defines a continuous function h.

The Lagrange multiplier correspondence $M^\circ(\alpha)$, and the function $h(\alpha, p)$ define a correspondence $G: C \Rightarrow C$, where $C = A \times \bar{M}$. The parameter set C is compact, convex, and, by theorem A.4.3, G is uppersemicontinuous in C, compact, and convex-valued. Therefore, by Kakutani's theorem A.4.2, there exists a fixed point $(\alpha^*, p^*) \in C$.

Step 2 The Fixed Point Is an Equilibrium

In the fixed point we have[4]

$$\alpha_i = \mu(\alpha_i + \delta s_i). \tag{3.2}$$

4. For notational convenience, we leave out the * superscript for denoting equilibrium values.

We show that all α_i's are positive starting with α_1. Suppose that $\alpha_1 = 0$; then by (3.2), $s_1 = 0$. We know that some price p_k will be positive as $0 \notin M^\circ(\alpha)$. Consumer 1 has thus positive income, since, by assumption S, his endowment is positive. By condition 1, since in the fixed point, prices p are Lagrange multipliers, $e_i(\alpha, p) = px_i$ holds. Hence, since $s_1 = 0$, it must be that $px_1 > 0$; however, $\alpha_1 = 0$ implies that $px_1 = 0$ (see condition 1 above), a contradiction. Therefore $\alpha_1 > 0$ in equilibrium, and by condition 7 above, all prices p_k will be positive. All consumers will have positive income, and by the same reasoning as for consumer 1, their welfare weight will be positive.

We now show that budget surplus s_i is zero for all i. By condition 6, commodity balances must hold with equality. Hence, since all consumers cannot have a budget surplus at prices p, some consumer must have a zero surplus. But then, in (3.2), μ must be equal to unity so that budget surpluses are zero for all i.

This proves the existence of a Negishi equilibrium, that is, of a solution to the welfare program in which budget deficits are zero. To prove that this is a competitive equilibrium, we invoke proposition 1.9 (a welfare optimum is a competitive equilibrium with transfers), and since budget deficits are zero, there is no need for lump-sum transfers. ■

The intuition behind this result is easy to grasp. For simplicity, assume there is no production, and consider the following two-consumer reformulation of the Negishi program:

$$\max \alpha_1 u_1(x_1) + \alpha_2 u_2(x_2),$$

$$x_1, x_2 \geqslant 0,$$

subject to

$$x_1 + x_2 \leqslant \omega_1 + \omega_2 \qquad (p),$$

where α_1, $\alpha_2 \geqslant 0$, $\alpha_1 + \alpha_2 = 1$, and p denotes the multipliers.

For each pair (α_1, α_2) the solution (x_1, x_2) will be unique. If we also suppose that the assumption made in assertion 10 of proposition 2.14 holds, then p is unique and $p(\alpha)$ is a continuous function for $\alpha_1 > 0$. Therefore the budget surplus of each consumer is

$$b_i(\alpha_1, \alpha_2) = p(\alpha_1, \alpha_2)\omega_i - p(\alpha_1, \alpha_2)x_i(\alpha_1, \alpha_2), \qquad i = 1, 2.$$

If $\alpha_1 = 0$, consumer 1 has no importance, and the solution of the program will be $px_1(\alpha) = 0$ (proposition 2.14, assertion 1) and $px_2(\alpha) = p(\omega_1 + \omega_2)$, and so $b_1(\alpha_1, \alpha_2) > 0$. By continuity, $b_1(\alpha_1, \alpha_2)$ will still be positive for α_1 positive, but close to zero. Likewise $b_1(\alpha_1, \alpha_2)$ will be negative for α_1 close to one. Since $b_1(\alpha_1, \alpha_2)$ is continuous, there must exist a value of (α_1, α_2) for which $b_1(\alpha_1, \alpha_2) = 0$. Since $p(x_1 + x_2) = p(\omega_1 + \omega_2)$, necessarily $b_2(\alpha_1, \alpha_2)$

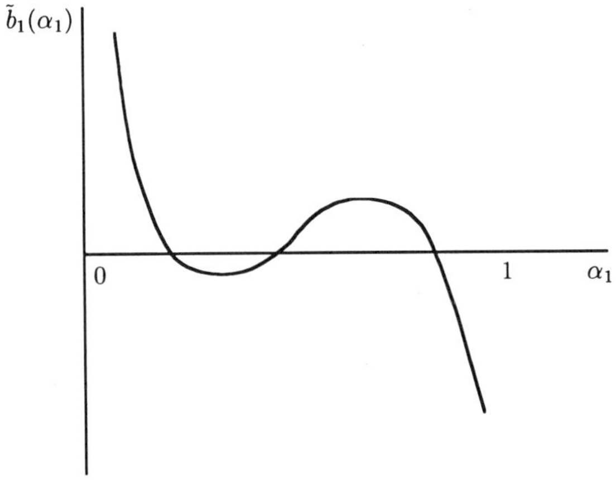

Figure 3.1
The excess budget function $\tilde{b}_1(\alpha_1)$ in the two-consumer case without production.

will also be equal to zero. This is illustrated in figure 3.1, which depicts the function $\tilde{b}_1(\alpha_1) = b_1(\alpha_1, 1 - \alpha_1)$. Hence the equilibria must lie on the bounded, open interval (0, 1).

3.1.2 Multiplicity of Equilibrium

As can be seen, the reasoning used to construct figure 3.1 is very similar to the one used in figure 1.3 for excess demand in the two-commodity economy. One difference is that when a price goes to zero, the budget deficit will remain bounded, whereas the boundedness of excess demand depends on particular assumptions on the utility functions. There is another difference that is more important in connection with the issue of multiplicity: When production sets are not strictly convex, the supply correspondences may not be single-valued, and this carries over to the excess demand correspondence, as shown in the shaded parts in figure 3.2, where we represent the excess demand correspondence in the two-commodity case.

Note, however, that in this figure the excess demand correspondence is single-valued at equilibrium. It is not easy to prove this property in the excess demand format, but in the Negishi format with budget surplus functions $s_i(\alpha, p(\alpha))$, the issue can be analyzed in the following, relatively straightforward manner which will also illustrate how an appropriate choice of format can facilitate the analysis.

Assume again that consumer 1 satisfies the condition of assertion 10 of proposition 2.14. Then, since α_1 is positive in equilibrium (see the proof of proposition 3.1), $p(\alpha)$ will be a continuous function around any equilibrium. Therefore the budget surplus function $b_i(\alpha) = p(\alpha)\omega_i + \Sigma_j \theta_{ij} \Pi_j(p(\alpha)) -$

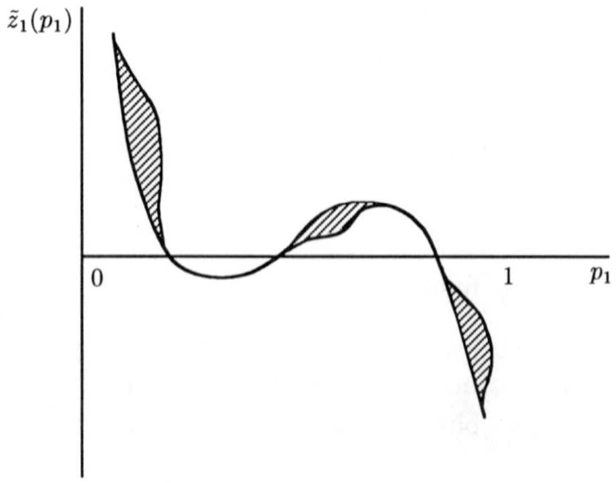

Figure 3.2
The excess demand correspondence $\tilde{z}_1(p_1)$ in the two-commodity case.

$p(\alpha)x_i^\circ(\alpha)$ is continuous for all α such that $\alpha_1 > 0$, though it may not be differentiable everywhere. If we are interested only in the points where it is continuously differentiable and has a regular Jacobian, we can immediately apply theorem A.7.6 and conclude that the number of regular equilibria is finite. Hence, since the Negishi equilibria coincide with excess demand equilibria, the same holds for excess demand equilibria, as reflected in figure 3.2. Therefore, though production sets are not strictly convex, the number of regular equilibria (i.e., equilibria where the budget surplus function is differentiable with nonsingular Jacobian) is finite.

3.2 The Full-format

Now we provide an alternative representation of a competitive equilibrium. This format will be useful in showing the impact of possible constraints on financial transactions (e.g., the budget constraints of the competitive equilibrium). For the competitive model it is not restrictive to leave the budget equations out of the constraint set of the Negishi program, since they are not binding, as can be verified in the following manner.

Assume that we have found equilibrium welfare weights α_i^* (such that $\Sigma_i \alpha_i^* = 1$), for the Negishi program as in proposition 3.1. Let $p^* \in M^\circ(\alpha^*)$ be associated equilibrium prices. We consider an extended version of the Negishi program with the budget equations in the constraints set:

$\max \Sigma_i \alpha_i^* u_i(x_i)$,

$x_i \geqslant 0$, all i, y_j all j,

subject to

$$\Sigma_i x_i - \Sigma_j y_j \leqslant \Sigma_i \omega_i \qquad (p),$$

$$y_j \in Y_j, \qquad\qquad (3.3)$$

$$p^* x_i - \Sigma_j \theta_{ij} p^* y_j \leqslant p^* \omega_i \qquad (\rho_i).$$

The following proposition follows naturally:

PROPOSITION 3.2 (Possibility of zero shadow prices associated with the budget constraints) Let assumption S hold for program (3.3). Then the budget constraints are not binding and $\rho_i^* = 0$ are equilibrium shadow prices.

Proof When the budget constraints are introduced in (3.3), the program remains feasible and bounded; since every consumer has positive income, Slater's constraint qualification is also satisfied. Therefore shadow prices p and ρ_i exist. Any equilibrium allocation corresponding to α_i^* is feasible in (3.3) also and supported by Lagrange multipliers p^*, $\rho_i^* = 0$. ■

This proposition shows that in the competitive equilibrium, the budget constraints do not impair on Pareto efficiency. The budget constraints can also be incorporated into the program from the outset, as suggested by Ginsburgh and Waelbroeck (1981), and the resulting program makes it possible to prove existence of a competitive equilibrium in a direct way. This will provide the basis for existence proofs in the model with overlapping generations (chapter 8) and when financial constraints do restrict welfare (chapter 12).

We define the full-format program as follows:

DEFINITION 3.3 (Full-format) The full format is defined as the constrained welfare optimum:

$$W(\alpha, \bar{p}) = \max \Sigma_i \alpha_i u_i(x_i),$$

$$x_i \geqslant 0, \text{ all } i, \ y_j \text{ all } j,$$

subject to

$$\Sigma_i x_i - \Sigma_j y_j \leqslant \Sigma_i \omega_i \qquad (p),$$

$$y_j \in Y_j,$$

$$\bar{p} x_i - \Sigma_i \theta_{ij} \bar{p} y_j \leqslant \bar{p} \omega_i \qquad (\rho_i),$$

for given $\alpha \in S^m$ and $\bar{p} \in S^r$ such that $\rho_i = 0$ is a shadow price and $\bar{p} = \lambda p$ for some scalar $\lambda > 0$.

Note that we normalize welfare weights and prices separately: At least one of the welfare weights and one of the prices should be positive, but the scaling itself does not affect allocations, since it amounts to multiplying the objective or the constraints by a positive constant.

PROPOSITION 3.3 (Existence of competitive equilibrium; the full-format proof) Under assumption S there exist values $\alpha^* \in S^m$ and $\bar{p}^*$ in the program of definition 3.3 such that the resulting allocation is a competitive equilibrium, as specified in definition 3.1.

Proof We consider the following variant of the program of definition 3.3 in which we add $\varepsilon > 0$ to the right-hand side of each budget constraint:

$$
\begin{aligned}
&\max \Sigma_i \alpha_i u_i(x_i), \\
&x_i \geqslant 0, \text{ all } i, \; y_j \text{ all } j, \\
&\text{subject to} \\
&\Sigma_i x_i - \Sigma_j y_j \leqslant \Sigma_i \omega_i \qquad (p), \\
&y_j \in Y_j, \\
&\bar{p} x_i - \Sigma_j \theta_{ij} \bar{p} y_j \leqslant \bar{p} \omega_i + \varepsilon \qquad (\rho_i),
\end{aligned}
\tag{3.4}
$$

for given α and $\bar{p}$ on the unit simplex: $\alpha \in S^m$ and $\bar{p} \in S^r$ and $\varepsilon > 0$.

For every such $(\alpha, \bar{p})$, program (3.4) is feasible, compact-valued, and satisfies Slater's constraint qualification (here we use $\varepsilon > 0$). Therefore it will always have a solution as well as Lagrange multipliers p and ρ_i. Following the method used in the proof of proposition 2.14, assertions 1–3 and 6–9, but applying it to program (3.4) instead of the welfare program, it is possible to show that in every optimum of (3.4), the following conditions hold:

1. $\alpha_i u_i(x_i^\circ) \geqslant (p + \rho_i \bar{p}) x_i^\circ$, and $\alpha_i = 0$ implies that $(p + \rho_i \bar{p}) x_i^\circ = 0$ for all $(p, \rho) \in M^\circ(\alpha, \bar{p})$, and any given $\bar{p}$.
2. $x_i^\circ(\alpha, \bar{p})$ is a continuous function for $\alpha_i > 0$ or $\alpha_1 > 0$.
3. $\alpha_i u_i^\circ(\alpha, \bar{p})$ is continuous for all $\alpha \in S^m$.
4. $W(\alpha, \bar{p})$ is homogeneous of degree one in α and homogeneous of degree zero in $\bar{p}$; $u_i^\circ(\alpha, \bar{p})$ and $x_i^\circ(\alpha, \bar{p})$ are homogeneous of degree zero in $(\alpha, \bar{p})$ (for $\alpha_i > 0$ or $\alpha_1 > 0$), and the correspondence $M^\circ(\alpha, \bar{p})$ of Lagrange multipliers is homogeneous of degree one in α.
5. $0 \notin M^\circ(\alpha, \bar{p})$.
6. If $\alpha_1 > 0$, then commodity balances will hold with equality.
7. If $\alpha_1 > 0$, then $(p + \rho_i \bar{p}) > 0$ for all $(p, \rho) \in M^\circ(\alpha, \bar{p})$, any given $\bar{p}$.

To obtain an equilibrium, we construct a fixed point mapping by defining one correspondence and two functions. The correspondence generates

Lagrange multipliers and the functions update welfare weights α and prices $\bar{p}$; we again proceed in two steps.

Step 1 Construction of the Fixed Point Correspondence

First, we consider the correspondence $M^\circ : A \Rightarrow \bar{M}, M^\circ(\alpha, \bar{p})$, where $A = S^m \times S^r$ is the set of parameters of the program (welfare weights α and prices $\bar{p}$) and $\bar{M}$ is the convex hull of $M^\circ(A)$ of Lagrange multipliers. By condition 4, constraining α and $\bar{p}$ to lie on their respective simplices does not affect allocations; by condition 5, $0 \notin M^\circ(\alpha, \bar{p})$. By the maximum theorem A.3.1, conditions 5a–5b, this correspondence is uppersemicontinuous in A and compact, convex-valued.

The continuous function $h_1 : A \rightarrow S^m$, $h_1(\alpha, \bar{p})$ adjusts welfare weights as follows: To simplify notation, we define $\hat{p}^i = (p + \rho_i \bar{p})$ and express budget surpluses in these prices. First, we define the "expenditure function" $e_i(\alpha, \bar{p}, p, \rho) = \min\{\hat{p}^i x_i^\circ(\alpha, \bar{p}),\ \alpha_i u_i(x_i^\circ(\alpha, \bar{p}))\}$, which by conditions 2 and 3 above is continuous for all $(\alpha, \bar{p})$. We store the current value α_i as $\hat{\alpha}_i = \alpha_i$ and obtain the new value of α_i by increasing it when there is a positive budget surplus $s_i(\alpha, \bar{p}, p, \rho) = \max\{\hat{p}^i \omega_i + \Sigma_j \theta_{ij} \Pi_j(\hat{p}^i) - e_i(.), 0\}$:

$$\alpha_i = \mu(\hat{\alpha}_i + \delta s_i), \tag{3.5}$$

where δ is a positive constant and μ is a scalar such that $\Sigma_i \alpha_i = 1$. Since at least one of the $\hat{\alpha}_i$'s is positive, such a scaling is always possible. Hence (3.5) defines a continuous function h_1.

We update $\bar{p}$ by defining a function $h_2 : S^r \times \bar{M} \rightarrow S^r$, $h_2(\bar{p}, p, \rho)$ as follows: We store the current value $\hat{p} = \bar{p}$ and obtain the new value of $\bar{p}$ as

$$\bar{p}_k = \lambda(p_k + \rho_0 \hat{p}_k), \tag{3.6}$$

where $\rho_0 = \max_i \rho_i$ and λ is such that $\Sigma_h \bar{p}_h = 1$. Since $0 \notin \bar{M}$, $(p + \rho_0 \hat{p})$ is nonzero, and such a scaling is possible.

The Lagrange multiplier correspondence $M^\circ(\alpha, \bar{p})$, and the functions $h_1(\alpha, \bar{p})$ and $h_2(\bar{p}, p, \rho)$ define a correspondence $G : C \Rightarrow C$, where $C = A \times \bar{M}$. The parameter set C is compact, convex, and, by theorem A.4.3, G is uppersemicontinuous in C, compact, and convex-valued. Therefore, by Kakutani's theorem A.4.2, there exists a fixed point $(\alpha^*, \bar{p}^*, p^*, \rho^*) \in C$.

Step 2 The Fixed Point Is an Equilibrium

In the fixed point we have[5]

$$\alpha_i = \mu(\alpha_i + \delta s_i). \tag{3.7}$$

We show that all α_i's are positive starting with α_1. Suppose that $\alpha_1 = 0$; then

5. For notational convenience, we leave out the * superscript for denoting equilibrium values.

by (3.7), $s_1 = 0$. We know that some price p_k will be positive as $0 \notin M^\circ(\alpha, \bar{p})$. Consumer 1 has thus positive income, since, by assumption S, his endowment is positive. In the fixed point, by condition 1, $e_i(.) = \hat{p}^i x_i$. Since $s_1 = 0$, it must be that $\hat{p}^1 x_1 > 0$; however, $\alpha_1 = 0$ implies that $\hat{p}^1 x_1 = 0$ (see condition 1 above), a contradiction. Therefore $\alpha_1 > 0$ in equilibrium and by condition 7 above, all prices $\hat{p}^i_k$ will be positive. All consumers will have positive income, and by the same reasoning as for consumer 1, their welfare weight will be positive.

We now show that budget surplus s_i is zero for all i. By condition 6, commodity balances must hold with equality. Hence, since all consumers cannot have a budget surplus at prices $\hat{p}^i$, some consumer must have a zero surplus. But then, in (3.7), μ must be equal to unity so that budget surpluses are zero for all i. Then, due to the introduction of ε, all budget constraints in program (3.4) will be nonbinding and ρ_i will be equal to zero for all i in any equilibrium. Hence $\rho_0 = \max_i \rho_i = 0$, and therefore, in equilibrium, $\hat{p}^i = p$, so budget surpluses are zero at prices p. Now (3.6) reduces to

$$\bar{p}_k = \lambda p_k \qquad \text{for all } k,$$

which shows that $\bar{p} \in S^r$ and is proportional to p.

Since $\rho_i = 0$ in equilibrium, proposition 3.2 applies and the solution of program (3.4) coincides with the solution of a Negishi equilibrium and therefore is a competitive equilibrium. Moreover (3.4) solves (3.3), since, in equilibrium, ε may be reduced to zero without affecting the optimal allocation. ■

We note that the main difficulty in this proof is to ensure Slater's constraint qualification. For the competitive model the proof only adds an unnecessary complication, when compared with the Negishi proof, because the welfare weights are adjusted as in the Negishi proof and a price adjustment is added.

3.3 The Open Economy Format

3.3.1 Existence of Equilibrium

Until now we have considered excess demand and welfare maximization as two alternative modes of obtaining equilibrium allocations and prices. We have seen that the excess demand format solves

$$z(p) \leqslant 0 \perp p \geqslant 0, \tag{3.8}$$

where z is excess demand, where we take $z(p)$ to be single valued. The welfare maximization (or Negishi) format solves

$$b(\alpha) \leqslant 0 \perp \alpha \geqslant 0, \tag{3.9}$$

where b is budget surplus, also taken to be single valued.[6]

To establish existence of a solution to (3.8) (respectively to equation (3.9)), we need to construct an adjustment rule for parameters (a part of the fixed point mapping) which does not belong to the model itself and which explicitly generates values for p and α, respectively. As we have seen, this can be cumbersome, since the artificial mapping would need to be specified and the fixed point so obtained proved to be an equilibrium.

We now follow Fischer et al. (1988) and consider an alternative format that does not require such a mapping; this format will be used for existence proofs and the examination of welfare properties in chapters 4 to 6. In this setting the closed economy is treated as a limiting case of an open one. This is implemented by allowing trade with the "outside world." The proof will show that this open economy will not trade when faced with low export prices, relative to import prices.[7]

Let "imports" and "exports" be denoted by m and e, respectively, and let $\bar{p}^m$ and $\bar{p}^e$ represent given import and export prices, and satisfy

$$\bar{p}^m \geqslant \bar{p}^e > 0.$$

We define a competitive equilibrium for this economy as follows:

DEFINITION 3.4 (Open and closed economy competitive equilibrium) The allocation y_j^*, all j, x_i^*, all i, m^*, and e^* supported by the price vector $p^* \geqslant 0$ is a competitive equilibrium for the open economy if:

1. for every producer j, $y_j^* \in \operatorname{argmax}\{p^* y_j$ subject to $y_j \in Y_j\}$;
2. for every consumer i, $x_i^* = \operatorname{argmax}\{u_i(x_i)$ subject to $p^* x_i \leqslant h_i^*$, $x_i \geqslant 0\}$, where $h_i^* = p^*\omega_i + \Sigma_j \theta_{ij} p^* y_j^*$;
3. all markets are in equilibrium

$$\Sigma_i x_i^* - \Sigma_j y_j^* - \Sigma_i \omega_i + e^* - m^* \leqslant 0 \perp p^* \geqslant 0,$$

$$e^* \geqslant 0 \perp p^* \geqslant \bar{p}^e,$$

$$m^* \geqslant 0 \perp p^* \leqslant \bar{p}^m.$$

6. Both (3.8) and (3.9) belong to the so-called class of nonlinear complementarity problems. There are two other well-known members in this class: The Kuhn-Tucker conditions resulting from a mathematical program and the equilibrium conditions of certain games. See section A.9 of the mathematical appendix for a further discussion.

7. A similar approach was followed by Morgenstern and Thompson (1976) to prove existence of equilibrium for the von Neumann model.

This equilibrium reduces to an equilibrium of a closed economy if $e^* - m^* = 0$.

The market equilibrium conditions state that domestic demand and exports should be met by domestic supply and imports, that a commodity will not be exported (or imported) unless the domestic price is equal to the export (or import) price, and that a commodity in surplus could be disposed of at zero price (though this will not happen here, since $\bar{p}^e$ is positive and $p \geqslant \bar{p}^e$).

Obviously, if the import price were to lie below the export price for a given commodity, there would be an infinite import and reexport (in the absence of further constraints and transaction costs); therefore this cannot be an equilibrium. If import and export prices are equal for some commodity, the model will only determine a net flow, though an open economy equilibrium will exist. For the closed economy, export and import prices are artificial parameters, and the assumption $\bar{p}^m > \bar{p}^e > 0$ is not restrictive.

To obtain such an equilibrium, we specify in definition 3.5 a mathematical program that maximizes the net revenue of the economy, under technological constraints and for given consumer demands, determined outside of the mathematical program through continuous feedback equations. We will show that this program with feedbacks has a solution that is an open economy competitive equilibrium that reduces to a closed economy equilibrium when export prices $\bar{p}^e$ are chosen sufficiently close to zero.

DEFINITION 3.5 (Open economy program with feedback) The model for the open economy is written as the following program, with given consumptions $\hat{x}_i$:

$$\max \bar{p}^e e - \bar{p}^m m,$$

$$m, e \geqslant 0,\ y_j \text{ all } j,$$

subject to

$$\Sigma_j y_j + m - e \geqslant \Sigma_i \hat{x}_i - \Sigma_i \omega_i \qquad (p),$$

$$y_j \in Y_j,$$

and with a feedback relation that sets $\hat{x}_i$ by solving the consumer problem:

$$\hat{x}_i = \operatorname{argmax}\{u_i(x_i) \text{ subject to } px_i \leqslant p\omega_i + \Sigma_j \theta_{ij} \Pi_j(p),\ x_i \geqslant 0\}, \qquad \text{all } i;$$

p is the vector of shadow prices associated to the balance constraints and $\Pi_j(p) = \max\{py_j, y_j \in Y_j\}$.

We can now use the format of this open economy program with feedback to state and prove:

PROPOSITION 3.4 (Existence of competitive equilibrium; the open economy format) Under assumption S:

1. if $\bar{p}^m \geqslant \bar{p}^e > 0$, there exists an allocation x_i^*, all i, y_j^*, all j, and a price vector p^* in the program of definition 3.5 that is an open economy competitive equilibrium;
2. if $\bar{p}^m > \bar{p}^e > 0$ and if $\bar{p}^e$ is sufficiently close to 0, the solution will be a (closed economy) competitive equilibrium, as in definition 3.1.

Proof To prove part 1, recall that Y_j is nonempty. Then, for given $\hat{x}_i$, the program is feasible by construction and satisfies Slater's constraint qualification. Although the constraint set is not bounded, compactness of Y_j, nonnegativity of m and e and $\bar{p}^m \geqslant \bar{p}^e > 0$ ensure that the program has a bounded objective.

Step 1 Construction of the Fixed Point Correspondence

The Kuhn-Tucker conditions (theorem A.2.1, assertion 8) are

$$\Sigma_j y_j + m - e \geqslant \Sigma_i \hat{x}_i - \Sigma_i \omega_i \perp p \geqslant 0, \tag{3.10}$$

$$e \geqslant 0 \perp p \geqslant \bar{p}^e, \tag{3.11}$$

$$m \geqslant 0 \perp p \leqslant \bar{p}^m. \tag{3.12}$$

Thus $p \in K = [\bar{p}^e, \bar{p}^m]$, a compact, convex set. This makes it possible to define a mapping $G: K \Rightarrow K, G(p) = P(\Sigma_i \hat{x}_i(p, h_i(p)))$, where $h_i(p) = p\omega_i + \Sigma_j \theta_{ij} \Pi_j(p)$ is the income of consumer i at given prices p, $\hat{x}_i(p, h_i)$ his utility maximizing consumption (a continuous function since p is positive) and $P(\Sigma_i \hat{x}_i)$ is a correspondence of Lagrange multipliers which, by the maximum theorem A.3.1, assertion 5, is uppersemicontinuous, compact, and convex-valued. Thus the properties of $P(\Sigma_i \hat{x}_i)$ carry over to $G(p)$, and we can invoke Kakutani's theorem A.4.2 to conclude that there exists a fixed point p^* such that $p^* \in G(p^*)$.

Step 2 The Fixed Point Is an Equilibrium

This can be verified as follows:

1. Producers maximize profits. This is a consequence of the Lagrange inequality of the program, and can be proved as in proposition 1.9.
2. Consumers maximize utility subject to their budget constraint. This is ensured by the feedback equation of definition 3.5.
3. Markets are in equilibrium by (3.10)–(3.12).

To prove part 2 of proposition 3.4, by (3.10), in equilibrium,

$$p^*\Sigma_i x_i^* = p^*\Sigma_j y_j^* + p^*\Sigma_i \omega_i + p^*m^* - p^*e^*.$$

Using (3.11) and (3.12), we can rewrite this expression as

$$p^*\Sigma_i x_i^* = p^*\Sigma_j y_j^* + p^*\Sigma_i \omega_i + \bar{p}^m m^* - \bar{p}^e e^*.$$

From summation of the budget equations, it follows that

$$p^*\Sigma_i x_i^* = p^*\Sigma_j y_j^* + p^*\Sigma_i \omega_i.$$

Therefore

$$\bar{p}^m m^* = \bar{p}^e e^*. \tag{3.13}$$

Now consider the equilibrium when $\bar{p}^e$ is set equal to a small positive value ε. There are two possibilities. If at this new value of $\bar{p}^e$ the equilibrium prices satisfy $p^* < \bar{p}^m$, then $m^* = 0$ by (3.12). Thus, by (3.13), $e^* = 0$, and we have a (closed economy) competitive equilibrium. Otherwise, it must be that $p_k^* = \bar{p}_k^m$ for some k. By assumption C2′, consumer 1 has an income at least equal to $\bar{p}_k^m \omega_{1k}$. Now suppose that export is positive for some commodity h (necessarily, by equations 3.11 and 3.12, $h \neq k$); now, $p_h^* = \varepsilon, e_h^* > 0$, and $m_h^* = 0$. By assumption C1′, consumer 1's utility is increasing, and for ε sufficiently small, by proposition 2.11, his demand will exceed domestic supply (which is bounded above by compactness in assumption P1′). This contradicts $e_h^* > 0$. Therefore $e^* = 0$ and by (3.13), $m^* = 0$. ■

Some remarks are in order. First, the introduction of positive $\bar{p}^e$ ensures positive equilibrium prices, since $p \geqslant \bar{p}^e$. This enables us to avoid the difficulties associated with zero prices, and continuity of demand is ensured without further complications. Result 2 for the closed economy excludes exports. However, if $\bar{p}^e$ is allowed to reach zero, exported commodities become free goods and positive export cannot be excluded, though the value of exports will go to zero. When import prices rise indefinitely, we cannot ensure that trade will actually become zero, but the quantity imported will move toward zero.

Second, we note that prices are generated by the open economy program, and there is no need to define artificial adjustment functions like for the price mapping in proposition 1.1 or for the welfare weight mapping in proposition 3.1. The price level is set by the fixed import and export prices in the rest of the world, as long as for one commodity at least, trade takes place. In equilibrium it is possible to renormalize all prices, including the export and import prices, say on the simplex. This renormalization is also meaningful in the closed economy case to avoid indeterminacy of the price level.

Finally, part 1 of proposition 3.4 will still hold if for some commodities we allow import and export prices to be equal, but then only net trade can be determined.

3.3.2 Extensions of the Open Economy Model

Under strict convexity of the production sets, supply is single-valued, so we can take suppliers decisions as part of the feedback and rewrite the model in definition 3.4 as a linear program for given productions $\hat{y}_j$ and consumptions $\hat{x}_i$:

$$\begin{aligned} &\max \bar{p}^e e - \bar{p}^m m, \\ &m, e \geqslant 0, \\ &\text{subject to} \\ &m - e \geqslant \Sigma_i \hat{x}_i - \Sigma_j \hat{y}_j - \Sigma_i \omega_i \qquad (p), \end{aligned} \tag{3.14a}$$

with feedback

$$\begin{aligned} &\hat{y}_j = \operatorname{argmax}\{p y_j \text{ subject to } y_j \in Y_j\}, \\ &\hat{x}_i = \operatorname{argmax}\{u_i(x_i) \text{ subject to } p x_i \leqslant p\omega_i + \Sigma_i \theta_{ij} \Pi_j(p),\ x_i \geqslant 0\}. \end{aligned} \tag{3.14b}$$

In this setup producers and consumers completely follow the specification of the existence proof in proposition 1.4. All the behavioral elements are shifted to the nonlinear continuous feedback, while the program has become linear and merely serves to generate the set of Lagrange multipliers to which the equilibrium price belongs.

The approach of treating all behavioral reactions as feedback functions can be extended to cover many other specifications, even if they do not result from optimizing behavior of agents. For example, due to aggregation problems, results from econometric estimation may violate the properties of demand or supply as specified in chapter 2. Likewise, instead of the principle of profit or utility maximization which may be unsuitable on a priori grounds, one could specify any continuous function, possibly including variables other than prices. This leads to the following program:

$$\begin{aligned} &\max \bar{p}^e e - \bar{p}^m m, \\ &m, e \geqslant 0, \\ &\text{subject to} \\ &m - e \geqslant \Sigma_i \hat{x}_i - \Sigma_j \hat{y}_j - \Sigma_i \omega_i \qquad (p), \end{aligned} \tag{3.15a}$$

with feedback

$$\begin{aligned} &\hat{y}_j = f_j(p), \\ &\hat{x}_i = g_i(p). \end{aligned} \tag{3.15b}$$

An open economy equilibrium will now exist as long as the following conditions are satisfied:[8]

1. Every commodity can be imported at positive and bounded import prices $\bar{p}^m$. Exports are possible for all commodities with positive export prices $\bar{p}^e$; as before $\bar{p}^m > \bar{p}^e$.

2. The feedback functions are continuous, and hence bounded, for all prices $p \in [\bar{p}^e, \bar{p}^m]$.

However, a closed economy solution may fail to exist unless appropriate budget restrictions are satisfied that make imports go to zero when export prices vanish; this will require replacing the function $g_i(p)$ for consumer i, say, by $g_i(p, \bar{p}^e e - \bar{p}^m m)$.

Finally, we mention that the open economy program (3.14) has a welfare-maximizing analog in Negishi format. This welfare program will be used extensively in chapters 4 to 6, and it is written

$$\max \Sigma_i \alpha_i u_i(x_i) + \alpha_0(\bar{p}^e e - \bar{p}^m m),$$

$$x_i \geqslant 0, \text{ all } i, \ y_j, \text{ all } j, \ e, \ m \geqslant 0,$$

$$\text{subject to} \qquad (3.16a)$$

$$\Sigma_i x_i - \Sigma_j y_j - m + e \leqslant \Sigma_i \omega_i \qquad (p),$$

$$y_j \in Y_j,$$

where welfare weights are on S^{m+1} and are such that all budget restrictions, including the trade balance are satisfied:

$$p x_i = p\omega_i + \Sigma_i \theta_{ij} \Pi_j(p), \qquad \text{all } i, \qquad (3.16b)$$

$$\bar{p}^e e = \bar{p}^m m. \qquad (3.16c)$$

3.4 The CGE Format

For models that satisfy assumptions P1 (strict convexity of production sets), C1′, and C2′, supply functions $y_j(p)$ and Marshallian demand functions $x_i(p, h_i)$ are continuous at positive prices. It is then straightforward to use the excess demand format of proposition 1.4:

$$\Sigma_i x_i(p, h_i) = \Sigma_j y_j(p) + \Sigma_i \omega_i, \qquad (3.17a)$$

8. This can be proved as follows: Since all commodities can be imported (possibly at a very high price), constraint qualification holds in (3.15a), so for $\bar{p}^m > \bar{p}^e$, program (3.15a) has a bounded optimal solution and bounded Lagrange multipliers $p \in [\bar{p}^e, \bar{p}^m]$. Since the feedback functions (3.15b) are continuous on the compact convex set $[\bar{p}^e, \bar{p}^m]$, the program (3.15a) with the feedback (3.15b) defines an uppersemicontinuous compact, convex-valued mapping that has a fixed point, which obviously is a solution of model (3.15).

with h_i defined by

$$h_i = p\omega_i + \Sigma_j \theta_{ij} p y_j(p), \qquad \text{all } i, \tag{3.17b}$$

and where the unknown is the price vector p, usually normalized so that $p_r = 1$. A solution can then be found by solving the system of nonlinear equations (3.17).

However, if constant returns to scale prevail for some producers, this approach cannot be used, since supplies are now correspondences instead of functions. The computable general equilibrium (or CGE) format overcomes this difficulty: It eliminates the set-valuedness by setting supply equal to aggregate demand and proceeds as follows.

3.4.1 The Basic CGE Model and the CGE Format

For the basic version of the CGE model, the following assumptions are made, some of which modify assumption S.

Goods and Factors

The set of commodities is partitioned into $G = \{1, 2, \ldots, n\}$ for goods and $F = \{n + 1, \ldots, r\}$ for factors. Goods and factors will be denoted by the superscripts g and f, respectively, and p^g, p^f will represent prices of goods and factors.

Production Sets

Every producer supplies a single good according to a constant-returns-to-scale production function as specified in proposition 2.4. Every good is produced by a single producer and requires inputs of factors, or of goods that require inputs of factors. Factors are not produced.

Endowments

Factors are owned by consumers in nonzero amounts and there are no endowments of goods. Consumer 1 owns positive endowments of all factors.

Utility Functions

Consumers have utility functions as specified in proposition 2.8 (continuous, strictly quasiconcave, and nonsatiated); consumer 1's utility is increasing in all factors (or in goods that together use all factors in production).

For producer j profit maximization implies cost minimization in accordance with proposition 2.4, and by Shephard's lemma, it leads to input demands $v_{jk}(p^g, p^f, q_j) = a_{kj}(p^g, p^f)q_j$, where q_j is the supply of the good produced by j. The coefficients $a_{kj}(.)$ define (price-dependent) matrices A^g and A^f of input-output coefficients with rows referring to demands for goods and factors, respectively (A^g is a square matrix). Since factors are not produced and constant returns to scale prevail, feasible supplies will be bounded. In addition, a viability requirement has to be fulfilled: There should exist a positive solution to the system of linear inequalities, $q^g > A^g(.)q^g$, $q^g \geqslant 0$. This means that at all positive prices, it is possible to supply the consumers with positive amounts of all goods.[9] In this case the inverse of $(I - A^g(.))$ exists and is nonnegative.

Consumer i's demand for goods and factors are $x_i^g(p^g, p^f, h_i)$ and $x_i^f(p^g, p^f, h_i)$, respectively. Since consumer 1's utility is increasing in all factors (or in goods that together use all factors in production), factor prices will be positive, and therefore every consumer's income will be positive.

Equations (3.17a) are now split between goods and factors; this amounts to replacing $\Sigma_j y_j(p)$ by $q^g - A^g(p^g, p^f)q^g$ for goods and by $-A^f(p^g, p^f)q^g$ for factors. The full model may be written

$$\Sigma_i x_i^g(p^g, p^f, h_i) + A^g(p^g, p^f)q^g = q^g, \tag{3.18a}$$

$$\Sigma_i x_i^f(p^g, p^f, h_i) + A^f(p^g, p^f)q^g = \Sigma_i \omega_i^f, \tag{3.18b}$$

$$h_i = p^f \omega_i^f. \tag{3.18c}$$

Moreover

$$p^g = p^g A^g(p^g, p^f) + p^f A^f(p^g, p^f). \tag{3.18d}$$

Equations (3.18a) and (3.18b) are the balances for goods and factors. Under constant returns to scale there are no profits distributed to consumer i in (3.18c), and the goods price in (3.18d) are set equal to unit cost (the value of goods and factors used as inputs).

Note that part of the model is recursive. Therefore the solution can be obtained as follows:

1. For given input-output coefficients and given p^f, one can solve for p^g from the linear equation (3.18d). This yields nonnegative prices for goods $p^g = p^f A^f (I - A^g)^{-1}$, and for positive p^f, p^g will be positive.

2. Income h_i follows from (3.18c), so consumption for goods $\Sigma_i x_i^g(p^g, p^f, h_i)$ can be computed.

3. Equation (3.18a) is solved for production of goods q^g, like in input-output analysis, thus eliminating the set-valuedness in q^g, by setting it equal to demand.

4. Factor input demand $A^f(p^g, p^f)q^g$ and $\Sigma_i x_i^f(p^g, p^f, h_i)$ in (3.18b) can finally be computed as well.

9. Note that if the viability condition holds, then by scaling production levels downward, it is possible to satisfy positive consumer demands for all factors, since total factor endowment is positive.

5. One checks whether demand equals supply in (3.18b). This equation is in fact the excess demand function, with its dimension reduced to the number of factors, and factor prices as unknowns.

6. Factor prices are adjusted and steps 1 to 5 are repeated until (3.18b) is satisfied.

The excess demand for factors depends only on the prices of factors, and it satisfies homogeneity and Walras's law. At positive prices it is continuous, and the assumption on consumer 1 implies desirability and, hence, positive prices. Therefore existence of equilibrium can be proved as in the model of proposition 1.4. But, if some of the assumptions do not hold, like viability, a closed economy solution may fail to exist; opening the economy to imports and allowing for positively priced exports will ensure existence of a solution under the conditions that were discussed for model (3.15). Similarly, if the demand functions $x_i(.)$ or $A(.)$ are not derived from optimizing behavior but are based on "institutional" or "structuralist" considerations,[10] an open economy solution will exist.

We will consider that a model is written in CGE format when it is specified as a system of simultaneous equations consisting of commodity balances (3.18a) and (3.18b) for goods and factors, price equations (3.18d) for goods, and budget equations (3.18c). This is a wide class that contains many cases; some of them will be discussed in section 3.6 and in subsequent chapters. We already note here that this class excludes models with inequality constraints, with an infinite number of commodities, agents, or periods, or with increasing returns to scale in production. We will now consider three immediate extensions: decreasing returns, closure rules, and markups; other extensions (e.g., international trade, taxes, dynamics) require developments of the general equilibrium model (to be covered in later chapters).

3.4.2 Extensions of the Basic CGE Model

Decreasing Returns

In the basic CGE model we have assumed a constant-returns-to-scale technology, so, in equilibrium, profit maximization leaves the scale of production indeterminate. Obviously, when one or more factors are firm specific (e.g., land or a capital good once installed), the indeterminacy disappears provided that the production function is increasing in such a factor. Since the firm-specific inputs will be utilized fully, they can be treated as a vector of fixed stocks k_j and the production function exhibits decreasing returns to scale in the remaining inputs. The fixed stocks can then be left

10. See, for example, Taylor (1990).

out of the commodity list.[11] The profit-maximizing formulation will now yield positive profits which have to be redistributed to consumers according to fixed ownership shares θ_{ij}. For these firms Hotelling's lemma of proposition 2.1 can now be used to derive net supplies as

$$y_j^g(p^g, p^f, k_j) = \frac{\partial \Pi_j}{\partial p^g},$$

$$y_j^f(p^g, p^f, k_j) = \frac{\partial \Pi_j}{\partial p^f}.$$

These net supply functions can be introduced as additional terms in the right-hand side of (3.18a)–(3.18b). In practice, it is not always possible to observe all the elements of vector k_j. In that case the production function exhibits decreasing returns in the observed factors. In virtually all applied CGE models, this formulation is used.

Closure Rules

Modelers often wish to incorporate macroeconomic mechanisms into general equilibrium models. Macroeconomic equations are then grafted onto these models by introducing for each new macroeconomic variable, a balancing constraint that does not necessarily represent a market equilibrium condition. The association of a particular endogenous variable to a constraint is commonly referred to as a closure rule.[12] Closure rules provide a flexibility that is operationally attractive, since they make possible to introduce many variables that are central to policy discussions.

For example, one could introduce a government that taxes consumers' income, through a proportional rate τ, to finance an exogenously given public consumption of goods ξ_0^g. The CGE model would then be written

$$\Sigma_i x_i^g(p^g, p^f, h_i) + \xi_0^g + A^g(p^g, p^f)q^g = q^g, \tag{3.19a}$$

$$\Sigma_i x_i^f(p^g, p^f, h_i) + A^f(p^g, p^f)q^g = \Sigma_i \omega_i^f, \tag{3.19b}$$

$$h_i = (1 - \tau)p^f \omega_i^f, \tag{3.19c}$$

$$p^g = p^g A^g(p^g, p^f) + p^f A^f(p^g, p^f). \tag{3.19d}$$

The government is also assumed to run a balanced budget, imposed as the additional constraint

$$p^g \xi_0^g = \tau \Sigma_i p^f \omega_i^f. \tag{3.19e}$$

11. Intersectorally mobile capital goods should be represented explicitly as commodities leased by "capital rental" firms.

12. See Sen (1963) and Taylor and Lysy (1979).

The tax rate τ is the closure variable, and the government budget is the associated constraint.[13]

One could think of other ways to finance the public expenditure, like a flat ad valorem tax σ on goods; this would only change equations (3.19c)–(3.19*e*) as follows:

$$h_i = p^f \omega_i^f, \tag{3.19c'}$$

$$p^g = (1+\sigma)(p^g A^g(p^g, p^f) + p^f A^f(p^g, p^f)), \tag{3.19d'}$$

$$p^g \xi_0^g = \sigma(p^g A^g(p^g, p^f) + p^f A^f(p^g, p^f) q^g), \tag{3.19e'}$$

where the last equation can also be written as

$$p^g \xi_0^g = \frac{\sigma}{1+\sigma} p^g q^g.$$

In this example, the choice of ξ_0^g is not related to the preferences of any consumer, and there could still have been other ways to finance public consumption. The approach is somewhat ad hoc and is mainly meant to represent institutional characteristics and policy scenarios in which some targets are set (here, public consumption) and a policy instrument adjusts endogenously (here, the tax rate). Closure rules have also been defined for investment, exports, and so on. These rules and their limitations will be treated further later on.[14]

Markups

When the taxes levied in (3.19d′) accrue directly to consumers rather than to the government, modelers refer to σ as a markup rate, which can be interpreted as a rent accruing to commodities not explicitly represented in the model.[15] Markups also appear in connection with imperfect competition, which will be considered in chapter 11.

3.5 Comparing the Formats

So far we have presented five alternative formats for a general equilibrium model: the excess demand format of proposition 1.4 (under more restrictive assumptions), the Negishi format, the full-format, the open economy program with feedback, and the CGE format. For the first four formats, we

13. Note that the exogenous setting of government consumption ξ_0^g may lead to an infeasibility and impair on existence of a solution. Equilibrium can be restored by scaling down ξ_0^g whenever the tax rate τ reaches a given value that is smaller than unity.

14. See also Dewatripont and Michel (1987) and Taylor (1990).

15. See also section 3.6.2.

have seen that under assumption S a competitive equilibrium exists and that all will yield the same equilibria.

Yet there are significant differences between the various proofs of existence of competitive equilibrium. For example, in proposition 1.4 (excess demand format) consumer demand can go to infinity when prices approach zero, causing a discontinuity. This discontinuity is avoided by imposing an artificial bound on demand, which has to be made ineffective in equilibrium. This condition is also needed for the CGE but not for the other three formats. In the Negishi and the full-format, consumptions and budget deficits remain bounded even when welfare weights go to zero; in the open economy format, export prices keep prices bounded away from zero. Another example is that in propositions 3.1 and 3.3 (the Negishi and the full-format) we need a rather involved construction to obtain excess budgets that are single valued for all welfare weights. As will become clear in later chapters, the choice of format for a particular proof is often dictated by the characteristics of the model itself. When the model is extended and some assumptions are relaxed, it can be easy to give a proof using one format and very difficult with another; for example, we have seen in section 3.1.2 that when supplies are set-valued, it is far more convenient to establish local uniqueness of equilibrium in the Negishi format than in the excess demand format.

As discussed more formally in mathematical appendix A.4, the existence proofs in the four formats (propositions 1.4, 3.1, 3.3, and 3.4) have in common that they construct a fixed point mapping by varying the parameters of one or several mathematical programs by adjustment functions which have the optimal solutions and the associated Lagrange multipliers as arguments; this is illustrated in figure 3.3.

In the Negishi format and in the full format, there is a single welfare program, and the parameters adjusted in the feedback are welfare weights and prices. In the excess demand format, every agent solves his own program, and the feedback is an adjustment on prices. In the open economy format, there is one central program, while each consumer solves his own program in the feedback, sending his consumption as a parameter to the central program. We summarize the relative merits of each format as follows.

The excess demand format is the natural point of reference whenever microfoundations are considered desirable. Therefore under any extension it is natural to investigate the implications for this format. In particular, one might want to identify the signals to the individual agents that are needed for decentralization of a specific welfare optimum.

We have already seen in the propositions of section 1.4, that the Negishi format provides a direct link to welfare analysis and that it makes it possible to use weaker assumptions on the production technology. Sometimes (e.g.,

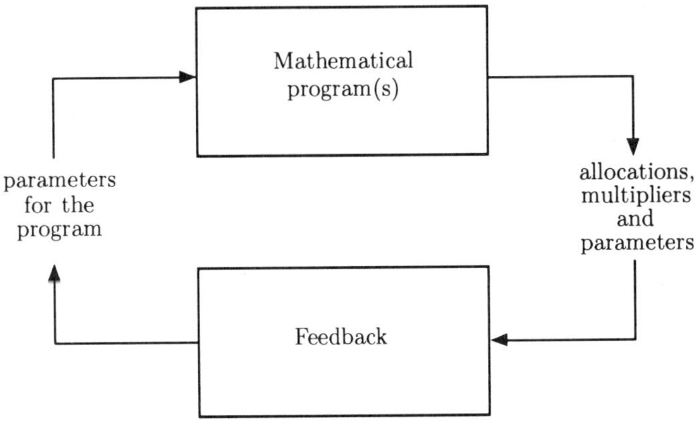

Figure 3.3
A representation of the fixed point mapping.

with externalities or nonconvexities) it is easier to formulate a centralized welfare program (the Negishi format) than to specify its decentralized counterpart (the excess demand or the CGE format).

Since the full-format is an extension of the Negishi format, it possesses the same advantages. It is particularly useful to represent financial constraints that define the opportunities for transactions (the "transaction technology"). Treating the financial and the physical constraints in the same program is, as we will see in chapter 12, helpful in clarifying the links between the real and the financial sphere.

The open economy format exploits duality: It can accommodate forms that are well suited for econometric estimation. Like the Negishi format and the full-format, it can handle inequality constraints to treat buffer stocks, trade regimes, transport, and rationing. The format can also incorporate nonoptimizing behavior, and it yields a solution for an open economy in cases where the closed economy model has no equilibrium.

Finally, the CGE model is the easiest for applications, but its assumptions are more restrictive. It can be used to represent nonoptimizing behavior, in the same way as the open economy format. Whenever possible, we will show how the theoretical models that use other formats can be brought into the CGE format.

Shoven and Whalley (1972) were the first to set up applied models in the Arrow-Debreu tradition according to the excess demand format and to solve them using globally convergent algorithms, following Scarf's (1967, 1973) work on the computation of equilibria.

Dixon (1975), Ginsburgh and Waelbroeck (1976, 1981), Makarov and

Rubinov (1977), Trzeciakowski (1971), and Manne, Chao, and Wilson (1980) have used the Negishi format to model multiconsumer (countries, regions) general equilibria. The single-consumer case (Sandee 1960, Bruno 1966, and many others) has a long-standing tradition in mathematical programming, though it was not clear from the beginning, that the shadow prices associated to commodity balances were actually equilibrium prices. The recent macroeconomic literature also follows the representative consumer approach, and in this sense, these macroeconomic models also fall in the AGE category.[16]

Starting from the nonlinear complementarity format, Keyzer (1977) introduced price rigidities and inequality constraints in linked national models of world agriculture; this approach eventually led to the open economy format.

CGE's may be looked as descendants of input-output and activity analysis that incorporate price-dependent input-output coefficients and final demand. Johansen (1960) was the first to build a multisectoral model that came very close to the CGE model. The first "true" CGE models were developed by Adelman and Robinson (1978). Keller (1980) developed the public and the foreign trade aspects in order to study the impact of tax policies on income distribution. Nowadays CGE-models have become a standard tool, in particular, due to the availability of specifically designed commercial software packages.[17]

3.6 Model Implementation

The construction of an AGE model still involves the following steps, once a mathematical specification has been obtained:

Step 1 Empirical Elaboration

1a. Choosing an exhaustive classification for agents and commodities.

1b. Collecting and organizing the data in a way compatible with the specification of the model and the classifications.

1c. Choosing an appropriate format.

1d. Specifying functional forms.

1e. Estimating the parameters.

16. See Blanchard and Fischer (1989), Farmer (1993), and Stokey and Lucas (1989).

17. See Brooke et al. (1992) and Codsi et al. (1992). All applied models are necessarily "computable." Computation did not start with competitive general equilibrium. The abstract presentation of general equilibrium theory and the nonconstructive fixed point arguments were thought to make computation intractable, until Scarf discovered his algorithm.

Step 2 Commodity Classification

In practice, it is often impossible to make the commodity classification exhaustive, and the issue of how to deal with missing commodities and markets must be addressed.

Step 3 Simulation Runs

This requires bringing the model within a computer program that can solve it, and running simulations under alternative values of exogenous variables and parameters.

3.6.1 Empirical Elaboration

Classifying Commodities

Proceeding on the assumption that the number of commodities with different physical characteristics is finite, we could attempt to list them exhaustively. In practice, this will lead to an unmanageably long list, and some type of commodity aggregation is unavoidable.

The modeler generally must decide to focus on a single characteristic and aggregate quantities in some unit of measurement (e.g., headcounts, weight, metric volume). The commodities in that group are then treated as perfect substitutes. For example, we could assume that soybean and sunflower cake are perfect substitutes as far as protein content is concerned and define a commodity called "protein feeds." At the other extreme, if commodities are used in the same fixed proportions everywhere, they are pure complements and can be treated in a single balance: For example, we do not consider separate markets for left- and right-hand gloves. Otherwise, every commodity needs a balance equation.

To keep the classification within a reasonable size, we could choose to use different classifications in various parts of the model and interconnect them through transition matrices that transform one classification into the other. The coefficients in such matrices are often fixed, but variable coefficients derived from functional separability in the consumer and producer models can be constructed. This approach was discussed under the heading of separability of the utility function for commodity groupings in section 2.2.5, where use was made of the assumption that a particular grouping of commodities generates a utility (the aggregate or composite commodity) according to a function that is homogeneous of degree one. Clearly we can derive the same aggregate, using unit cost rather than expenditure functions. Note that prices have to clear markets at the disaggregated level and not at the level of the composite commodity. The idea of introducing composite goods was pioneered by Armington (1969) for aggregating a domestic and a foreign good or goods imported from different origins.[18]

Classifying Agents

Agent classification must depend also on the purposes for which the model is built. The modeler may choose to distinguish rural and urban areas, specific socioeconomic groups, firms with different technologies, or adopt an explicit geographical or administrative breakdown (regions, countries).

For consumers, two criteria should be taken into account. One important consideration is whether groups are expected to be differentially affected by policy changes. On this ground the modeler may want to single out groups such as landless laborers or rural households with access to remittances. Vulnerable groups (like the unemployed) might also be distinguished. A second criterion is that within each group, heterogeneity should be as limited as possible. Income distribution within classes can be represented by imposing a fixed (discrete or continuous) income distribution. It must be recognized that all applied general equilibrium models suffer from problems involved in the representative agent construction: Aggregation of individual net trades does not preserve microeconomic properties other than continuity, homogeneity, and adding up. This is a fundamental problem from which any level of disaggregation suffers, as discussed in section 2.2.5.

The classification of consumers is often a decisive factor for producer classification, since profits of specific firms are distributed to specific consumers: Small farmers do not have shares in large farms or in industry. Of course, under constant returns, profits will be zero, so ownership of firms does not matter. Moreover differences in technology could require separate representation of firms, even if they produce the same commodities: The modeler may want to distinguish large and small farmers, cotton mills and handlooms, or more generally large-scale and cottage industries. In the CGE format (where goods are produced under constant returns), there will be a single sector for each good. In other formats, or if decreasing returns to scale prevail, it is possible to have several production functions for the same commodity, leading to several columns in the input-output matrix; multiple output firms can be represented through one or several profit functions, or by activity analysis (section 2.1.2).

18. See also section 4.4.2 where this approach is further discussed.

Organizing the Data: The Social Accounting Matrix (SAM)

An applied general equilibrium model needs data from many sources. These data will usually be in contradiction with each other and incomplete. Therefore the first step in building an AGE model is to set up a consistent and comprehensive database for one or several reference years. For this the social accounting matrix (SAM) provides a convenient structure, since it contains, for every agent of the model, the revenue and expenditure (as well as sales and purchases in quantity terms) for every commodity.[19] It provides the data on exogenous and lagged variables (initial endowments, capital stocks) and possibly on model parameters (shares in profits) that are needed by the model itself for the base year; it also provides a consistent set of data on all endogenous variables, which can be used to test the base year outcome of the model.

To illustrate how a SAM structure follows from the specification of a model, we consider a general equilibrium solution y_j^* all j, x_i^*, all i, a price vector p^*, and the market equilibrium conditions

$$\Sigma_i x_{ik}^* - \Sigma_j y_{jk}^* - \Sigma_i \omega_{ik} \leqslant 0 \perp p_k^* \geqslant 0, \qquad \text{all } k, \tag{3.20a}$$

the profits of producers

$$\Pi_j^* = p^* y_j^*, \qquad \text{all } j, \tag{3.20b}$$

and the consumer budget equations

$$p^* x_i^* = p^* \omega_i + \Sigma_j \theta_{ij} \Pi_j^*, \qquad \text{all } i. \tag{3.20c}$$

The left-hand side of (3.20a) is the commodity balance for commodity k in quantity terms, and it holds with equality under our standard assumptions. The complementarity conditions imply that

$$p_k^* \Sigma_i x_{ik}^* = p_k^* \Sigma_j y_{jk}^* + p_k^* \Sigma_i \omega_{ik}. \tag{3.20d}$$

In organizing a database for this model, we need to compile data for every commodity k, producer j, and consumer i. These data can be presented in matrix form as shown in table 3.1 where, as in the CGE format, we introduce a distinction between goods and factors and differentiate between production and input demand.[20] The correspondence of this SAM with equations (3.20) may be explained as follows. Let k denote a good or a factor. In value terms, its balance is (3.20d). The left-hand side of each such

19. See Pyatt and Round (1979) and the papers in Pyatt and Round (1985) for further details.

20. The term "SAM-based model" is sometimes used for the class of all models that are based on a SAM. This is misleading since a SAM does not define a model. Rather the structure of a SAM can be derived from any model that has commodity balances and budget constraints, and the solution of the model provides the numerical value for the SAM.

Table 3.1
Structure of the SAM for an AGE model

	Goods	Factors	Firms	Consumers
Goods			Input	Consumption
Factors			Input	Consumption
Firms	Production	(Production)		
Consumers	(Endowments)	Endowments	(Profits)	

Note: In the SAM for the basic CGE model, the terms in parentheses are zero.

balance represents demand and appears in the SAM as one of the goods or factors rows. The two expenditure components appear in the columns for firms and consumers. The demand of commodity k by sectors is booked as input; similarly the demand of that commodity by consumers ($p_k x_{ik}(.)$) is booked as consumption. The right-hand side of equation (3.20d) represents supply from production and endowments and appears in the kth goods and factors column, respectively. For agents, the columns relate to outlays and the rows to receipts, while for goods and factors, the columns relate to the right-hand side of (3.20d) and rows to its left-hand side. Obviously, since the entries are derived from a balance equation, each column total will be equal to the associated row total.

Once a SAM (in values) has been obtained, estimates of prices are derived. This is feasible if all data for balances (3.20a) in quantity terms are available, as is often the case for agricultural products, or fuels expressed in energy content; for many commodities, however, one can hardly avoid setting prices equal to one, which implies that all quantities are measured in terms of base-year prices.[21]

Choice of Model Format and of Functional Forms

The format has important consequences for the choice between primal and dual formulation of the consumer and producer problems. For example, choosing the Negishi format implies that only primal forms can be used. In the excess demand format it is possible to use primal formulations for each agent, but this is cumbersome, in particular, from an econometric point of view (unless one opts for CES-related forms).[22]

21. The disadvantage of the latter procedure is that price and quantity results cannot be compared directly with published data. A more serious drawback is that the information contained in price differences, such as on transportation costs, is likely to be lost. For example, if rice is produced in different locations, then interregional price differences will be attributed to heterogeneity rather than to transportation costs, even if the product is homogeneous. We return to this in chapter 4, section 4.4.

22. Therefore modelers often avoid using the mathematical programs and work with demand and supply functions. See also the discussion in section 2.2.5.

The choice of functional forms largely depends on data availability, since some specifications, like flexible forms require estimation of a large number of parameters.[23] This number can be reduced by using a multi-level formulation[24] or by setting a priori values for selected coefficients.[25] In practice, the choice of functional form also depends on the amount of effort the modeler is willing to invest in econometric estimation.

It is fair to add that so far, the more refined econometrics are only performed on model components such as consumer demand systems or input demand functions, without taking into account that equilibrium imposes additional restrictions. The coefficients resulting from econometric estimation are subsequently calibrated to ensure satisfaction of these restrictions in the reference case (in the base year for which a SAM is available).[26]

Uniqueness of Equilibrium

As we have seen in sections 1.2.3 and 3.1.2, there is in general no guarantee for uniqueness of equilibrium. However, in the case of perfect competition, if all consumers have the same homothetic utility function, they can be represented by a single consumer, as in section 1.4.5. This ensures that equilibria are solutions of a convex program and that if there are several, their convex combinations will also be equilibria, so that there are no multiple, isolated equilibria. More important, if, as is often the case, the common utility function also has strictly convex indifference curves (e.g., CES or Cobb-Douglas), equilibrium will be unique. Yet this only holds for the purely competitive model. As discussed in section 1.4.6, most market imperfections (e.g., taxes or monopoly premia) can, in the Negishi format, be represented through the utility function of an "additional consumer," which will be different from the one of "true" consumers. Therefore the representative consumer construction will no longer hold, and uniqueness becomes an empirical matter.[27]

Parameter Estimation and Model Calibration

Full system econometric estimation of parameters in equations (3.17) is usually impossible, mainly for two reasons. First, there are identification

23. See also sections 2.1.4 and 2.2.5 where some functional forms are presented.

24. See section 2.1.4.

25. See, for example, Condon et al. (1990).

26. See also section A.9 of the mathematical appendix for a brief discussion on the introduction of calibration procedures in a GAMS program.

27. See, for example, Kehoe and Whalley (1985). See also Ginsburgh and Van der Heyden (1988).

problems, since the number of endogenous variables is very large.[28] Second, full system estimation is hampered by lack of data. For consumption, household budget surveys may be available for a few years, possibly supplemented by aggregate time series data. For inputs, one rarely has access to more than one or two input-output tables, supplemented with time series of a more specialized nature (e.g., agricultural production series). This severely limits the scope for a more than rudimentary treatment of autocorrelation of disturbances. Modelers react to these problems by proceeding in two steps:

1. Using econometric estimation component by component, that is, separately for consumption and production, with some parameters borrowed from the literature.

2. Setting the remaining parameters through calibration methods, also called benchmarking methods.

The econometrics were briefly discussed in sections 2.1.4 for producers and 2.2.5 for consumers, and a further coverage falls outside the scope of this book.

Calibration is defined as choosing numerical values of the parameters for consumer demand functions $x_i(p, h_i)$ as well as for net supply functions $y_j(p)$ (or unit input demand $A(p)$ functions for the CGE model). This can be expressed in terms of the SAM variables x_i^*, h_i^*, y_j^*, p^* as solving the following system of equations:

$$x_{ik}(p^*, h_i^*; \beta, \gamma) = x_{ik}^*, \qquad \text{all } i \text{ and } k, \tag{3.21a}$$

$$y_{jk}(p^*; \beta, \gamma) = y_{jk}^*, \qquad \text{all } j \text{ and } k. \tag{3.21b}$$

For the goods in the CGE model, this is

$$A_{kh}(p^*; \beta, \gamma) = A_{kh}^*, \qquad \text{all } k \text{ and } h. \tag{3.21c}$$

In (3.21), β and γ are vectors of parameters and A^* is the SAM-value of the input-output matrix. Often the crucial parameter values (e.g., substitution elasticities or the rate of returns to scale), here the γ's, are taken as exogenous, originating either from econometric estimation with the data of the model or from the literature. The vector β of other parameters is the unknown in (3.21), for which the system is solved. In principle, the dimension of the vector β should not be less than the number of equations in (3.21), but even then no solution may exist. In practice, one often associates a single, multiplicative parameter β_h to every equation, and this

28. One could of course attempt to instrumentalize at least some of these endogenous variables, but this is rarely seen.

obviously makes it possible to compute its value for every equation separately.

In many modeling applications not only are the levels of variables calibrated to their SAM-value, but also function elasticities are calibrated to values that are given a priori. This can be done, for example, by extending (3.21) with elasticity equations for demand:

$$\frac{\partial \ln x_{ik}(p^*, h_i^*; \beta, \gamma)}{\partial \ln p_k} = \varepsilon_{ik}^*, \tag{3.21d}$$

with given ε_{ik}^*. Note that to satisfy both (3.21a) and (3.21d), these equations have to be solved simultaneously.

When time series are available, equations (3.21a)–(3.21c) and possibly (3.21d) should be solved simultaneously for all years. Note that when the dimension of β is less than the number of equations to be satisfied, the problem becomes a standard econometric one, since the model outcomes will deviate from the observations. In many applications the estimation of β is not done via formal algorithms, but by trial and error, since the number of parameters is very large and outcomes that are considered "unrealistic" have to be avoided.[29]

Finally, note that since prices are known to be positive in equilibrium, the balance equations (3.20a) hold with equality, and there is no need to impose the complementarity conditions in calibration. We will return to this question in section 4.4.2, where the complementarity conditions will matter.

After calibration, the model will fully replicate the SAM(s), for the base year or for the reference period. Clearly such a solution does not add any new information for these years, but it does provide a consistency check.[30]

3.6.2 The Treatment of Missing Commodities and Markets

To achieve Pareto-efficient decentralization, agents need to receive signals that they cannot affect in a direct way. Until now prices and lump-sum transfers have performed this role.[31] Prices have to be generated either by competitive markets or by a central "Negishi planner," who computes them as dual variables associated with the commodity balances.

Though whenever a commodity exists, there is a physical balance that must be satisfied, there may be no price quoted. Managerial skills or labor

29. See also Mansur and Whalley (1984).

30. When the outcomes of the model do not coincide with those of the SAM, one may be able to identify the source of the error by comparing all row and column totals. There are two possible sources of differences: Either the SAM violates model restrictions (e.g., nonnegativity of income or demand) or the model does not follow the SAM structure (this may be due to a programming error or to a difference in design).

31. See propositions 1.9 and 1.10.

of self-employed are so specific that it would be very cumbersome to value them. There is no price for the Mona Lisa, since it is not for sale and considered invaluable; likewise there is no price for effort that is difficult to measure, fresh air has a zero price, and no trader is interested to operate on that market, and so on. Sometimes the price is expressed as a percentage of the value of some other good (expertise fees for paintings are usually charged in that way). This indicates that the markets for these commodities are either missing or imperfect.

An important reason for the set of commodity markets to be incomplete is that in the real world economy, it is too costly to organize all of them because this involves (1) setup costs (building a marketplace in a village), (2) measuring the quantity of a commodity, service, or good (number of hours needed for an expertise or for grading the varieties of a commodity), (3) defining and recognizing the property rights over it, (4) monitoring transactions (protect the property rights).

If the commodity for which the market is missing has to be produced at positive cost, no firm will be willing to supply it; if the commodity is a natural resource, it will be supplied freely, and its use will be wasteful. In both cases Pareto-efficiency will be lost. To restore it, new institutions have to be created. Further discussion of this topic is postponed to chapters 9 and 10, since it involves external effects and possibly nonconvex technologies. The issue is also relevant in dynamics and even more so under uncertainty and we return to it in sections 7.2.2 and 7.2.3 for dynamics and in section 12.2 for uncertainty. Here we merely indicate how to represent commodities that do play a role in the static model under certainty but for which markets are not fully organized.

In applied models the set of commodity markets may be incomplete as a representation of a property of the economy under study but also because data are missing or because, for instance, detailed information on the nature of competition on given markets is unknown so that limited research time makes shortcuts unavoidable. We give a typology of incomplete commodity markets and ways of formulating such shortcuts.

Rough Markets

When there are many varieties or qualities of a given product, the market (or the modeler) may not always be able to organize competitive pricing for each of them, and the products will be treated without distinction on an aggregate market. Grading systems vary across countries and products; the distinction may be by regions of origin, modes of production, and so on. When there is no grading, there will be only one price, though consumers value the varieties differently. If they can choose and strictly prefer one variety to the other, they wil pick the better qualities, and leave the others on the shelves; then apparently for the product as a whole, there will be excess supply at positive price. The sellers may be able to restrict their choice, for example, by bundling qualities or rationing high quality. For producers, lack of recognized product differentiation will be an incentive to supply the variety that is cheapest to manufacture; this often will be the one of lowest quality. All this illustrates that deficient grading generates welfare losses.

When the roughness of markets is diminished, and new grading systems are put in place, the number of channels between producers and consumers is augmented, but this does not change the specification of the producer and consumer models themselves; it merely adds new prices to the model.

Markets for Characteristics

There are instances in which consumers care for the characteristics commodities bring, more than for the commodities themselves: In these cases the service is being provided by the characteristics not by the commodity. Car makes differ widely, and in the end, a car is chosen for its specifications, such as speed, mileage, or number of doors. Modularity in design increasingly makes it possible to freely compose a mix of characteristics, which the manufacturer will combine into a commodity. This means that a virtually infinite range of commodities can be obtained from a limited number of characteristics.[32]

If both consumers and producers are able to generate their demands and supplies in terms of characteristics, and take the prices of characteristics as given, the model is competitive in terms of characteristics, and the only difference with the usual competitive model is that there is an intermediate mapping from commodities to characteristics in the producer model. Then consumer demand will be expressed in characteristics only. Alternatively, if the producer supplies commodities and the consumer cares for characteristics, the consumer model will be

$$\max u_i(c_i),$$

$$c_i, x_i \geqslant 0,$$

subject to

$$c_i = Tx_i,$$

$$px_i \leqslant h_i,$$

32. See Griliches (1971) for one of the first applications. Many authors only estimate the reduced form of these models and specify a so-called hedonic regression where prices of commodities are related to characteristics. Rosen (1974) develops a full general equilibrium model in terms of characteristics and points to limitations of reduced form approaches. See also Bartik (1987) and Epple (1987) on related econometric issues.

where T is a matrix, assumed common to all consumers, that maps[33] commodities to characteristics. This is the so-called Lancaster (1966, 1971) model of characteristics.[34] It cannot be expected that technology sets are always sufficiently flexible to ensure that any combination of characteristics can be accommodated by a finite set of commodities.[35] In this case the market will not be able to provide the ideal blend of characteristics, and we can again characterize the markets as rough.

In applications, particularly for food demand, the characteristics approach is often implemented by distinguishing a raw product and linked to it, a demand for processing or quality, so that each product has two markets: one for the raw version and one for the additives, which enter as such in the utility function (sometimes the firms producing additives are grouped under meat processing, baking, etc.).

Commodities Not Traded between Agents

A commodity that is supplied and used within a household or a firm without being traded (a flat that is not for sale, the building of a factory) can be left out from the budget constraint of the household and from the commodity balance of the economy at large. Thus there is no need to price this commodity, and it is sufficient to represent it within the relevant programs of consumers and producers.

It may also happen that such a commodity does not appear at all in the model but is a fixed input in production that has an indirect effect via the decreasing-returns-to-scale property. This will allow the firm to make a profit, which is the rent on this fixed input. For consumers the only commodities that can be treated are those that are not only nontraded but also in fixed use within the household.

Commodities Traded between Agents but Not Fully Represented

Goods that currently have a zero price may become scarce in the future; historically this happened for several resources like pasture grounds and minerals of various kinds. The policymaker may find it wise to anticipate this and establish property rights on these resources before they become scarce.[36] In such circumstances general equilibrium modeling can be of

33. Note that this is nothing more than writing the utility function in nested form, as $u_i(c_i(x_i))$.

34. See also Drèze and Hagen (1978).

35. Consider the system of linear equations $c = Tx$, where c is a vector of characteristics, x a vector of consumptions, and T a transformation matrix from consumptions to characteristics. The technology is flexible if for any given $c \geqslant 0$ one can find $x \geqslant 0$. This will depend on the properties of the matrix T.

36. This will be discussed further in section 9.2.

special relevance by representing the allocation of such commodities and comparing the outcomes of scenarios in which these commodities go unpriced with those in which prices are assumed to be competitive, since these can only be imputed from a model and not observed on the market.

Markups

Markups were introduced earlier in section 3.4.2 as taxes that accrue to consumers. In theoretical models they appear in connection with imperfect competition, as a representation of producer market power, and section 11.4.3 will deal with this aspect in detail. The concept is also used to represent the cost of missing commodities or commodities for which no markets exist. The approach originates from Kalecki (1943), was pursued by Robinson and Eatwell (1973), and later on, extensively used by Taylor (1979, 1983, 1990), and many others.[37]

As long as markups are interpreted as rewarding the true cost of commodities absent from the model, they do not cause any distortion in principle. However, since they are generated as functions of other prices (and maybe other variables also), they can hardly be interpreted as "true" prices, and inefficiency is inherent.

When specifying the markup functions, one should keep in mind that they represent the price of a missing commodity. It may therefore be reasonable to treat the markup as a function of the prices of the commodities needed to produce the missing commodity: For labor, this could be a cost of living index; for capital, an index of the cost of replacement of capacity (equipment and buildings). Omitted from the model in this representation are the (missing) factors covered by the markup. The modeler sometimes appends an indicator that measures the demand for these factors once the model has been solved; this does not affect allocations and amounts to assuming that the supply of these factors is not constraining.

Markups and Bounds in the CGE Model

Introducing markups in the CGE model (3.18) amounts to replacing (3.18d) by

$$p^g = p^g A^g(p^g, p^f) + p^f A^f(p^g, p^f) + \mu^g,$$

where $p^g A^g(p^g, p^f) + p^f A^f(p^g, p^f) = c^g$ is the unit cost (recall that f and g denote factors and goods, respectively) and μ^g is the markup level. Alternatively, the markup can be expressed as a percentage η of c^g, so that $p_k^g = c_k^g(1 + \eta_k)$ for commodity k. In some way or another most CGE

37. Markups will be discussed in more detail in chapter 11, especially in section 11.4.

models, starting with Adelman and Robinson (1978), make use of such markup rules.

3.6.3 Simulations

Solving the Model

The choice of format also has consequences for computation. The mathematical programming format has the advantage that for given welfare weights, the model can be solved by standard programming algorithms. This is especially relevant when the number of consumers is small, relative to the number of commodities, as will be the case if production is represented through large-scale systems that involve many intermediate commodities. The choice of format also depends on whether functional forms for consumer demand and net supply are available in primal or in dual form.[38] Clearly the use of dual forms rules out the option of using the Negishi and the full-format.[39]

A brief technical discussion of computation is given in appendixes A.7–A.9, where we describe how convergent algorithms can be designed. These procedures follow the path of a differential equation that is approximated through a difference equation. However, applied modelers often try to avoid such an approximation and find it more convenient to use well-established algorithms because these are widely available as software packages.

Although for many of these algorithms, convergence is not ensured theoretically, this does not seem to pose serious problems in practice. Equality-constrained problems are often solved through Gauss-Seidel algorithms.[40] Codsi et al. (1992) solve such problems through successive linear approximation. Ginsburgh and Waelbroeck (1976, 1981) use a linear programming package to solve a Negishi welfare program with inequality constraints, adjusting the welfare weights through a tatonnement rule, so as to bring budget deficits to zero.[41] Matthiesen (1985) approximates a nonlinear complementarity problem by a sequence of linear complementarity problems. Finally, we show in appendixes A.9 and B.1–B.6 and in section 3.6.4 how to use the general algebraic modeling system (GAMS) developed by Brooke et al. (1992) to solve nonlinear complementarity problems via nonlinear programming methods. The experience is that computation is no longer a problem for small to medium-sized AGE models.

38. See also sections 2.1.4 and 2.2.5.

39. The formulation with infinitely lived consumers to be discussed in section 8.1 will need primal formulation if one relies on dynamic programming to compute a solution.

40. See Ortega and Rheinboldt (1970).

41. Such an approach is often followed in energy modeling. See also Dixon (1975) and Makarov and Rubinov (1977).

Running Simulations

The AGE model as formulated so far is basically static. In many applications the model indeed applies to a single period, which may range from a year to a lifetime. The model is then used in the so-called comparative static mode in which some parameters are changed and, for the same period, a new, "counterfactual" solution is computed. Long time periods have the attraction that short-run effects can be disregarded, but they introduce three problems. First, there is time aggregation, which has effects on substitution possibilities. Second, very few data points will be available if time series have to be aggregated over long periods. Third, time aggregation obscures the process of accumulation. Hence the approach seriously limits the scope for empirical implementation: A lifetime SAM is required, and very few degrees are left to validate the model for years other than the base period validation. Therefore it seems advisable to choose as short a time period as the data allow and to reserve the use of static models to single, short period applications.

We will discuss several representations where time enters explicitly: recursive dynamics (chapter 7) and models with intertemporal optimization (chapters 7 and 8). For these models there is more scope for validation, since time series observations can be used. The differences between simulation results and observed values are then used to adjust the values of parameters for which only a base-year observation is available. This procedure is necessarily informal and has the advantage that the performance of the model over time can be reported. This appears to be a reasonable minimal requirement, but in current modeling it is unfortunately not a common practice to publish such tests.

Policy Analysis with the AGE Model

AGE models have been used to evaluate a wide range of policy issues, including changes in indirect taxes, trade policy, income redistribution, and public investments. Well-known examples include Adelman and Robinson (1978), Dervis et al. (1982), Dixon et al. (1977), and Whalley (1985). Every national AGE model has at least once been used for running policy simulations.

Usually the changes modeled are based on existing reform proposals. The evaluation is commonly in terms of national income, incomes of various socioeconomic groups, or formal welfare criteria as well as nutrition, market shares, production in various sectors, and the like. Applied general equilibrium models can be particularly useful in this sort of application, since the welfare effects of the policy proposals cannot be predicted on theoretical grounds only. This is so because the reforms typically do not involve full compensation of the losers and because they do not remove all distortions.

Many of these studies use models that belong to the class of CGE models, but some contain inequality constraints to represent downward real wage rigidity or import quota. Policy packages typically involve simultaneous changes in variables under the control of the government such as taxes and public investments. The models have the advantage that specific policy measures can be accommodated without excessive simplification and aggregation. The classification can correspond closely to the one the policymaker is used to. Using such models not only has the advantage that general equilibrium effects are taken into account but also that interaction of different policy measures can be studied. For example, Narayana et al. (1990) show that the combination of infrastructural investment with food-for-work programs is a more effective way of reducing poverty than food subsidies. Bevan et al. (1990) analyze the effect of world coffee price increases combined with partial trade liberalization in Kenya; they conclude that a large part of the benefit to smallholder producers accrues as rents to owners of trade quota. Perhaps most important, when used in the recursive dynamic mode, AGE models make it possible to evaluate lagged effects of policies, in particular, of public investment. This can be done by distinguishing investment projects of various degrees of completion (e.g., in irrigation; see Salauddin Ahmad et al. 1986).

AGE models have been constructed mainly for developing countries. Developed economies have been represented as part of multi-country international trade models and as independent national models in connection with tax incidence, unemployment, reforms of the social security system, trade and energy policy. There have been several surveys of applications, such as Shoven and Whalley (1984), Manne (1985), Devarajan et al. (1986), de Melo (1988), Decaluwé and Martens (1989), Pereira and Shoven (1988), Robinson (1989), and Gunning and Keyzer (1995). In the application sections of the following chapters we will refer to typical examples, without any claim to comprehensiveness.

3.6.4 A GAMS Application of the Basic CGE Model

To show that the basic CGE-model (3.18) can readily be transformed into a computer program, we list in this section the full GAMS representation as well as the results from computation. We consider three classes of consumers (farmers, workers, employers) and four commodities, two goods (food and textile) and two factors (labor and equipment). The model is only a prototype, and the numerical values of the parameters and initial values are arbitrary. It is solved by making use of a nonlinear programming algorithm that minimizes the infeasibility on two nonlinear equations (variable input-output coefficients and consumer demand). The computational approach is described in more detail in appendix A.9. A more elaborate application is considered in appendix B, where the GAMS statements used here are also explained. Therefore, and since the GAMS notation is largely self-explanatory, the comments can be brief in this section.[42]

The only addition to model (3.18) is that prices are normalized explicitly: To avoid scaling problems, it is often practical to normalize so as to keep the value of endowments equal to unity rather than the sum of prices itself. The index sets of the model are I for consumer groups, K for commodities, G for goods, and F for factors. Goods are produced with a CES-type technology.[43] Consumer demand follows a linear expenditure system. The complete GAMS program is written as follows:[44]

```
 *== Dollar control commands ==*

$TITLE Basic CGE Model (equations (3.18a-d))
* suppress some print
$OFFSYMXREF
$OFFDOLLAR
$INLINECOM { }

*== 1. Declare and initialize sets ==*

 SET  IT      CONSUMERS      / FARMERS, WORKERS,
                               EMPLOYERS, NATIONAL /
      I(IT)   CONSUMERS      / FARMERS, WORKERS,
                               EMPLOYERS /

      KT      COMMODITIES    / FOOD, TEXTILE, LABOR,
                               EQUIPMENT, TOTAL /
      K(KT)   COMMODITIES    / FOOD, TEXTILE, LABOR, EQUIPMENT /
      KN(K)                  / FOOD, TEXTILE, LABOR /

      G(K)    GOODS          / FOOD, TEXTILE /

      F(K)    FACTORS        / LABOR, EQUIPMENT /
 ;
      ALIAS  (G,  GP);
      ALIAS  (K,  KP);
```

42. For further details, see Brooke et al. (1992).

43. Under the CES technology, the optimal input mix is a solution of $\min\{\Sigma_k p_k a_{kg} \,|\, a_{kg} \geqslant 0$, all $k, \Sigma_k(\gamma_{kg} a_{kg}^{-\rho_g})^{-1/\rho_g} = 1\}$. The associated first-order condition is $p_k = \lambda_g \gamma_{kg} a_{kg}^{-(1+\rho_g)}$. Since under profit maximization, the price is equal to the marginal cost, it follows that $\lambda_g = \mathrm{p}_g$, and for $\rho_g > -1$, the expression reduces to $a_{kg} = \kappa_{kg}(p_g/p_k)^{\sigma_g}$, where $\kappa_{kg} = (\gamma_{kg})^{\sigma_g}$ and $\sigma_g = 1/(1+\rho_g)$. This is the form used in equation ACAL(K,G) of the program.

44. In this program, comment lines are denoted by an * in the first column. A $ sign is used for the so-called $-control commands (such as $TITLE) at the beginning of the program.

```
*== 2. Parameters and Tables ==*

 PARAMETERS
        AO(K,G)          input output constants
        BETA(I,K)        budget shares
        OMEGA(I,F)       endowments
        OMEGTOT(F)       total endowments
        PNORM            price norm
        SIGMA(G)         input output substitution elasticities
;
 TABLE AO(K,G)   Input output coefficients

                 FOOD        TEXTILE
 FOOD            0.          .2
 TEXTILE         .1          .0
 LABOR           .5          .29
 EQUIPMENT       .3          .5
;

PARAMETER SIGMA(G)   Input output substitution parameter

                     / FOOD 2., TEXTILE .2/
;

 TABLE BETA(I,K)    Budget shares

                 FOOD        TEXTILE       LABOR        EQUIPMENT
 FARMERS         .461        .239          .130         .170
 WORKERS         .532        .128          .180         .160
 EMPLOYERS       .263        .137          .080         .520
;

 TABLE OMEGA(I,F)   Factor endowments

                 LABOR   EQUIPMENT
 FARMERS         10.0    30.0
 WORKERS         20.0    0.0
 EMPLOYERS       10.0    20.0
;
* Total endowments
  OMEGTOT(F) = SUM(I,OMEGA(I,F)) ;

* Price norm
  PNORM = 100.   ;

*== 3. Variables, equations and model definition ==*

 POSITIVE VARIABLES
  A(K,G)                 input output matrix
  H(I)                   income
  P(K)                   price
  Q(G)                   production
  X(I,K)                 consumption
  ZAM(K,G)               deficit input quantity (slack)
```

```
 ZAP(K,G)            excess input quantity (slack)
 ZXM(I,K)            deficit consumption quantity (slack)
 ZXP(I,K)            excess consumption quantity (slack)
 ;
VARIABLES
 TMAX                objective
 ;
* Equations: declaration

EQUATIONS
 ACAL(K,G)         first-order condition of cost minimization
                   { CES demand for inputs per unit of output }

 BALGOOD(G)        balance of goods

 BALFAC(F)         balance of factors

 INC(I)            income
                   { value of endowments }

 PRIGOOD(G)        pricing of goods
                   { price of a good equal to unit cost }

 PRINORM           price normalization
                   { total value of endowments fixed }

 T1OBJ(K,G)        maximal excess input

 T2OBJ(I,K)        maximal excess consumption

 XCAL(I,K)         calculate consumption
                   { fixed budget shares }
 ;

* Equations: specification
 ACAL(K,G)     ..  A(K,G) + ZAP(K,G)  =E=  ZAM(K,G)
                     + AO(K,G) * (P(G)/P(K))**SIGMA(G);

 XCAL(I,K)     ..  X(I,K) + ZXP(I,K)  =E=  ZXM(I,K)
                     + BETA(I,K)*H(I)/P(K);

 BALGOOD(G)    ..  SUM(I, X(I,G)) + SUM(GP, A(G,GP)*Q(GP))  =E=  Q(G);

 BALFAC(F)     ..  SUM(I, X(I,F)) + SUM(GP, A(F,GP)*Q(GP))
                     =E=  OMEGTOT(F);

 INC(I)        ..  H(I)  =E=  SUM(F, P(F)*OMEGA(I,F));

 PRIGOOD(G)    ..  P(G)  =E=  SUM(GP, P(GP)*A(GP,G))
                     + SUM(F, P(F)*A(F,G));

 PRINORM       ..  SUM(F, P(F)*OMEGTOT(F))  =E=  PNORM;

 T1OBJ(K,G)    ..  TMAX  =G=  10*ZAP(K,G) + ZAM(K,G);

 T2OBJ(I,K)    ..  TMAX  =G=  10*ZXP(I,K) + ZXM(I, K);
```

```
* Define the model

  MODEL CGE /ACAL, XCAL, BALGOOD, BALFAC, INC, PRIGOOD,
              PRINORM, T1OBJ, T2OBJ/;

* Impose lower bounds for safety during iterations

  A.LO(K,G) = .001$(A0(K,G) GT 0);
  H.LO(I) = .001;
  P.LO(K) = .001; Q.LO(G) = .001;
  X.LO(I,K) = .001;
* Initialization

  A.L(KN,G) = 1.;
  H.L(I) = 1.;
  P.L(K) = 1.;
  Q.L(G) = 1.;
  TMAX.L = 100;
  ZAM.L(K,G) = 10; ZXM.L(I,K) = 10.;
  ZAP.L(K,G) = 10; ZXP.L(I,K) = 10.;

*== 4. Computation of equilibrium ==*

  OPTION ITERLIM = 2000;
  SOLVE CGE USING NLP MINIMIZING TMAX;

*== 5. Accounting (post-optimal calculations) and reporting ==*

* Declare sets for mapping

  SET GG        Sectors /FOOD, TEXTILE, HOUSEHOLD, NATIONAL /

  CORR(GG,G) Map sector to goods
                        /FOOD.FOOD, TEXTILE.TEXTILE /

  NM           Items   /'CONSUMPT.', ENDOWMENT/
;
* Declare parameters (post-optimal variables)

 PARAMETERS
  GDP            TOTAL GDP
  GDPS(G)        SECTORAL GDP
  QCONS(K)       CONSUMER DEMAND BY COMMODITY
  QINT(K)        INTERMEDIATE DEMAND BY COMMODITY
  QQ(K)          PRODUCTION QUANTITY BY COMMODITY
;

  QCONS(K) = SUM(I, X.L(I,K));
  QINT(K) = SUM(G, A.L(K,G)*Q.L(G));
  QQ(F) = OMEGTOT(F);
  QQ(G) = Q.L(G);
  GDP = SUM(K, P.L(K)*(QQ(K)-QINT(K)));
  GDPS(G) = P.L(G)*Q.L(G) - SUM(GP,P.L(GP)*A.L(GP,G))*Q.L(G);

* Prepare tables
```

```
* Table 1: Commodity account in quantity terms and equilibrium price

 PARAMETER TABLE1(K,*);
  TABLE1(K, 'CONSUMPT.') = QCONS(K);
  TABLE1(K, 'INPUT') = QINT(K);
  TABLE1(K, 'SUPPLY') = QQ(K);
  TABLE1(K, 'IMBALANCE') = QCONS(K)+QINT(K)-QQ(K);
  TABLE1(K, 'PRICE') = P.L(K);
  DISPLAY "COMMODITY ACCOUNT IN QUANTITY TERMS", TABLE1;

* Table 2: Commodity account in value terms

 PARAMETER TABLE2(KT,*);
  TABLE2(K, 'CONSUMPT.') = P.L(K)*QCONS(K);
  TABLE2(K, 'INPUT') = P.L(K)*QINT(K);
  TABLE2(K, 'SUPPLY') = P.L(K)*QQ(K);
  TABLE2('TOTAL', 'CONSUMPT.') = SUM(K, TABLE2(K, 'CONSUMPT.'));
  TABLE2('TOTAL', 'INPUT') = SUM(K, TABLE2(K, 'INPUT'));
  TABLE2('TOTAL', 'SUPPLY') = SUM(K, TABLE2(K, 'SUPPLY'));
  DISPLAY "COMMODITY ACCOUNT IN VALUE TERMS",TABLE2;

* Table 3: Consumption and revenue by class

 PARAMETER TABLE3(NM,IT,KT);
  TABLE3('CONSUMPT.',I,K) = P.L(K)*X.L(I,K);
  TABLE3('ENDOWMENT',I,F) = P.L(F)*OMEGA(I,F);
  TABLE3('CONSUMPT.',I,'TOTAL') = SUM(K, TABLE3('CONSUMPT.',I,K));
  TABLE3('ENDOWMENT',I,'TOTAL') = SUM(K, TABLE3('ENDOWMENT',I,K));
  TABLE3('CONSUMPT.','NATIONAL',K)
          = SUM(I, TABLE3('CONSUMPT.',I,K));
  TABLE3('ENDOWMENT','NATIONAL',K)
          = SUM(I, TABLE3('ENDOWMENT',I,K));
  TABLE3('CONSUMPT.','NATIONAL','TOTAL')
          = SUM(I, TABLE3('CONSUMPT.',I,'TOTAL'));
  TABLE3('ENDOWMENT','NATIONAL','TOTAL')
          = SUM(I, TABLE3('ENDOWMENT',I,'TOTAL'));
* option with three decimals 1 index as row (NM), 2 as columns (IT,KT)
  OPTION TABLE3:3:1:2;
  DISPLAY "CONSUMPTION AND REVENUE BY CLASS",TABLE3;

* Table 4: Value added by sector

 PARAMETER TABLE4(GG);
  TABLE4(GG) = SUM(G $ CORR(GG,G), GDPS(G));
  TABLE4('HOUSEHOLD') = GDP - SUM(G,GDPS(G));
  TABLE4('NATIONAL') = GDP;
  OPTION TABLE4:3:0:1;
  DISPLAY "VALUE ADDED BY SECTOR",TABLE4;
```

Results from tabulation
(layout as printed by the program)

---- 190 COMMODITY ACCOUNT IN QUANTITY TERMS

---- 190 PARAMETER TABLE1

	CONSUMPT.	INPUT	SUPPLY	IMBALANCE	PRICE
FOOD	33.202	4.040	37.242	−7.3186E-13	1.238
TEXTILE	16.070	4.535	20.604	−4.0146E-13	1.122
LABOR	11.150	28.850	40.000	−7.1054E-15	1.117
EQUIPMENT	25.693	24.307	50.000		1.107

---- 201 COMMODITY ACCOUNT IN VALUE TERMS

---- 201 PARAMETER TABLE2

	CONSUMPT.	INPUT	SUPPLY
FOOD	41.093	5.001	46.093
TEXTILE	18.024	5.086	23.110
LABOR	12.452	32.220	44.672
EQUIPMENT	28.431	26.897	55.328
TOTAL	100.000	69.204	169.204

---- 218 CONSUMPTION AND REVENUE BY CLASS

---- 218 PARAMETER TABLE3

	FARMERS FOOD	FARMERS TEXTILE	FARMERS LABOR	FARMERS EQUIPMENT	FARMERS TOTAL
CONSUMPT.	20.452	10.603	5.767	7.542	44.365
ENDOWMENT			11.168	33.197	44.365

+	WORKERS FOOD	WORKERS TEXTILE	WORKERS LABOR	WORKERS EQUIPMENT	WORKERS TOTAL
CONSUMPT.	11.883	2.859	4.020	3.574	22.336
ENDOWMENT			22.336		22.336

+	EMPLOYERS FOOD	EMPLOYERS TEXTILE	EMPLOYERS LABOR	EMPLOYERS EQUIPMENT	EMPLOYERS TOTAL
CONSUMPT.	8.758	4.562	2.664	17.316	33.299
ENDOWMENT			11.168	22.131	33.299

+	NATIONAL FOOD	NATIONAL TEXTILE	NATIONAL LABOR	NATIONAL EQUIPMENT	NATIONAL TOTAL
CONSUMPT.	41.093	18.024	12.452	28.431	100.000
ENDOWMENT			44.672	55.328	100.000

---- 227 VALUE ADDED BY SECTOR

---- 227 PARAMETER TABLE4

NATIONAL	100.000
FOOD	41.007
TEXTILE	18.110
HOUSEHOLD	40.883

4 International Trade

An important body of the literature on applied general equilibrium models is devoted to the consequences of international tariff and trade reforms. This chapter studies various representations of international trade in general equilibrium models. The introduction of tariffs (and other taxes) is postponed until chapter 5. As will become clear from the discussion in sections 4.1 and 4.2, international trade models can be obtained from a reinterpretation of the competitive model of the previous chapters. Section 4.3 briefly discusses the welfare properties of such models. Section 4.4 describes modeling techniques used in applied work, such as the CGE representation, ways to avoid extreme specialization, aggregation of imports from different origins, and simplified representations of the rest of the world.

4.1 Reinterpreting the Competitive Equilibrium Model as a World Model

The reinterpretation of the competitive equilibrium model as a world model of international trade starts with the grouping of consumers and producers into countries or regions. Having grouped agents in this way, the modeler has to make decisions of three kinds.

First, it may be necessary to distinguish some commodities by location (houses and oranges in Marseilles are different from those in New York), while others may be treated as homogeneous with respect to location (e.g., French wheat may be considered as a perfect substitute for American wheat). A further distinction is whether the commodity can be traded internationally, in the sense of being shipped in and out of the country the way oranges and wheat can but houses cannot.

Second, one has to describe the set of actions open to agents. Should the modeler account, say, for French residents consuming in the United States while on vacation? This consumption is an export from the United States to France, though, physically, it remains in the United States. Similarly an American resident may earn income from selling a house he owns in France. French consumers may also own shares in American firms which would provide profit income. Hence the total value of imports shipped into France may exceed the value of exports leaving it, though there is no French payment deficit. In the competitive model discussed so far, every agent has the option to buy every commodity. In the international trade model, these actions are usually restricted by assuming, say, that French consumers can only buy commodities in France and that they do not own shares in American firms. Though theory does not impose such limitations, applications often do, both for statistical and computational reasons.

Third, the modeler has to decide on how to represent the various countries. He has the possibility of representing all of them as (groups of) agents, and this will lead to a model in which all prices are endogenous.

Alternatively, he may opt for a "small country" representation in which the model only covers a group of countries (or a single country) that takes the prices at which it trades with the outside world as given.

In applied work, one will in every country model restrict attention to commodities that are available to the agents of the country, and this defines country-specific commodity sets. However, to simplify notation, we will work with a full commodity set. This does not cause difficulties because, by assumption S, nonsatiation applies to the utility functions defined on the full commodity space; similarly convexity of production sets is defined on the full commodity space. Therefore we may express all supply, demand, and price vectors in the full commodity space, where traded as well as nontraded commodities appear separately.

4.2 International Trade in the General Equilibrium Model

The grouping of agents can be represented as follows. The index set $I = \{1, 2, \ldots, m\}$ of consumers is subdivided into C country-specific sets I_c (where $\Sigma I_c = I$). Similarly the index set $J = \{1, 2, \ldots, n\}$ of firms is subdivided into sets J_c (where $\Sigma J_c = J$). Hence x_c will denote[1] $\Sigma_{i \in I_c} x_i$, where x_i refers to the consumption by consumer i in country c (but this consumption will not always physically take place in country c). The commodity index set $K = \{1, 2, \ldots, r\}$ is valid in all countries and includes location-specific as well as homogeneous commodities. Nontradability of a particular commodity is represented by restricting the possibility for agents to demand from or supply to other countries.

4.2.1 The Multi-country Case

To construct a world model, we define country-specific (or national) demand and supply functions[2] as well as national endowment in country c as

$$x_c(p^w) = \Sigma_{i \in I_c} x_i(p^w),$$

$$y_c(p^w) = \Sigma_{j \in J_c} y_j(p^w),$$

$$\omega_c = \Sigma_{i \in I_c} \omega_i,$$

where p^w is the vector of world prices; accordingly the national commodity balance can be written

$$x_c + e_c = y_c + \omega_c + m_c, \tag{4.1a}$$

1. This notation is introduced for simplicity. It could lead to confusion, since, say, x_1 could refer to a consumer as well as to a country. We will use numerical indexes only when no confusion can arise.

2. Here, for simplicity, we use functions, but we will allow for correspondences later on.

where e_c and m_c are exports and imports, respectively. Since each element in these vectors concerns one homogeneous commodity, only net imports (net trades)

$$z_c = m_c - e_c \tag{4.1b}$$

can be determined. For this reason we will often denote trade flows of country c by z_c.

Consumers in country c may obtain income (e.g., from labor, resource endowments, or profits) from abroad. Assuming, as before, that each consumer spends all his income on consumption, this will also hold at country level so that each national budget is balanced and

$$p^w x_c = p^w \omega_c + \Sigma_j \theta_{cj} p^w y_j, \tag{4.1c}$$

where $\theta_{cj} = \Sigma_{i \in I_c} \theta_{ij}$. Walras's law holds at world level. Indeed, since

$$p^w \Sigma_c x_c = p^w \Sigma_c \omega_c + p^w \Sigma_c \Sigma_j \theta_{cj} y_j,$$

and since $\Sigma_c \theta_{cj} = 1$ for all j, this reduces to

$$p^w \Sigma_c x_c = p^w \Sigma_c \omega_c + p^w \Sigma_j y_j.$$

By (4.1a)–(4.1b),

$$\Sigma_c p^w z_c = \Sigma_c p^w (m_c - e_c) = 0.$$

Note that $p^w z_c = p^w(m_c - e_c)$, the value of national net import by country c may be nonzero if consumers in country c earn, say, profit income from abroad or receive income from labor originating from relatives abroad.[3]

The value $p^w z_c$ is the trade deficit, or deficit on the trade balance. The theory presented so far allows for nonzero trade deficits. Obviously, if profits accrue only to consumers of the country where firms are located, then $p^w z_c$ will be zero. If consumers only own endowments located in this country and do not consume abroad, then the trade balance coincides with the balance of payments.[4]

We can now define an international competitive equilibrium as

DEFINITION 4.1 (International competitive equilibrium) The allocation y_j^*, all j, x_i^*, all i, e_c^* m_c^*, all c, supported by the price vector $p^{w*} \geqslant 0$, $p^{w*} \neq 0$ is

3. If x_c, y_c, and ω_c have nonzero entries only for commodities that are geographically located in country c, then z_c refers to a net inflow of commodities into c. Otherwise, it denotes the net flow controlled by agents of country c.

4. In models without financial assets, the trade balance refers to flows of commodities (in value terms) that enter or leave the country. The balance of payments is the budget equation of the country and includes payments and revenues from transactions abroad, like transfers by migrant workers or by firms.

a general competitive equilibrium if the following conditions are satisfied:

1. In every country c:

a. for every producer $j \in J_c$, y_j^* solves $\max_{y_j}\{p^{w*}y_j \mid y_j \in Y_j\}$;

b. for every consumer $i \in I_c$, x_i^* solves $\max_{x_i \geqslant 0}\{u_i(x_i) \mid p^{w*}x_i \leqslant h_i^*\}$, where $h_i^* = p^{w*}\omega_i + \Sigma_j \theta_{ij} p^{w*} y_j^*$;

c. the commodity balance is satisfied, $\Sigma_{i \in I_c} x_i^* - \Sigma_{j \in J_c} y_j^* - \Sigma_{i \in I_c}\omega_i - m_c^* + e_c^* = 0$.

2. At the international level, trade balances, $\Sigma_c m_c^* - \Sigma_c e_c^* \leqslant 0$.

An alternative representation is based on the open economy model of definition 3.5. Using this framework and recalling that $\bar{p}^m \geqslant \bar{p}^e > 0$, the international model can be written

$$\max_{m, e \geqslant 0,\, y_j,\, \text{all } j,\, z_c,\, \text{all } c} \bar{p}^e e - \bar{p}^m m$$

subject to

$$m - e \geqslant \Sigma_c z_c \qquad (p^w), \tag{4.2a}$$

for every c,

$$\Sigma_{j \in J_c} y_j + z_c \geqslant \Sigma_{i \in I_c} \hat{x}_i - \Sigma_{i \in I_c} \omega_i \qquad (p_c), \tag{4.2b}$$

$$y_j \in Y_j, \text{ all } j \in J_c$$

for given

$$\hat{x}_i = \operatorname{argmax}\{u_i(x_i) \text{ subject to } p_c x_i \leqslant p_c \omega_i + \Sigma_{j \in J_c} \theta_{ij} \Pi_j(p_c) + \Sigma_{c' \neq c} \Sigma_{j \in J_{c'}} \theta_{ij} \Pi_j(p_{c'}),\ x_i \geqslant 0\}, \text{ all } i \in I_c, \text{ all } c. \tag{4.3}$$

In (4.2b) and (4.3) p_c is the vector of shadow prices associated to the commodity balances of country c, $\Sigma_{j \in J_c} \theta_{ij} \Pi_j(p_c)$ represents profits collected from firms operating in country c, and $\Sigma_{c' \neq c} \Sigma_{j \in J_{c'}} \theta_{ij} \Pi_j(p_{c'})$ denotes profits from abroad.

Of course, if the model covers the whole world, the possibility of external trade (e and m) is only a mathematical artefact, and to obtain a meaningful equilibrium, prices $\bar{p}^e$ have to be chosen sufficiently close to 0 to make e and m vanish. We use this formulation because it makes it possible to represent at the same time the world model and the small-country specification to which we will turn in section 4.2.2.

The following proposition of existence of equilibrium, based on (4.2)–(4.3), is now immediate:

PROPOSITION 4.1 (Existence of an international equilibrium) Under assumption S and for positive $\bar{p}^e$ sufficiently close to 0, the solution of

(4.2)–(4.3) exists and is an international competitive equilibrium as specified in definition 4.1.

Proof The proof is identical to the one of proposition 3.4. ■

Note that in equilibrium, domestic prices p_c will be equal to international prices p^w, since the net imports of the various countries are perfect substitutes.

In this framework the constraints of the mathematical program can be decomposed into two groups: world markets (equations 4.2a) and country markets (equations 4.2b). This distinction facilitates the analysis, since it separates the international from the national components. Moreover the open economy mathematical program can itself be decomposed into the international linkage and national programs as follows:

1. *International linkage*. For given m_c, e_c transmitted by the countries, the linkage solves

$$\begin{aligned}&\max \bar{p}^e e - \bar{p}^m m,\\ &m, e \geqslant 0,\\ &\text{subject to}\\ &m - e \geqslant \Sigma_c(m_c - e_c) \qquad (p^w).\end{aligned} \tag{4.4}$$

2. *National programs*. For given p^w transmitted by the linkage and for given consumptions $\hat{x}_i$, each country c solves

$$\begin{aligned}&\max p^w e_c - p^w m_c,\\ &m_c, e_c \geqslant 0,\ y_j, \text{ all } j \in J_c,\\ &\text{subject to}\\ &\Sigma_{j\in J_c} y_j + m_c - e_c \geqslant \Sigma_{i\in I_c}\hat{x}_i - \Sigma_{i\in I_c}\omega_i \qquad (p_c),\\ &y_j \in Y_j,\end{aligned} \tag{4.5a}$$

for given

$$\begin{aligned}\hat{x}_i = \operatorname{argmax}\{u_i(x_i) \text{ subject to } p_c x_i \leqslant p_c\omega_i + \Sigma_{j\in J_c}\theta_{ij}\Pi_j(p_c)\\ + \Sigma_{c'\neq c}\Sigma_{j\in J_{c'}}\theta_{ij}\Pi_j(p_{c'}),\ x_i \geqslant 0\}, \text{ all } i \in I_c, \text{ all } c.\end{aligned} \tag{4.5b}$$

In a world model both (4.4) and (4.5) need to be solved. Note that since here p^w is both the import price and the export price, every program (4.5a) with feedback (4.5b) will only determine net trades $z_c = m_c - e_c$ and not exports and imports separately.

Recall that commodities that are not traded between countries are nonetheless elements of the vectors e, m, and p^w; they are only supplied and demanded by domestic agents.[5] Hence each country vector has zero entries for nontraded commodities of other countries.

4.2.2 The Two-Country Case and the Small-Country Assumption

The two-country case is readily obtained by setting $C = 2$. This may seem attractive for applications, since it significantly reduces the size of the model, but it means that one has to represent the country of interest and the rest of the world by a full "country" model. This full representation of the rest of the world is problematic, and modelers have avoided this, using shortcuts that will be described in section 4.4.

If country $c = 1$, say, has an impact on world prices, the rest of the world has to be represented. If this impact is negligible, one can make the small-country assumption and solve (4.5a)–(4.5b), where p^w is the given international price to which the domestic price will adjust;[6] in the feedback (4.5b) profits from abroad will then become exogenous. To allow for nontraded commodities, one could replace the objective of (4.5a) by $(\bar{p}_c^e e_c - \bar{p}_c^m m_c)$; for nontraded commodities the export price will have to be set sufficiently low and the import price sufficiently high. For traded commodities they can be set equal to the international prices. This leads to the following reformulation of the open economy program with feedback (definition 3.5):

DEFINITION 4.2 (Competitive equilibrium under the small country assumption) In the open economy format, the competitive equilibrium is

$$\max \bar{p}_c^e e_c - \bar{p}_c^m m_c,$$

$$m_c, e_c \geqslant 0, \; y_j, \text{ all } j \in J_c,$$

subject to (4.5a′)

$$\Sigma_{j \in J_c} y_j + m_c - e_c \geqslant \Sigma_{i \in I_c} \hat{x}_i - \Sigma_{i \in I_c} \omega_i \qquad (p_c),$$

$$y_j \in Y_j,$$

and with a feedback relation that sets $\hat{x}_i$ by solving the consumer problem:

$$\hat{x}_i = \operatorname{argmax}\{u_i(x_i) \text{ subject to } p_c x_i \leqslant p_c \omega_i + \Sigma_{j \in J_c} \theta_{ij} \Pi_j(p_c) + \Pi_i^w, \; x_i \geqslant 0\},$$

all $i \in I_c$. (4.5b′)

In (4.5a′) and (4.5b′), p_c is the vector of shadow prices associated to the

5. Under assumption S there is an exception: Consumer 1 has increasing utility for all commodities and may therefore consume all commodities in all countries, but since his endowments may be taken to be arbitrarily small, this is innocuous.

6. Domestic prices will then follow the normalization imposed by world prices, but this has no effect on equilibrium allocations as long as demand and supply are homogeneous of degree zero in prices.

balance constraints and $\Pi_i^w = \Sigma_{c' \neq c} \Sigma_{j \in J_{c'}} \theta_{ij} \Pi_j(p_{c'})$ is the given profit from abroad received by consumer i. For nontraded commodity k, $\bar{p}_{ck}^e = \bar{p}_k^e$, and $\bar{p}_{ck}^m = \bar{p}_k^m$, where $\bar{p}_k^e$ and $\bar{p}_k^m$ are low and high values for prices; for a traded commodity k, $\bar{p}_{ck}^e = \bar{p}_{ck}^m = p_k^w$.

Like in definition 3.4, it follows from the Kuhn-Tucker conditions for this program that, in equilibrium,

$$\Sigma_{i \in I_c} x_i^* - \Sigma_{j \in J_c} y_j^* - m_c^* + e_c^* + \Sigma_{i \in I_c} \omega_i \leqslant 0 \perp p_c^* \geqslant 0, \tag{4.6a}$$

$$e_c^* \geqslant 0 \perp p_c^* \geqslant \bar{p}_c^e, \tag{4.6b}$$

$$m_c^* \geqslant 0 \perp p_c^* \leqslant \bar{p}_c^m. \tag{4.6c}$$

Existence of equilibrium is obvious:

PROPOSITION 4.2 (Existence of a competitive equilibrium under the small-country assumption) Under assumption S and for given export prices $\bar{p}_c^e$ and import prices $\bar{p}_c^m$, such that $\bar{p}_c^m \geqslant \bar{p}_c^e > 0$, the small-country assumption model of definition 4.2 possesses an equilibrium.

Proof The proof is identical to the one of proposition 3.4, part 1. ■

4.3 Welfare Implications

4.3.1 The International Competitive Equilibrium

Since the international competitive equilibrium of definition 4.1 is a straightforward reinterpretation of the competitive equilibrium, the welfare theorems apply, and we may conclude that the international trade equilibrium is Pareto-efficient.

Many trade studies are concerned with the welfare gains from reducing tariffs and relaxing quotas. This will be discussed in chapter 5, where we will see that gains do not necessarily accrue to the country that liberalizes its trade and that, even if they do, some of its consumers may lose. The second welfare theorem states that if compensating transfers are given, no one needs to lose. What matters for the country concerned is the price change it will face on the international market as a result of its action. In particular, if the small-country assumption holds, it cannot lose.

4.3.2 The Competitive Equilibrium under the Small-Country Assumption

Under the small-country assumption, import and export prices $\bar{p}_c^m$ and $\bar{p}_c^e$ as well as the profits $\Pi_c^w = \Sigma_i \Pi_i^w$ earned from foreign firms are given for the individual country. Using the Negishi format (3.16a), the welfare program can be written

$$\max \Sigma_{i \in I_c} \alpha_i u_i(x_i) + \rho_c(\bar{p}_c^e e_c - \bar{p}_c^m m_c),$$

$$x_i \geqslant 0, \text{ all } i \in I_c, m_c, e_c \geqslant 0, y_j, \text{ all } j \in J_c,$$

subject to (4.7)

$$\Sigma_{j \in J_c} y_j + m_c - e_c \geqslant \Sigma_{i \in I_c} x_i - \Sigma_{i \in I_c} \omega_i \qquad (p_c),$$

$$y_j \in Y_j,$$

where the α_i's and ρ_c are equilibrium welfare weights.

This program shows that under the small-country assumption, the rest of the world can be thought of as having a linear utility function $(\bar{p}_c^e e_c - \bar{p}_c^m m_c)$, often called trade welfare function. Obviously the equilibrium allocation associated with this program is Pareto-efficient.

In equilibrium ρ_c can also be viewed as the Lagrange multiplier[7] associated to the balance-of-payments constraint:

$$(\bar{p}_c^e e_c - \bar{p}_c^m m_c) \leqslant \Pi_c^w.$$

By entering this constraint into the program, we can eliminate the associated welfare term from the objective and obtain a national welfare program, with the balance of payments as a constraint (there is then no need to adjust the welfare weight for the rest of the world, since it is now a Lagrange multiplier). This welfare program represents all possible efficient allocations and shows that any additional constraint on international trade, say, a quota on imports $m_c \leqslant \bar{m}_c$, will reduce welfare for the country concerned.

4.4 Practices in Applied Modeling

In principle, the two types of models (international equilibrium and small-country model) of section 4.2 can be applied without modification. They closely follow the traditional model of international trade and will, if constant returns prevail in production, lead to specialization. If each firm produces a single output under constant returns to scale, and there are several firms producing the same output, then only the most efficient firm will produce, and specialization will occur in the economy. In the context of a world model the same will happen, and countries will end up producing

7. Note that by the Kuhn-Tucker conditions of the resulting program, for commodities for which imports or exports are positive, domestic prices will be proportional to the given world import and export prices, respectively. The multiplier ρ_c only bridges the normalization of welfare weights with that of import and export prices. It should not be viewed as an exchange rate.

only the commodities for which they are more efficient than others. This is the well-known specialization effect induced by international trade.

However, this is a long-run effect only, which applied modelers have tried to avoid because it is not a realistic representation of economies in the short run and because it may lead to widely varying patterns of specializations under minor changes in the parameters of a model. It is the way in which this problem is dealt with in applied general equilibrium models with international trade that gives them a special structure.

In reality, specialization is more gradual than the classical trade model suggests:

1. Capacity constraints may be effective, and if this is so for the most efficient producer, the less efficient ones have an opportunity to produce.

2. If producers operate under decreasing returns to scale, less efficient producers may stay in the market by accepting lower (but still positive) profits than the more efficient ones.

3. Transportation systems create wedges between producers located in different countries; therefore an inefficient producer in an importing country located far away from the efficient producers may be able to survive.

4. Single output production processes are rare in reality; they are intoduced to facilitate modeling by breaking the production system into single output firms. If the whole production system were represented through a transformation function, specialization would rarely come about, though some commodities may not be supplied.

5. Foreign and domestic goods are not homogeneous, and this also reduces the tendency toward specialization.

In the CGE model of section 4.4.1, specialization is avoided via assumptions on heterogeneity of commodities. In section 4.4.2 the problems caused by specialization are treated in other ways. Sections 4.4.3 and 4.4.4 briefly address issues in data organization and simulation.

4.4.1 International Trade in the CGE Model

The representation of international trade in a CGE framework requires a reinterpretation of what was done in section 3.4.1, so here we only present the small-country version of the basic CGE model. It calls for some modifications of the small-country model, since in many applications the country can be assumed to be "small" only as an importer (and to take import prices as given) but to export at domestic prices.[8]

8. These assumptions can and have been modified: Export prices can be taken as given and import prices can be made endogenous.

The basic CGE model is modified by adding imports and exports to the balances for goods (factors are assumed not to be traded). This leads to

$$\Sigma_i x_i^g(p^g, p^f, h_i) + A^g(p^g, p^f)q^g + e^g = q^g + m^g \qquad \text{for goods}$$

and

$$\Sigma_i x_i^f(p^g, p^f, h_i) + A^f(p^g, p^f)q^g = \Sigma_i \omega_i^f \qquad \text{for factors.}$$

This setting implies that imports and exports are perfect substitutes for domestic goods. It is usual to assume that this is not the case and to model imports like intermediate demand for factors, with given prices $\bar{p}^m$ and perfectly elastic supply in the rest of the world. The exported goods do not have perfect substitutes in the rest of the world. The equations become

$$\Sigma_i x_i^g(p^g, p^f, h_i) + A^g(p^g, p^f, \bar{p}^m)q^g + e^g = q^g \qquad \text{for goods,} \tag{4.8a}$$

$$\Sigma_i x_i^f(p^g, p^f, h_i) + A^f(p^g, p^f, \bar{p}^m)q^g = \Sigma_i \omega_i^f \qquad \text{for factors,} \tag{4.8b}$$

$$A^m(p^g, p^f, \bar{p}^m)q^g = m \qquad \text{for imports.} \tag{4.8c}$$

The matrices A^g, A^f, and A^m are intermediate demand coefficients derived from unit cost functions, and are thus functions of p^g, p^f, and $\bar{p}^m$.

To complete the model, three further equations are needed, for income, price formation and exports, respectively:

$$h_i = p^f \omega_i^f + \Pi_i^w, \tag{4.8d}$$

$$p^g = p^g A^g(p^g, p^f, \bar{p}^m) + p^f A^f(p^g, p^f, \bar{p}^m) + p^m A^m(p^g, p^f, \bar{p}^m), \tag{4.8e}$$

$$e^g = \text{argmax}_{e^g \geqslant 0}\{u^w(e^g) \mid p^g e^g = \bar{p}^m m - \Sigma_i \Pi_i^w\}. \tag{4.8f}$$

Equation (4.8d) determines consumer i's income as income from his factor endowment plus income from abroad, which is given. Equation (4.8e) computes the prices for goods as the sum of unit costs from three components: intermediate goods, factors, and imports. Finally, equation (4.8f) determines exports as if the rest of the world were a consumer who only consumes exports and sells imports, with the balance of payments as his budget constraint. Note that we use domestic prices of goods p^g to value exports e^g because it is assumed that the exported goods are not produced by other countries. In practice, demand for exports is often specified in an ad hoc way, as a function, say, $e_k^g(\bar{p}_k^g/\bar{p}_k^m)$ where $\bar{p}_k^m$ are given prices of competing countries.

In applications this approach has two important limitations: First, it disregards boundedness of supply of the rest of the world that is thus able to supply unlimited amounts; second, this formulation can create unboundedness of the model itself because p^g, the price for exports is not fixed;

hence it may happen that the country exports a small amount at an infinitely large price, while importing infinitely large quantities so that no bounded equilibrium exists.

4.4.2 Issues in Model Specification

In principle, the representations in sections 4.2 and 4.4.1 already constitute a complete model of international trade, and many applied models do not go beyond this level of elaboration of international trade. However, such a general approach may bypass characteristics that often need to be addressed in empirical models, in particular, the heterogeneity of commodities produced in different countries. Also there is the difficulty that in the competitive model, consumers and producers do not have to know to whom they sell and from whom they buy. Similarly the multicountry model is unable to explain the commodity flows between countries and assumes that all routes are open, that traffic is not congested, and that transportation costs can be disregarded.

Such assumptions are of course not realistic, and this has led to the development of models with explicit transportation systems. These models can be applied to study the welfare gains resulting from improvements of the transportation system, such as opening new road links or improving capacities of existing roads.

Trade without Transportation Cost

Consider a commodity such as an orange, which can be defined by a bundle of physical characteristics (taste, color, shape), a location of origin (Morocco, Spain), and a location of use (France, the United States). Unless there exist technologies to reshape, say, Moroccan into Spanish oranges, the French consumer is faced with two different commodities, and there is nothing one can do to reduce this list without imposing additional restrictions on consumer behavior.

However, if Moroccan and Spanish oranges offer the same characteristics to the French consumer, all that differentiates them is location of production, and this is overcome once transportation to France has taken place. Transportation can thus be considered as a reshaping technology, and it requires distinct treatment. In applications where one might wish to maintain product heterogeneity and keep track of bilateral trade flows between countries, transportation will be of particular relevance, since trade agreements such as barter trade contracts or preferential treatment of some partners have to be accounted for.[9]

9. Product heterogeneity in characteristics and location can obviously occur within a national model as well. The discussion that follows on the representation of transportation flows also applies within a national context.

For illustration, we consider a three-country example and assume that each country c can obtain its net imports from any other country. Then net imports are given by

$$
\begin{aligned}
z_1 &= \quad\quad +s_{21} \quad +s_{31} \quad\quad\quad -s_{12} \quad -s_{13}\\
z_2 &= +s_{12} \quad\quad\quad +s_{32} \quad -s_{21} \quad\quad\quad -s_{23}\\
z_3 &= +s_{13} \quad +s_{23} \quad\quad\quad -s_{31} \quad -s_{32}
\end{aligned}
$$

where $s_{cc'} \geqslant 0$ is the flow (a vector whose dimension is the number of commodities r) from c to c'.[10] This can trivially be written

$$z_1 = 0s_{11} + \mathrm{I}s_{21} + \mathrm{I}s_{31} - 0s_{11} - \mathrm{I}s_{12} - \mathrm{I}s_{13},$$

$$z_2 = \mathrm{I}s_{12} + 0s_{22} + \mathrm{I}s_{32} - \mathrm{I}s_{21} - 0s_{22} - \mathrm{I}s_{23},$$

$$z_3 = \mathrm{I}s_{13} + \mathrm{I}s_{23} + 0s_{33} - \mathrm{I}s_{31} - \mathrm{I}s_{32} - 0s_{33},$$

where 0 and I are zero and unit matrices, respectively (of dimension r).

Constrained routing sets bounds on $s_{cc'}$. In the extreme case $s_{cc'}$ is set to zero, which is a means of barring route cc', reflecting the technical impossibility or the undesirability of the corresponding flow. If, for instance, the route between 1 and 2 becomes a one-way street, we simply drop s_{12} from the above equations, which amounts to replacing an I matrix by a 0 matrix. Alternatively, and more generally, we can write

$$z_c = \Sigma_{c'} R_{c'c} s_{c'c} - \Sigma_{c'} R_{cc'} s_{cc'},$$

$$0 \leqslant s_{cc'} \leqslant \bar{s}_{cc'}, \text{ all } c, c',$$

where $R_{cc'}$ is a diagonal matrix with elements 0 or 1.

Trade with Transportation Cost

In the real world, transportation of a commodity always requires inputs (at least energy); let $T^m_{c'c}$ and $T^e_{cc'}$ represent country c's matrices of inputs associated to its import and export vectors; such matrices can contain several techniques (columns) for a single route (air, rail, etc.), implying that they may be rectangular, though of the same dimension. In our earlier example, a column of R would contain at most one nonzero entry, but it may be realistic to allow for joint transportation, if only because the vehicle always travels with the transported commodity. This may be formalized as

$$z_c = \Sigma_{c'}(R_{c'c} + T^m_{c'c}) s_{c'c} - \Sigma_{c'}(R_{cc'} - T^e_{cc'}) s_{cc'},$$

$$0 \leqslant s_{cc'} \leqslant \bar{s}_{cc'}, \text{ all } c, c'.$$

10. Note that this flow may again leave c' for another destination (transit trade).

It is now possible to incorporate transportation into the international model. This amounts to replacing the international balances (4.2a) by

$$z_c \leqslant \Sigma_{c'}(R_{c'c} + T^m_{c'c})s_{c'c} - \Sigma_{c'}(R_{cc'} - T^e_{cc'})s_{cc'} + \sigma^m_c - \sigma^e_c,$$

$$0 \leqslant s_{cc'} \leqslant \bar{s}_{cc'},\ \sigma^m_c,\ \sigma^e_c \geqslant 0,\ \text{all } c,\ c', \tag{4.2a$'$}$$

$$m = \Sigma_c \sigma^m_c;\ e = \Sigma_c \sigma^e_c,$$

where σ^m_c and σ^e_c are slack variables that simply ensure feasibility and nonnegativity of the dual prices in the program;[11] for $\bar{p}^e$ chosen sufficiently close to zero, they will be equal to zero in equilibrium.[12] The equations that define m and e link the constraints of the program to the objective function $(\bar{p}^e e - \bar{p}^m m)$. Transportation costs will make country-specific prices p_c differ. Note that excessive input requirements and bounds on trade flows may bring a country in autarky for some commodities.

In (4.2a′) the R and T matrices contain fixed parameters, implying constant returns to scale and no input substitution within each transportation activity. This can readily be extended to cases with decreasing returns to scale and substitution. The matrices would then be functions of scale (here, trade flows) and prices.

The Kuhn-Tucker conditions of the mathematical program with (4.2a′) yield the following costing and pricing restrictions for every commodity:

1. Under constant returns to scale in transportation, the cost of transportation will be equal to the cost of inputs involved (zero profits).

2. For any positive flow the price at the destination will be equal to the price at the source, incremented by the cost of the transportation activity.

3. At any location, for a given commodity, there will be a single price for each incoming and outgoing flow.

4. Under constant returns to scale, transportation flows may become indeterminate (set-valued); for instance, since all modes of transportation will be competitively priced, all modes used will make zero profit, and it may be impossible to determine a unique route or mode of transportation.[13]

These implications are disappointing: Competition among suppliers of transportation services may equalize the price of a commodity from several origins, making it impossible for an importing country to choose between

11. For existence of an equilibrium with zero slacks, some viability assumption is required. An admittedly overrestrictive but convenient assumption would be that each country's economy is viable in complete autarky. To alleviate this restriction, one would need assumptions on the R and T matrices so as to make international trade possible.

12. See proposition 4.1.

13. See Elbers (1992).

suppliers. The outcome may thus not be more informative than the model without transportation matrices.

The transportation framework is seldom used in applications: It requires a vast amount of information, and the modeler may find it unimportant or impossible to represent bilateral trade. Three different techniques (which may complement each other) have been developed to circumvent these difficulties. The first is the "trade-pool approach," which gives a simple structure to the transportation system itself; the second differentiates commodities by location, and the third focuses on a particular set of countries (often a single one) and aggregates the rest of the world via a trade welfare function.

The Trade-Pool Approach

To circumvent the indeterminacy of trade flows mentioned above, the most common approach is to consider a pool into which all exports flow and from which all imports originate; this trivially eliminates indeterminacy, giving up the representation of bilateral flows. Still transportation routes and costs as well as constraints to and from the pool can be considered. The specification can be simplified further by ignoring alternative modes of transportation, so that the R, T^e, and T^m matrices of (4.2a′) become square. The transportation system can then be written

$$\begin{aligned}
&\Sigma_c m_c \leqslant \Sigma_c e_c \qquad (p^w),\\
&z_c \leqslant (R_c + T_c^m)m_c - (R_c - T_c^e)e_c + \sigma_c^m - \sigma_c^e\\
&0 \leqslant m_c \leqslant \bar{m}_c,\ 0 \leqslant e_c \leqslant \bar{e}_c,\ \sigma_c^m,\ \sigma_c^e \geqslant 0, \text{ all } c,\\
&m = \Sigma_c \sigma_c^m;\ e = \Sigma_c \sigma_c^e.
\end{aligned} \qquad (4.2a'')$$

Equation (4.2a″) replaces (4.2a′). Since all variables in the second and third equations of (4.2a″) relate only to country c, these equations can be treated as part of the country component of (4.2).

Under the trade-pool approach, each national model can now be obtained by introducing the transportation system (dropping the slack variables) into (4.5). Building on definition 4.2, this yields the following mathematical program with feedback:

$$\max \bar{p}_c^e e_c - \bar{p}_c^m m_c,$$

$$m_c,\ e_c \geqslant 0,\ y_j \text{ all } j \in J_c,$$

subject to

$$\Sigma_{j\in J_c} y_j + (R_c + T_c^m)m_c - (R_c - T_c^e)e_c \geqslant \Sigma_{i\in I_c} \hat{x}_i - \Sigma_{i\in I_c} \omega_i \qquad (p_c),$$

$$y_j \in Y_j,$$

$$0 \leqslant m_c \leqslant \bar{m}_c,\ 0 \leqslant e_c \leqslant \bar{e}_c.$$

A feedback relation sets $\hat{x}_i$ by solving the consumer problem

$$\hat{x}_i = \text{argmax}\{u_i(x_i) \text{ subject to } p_c x_i \leqslant p_c \omega_i + \Sigma_{j \in J_c} \theta_{ij} \Pi_j(p_c) + \Pi_i^w,\ x_i \geqslant 0\},$$

all $i \in I_c$.

This approach also applies to the two-country and, obviously, to the small country settings.

Heterogeneous Commodities

As mentioned earlier, transportation is a means of homogenizing commodities that only differ by location and are otherwise physically identical: Liquid methane from Algeria and from Indonesia should fetch the same price in Marseilles. However, Guatemalan and Brazilian coffee will be priced differently. And this is the type of heterogeneity we focus on here.

The modeler can of course disregard heterogeneity altogether and treat foreign and domestic commodities as perfect substitutes. However, if he decides to account for it, he may find it practical to distinguish different commodity lists in various parts of the model, using nested utility and cost functions.[14] In the context of international trade, this nesting is known as the Armington approach.[15]

To illustrate the approach, we consider a three-country world in which each country produces a specific commodity that it may use domestically or export. On the demand side, country 1 can use its own (domestic) as well as the two other (foreign) goods in consumption and/or production. We can often assume that the domestic and the foreign goods are used in the same proportions everywhere in the national economy; hence they constitute a composite commodity generated through an aggregation function $f(p)$. For the two-level CES this is

$$f(p_1, p_f) = [\alpha_i p_1^{-\rho} + \alpha_f p_f^{-\rho}]^{-1/\rho},$$

where

$$p_f = [\alpha_2 p_2^{-\sigma} + \alpha_3 p_3^{-\sigma}]^{-1/\sigma}.$$

The quantities of each of the three goods per unit of the composite good can be computed by applying Shephard's lemma,[16] that is, by taking derivatives of the aggregator function $f(p_1, p_f(p_2, p_3))$ with respect to prices. Within country 1, we now work with the composite commodity; as soon as we turn to imports and exports, the commodities must again be specified separately.

14. See sections 2.1.4 and 2.2.5.

15. See Armington (1969).

16. See proposition 2.4, assertion 5.

On the supply side, the country clearly can only produce its own commodity,[17] and it is the domestic (not the composite) commodity that is exported. Hence exports and imports are treated as different commodities and will not appear in the same commodity balance.

The decisions on the representation of heterogeneity have important consequences for the way in which data have to be organized and processed. When there is heterogeneity, it becomes meaningless to calculate price differences between goods; if transportation (or processing) is the only differentiating factor, it is possible to measure the flows in a common unit and interpret price differences as trade or transportation margins.

On the basis of statistical observations alone, one cannot determine which part of a price difference is attributable to heterogeneity and which to transportation costs. However, time series may help clarify this choice. We can indeed test whether the assumption of nonheterogeneity holds on the basis of a cost function that includes transportation (and processing) costs.

Often the modeler has little choice in his treatment of heterogeneity because he has to accept the assumptions made by the statistician at the time the data were compiled. Quantity data are often presented in the form of commodity balances (supply-utilization accounts) where supply, demand, imports, and exports match. Thus imports and exports are additive with respect to demand and production, which implies perfect substitution (these data are already the result of an aggregation, e.g., in metric tons, performed by the statistical offices). The database may also consist of time series of price indexes and values that balance supply and demand, including foreign trade. Aggregate quantities are then obtained through division of values by appropriate price indexes, which take the form of a cost function[18] and which reflect the assumptions made on heterogeneity.

Representation of the Rest of the World by a Trade Welfare Function

To simplify the world model, we can describe the rest of the world by representative agents: one or several producers and one consumer who maximizes utility under his budget constraint. In such a simplified context, neither the production system nor the preferences of the consumer can be modeled explicitly in a realistic way. This has led to the more implicit

17. Its consumers may own shares of foreign firms, but the commodities produced by those firms will be "foreign."

18. The indexes should be defined in a consistent way. The value index is always $\Sigma_k p_{kt} q_{kt} / \Sigma_k p_{ko} q_{ko}$; the price index multiplied by the quantity index should generate this. For the price index one could choose the so-called Paasche index $\Sigma_k p_{kt} q_{kt} / \Sigma_k p_{ko} q_{kt}$, while the quantity index would be Laspeyres: $\Sigma_k p_{ko} q_{kt} / \Sigma_k p_{ko} q_{ko}$. In this setup initial prices can be set to one without loss of generality. Observe that this means that perfect substitution is assumed within the aggregate, using the base year prices as weights. Assuming this means that one attributes physical properties to base-year prices, treating them as if they would reflect a product characteristic like calorie or protein content.

treatment through the trade welfare function,[19] which is obtained as the value function $W_0(\omega_0 + m_0 - e_0)$ of the program

$$\max\{u_0(x_0) \mid x_0 - \Sigma_{j\in J_0} y_j \leqslant \omega_0 + m_0 - e_0, y_j \in Y_j\},$$

$$e_0, m_0, x_0 \geqslant 0,\ y_j, \text{ all } j \in J_0.$$

If the utility function $u_0(x_0)$ satisfies assumption C1′, if ω_0 is nonnegative and nonzero, and if the production sets Y_j satisfy assumption P, then the value function $W_0(.)$ will be continuous, strictly concave, and nonsatiated[20] and can be used as a "utility" function for the rest of the world. The budget constraint of the rest of the world is

$$p^w m_0 - \Pi_0(p^w) \leqslant p^w e_0,$$

where $\Pi_0(p^w)$ is the given net transfer of profits originating from the rest of the world.

Net import demand by the rest of the world is now

$$(m_0 - e_0) = \operatorname{argmax}_{m_0, e_0 \geqslant 0}\{W_0(\omega_0 + m_0 - e_0) \mid p^w m_0 - \Pi_0(p^w) \leqslant p^w e_0\}.$$

As can be seen from these expressions, the trade welfare function only generates net import $(m_0 - e_0) = z_0$ and not imports and exports separately.

The trade welfare function has been used in various applications but many researchers have taken the more pragmatic approach of specifying imports and exports directly as ad hoc demand and supply functions.[21]

4.4.3 Empirical Elaboration: The SAM for a Country Model with International Trade

We only describe the structure of the SAM for the small country model of definition 4.2, since for the international trade model of definition 4.1 it can be obtained by reinterpreting the SAM of section 3.6.1, with agents grouped by country. We follow the steps of section 3.6.1, starting from the analogue of equations (3.20).[22] These equations are

For market equilibrium conditions,

$$\Sigma_{i\in I_c} x^*_{ik} - \Sigma_{j\in J_c} y^*_{jk} - m^*_{ck} + e^*_{ck} - \Sigma_{i\in I_c} \omega_{ik} \leqslant 0 \perp p^*_{ck} \geqslant 0, \qquad \text{all } k. \tag{4.9a}$$

For profits of domestic producers,

$$\Pi^*_j = p^*_c y^*_j, \qquad \text{all } j \in J_c. \tag{4.9b}$$

19. See, for example, Negishi (1972).

20. See theorems A.3.1 and A.3.2.

21. It may be noted that the concept of the trade welfare function can be avoided if one estimates directly the net demand functions derived from $\max_{x_0 \geqslant 0}\{u_0(x_0) \mid p^w x_0 \leqslant p^w \omega_0\}$.

22. Note that this model abstracts from transportation costs.

For budget equations of domestic consumers,

$$p_c^* x_i^* = p_c^* \omega_i + \Sigma_{j \in J_c} \theta_{ij} \Pi_j^* + \Pi_i^w, \qquad \text{all } i \in I_c. \tag{4.9c}$$

For the budget equation of the rest of the world (foreign sector),

$$p_c^* e_c^* + \Sigma_{i \in I_c} \Pi_i^w = p_c^* m_c^*. \tag{4.9d}$$

The complementarity conditions (4.9a) imply that

$$p_{ck}^* \Sigma_{i \in I_c} x_{ik}^* + p_{ck}^* e_{ck}^* = p_{ck}^* \Sigma_{j \in I_c} y_{jk}^* + p_{ck}^* \Sigma_{i \in J_c} \omega_{ik} + p_{ck}^* m_{ck}^*, \tag{4.9e}$$

so summing (4.9c) over consumers and subtracting (4.9e) implies (4.9d). In other words, satisfaction of budgets for every domestic consumer implies that the budget equation for the rest of the world (the balance of payments) is satisfied.

To obtain the SAM in tabular form, we refer to the SAM of section 3.6.1, extending it with a column and a row for the foreign sector in accordance with equation (4.9d), which of course affects the other columns and rows, as shown in Table 4.1. The result is a SAM in value terms. The commodity balances are in quantity terms, so unit values p_c^* can be computed. However, we must still account for the complementarity conditions (4.6b)–(4.6c):

$$e_{ck}^* \geqslant 0 \perp p_{ck}^* \geqslant \bar{p}_{ck}^e, \tag{4.9f}$$

$$m_{ck}^* \geqslant 0 \perp p_{ck}^* \leqslant \bar{p}_{ck}^m. \tag{4.9g}$$

If separate information is available for the three prices in (4.9f)–(4.9g), and statistics are recorded for both exports and imports (and not only for net imports), then these two equations can lead to conflict between the data and model structure adopted which can be resolved as follows: As is often the case, both imports and exports are positive for a given commodity, especially if the commodity is highly aggregated. To avoid the conflict, one could either net out import and exports or treat imports and exports under two different commodities as in the CGE model. Then, if price constraints are still violated, the foreign price should be set equal to the domestic price for the positive flow and the other foreign price set away from its bound.

However, either effort could lead to unrealistic foreign prices, especially over time. If time series are available for prices, one could introduce trade and processing margins (booked in the SAM as services provided by the trade and transportation sectors) to account for the discrepancies. In practice, this is often not satisfactory because these margins should not fluctuate excessively over time, and such consistency checks often lead to several rounds of revisions of other parts of the SAM.

Table 4.1
Structure of the SAM for the small-country AGE model

	Goods	Factors	Firms	Consumers	Foreign sector
Goods			Input	Consumption	Exports
Factors			Input	Consumption	(Exports)
Firms	Production	(Production)			
Consumers	(Endowments)	Endowments	(Profits)		Foreign profits
Foreign sector	(Imports)	Imports			

Note: In the SAM for the CGE model, the terms in parentheses are zero.

4.4.4 Simulations

Solving the Model

In applications the national models are well-identified entities, and the international model is no more than the linkage of these models through international commodity balances that clear by adjustment of international prices. It is tempting to solve each national model separately in (4.5a′–4.5b′), calculate the prices of nontradables for given prices of internationally traded commodities, compute the total net excess demand (here net imports), and apply a fixed point algorithm to adjust international prices until, in (4.4) the excess demand vanishes, thus treating each country's net import as the net demand of an agent. The approach would be appropriate if the net demands were known to be continuous functions (or uppersemicontinuous compact and convex-valued correspondences). This condition is fulfilled if there is only one consumer for each country, since the net demand is then obtained as the solution of a single mathematical program,[23] which, by the maximum theorem A.3.1, has these properties. When there are several consumers, however, this property does not hold unless each national model has a unique equilibrium at all possible international prices. As discussed in section 1.2.3, this can only be ensured under very restrictive conditions. Therefore, to compute the equilibrium of an international model with many consumers in any country, one has to adjust simultaneously all prices for tradables and nontradables.

Running Simulations

Applied models of international trade are always used in conjunction with trade policies, which involve quota, tariffs and other taxes, and subsidies. The discussion of applications and simulations of policy alternatives is thus postponed to section 5.4 of the next chapter.

23. See section 1.4.5.

5 Taxes, Tariffs, and Quotas

Applied general equilibrium models are often used to analyze the consequences of tax and trade reforms and the restructuring of public spending. Taxes can be levied on various types of transactions, and section 5.1 provides a short typology. Equilibrium under taxes (on consumption and production) and tariffs (on exports and imports) is studied in section 5.2, where we apply the open economy format introduced in section 3.3; this makes it possible to study existence for all possible combinations of taxes and tariffs within the same setting. Taxes have many effects that can hardly be foreseen on theoretical grounds only and require numerical exercises. Theory provides some clearcut answers on efficiency, but only for very drastic or specific tax reforms. This is our focus in section 5.3, where we use the Negishi format. By proposition 1.11 we obtain the standard results on welfare gains from abolishing taxes. More important, proposition 1.11 will also lead to the less widely known formulation of gradual Pareto-improving reforms. Yet although such theoretical results give important information on what to expect from such reforms, applied policy analysis, treated in section 5.4, can give answers in cases not covered by the theorems. The taxes we consider are levied on consumers, producers, and international trade and call for a specific treatment in each case, but some repetition is unavoidable.

5.1 Government and Taxation

5.1.1 Government

In this chapter we assume that one of the consumers is the national government. The government purchases and sells at given prices, receives income from its endowment and from profits, like any other consumer; hence it may own public firms, but these firms will operate like private ones and not as price setters. Moreover the government is endowed with the power of imposing taxes, subsidies, and quantity constraints (quotas). It uses the proceeds of taxation to finance its own public consumption and to redistribute the remainder to private consumers. The preferences of the government can be given by a utility function $u_g(x_g)$ independently from the preferences of the other consumers, and government maximizes this utility subject to its budget constraint. In chapter 9, where we consider the case of nonrival goods, public demand is an argument of the consumers' utility functions: Consumer i will have a utility function $u_i(x_i, x_g)$ and choose x_i as well as x_g (note that x_g is the same for all consumers and will have to be agreed upon collectively).

5.1.2 A Short Typology of Taxes

A first common distinction is between indirect and direct taxes. Indirect taxes (resp. subsidies) introduce a wedge between prices at the various stages

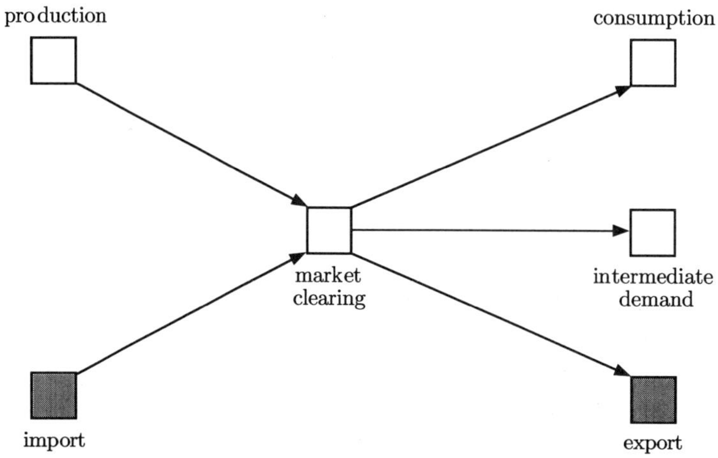

Figure 5.1
Price wedges between supply and utilization.

of supply and utilization of a commodity. Direct taxes (resp. payments) take away (give) part of income and profits from (to) the agents; thus direct payments can be used to distribute the proceeds from indirect taxes. The price wedges generated by indirect taxes are indicated as the boxes in figure 5.1.

As is seen from figure 5.1, price wedges can be differentiated according to the point at which they occur. When they impact on exports or imports (the shaded boxes), we call them tariffs (or subsidies); otherwise, we call them indirect taxes (or subsidies). Price wedges can result from various functional specifications, which do usually not follow from theory but merely reflect institutional settings. For example, when the tax rate is fixed, the consumer price p_k^c of commodity k is (in the absence of any other processing cost) related to the clearing price p_k (i.e., the price associated with the market equilibrium condition) according to

$$p_k^c = (1 + \tau_k^c)p_k;$$

the rate τ_k^c is positive for a tax and negative for a subsidy (with $\tau_k^c > -1$, for otherwise the consumer price of commodity k would be zero or negative). In the remainder of this chapter, a subsidy will be a tax with a negative value, and we will no longer distinguish between both.[1]

1. Subsidies may cause special problems because they are an expenditure item of government that has to be financed through revenue from other sources, in particular, from direct taxes collected from consumers. The problem is that one has to verify that consumers' income after taxes remains positive.

The budget constraint of consumer i will thus read as

$$\Sigma_k (1 + \tau_k^c) p_k x_{ik} \leqslant h_i.$$

The tax is ad valorem if τ_k^c is fixed. It is nominal if $\tau_k^c p_k$ is fixed, and it is a variable levy if τ_k^c is adjusting so as to keep the consumer price fixed. In short, τ_k^c can be specified as an arbitrary continuous function of, say, prices and quantities consumed. However, to maintain homogeneity of demand in terms of clearing prices, this function has to be homogeneous of degree zero in prices. Note that nominal taxes and variable levies do not fulfill this requirement.

In this setup taxes on commodity k are imposed on total consumption at the same rate; it is also possible to distinguish different tax rates, say, a zero or a subsidized rate for consumption below some ration $\bar{x}_k$ and a positive rate beyond. The consumer problem is then

$$\max u(x^1 + x^2),$$

$$x^1, x^2 \geqslant 0,$$

subject to

$$p^1 x^1 + p^2 x^2 \leqslant h,$$

$$x^1 \leqslant \bar{x}^1.$$

Here p^1 is the price for the untaxed or the subsidized part of consumption and p^2 the tax-ridden price. For example, $\bar{x}^1$ could be equal to the endowment, so the consumption out of own endowment would be untaxed.

Taxes can also be made consumer (or producer) specific, for example, to subsidize old or disabled consumers or to help a specific group of producers, say, farmers or steel mills.

Sometimes government will impose quantity constraints on transactions, such as bounds on imports (import quotas). In a competitive setting these will create wedges between the world and the domestic price; the quota can be interpreted as rights to import and can be distributed among importers through import licenses, to which importers can subscribe. The revenue of the bids will accrue to government and may be treated as a tariff revenue. Hence in equilibrium there will be a tariff (more generally a tax) equivalent to such a quota.[2]

Indirect taxes and quotas are introduced as exogenous parameters in order to create price wedges; they do not follow from any principle of efficient allocation, and as we will see, they will tend to reduce welfare. Since

2. In many real world situations these quotas are not auctioned on a regular basis, and the revenue so generated accrues to the owners rather than to the government.

they are nonetheless so common in real life, we have to allow for them in general equilibrium models. Among the reasons for imposing them, we list the following: supporting income of agents by means of prices rather than by income transfers, protecting a particular industry from the effects of international competition, avoiding or promoting the use of specific commodities (e.g., alcoholic beverages vs. primary education), affecting international prices by discouraging domestic production of a commodity in which the country has a large market share, and financing government spending.

Government will use the proceeds of taxes to cover public consumption x_g, price subsidies $-\tau_k^c$, and lump-sum transfers T_i. The income h_i of consumer i now contains a lump-sum transfer T_i and may be written

$$h_i = h_i^0 + T_i, \tag{5.1}$$

where $h_i^0 = p\omega_i + \Sigma_j \theta_{ij} \Pi_j(p)$ is the income before taxation. Transfers T_i are distributed via functions that represent the income tax schedule (a negative T_i is an income tax) and satisfy the following assumption:

ASSUMPTION T (Transfer distribution) Lump-sum transfers (also called direct taxes) are distributed according to $T_i(p, T, h_1^0, h_2^0, \ldots, h_m^0)$ for all i. This function satisfies the following properties

1. Continuity.
2. Homogeneity of degree 1 in $(p, T, h_1^0, h_2^0, \ldots, h_m^0)$.
3. Adding-up: $\Sigma_i T_i(.) \geqslant T$, with equality whenever $T + \Sigma_i h_i^0 > 0$.
4. Positiveness: $h_i^0 + T_i > 0$.
5. There exists a constant $\bar{\eta} \in (0, 1)$ such that if $T + (1 - \bar{\eta})\Sigma_i h_i^0 \leqslant 0$, then all consumers are taxed at the same proportional rate so that for all i, $h_i = \eta h_i^0$ for $\eta \in [0, \bar{\eta}]$.

Requirements 1, 2, and 4 ensure continuity and homogeneity of demands in prices, requirement 3 should hold with equality in equilibrium, to have the government redistribute what it collects, while out of equilibrium, positiveness of income may necessitate the restriction to hold with strict inequality. If there are only taxes, conditions 3 and 4 are always satisfied, and condition 5 is not needed. All that condition 5 does is specifying how the positiveness in after-tax income is maintained if almost all consumers' income has to be taxed away to finance price subsidies or public consumption.

We can now rewrite (5.1) as

$$h_i = p\omega_i + \Sigma_j \theta_{ij} \Pi_j(p) + T_i(.). \tag{5.1$'$}$$

Note that in the absence of taxes, the "income taking" behavior of the consumer is an immediate consequence of price-taking behavior, so it does not require separate discussion. When taxes are imposed, we have to make the specific assumption that consumers, including government, do not anticipate the impact on T (and on T_i) of their own consumption, and take T_i and thus h_i as given.

Functions satisfying the properties in assumption T can be constructed in an ad hoc way, but they can also be derived from maximization of a social welfare function $W(u_1, u_2, \ldots, u_m)$, where government is one of the agents and the contribution of public consumption to social welfare is postulated a priori[3] as in section 2.2.5, definition 2.2.

As shown in figure 5.1, indirect taxes can also be levied on inputs, outputs, imports, or exports. Direct taxes (transfers) can be levied on producers, but as long as they are lump sum, they will not affect production decisions and can therefore be treated as lump-sum taxes on the consumers that own the firms.

5.1.3 Taxes May Affect Allocations

Direct taxes on producers affect profits. Since the objective of the producer is profit maximization, one may think that direct taxes will not affect his plans. Indeed, as long as the marginal tax rate is less than unity, the producer problem $\max\{py_j - \tau(py_j, p),\ y_j \in Y_j\}$ yields the same optimal production plan y_j as $\max\{py_j,\ y_j \in Y_j\}$. However, if a producer operates several firms and the marginal tax rates are not the same for all firms, then $\max\{\Sigma_j py_j - \tau(\Sigma_j py_j, p),\ y_j \in Y_j\}$ may lead to another allocation than profit maximization by each firm separately.[4] This indicates that if direct taxes are imposed in a non lump-sum way, the decentralization of producers into firms may affect allocations and hence lead to inefficiencies. Similar difficulties emerge for the consumer himself and for the decentralization between consumer and producer decisions.

Consider a (private) consumer whose only endowment consists of labor; let x_{i1} be his demand for leisure, ω_{i1} the labor endowment, p_1 the wage rate, and τ_1 the rate of direct tax. Ignoring income from profits, the consumer earns $h_i = (1 - \tau_1)p_1(\omega_{i1} - x_{i1})$, where $(\omega_{i1} - x_{i1})$ is his labor supply. If the consumer is unaware of this relation, his decision problem amounts to solving $\max\{u_i(x_i)$ subject to $\Sigma_{k \neq 1} p_k x_{ik} \leqslant h_i\}$; if, however, this relation is anticipated, his budget constraint must be rewritten as

$$\Sigma_{k \neq 1} p_k x_{ik} \leqslant (1 - \tau_1)p_1(\omega_{i1} - x_{i1})$$

3. As can be checked, these functions satisfy the conditions in assumption T, where positiveness can be maintained by assuming that $W(\,.\,) = 0$ whenever $u_i(\,.\,) = 0$ for some i.

4. This will of course not happen if tax rates on profits are common to all firms, or if every producer operates one firm only.

or

$$(1 - \tau_1)p_1 x_{i1} + \Sigma_{k \neq 1} p_k x_{ik} \leqslant (1 - \tau_1)p_1 \omega_{i1}.$$

As this equation shows, the direct tax contains an element that, on the left-hand side of the budget constraint, operates as an indirect tax, so an agent can "avoid" taxes: Direct taxes can have an influence on allocations. If the tax affects variables that cannot be influenced by the agent, like his endowment (e.g., land or real estate), then taxation will not affect allocations and will act like a lump-sum transfer.[5]

To illustrate the effect of decentralization, assume that a consumer owns a farm, whose profit is taxed. He has the choice between buying his food on the market (x^c) or consuming his own crop (x^s). The problem can be written

$$\max u(x^c + x^s),$$

$$x^c, x^s \geqslant 0, y,$$

subject to

$$px^c \leqslant p\omega + \Pi,$$

$$\Pi = (1 - \tau)p(y - x^s),$$

$$x^s \leqslant y + \omega,$$

$$y \in Y.$$

For $\tau = 0$ the producer and consumer decisions can be decentralized in the usual way, since the linking constraint $x^s \leqslant y + \omega$ will always have a zero shadow price as the consumer can buy outside at the same price. For $\tau > 0$, however, it is evident that the farmer will prefer consuming from own production, and this will affect his production plan. In this case decisions are not decentralizable through prices only.[6]

Applied models combine consumer and producer taxes with tariffs and sometimes with quotas. For expository purposes, in sections 5.2 and 5.3 we treat indirect taxes on consumption, production, and international trade separately. They will be combined in section 5.4 where model implementation is discussed.

5.2 Taxes and Quotas in General Equilibrium

It is intuitive that a general equilibrium with taxes, tariffs, and quotas will exist, provided that the proceeds are redistributed among consumers and

5. As long as consumers do not anticipate this, namely in a single-period context.

6. If the shadow price on the bound $x^s \leqslant y + \omega$ is treated as a market price, then decentralization is possible in a mathematical sense, but clearly such a bound cannot be considered as a balance constraint of a market that is competitive in the conventional sense.

leave them with a positive income. Since for taxes, the theoretical discussion is mainly concerned with efficiency properties of tax reforms, the reader can proceed to section 5.2.4 where the main specifications are summarized. For the sake of completeness and to suggest computational schemes, we nonetheless give the full definitions of equilibrium in every case as well as an existence proof for consumer taxes; we sketch how this proof can be amended for the other cases. The issue that needs special attention in the proof is preventing consumers' income from becoming negative when lump-sum transfers are excessively negative (due to the financing of price subsidies).

5.2.1 Indirect Taxes on Consumption

We consider fixed[7] indirect tax rates τ_k^c on the consumption of commodity k and define tax-ridden prices $p_k^c = (1 + \tau_k^c)p_k$. These enter into the budget constraint of each consumer as $p^c x_i \leqslant h_i$, where h_i is after tax income:

$$p^c x_i \leqslant h_i^0 + T_i(p, T, h_1^0, h_2^0, \dots, h_m^0); \tag{5.2}$$

as before, $h_i^0 = p\omega_i + \Sigma_j \theta_{ij} \Pi_j(p)$ is pre-tax income.

We assume that government operates as a consumer in accordance with (5.2); it also ensures the proceeds from consumer taxes $\Sigma_k \tau_k^c p_k \Sigma_i x_{ik}$ to be fully redistributed to consumers so that all budgets balance:

$$T = (p^c - p)\Sigma_i x_i. \tag{5.3}$$

We now define an open (and closed) economy equilibrium with consumer taxes:

DEFINITION 5.1 (Equilibrium with consumer taxes) The allocation y_j^*, all j, x_i^*, all i, m^* and e^* supported by the price vector $p^* \geqslant 0$ and a level of transfers T^* is an equilibrium with consumer taxes for the open economy if:

1. for every producer j, $y_j^* \in \operatorname{argmax}\{p^* y_j \text{ subject to } y_j \in Y_j\}$ and profit is Π_j^*;
2. for every consumer i, $x_i^* = \operatorname{argmax}\{u_i(x_i) \text{ subject to } p^{c*} x_i \leqslant h_i^*, x_i \geqslant 0\}$, where $h_i^* = h_i^{0*} + T_i^*$ is after tax income and $h_i^{0*} = p^* \omega_i + \Sigma_j \theta_{ij} \Pi_j^*$ is pre-tax income;
3. for every commodity k there is a proportional consumer tax $\tau_k^c > -1$ so that $p_k^{c*} = (1 + \tau_k^c) p_k^*$;
4. proceeds from consumer taxes are fully redistributed: $T_i^* = T_i(p^*, T^*, h_1^{0*}, h_2^{0*}, \dots, h_m^{0*})$, where $T^* = (p^{c*} - p^*)\Sigma_i x_i^*$;

7. These are fixed ad valorem indirect taxes. Any continuous tax function that is homogeneous of degree zero in prices could be used, and proposition 5.1 given below would apply.

5. all markets are in equilibrium:

$$\Sigma_i x_i^* - \Sigma_j y_j^* - \Sigma_i \omega_i + e^* - m^* \leqslant 0 \perp p^* \geqslant 0,$$

$$e^* \geqslant 0 \perp p^* \geqslant \bar{p}^e,$$

$$m^* \geqslant 0 \perp p^* \leqslant \bar{p}^m.$$

This equilibrium reduces to an equilibrium of a closed economy if $e^* - m^* = 0$.

Existence of equilibrium[8] is stated and proved as follows:

PROPOSITION 5.1 (Existence of equilibrium with consumer taxes) Under assumptions S and T:

1. if $\bar{p}^m \geqslant \bar{p}^e > 0$, then there exist allocations x_i^*, all i, y_j^*, all j, a price vector p^* and transfers T^*, that define an open economy equilibrium with consumer taxes as in definition 5.1;

2. all solutions in part 1 will satisfy the condition of trade balance equilibrium: $\bar{p}^m m^* - \bar{p}^e e^* = 0$;

3. if $\bar{p}^m > \bar{p}^e > 0$ and if $\bar{p}^e$ is sufficiently close to 0, the solution will be a closed economy equilibrium with consumer taxes.

Proof *Part 1:* As with proposition 3.4 we start from the open economy format of definition 3.5. Introducing consumer taxes only affects the budgets (5.2) of the consumer problem (the feedback equation) and not the program itself; hence the open economy program is feasible, bounded, and satisfies Slater's constraint qualification.

Step 1. Construction of the Fixed Point Correspondence

The mathematical program of definition 3.5 generates net supplies y_j as primal variables and prices p as dual variables; it may also be viewed as transmitting its demand parameters $\mathring{x}_i$ without changing them. This is illustrated in figure 5.2, which describes the fixed point mapping.

The feedback equation is defined as

$$\mathring{x}_i = \min(\hat{x}_i, \bar{x}_i), \tag{5.4}$$

where

1. $\hat{x}_i = \operatorname{argmax}\{u_i(x_i) \mid p^c x_i \leqslant h_i^0 + T_i(p, T, h_1^0, h_2^0, \ldots, h_m^0), x_i \geqslant 0\}$, with $h_i^0 = p\omega_i + \Sigma_j \theta_{ih} \Pi_j(p)$;

8. Early proofs of existence of equilibrium with consumer taxes have been given by Shoven (1972) and Shoven and Whalley (1974a).

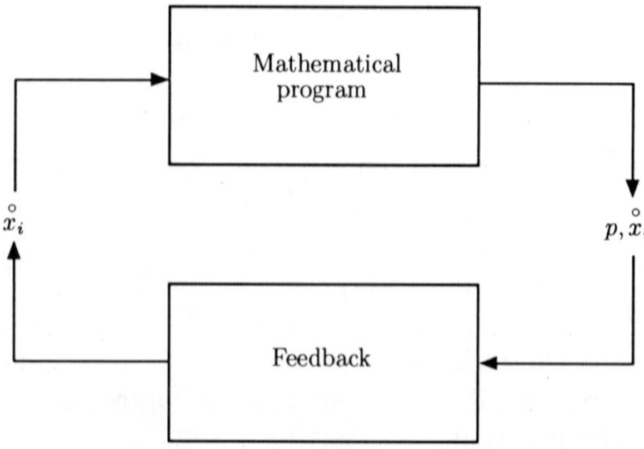

Figure 5.2
Fixed point mapping for equilibrium with consumer taxes.

2. $T = \max\{\Sigma_k \tau_k^c p_k \Sigma_i \mathring{x}_{ik}, T_0\}$, with $T_0 = (1 - \eta)\Sigma_i h_i^0$, for fixed $\eta \in (0, \bar{\eta}]$ and $\eta < (1 + \min_k \tau_k^c)$ (such a value for η exists since $\tau_k^c > -1$);

3. $\bar{x}_i$ is a fixed upper bound for the feasible consumption of i, which may be obtained as follows. We start with the sum of the budget constraints $p^c \Sigma_i x_i \leqslant \Sigma_i h_i$, and substitute (5.2) and (5.3). This yields $p^c \Sigma_i x_i \leqslant p\Sigma_i \omega_i + \Sigma_j \Pi_j(p) + (p^c - p)\Sigma_i x_i$, which reduces to $p\Sigma_i x_i \leqslant p\Sigma_i \omega_i + \Sigma_j \Pi_j(p)$. Since $\Pi_j(p)$ is continuous in p and $\bar{p}^e \leqslant p \leqslant \bar{p}^m$, there is a maximum value $\bar{\Pi}_j$; hence $\bar{p}^e \Sigma_i x_i \leqslant \bar{p}^m \Sigma_i \omega_i + \Sigma_j \bar{\Pi}_j$, and we can set $\bar{x}_{ik} > [\bar{p}^m \Sigma_i \omega_i + \Sigma_j \bar{\Pi}_j]/\bar{p}_k^e$, the required upper bound.

The function (5.4) is continuous. Therefore the remainder of the proof is as in proposition 3.4, part 1.

Step 2. The Fixed Point Is an Equilibrium

By construction, the upper bound $\bar{x}_i$ is nonbinding in equilibrium. To prove that the fixed point is an equilibrium, we need to show that the consumer is on his budget, namely that $\Sigma_i T_i = T$ which is ensured if $T > T_0$.

Suppose, on the contrary, that $T = T_0$ in the fixed point. Then, by assumption T, part 5,

$$p^c \Sigma_i x_i = \eta \Sigma_i h_i^0. \tag{5.5}$$

By the proportionality of consumer taxes,

$$p\Sigma_i x_i \leqslant \delta p^c \Sigma_i x_i, \tag{5.6}$$

where $\delta = 1/(1 + \min_k \tau_k^c)$, which will be positive and bounded since $\tau_k^c > -1$.

Commodity balances and the definition of pre-tax income imply that, in the fixed point,

$$p\Sigma_i x_i = pm - pe + \Sigma_i h_i^0, \tag{5.7}$$

so after substituting (5.5) in (5.6) and then in (5.7), we obtain

$$\delta\eta\Sigma_i h_i^0 \geqslant pm - pe + \Sigma_i h_i^0,$$

which implies that $pm - pe < 0$ since $\delta\eta < 1$. However, from consumer budgets one finds that $p^c\Sigma_i x_i \geqslant (p^c - p)\Sigma_i x_i + \Sigma_i h_i^0$. Hence $p\Sigma_i x_i \geqslant \Sigma_i h_i^0$, so in (5.7), $pm \geqslant pe$, a contradiction. Therefore $T > T_0$, so in assumption T, part 3, equality holds and $T = \Sigma_i T_i$. The remainder of the proof is identical to the one given for proposition 3.4, part 1.

Part 2: In equilibrium Kuhn-Tucker conditions of the open economy program imply that $p^*\Sigma_i x_i^* - p^*\Sigma_j y_j^* - p^*\Sigma_i\omega_i + p^*e^* - p^*m^* = 0$, $p^*e^* = \bar{p}^{e*}e$ and $p^*m^* = \bar{p}^m m^*$ so that

$$p^*\Sigma_i x_i^* = \bar{p}^m m^* - \bar{p}^e e^* + p^*\Sigma_j y_j^* + p^*\Sigma_i\omega_i. \tag{5.8}$$

After substitution of (5.3), the consumer budget equations (5.2) imply that $p^{c*}\Sigma_i x_i^* = p^*\Sigma_j y_j^* + p^*\Sigma_i\omega_i + (p^{c*} - p^*)\Sigma_i x_i^*$, which reduces to

$$p^*\Sigma_i x_i^* = p^*\Sigma_j y_j^* + p^*\Sigma_i\omega_i. \tag{5.9}$$

From (5.8) and (5.9) it follows that

$$\bar{p}^m m^* = \bar{p}^e e^*,$$

proving part 2.

Part 3: When export of commodity k is positive, $p_k = \bar{p}_k^e$. Hence, if $\bar{p}_k^e$ is reduced towards zero, either e_k has to become zero or p_k^c will approach zero together with $\bar{p}_k^e$ since τ_k^c is fixed. As long as there is import for some other commodity h, this commodity generates a pre-tax income so that $h_i^0 \geqslant \bar{p}_h^m\omega_{ih}$, and by assumption T, part 4, after-tax income $h_i \geqslant \eta\bar{p}_h^m\omega_{ih}$. Hence this income does not go to zero as $\bar{p}^e$ approaches zero. By the same arguments as in the proof of proposition 3.4, there can be no exports, but by part 2, this also means that imports vanish and a closed economy equilibrium is obtained. ■

Parts 1 and 2 of the proof are technical, and the complication in part 1 is due to the possibility of T becoming negative if subsidies become large. Part 2 is of direct relevance for applied modeling, since it establishes that trade balance equilibrium follows if there is equilibrium on commodity markets and if both consumers and government satisfy their budget constraints. The proof also shows that whenever commodity markets clear, and two out of the three types of budget equations (consumer budgets, government budget and trade balance) are satisfied, the third follows; we may thus impose balance of trade and consumer budgets, and the government budget will be in equilibrium. This is as in section 4.4.3, where market equilibrium and the satisfaction of budget constraints for every domestic consumer (now including government) implies that the balance of payment equation (i.e., the budget equation of the foreign consumer) is also satisfied.

Proposition 5.1 proves that an equilibrium exists with consumer prices p^c for any set of tax rates (as long as they exceed -1). One could ask the reverse question as to whether there are equilibrium tax rates that support an arbitrarily given set of consumer prices.[9] The proof does not allow us to answer this question in the affirmative. In part 2 we have used the property that p_k^c goes to zero whenever p_k does. Hence it may not be possible to implement an equilibrium with a variable tax rate such that it maintains a given fixed consumer price (i.e., a price rigidity). To ensure existence of such an equilibrium, we need quantity adjustments on the international trade side and, in a closed economy, buffer stocks or rationing. This will be covered in chapter 6.

Tax-ridden prices p^c can also be generated directly through the mathematical program, by transforming the open economy program of definition 3.5 as follows:

$$\max \bar{p}^e e - \bar{p}^m m - \xi^c x,$$

$$m, e \geqslant 0, x, y_j, \text{ all } j,$$

$$\text{subject to} \qquad (5.10)$$

$$x - \Sigma_j y_j - m + e \leqslant \Sigma_i \omega_i \qquad (p),$$

$$x \geqslant \Sigma_i \hat{x}_i \qquad (p^c),$$

$$y_j \in Y_j.$$

In (5.10), x stands for total consumption demand and is introduced to isolate the consumer price p^c. The consumption feedback $\hat{x}_i$ is defined like in part 2 of definition 5.1,[10] and ξ_k^c is determined via the tax feedback equation

$$\xi_k^c = \tau_k^c p_k$$

for $\tau_k^c > -1$.[11]

9. See also proposition 11.1, where we prove that an arbitrary allocation can be supported by markup-ridden prices.

10. When taxes are consumer specific (τ_{ik}^c), there will be separate constraints of the form $x_{ik} \geqslant \hat{x}_{ik}$ in the mathematical program, with dual prices p_{ik}^c associated to each constraint. In the objective function, parameters ξ_{ik}^c will be differentiated by consumer.

11. To keep program (5.10) bounded, the parameter values of ξ^c must be constrained so that $\bar{p}^m + \xi^c > \bar{p}^e$. In equilibrium this condition is equivalent to $\tau_k^c > -1$.

Among the Kuhn-Tucker conditions of program (5.10), we find

$$p \geqslant \bar{p}^e \perp e \geqslant 0,$$

$$p \leqslant \bar{p}^m \perp m \geqslant 0, \tag{5.11}$$

$$p^c \leqslant p + \xi^c \perp x \geqslant 0.$$

By construction, this program together with the feedbacks has a solution that is an equilibrium with consumer taxes as specified in definition 5.1. We will use this representation in relation to welfare analysis.

5.2.2 Indirect Taxes on Production and Intermediate Demand

Indirect taxes on production and intermediate demand can be introduced along similar lines. For this, it is useful to distinguish outputs y_j^+ and inputs y_j^- in firm j's net output vector y_j. We now define an open (and closed) economy equilibrium with taxes on production and intermediate demand (producer taxes for short):

DEFINITION 5.2 (Equilibrium with producer taxes) The allocation y_j^*, all j, x_i^*, all i, m^*, and e^* supported by the price vector $p^* \geqslant 0$ and a level of transfers T^* is an equilibrium with producer taxes for the open economy if

1. for every producer j, y_j^{+*} and y_j^{-*} maximize $(p^{+*}y_j^+ - p^{-*}y_j^-)$ subject to $(y_j^+ - y_j^-) \in Y_j$, $y_j^+, y_j^- \geqslant 0$, with optimal profit denoted by Π_j^*;

2. for every consumer i, $x_i^* = \text{argmax}\{u_i(x_i)$ subject to $p^* x_i \leqslant h_i^*$, $x_i \geqslant 0\}$, where $h_i^* = h_i^{0*} + T_i^*$ is after tax income and $h_i^{0*} = p^*\omega_i + \Sigma_j \theta_{ij} \Pi_j^*$ is pre-tax income;

3. for every commodity k, there are proportional taxes on outputs τ_k^+ (with $\tau_k^+ < 1$) and on inputs τ_k^- (with $\tau_k^- > -1$) so that $p_k^{+*} = (1 - \tau_k^+)p_k^*$ and $p_k^{-*} = (1 + \tau_k^-)p_k^*$;

4. proceeds from producer taxes are fully redistributed:

$$T_i^* = T_i(p^*, T^*, h_1^{0*}, h_2^{0*}, \ldots, h_m^{0*}),$$

$$\text{where} \quad T^* = (p^* - p^{+*})y_j^{+*} + (p^{-*} - p^*)y_j^{-*};$$

5. all markets are in equilibrium:

$$\Sigma_i x_i^* - \Sigma_j y_j^* - \Sigma_i \omega_i + e^* - m^* \leqslant 0 \perp p^* \geqslant 0,$$

$$e^* \geqslant 0 \perp p^* \geqslant \bar{p}^e,$$

$$m^* \geqslant 0 \perp p^* \leqslant \bar{p}^m.$$

This equilibrium reduces to an equilibrium of a closed economy if $e^* - m^* = 0$.

In this definition the proceeds from producer taxes accrue to the government, which redistributes them. They may also accrue to the owner of the firm to represent a "self-imposed" reduction of marginal returns, which is a virtual tax when the wedge arises through a rationing scheme,[12] and a markup to represent missing cost items or imperfect competition.[13]

Existence of equilibrium can be proved along the lines of the proof of proposition 5.1.[14]

Tax-ridden prices p^+ and p^- can, as in (5.10) and (5.11), be obtained by transforming the open economy program of definition 3.5 as follows:

$$\max \bar{p}^e e - \bar{p}^m m - \xi^+ \Sigma_j y_j^+ - \xi^- \Sigma_j y_j^-,$$

$$m, e \geqslant 0,\ y_j^+, y_j^- \geqslant 0,\ y_j \text{ all } j,$$

$$\text{subject to} \tag{5.12}$$

$$\Sigma_j(y_j^+ - y_j^-) + m - e \geqslant \Sigma_i \hat{x}_i - \Sigma_i \omega_i \qquad (p),$$

$$y_j^+ - y_j^- = y_j \qquad (p^f),$$

$$y_j \in Y_j,$$

where the consumption feedback $\hat{x}_i$ is defined like in part 2 of definition 5.1, and ξ_k^+ and ξ_k^- are determined[15] via the tax feedback equations

$$\xi_k^+ = \tau_k^+ p_k, \quad \xi_k^- = \tau_k^- p_k.$$

Among the first-order conditions, we find

$$\begin{aligned} &p \geqslant \bar{p}^e \perp e \geqslant 0,\\ &p \leqslant \bar{p}^m \perp m \geqslant 0,\\ &p^f \geqslant (p - \xi^+) \perp y_j^+ \geqslant 0,\\ &p^f \leqslant (p + \xi^-) \perp y_j^- \geqslant 0. \end{aligned} \tag{5.13}$$

We note that taxes are treated as costs in the objective function of the mathematical program, which, as before, may be viewed as maximizing the net revenue of the producers in the economy. This program with its

12. See Chapter 6.

13. See sections 3.4.2, 3.6.2, and chapter 11.

14. The main difference is that the identity mapping for consumption has to be replaced by a mapping for net outputs. Compact valuedness of this mapping follows from compactness of the production sets, from $\bar{p}^m \geqslant \bar{p}^e > 0$ and from $\tau^+ < 1$ and $\tau^- > -1$. These assumptions also make it possible to prove, as in part 2 of proposition 5.1, that in the fixed point $T > T_0$.

15. To keep program (5.12) bounded, the parameter values of ξ^- must be constrained so that $\bar{p}^m + \xi^- \geqslant \bar{p}^e$.

feedbacks has a solution that is an equilibrium with producer taxes specified in definition 5.2 and will be used in welfare analysis.

5.2.3 Tariffs and Quotas in International Trade

Tariffs in the Small-Country Model

We first consider the open economy model under the small-country assumption. Let then $\bar{p}^e$ and $\bar{p}^m$ be the exogenously given trade prices of exports and imports, respectively, with $\bar{p}_k^e$ small and $\bar{p}_k^m$ large for nontraded commodities. We define tariff-ridden (or border prices) as

$$\hat{p}_k^e = (1 - \tau_k^e)\bar{p}_k^e,$$

$$\hat{p}_k^m = (1 + \tau_k^m)\bar{p}_k^m,$$

with τ_k^e and τ_k^m such that tax-ridden prices satisfy $\hat{p}_k^m \geqslant \hat{p}_k^e > 0$. This defines the tariff wedges as

$$\xi_k^e = \tau_k^e \bar{p}_k^e,$$

$$\xi_k^m = \tau_k^m \bar{p}_k^m,$$

which, unlike the ξ's for consumption and production, do not depend on domestic prices and are therefore fixed.

DEFINITION 5.3 (Equilibrium with tariffs under the small-country assumption) The allocation y_j^*, all j, x_i^*, all i, m^*, and e^* supported by the price vector $p^* \geqslant 0$ and a level of transfers T^* is an equilibrium with tariffs if:

1. for every producer j, $y_j^* \in \operatorname{argmax}\{p^* y_j \text{ subject to } y_j \in Y_j\}$ with optimal profits denoted as Π_j^*;

2. for every consumer i, $x_i^* = \operatorname{argmax}\{u_i(x_i) \text{ subject to } p^* x_i \leqslant h_i^*, x_i \geqslant 0\}$, where $h_i^* = h_i^{0*} + T_i^*$ is after tax income and $h_i^{0*} = p^*\omega_i + \Sigma_j \theta_{ij} \Pi_j^*$ is pre-tax income;

3. for every commodity k, the tariff-ridden export and import prices are $\hat{p}_k^e = (1 - \tau_k^e)\bar{p}_k^e$ and $\hat{p}_k^m = (1 + \tau_k^m)\bar{p}_k^m$, with τ_k^e and τ_k^m such that $\hat{p}_k^m \geqslant \hat{p}_k^e > 0$;

4. proceeds from tariffs are fully redistributed: $T_i^* = T_i(p^*, T^*, h_1^{0*}, h_2^{0*}, \ldots, h_m^{0*})$, where $T^* = (\hat{p}^m - \bar{p}^m)m^* + (\bar{p}^e - \hat{p}^e)e^*$;

5. all markets are in equilibrium:

$$\Sigma_i x_i^* - \Sigma_j y_j^* - \Sigma_i \omega_i + e^* - m^* \leqslant 0 \perp p^* \geqslant 0,$$

$$e^* \geqslant 0 \perp p^* \geqslant \hat{p}^e,$$

$$m^* \geqslant 0 \perp p^* \leqslant \hat{p}^m.$$

Existence of equilibrium can be proved along the lines of the proof of proposition 5.1.[16]

The associated mathematical program is

$$\max \bar{p}^e e - \bar{p}^m m - \xi^e e - \xi^m m,$$

$$m, e \geqslant 0, y_j, \text{ all } j,$$

subject to (5.14)

$$\Sigma_j y_j + m - e \geqslant \Sigma_i \hat{x}_i - \Sigma_i \omega_i \qquad (p),$$

$$y_j \in Y_j,$$

where $\hat{x}_i$ is obtained from the usual feedback equation.

The Kuhn-Tucker conditions on import and export prices are as in part 5 of definition 5.3. Like in part 2 of proposition 5.1, the trade deficit will be zero in equilibrium:

$$\bar{p}^m m^* - \bar{p}^e e^* = 0.$$

Note that since this is the budget constraint of the rest of the world, the valuation is at international prices and not at tariff-ridden prices.

Quotas on exports and imports can be introduced directly into the constraint set of the mathematical program as

$$e \leqslant \bar{e} \qquad (\mu^e),$$

$$m \leqslant \bar{m} \qquad (\mu^m),$$

where μ^e and μ^m are the quota rents (shadow prices). These can be interpreted as tariff proceeds and redistributed to consumers via the government budget, although in the real world these rents usually remain with the traders who own the import and export licenses. This can be handled by redistributing them directly to specific consumers; they are then called rents or virtual taxes.

Tariffs in the International Model

The international model can be considered as consisting of national models (5.14), linked as in (4.4), through international commodity balances and

16. Here the identity mapping for consumption has to be replaced by a mapping for net imports. Compact valuedness of this mapping follows from compactness of the production sets and from the assumption that $\hat{p}_k^m \geqslant \hat{p}_k^e > 0$. This also makes it possible to prove, as in part 2 of proposition 5.1, that $T > T_0$ in the fixed point. See Shoven and Whalley (1974b) for an early proof and an algorithm.

with import and export prices p_k^w that are now endogenous[17] and should in equilibrium clear markets for traded as well as for nontraded goods. Here we assume that tariff receipts levied in country c are redistributed to consumers in country c only. Then, in the open exchange format, the international model may be written as

$$\max \bar{p}^e e - \bar{p}^m m - \Sigma_c \xi_c^e e_c - \Sigma_c \xi_c^m m_c,$$

$$m, e \geqslant 0, m_c, e_c \geqslant 0, \text{ all } c, y_j, \text{ all } j,$$

subject to (5.15a)

$$m - e \geqslant \Sigma_c (m_c - e_c) \qquad (p^w),$$

for every c:

$$\Sigma_{j \in J_c} y_j + (m_c - e_c) \geqslant \Sigma_{i \in I_c} \hat{x}_i - \Sigma_{i \in I_c} \omega_i \qquad (p_c),$$

$$y_j \in Y_j, \text{ all } j \in J_c,$$

for given consumptions

$$\hat{x}_i = \operatorname{argmax}\{u_i(x_i) \text{ subject to } p_c x_i \leqslant h_i^0 + T_i(p_c, T_c, \{h_i^0\}_{i \in I_c}), x_i \geqslant 0\}, \tag{5.15b}$$

where

$$h_i^0 = p_c \omega_i + \Sigma_{j \in J_c} \theta_{ij} \Pi_j(p_c) + \Sigma_{c' \neq c} \Sigma_{j \in J_{c'}} \theta_{ij} \Pi_j(p_{c'}), \text{ all } i \in I_c, \text{ all } c,$$

for given tariff wedges in country c,

$$\xi_{ck}^e = \tau_{ck}^e p_k^w \quad \text{and} \quad \xi_{ck}^m = \tau_{ck}^m p_k^w, \text{ all } c, \text{ all } k \tag{5.15c}$$

and for given tariff receipts in country c,

$$T_c = \xi_c^e e_c + \xi_c^m m_c, \text{ all } c. \tag{5.15d}$$

The term $\Sigma_{c' \neq c} \Sigma_{j \in J_{c'}} \theta_{ij} \Pi_j(p_{c'})$ represents profits from abroad and will be treated as part of remittances from abroad, to be denoted as Π_i^w as in chapter 4.

Alternatively, tariffs can also be interpreted as indirect taxes on sales and purchases of a production activity called trade, in a model in which the agents of the country can only engage in transactions with this trade sector and not directly with foreign countries. Then the model has the structure of the closed model in definition 5.2.

17. Recall that commodities not traded between countries are nonetheless elements of the vectors e, m, and p^w. They are only supplied and demanded by the agents in this country. Hence each country vector has zero entries for nontraded commodities of other countries.

Existence of equilibrium in this model can be proved along the lines of proposition 5.1 with a modification to account for the country-specific government budget equilibrium condition (5.15d).

5.2.4 Government Activity: A Summary

We have assumed that the government is a consumer with an initial endowment and shares in firms. It imposes indirect taxes and tariffs (or quotas), whose proceeds it partly redistributes to the other consumers so as to maintain its own as well as the overall financial balance. As we have seen, this also ensures satisfaction of the trade balance. As before, we refer to government as consumer g. The activities and transactions in which it is involved can be summarized as follows:

Public consumption,

$$\hat{x}_g = \operatorname{argmax}\{u_g(x_g) \text{ subject to } p^c x_g \leqslant h_g^0 + T_g,\ x_g \geqslant 0\}.$$

Income from endowment and production and remittances from abroad,

$$h_g^0 = p\omega_g + \Sigma_j \theta_{gj}(p^+ y_j^+ - p^- y_j^-) + \Pi_g^w.$$

Redistributed income,

$$T_g = T_g(p, T, h_1^0, h_2^0, \ldots, h_g^0, \ldots, h_m^0).$$

Transfers to private consumers,

$$T_i = T_i(p, T, h_1^0, h_2^0, \ldots, h_g^0, \ldots, h_m^0) \qquad \text{for all } i \text{ and } i \neq g.$$

Proceeds from indirect taxation of consumers and producers, from tariffs,

$$T = (p^c - p)\Sigma_i x_i + (p - p^+)y_j^+ + (p^- - p)y_j^- + (\hat{p}^m - \bar{p}^m)m + (\bar{p}^e - \hat{p}^e)e.$$

Budget balancing,

$$\Sigma_{i \neq g} T_i + T_g = T.$$

Trade balancing,

$$\bar{p}^m m - \bar{p}^e e = \Sigma_i \Pi_i^w.$$

Recall that the budget and trade balances are equilibrium conditions.

5.3 Welfare Implications

The specifications introduced above suggest that government consumption and taxes may affect allocations in many ways. Government consumption does not provide any utility to private consumers and merely uses resources, as mentioned in section 5.1.1. Indirect taxes create price wedges that tend to benefit certain agents at the expense of other agents. This naturally calls for a Pareto-efficiency study of the allocations under these policies, but we note from the outset that indirect taxes on endowments or a single-rate tax on all consumer expenditures will act like lump-sum taxes and not affect efficiency.

Since public consumption does not benefit private consumers, maximization of private welfare would require its suppression. However, if one accepts that government consumption contributes to social welfare, it has to appear in the social welfare function, and the issue is only how much weight it should carry. Throughout this chapter we treat public consumption like private consumption, so the weight on government's utility is determined in the same way as for other consumers, by the income it can spend.

When studying the consequences of reforms, we will pursue the approach introduced in section 1.4.6 and apply proposition 1.11 to establish the main results. This proposition establishes that aggregate welfare of one out of two groups of consumers cannot decrease when the welfare weight of the other group is reduced. We will now see that, as in (5.10), (5.12) and (5.14), the objective of the welfare program can be extended with tax terms that operate like "costs" on the production allocation. These costs will be viewed as the utility of an additional agent who competes for welfare with the collective of consumers, suggesting tax-reforms study by varying this agent's welfare weight as in proposition 1.11. Hence we will establish not only the classical distortionary effect of taxes and tariffs but also suggest Pareto-improving piecemeal reforms.

We first deal with consumer taxes. The main arguments for efficiency gains under (full and partial) reforms with compensation of losers can be understood from the treatment given in the first section. As before, the model with international trade needs a separate discussion for the small open economy because it does not require compensating transfers from (or to) other countries. Finally, we also consider the more realistic situation of a reform without compensating payments.

5.3.1 Indirect Taxes on Consumption in the Closed Economy

We start with indirect taxes on consumption in the closed economy of proposition 5.1. We also make use of the equivalence between the closed economy model of proposition 3.4 and the Negishi format of proposition 3.1, while maintaining the representation of consumer taxes in the objective, as in (5.10).[18] This is written

$$\max \Sigma_i \alpha_i u_i(x_i) - \xi^c x,$$

$$x,\ x_i \geqslant 0 \text{ all } i,\ y_j \text{ all } j,$$

18. Recall that the constraint $x \geqslant \Sigma_i x_i$ serves to generate consumer prices p^c, as in (5.10).

subject to (5.16)

$$x - \Sigma_j y_j \leqslant \Sigma_i \omega_i \qquad (p),$$

$$x \geqslant \Sigma_i x_i \qquad (p^c),$$

$$y_j \in Y_j.$$

The α_i are set in such a way that every consumer (private or public) is on his budget constraint:

$$p^c x_i = h_i^0 + T_i(p, T, h_1^0, h_2^0, \ldots, h_m^0),$$

$$h_i^0 = p\omega_i + \Sigma_j \theta_{ij} \Pi_j(\mathrm{p}), \qquad (5.17)$$

$$T = \xi^c \Sigma_i x_i,$$

$$\xi_k^c = \tau_k^c p_k.$$

Any equilibrium solution of the model in proposition 5.1 will be optimal in problem (5.16) and satisfy (5.17). Note, however, that when $\xi^c \neq 0$, the α_i may no longer be normalized to add up to 1.[19]

Proposition 5.2 shows that abolishment of all indirect taxes cannot reduce welfare if compensating payments are given so as to maintain constant relative welfare weights among consumers.

PROPOSITION 5.2 (Welfare loss from a consumption tax) Let $W(\alpha, \xi^c) = \Sigma_i \alpha_i u_i(x_i(\alpha, \xi^c))$, with $x_i(.)$ determined as in (5.16). Then $W(\alpha, \xi^c) \leqslant W(\alpha, 0)$.

Proof By strict concavity of utility and the maximum theorem A.3.1, $x_i(\alpha, \xi^c)$ is a continuous function of ξ^c. The constraint set is unchanged when ξ^c (i.e., τ^c) is set to zero, but it is $W(.)$ that is now maximized in (5.16). ■

This shows that removing taxes cannot reduce total welfare.[20] The proposition is weak, however. It may happen that no agent gains, so any reform would be useless, as illustrated by the following example: Assume that all commodities are subject to the same tax rate; assume also that tax redistribution is such that every consumer receives a tax rebate (as a lump-sum transfer) equal to the amount of indirect taxes he has paid. Consumers' welfare is unaffected by taxation, since the only effect of taxation

19. Multiplication of the objective function by a positive scalar may affect allocations unless $\xi^c = 0$ or is being multiplied by the same factor as the welfare weights.

20. Note that to establish Pareto-efficiency, the proposition should be strengthened to show that under appropriate transfers no consumer will be worse off. This can be done by adding to (5.16) constraints of the form $u_i(x_i) \geqslant u_i^0$, where u_i^0 is the utility of consumer i at the original level of ξ^c (see proposition 1.8). Since, again the constraint set will remain unaffected under elimination of ξ^c, aggregate welfare would not fall, and no consumer will have to lose.

would be to multiply both sides of the budget constraints by a common factor.

Proposition 5.2 suggests that all indirect taxes should be abolished in a single reform. This is rarely feasible and a more interesting question is what happens under a gradual reduction of ξ^c. We state and prove two results: The first (proposition 5.3) is global and shows that a reduction of wedges will not be welfare decreasing if all these wedges fall at the same rate and consumers are compensated. The second result (proposition 5.4) states a similar property for an arbitrary tax reduction when tax rates are already sufficiently close to zero.

PROPOSITION 5.3 (Welfare gains from reducing consumer taxes in the closed economy) Let $W(\alpha, \xi^c)$ be as in proposition 5.2. Then $W(\alpha, \xi^c) \leqslant W(\alpha, \alpha_0 \xi^c)$ for any $\alpha_0 \in [0, 1]$, where α is the vector of welfare weights of consumers 1 to n.

Proof We start by rewriting (5.16) as

$$\max \Sigma_i \alpha_i v_i + \alpha_0 v_0,$$

$$x, x_i \geqslant 0, \text{ all } i, v_0, v_i \text{ all } i, y_j \text{ all } j,$$

$$\text{subject to} \qquad (5.18)$$

$$u_i(x_i) \geqslant v_i,$$

$$-\xi^c x \geqslant v_0,$$

$$x - \Sigma_j y_j \leqslant \Sigma_i \omega_i \qquad (p),$$

$$x \geqslant \Sigma_i x_i \qquad (p^c),$$

$$y_j \in Y_j,$$

where $\alpha_0 = 1$. Since the α_i are inverse marginal utilities of income evaluated at equilibrium, they are positive. Hence the inequalities on v_i and v_0 will be binding in any optimum, and the solution of (5.18) will coincide with the solution of (5.16). Note that consumer prices p^c satisfy

$$p^c \leqslant p + \alpha_0 \xi^c \perp x \geqslant 0,$$

which shows that a reduction of α_0 amounts to a proportional reduction of all positive or negative wedges between consumer prices p^c and clearing prices p.

To establish the result, we note that the constraint set is a compact, convex possibility set for "utilities" $v_0, v_1, \ldots, v_m$. We can now apply proposition 1.11, to show that $W(\alpha, \alpha_0 \xi^c)$ is nonincreasing in α_0 and cannot fall if α_0 is being reduced. ■

Proposition 5.3 needs some further explanation. First, it establishes that there exists an equilibrium with reduced consumer taxes $\tau_k^c p_k$, where no consumer would be worse off, provided transfers are given to those who would lose as a consequence of the change in prices. The proposition fails to ensure that some consumer will gain, but if the utility possibility set has a smooth northeast boundary, there will be some gain (see figure 1.4). Second, the proposition only holds when the reduction is proportional, so reducing only some of the tax wedges ξ_k^c could lead to welfare losses, though such a reduction may be seen as a step in a sequence of reforms that will eventually lead to a proportional reduction. Finally, we must emphasize that a proportional reduction of the wedges ξ_k^c is not the same as a proportional reducton of tax rates τ_k^c; reducing τ_k^c will affect clearing prices, so $\xi_k^c = \tau_k^c p_k$ will not adjust by the same factor for all k.

We now turn to a local result that does not require a uniform reduction in all the wedges:

PROPOSITION 5.4 (Eventual welfare gains from reducing taxes in the closed economy) Let $W(\alpha, \xi^c)$ be as in proposition 5.2. Let $\xi(s)$ be a continuous function of a scalar s such that $\xi(0) = 0$ and $\xi(1) = \xi^c$. Then, if $W(\alpha, \xi^c) < W(\alpha, 0)$, there will exist a value s_0 such that for all $s \in (0, s_0]$, $W(\alpha, \xi(s))$ is nonincreasing in s and $W(\alpha, s_0) < W(\alpha, 0)$.

Proof By the maximum theorem A.3.1, $W(.)$ is continuous in s. Consider then figure 5.3, where $W(.)$ is depicted as a continuous function of s, with maximum at $s = 0$ and the result is obvious: Then, if $W(\alpha, \xi^c) < W(\alpha, 0)$ at that point, W cannot decrease when s is close to zero. ■

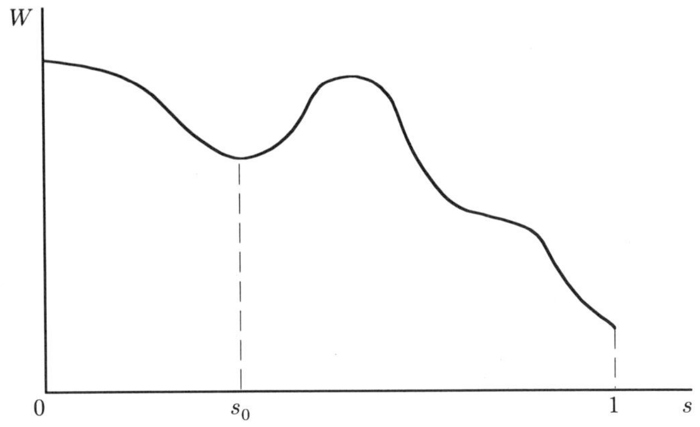

Figure 5.3
Welfare changes under tax reform.

Combining both propositions indicates that to prevent a drop in welfare, tax reductions could be performed in two phases. In the first phase (when s goes from 1 to s_0), the reduction should be simultaneous and proportional; in phase two (between s_0 and 0), it could proceed stepwise: No tax can be raised in absolute value, though some can be set to zero leading to piecemeal adjustments. Such a program works as long as the adjustments are not anticipated by the agents.

5.3.2 Indirect Taxes on Production and Intermediate Demand in the Closed Economy

Indirect taxes on producers can be treated by replacing the $-\xi^c x$ term by $-(\xi^+ y^+ + \xi^- y^-)$ in the objective and thus obtaining the counterparts of propositions 5.2 to 5.4. Obviously consumer and producer taxes can be combined.

5.3.3 Tariffs and Quotas in the International Trade Model

We now consider the welfare effects of tariffs in the international trade model (5.15). The resulting mathematical program may, analogously to (5.15), be written[21]

$$\max \Sigma_c \Sigma_{i \in I_c} \alpha_i u_i(x_i) - \Sigma_c \xi_c^e e_c - \Sigma_c \xi_c^m m_c,$$

$$m_c,\ e_c \geqslant 0, \text{ all } c,\ y_j, \text{ all } j,\ x_i \geqslant 0 \text{ all } i,$$

$$\text{subject to} \qquad (5.19)$$

$$\Sigma_c(m_c - e_c) \leqslant 0 \qquad (p^w),$$

for every c:

$$\Sigma_{j \in J_c} y_j + (m_c - e_c) \geqslant \Sigma_{i \in I_c} \hat{x}_i - \Sigma_{i \in I_c} \omega_i \qquad (p_c),$$

$$y_j \in Y_j, \text{ all } j \in J_c.$$

PROPOSITION 5.5 (Welfare gains from tariff reductions) A proportional reduction of tariff wedges in all countries and for all (traded) commodities cannot reduce world welfare when compensating transfers are given to the losers. In the neighborhood of $\xi^e = \xi^m = 0$, piecemeal reductions will work.

Proof The proof is analogous to that of propositions 5.3 and 5.4. ■

The following two remarks may be added. First, as in the consumer tax case, Propositon 5.5 assumes that transfers are given and that the result does not necessarily hold in their absence. Second, a reduction of quotas will enlarge the feasible set but not necessarily lead to welfare increases if there

21. Quotas can be introduced as bounds on e_c and m_c.

are tariffs present. However, the proposition does carry through if tariff wedges are reduced while quotas are present.

Proposition 5.5 only establishes a result at world level. It does not claim that a unilateral reduction of tariffs by one country can be performed without welfare loss for the world, unless the others pay some compensation and do not levy tariffs or reduce wedges at the same rate. But even if the other countries have no tariffs, a welfare loss may result for the country that is considering the reduction. We will show in the next section that on the contrary, for a country that faces given world prices (small-country assumption), no such welfare loss can occur provided that the proceeds from tariffs are redistributed in an appropriate way within the country.

5.3.4 Indirect Taxes and Tariffs in the Small Open Economy

Indirect Taxes

We introduce taxes on consumption and write (5.10) in the welfare maximizing format, which then reads (dropping the country index c):

$$\max \Sigma_i \alpha_i u_i(x_i) + (\bar{p}^e e - \bar{p}^m m) - \xi^c x,$$

$$m, e \geqslant 0, \; x_i \geqslant 0, \text{ all } i, \; x, \; y_j \text{ all } j,$$

$$\text{subject to} \tag{5.20}$$

$$x - \Sigma_j y_j - m + e \leqslant \Sigma_i \omega_i \qquad (p),$$

$$x \geqslant \Sigma_i x_i \qquad (p^c),$$

$$y_j \in Y_j,$$

where $\xi_k^c = \tau_k^c p_k$; the α_i are set in such a way that the consumers' budget constraints are met and the aggregate budget must satisfy

$$T = (p^c - p)\Sigma_i x_i.$$

Recall that imposing this condition ensures balance of trade equilibrium $(\bar{p}^e e - \bar{p}^m m) = \Pi_0$, where $\Pi_0 = \Sigma_i \Pi_i^w$ are remittances from abroad which can be zero and even negative provided that feasibility is maintained. Hence dropping the term $(\bar{p}^e e - \bar{p}^m m)$ from the objective and introducing the balance of trade as an extra constraint in the previous model at the given equilibrium ξ^c and α_i will generate the same equilibrium allocation. This leads to the format that is suited for welfare analysis of the small open economy:

$$\max \Sigma_i \alpha_i u_i(x_i) - \xi^c x,$$

$$m, e \geqslant 0, \; x_i \geqslant 0, \text{ all } i, \; x, \; y_j \text{ all } j,$$

$$\text{subject to} \tag{5.21}$$

$$x - \Sigma_j y_j - m + e \leqslant \Sigma_i \omega_i \qquad (p),$$

$$x \geqslant \Sigma_i x_i \qquad (p^c),$$

$$\bar{p}^m m - \bar{p}^e e \leqslant \Pi_0 \qquad (\rho),$$

$$y_j \in Y_j.$$

The trade balance now works like a technical constraint that restricts the trade possibilities of the country; it can also be thought of as a lower bound on the utility of the rest of the world $(\bar{p}^e e - \bar{p}^m m)$.[22] As can be verified from the Lagrangean of this problem, $\rho = 1$ will be an equilibrium shadow price before tax reduction. The question is now whether the country can gain by decreasing taxes. This is answered in the following proposition:

PROPOSITION 5.6 (Welfare gains from reducing consumer taxes in the small open economy) A proportional reduction of tax wedges on all commodities cannot reduce the country's welfare when compensating transfers are given to the losers. In the neighborhood of $\xi^c = 0$, piecemeal reductions will work.

Proof The proof is analogous to that of propositions 5.3 and 5.4 ∎

The same proposition can obviously be derived for producer taxes.

Tariffs

We proceed to the welfare program with tariffs, obtaining

$$\max \Sigma_i \alpha_i u_i(x_i) - \xi^e e - \xi^m m,$$

$$m, e \geqslant 0, x_i \geqslant 0, \text{ all } i, y_j \text{ all } j,$$

$$\text{subject to} \tag{5.22}$$

$$\Sigma_i x_i - \Sigma_j y_j - m + e \leqslant \Sigma_i \omega_i \qquad (p),$$

$$\bar{p}^m m - \bar{p}^e e \leqslant \Pi_0 \qquad (\rho),$$

$$y_j \in Y_j.$$

PROPOSITION 5.7 (Welfare gains from reducing tariffs in the small open economy) A proportional reduction of tariff wedges on all commodities cannot reduce the country's welfare when compensating transfers are given to the losers within the country. In the neighborhood of $\xi^e = \xi^m = 0$, piecemeal reductions will work.

Proof The proof is analogous to that of propositions 5.3 and 5.4. ∎

22. See also section 4.3.2.

After a reduction of tariffs, ρ will generally become different from unity, and the associated tariff revenue will be

$$T = \xi^m m + \xi^e e,$$

while the first-order conditions show that border prices p will not change proportionally:

$$p \geqslant \rho p^{-e} - \xi^e \perp e \geqslant 0,$$

$$p \leqslant \rho p^{-m} + \xi^m \perp m \geqslant 0.$$

There are two interpretations to these conditions. One could look at the wedge, say, for imports $p - p^{-m}$ as the actual tariff. This is equal to $(\rho - 1)p^{-m} + \xi^m$; the adjustment of ρ is then seen to change actual wedges in a nonproportional way. Alternatively, $(\rho - 1)$ may be viewed as the price paid for an import license by the importers to the exporters (for whom the price is being raised) and to the foreign lenders who provide Π_0. However, it should in no way be interpreted as an exchange rate.

5.3.5 Welfare Analysis in the Absence of Compensation

All the results discussed so far show that (weakly) Pareto-superior tax reforms can be designed provided that they are supported by compensating transfers. The following example shows that even full elimination of taxes does not necessarily lead to welfare increases in the absence of compensation.

Consider a closed economy with two consumers, a single commodity and no production. Endowments are ω_1 for consumer 1 and 0 for consumer 2. There is a 100 percent consumption tax whose proceeds are given to consumer 2; both consumers have the same strictly concave, increasing, and continuously differentiable utility function $u(x_i)$. Their budget equations are $p^c x_1 = p\omega_1$ and $p^c x_2 = T$, with $T = (p^c - p)(x_1 + x_2) = \xi^c(x_1 + x_2)$. For a 100% consumption tax, $\xi^c = p$ so that $p^c x_1 = p^c x_2 = p\omega_1$. Consequently both consumers have the same income, and since there is only one good, $x_1 = x_2 = \omega_1/2$. Since both consumers have the same utility and the same income, they have equal marginal utilities of income and thus equal welfare weights α_1 and $\alpha_2 = \alpha$. We now consider reducing ξ^c in the following welfare-maximizing program with equilibrium welfare weights:

$$\max \alpha u(x_1) + \alpha u(x_2) - \xi^c(x_1 + x_2),$$

$$x_1, x_2 \geqslant 0,$$

subject to

$$x_1 + x_2 \leqslant \omega_1,$$

where $\xi^c = 1$. By increasingness of $u(x_i)$, equality will hold in the commodity balance. Therefore we can substitute it into the objective function and obtain

$$\max \alpha u(x_1) + \alpha u(\omega_1 - x_1) - \xi^c \omega_1, \qquad x_1, x_2 \geqslant 0. \tag{5.23}$$

In the optimum

$$\frac{\alpha \partial u}{\partial x_1} = \frac{\alpha \partial u}{\partial(\omega_1 - x_1)} = \frac{\alpha \partial u}{\partial x_2}.$$

Thus, under a reduction of ξ^c, say, from 1 to 0, and compensating transfers, the equality $x_1 = x_2 = \omega_1/2$ will be maintained, and no consumer will be worse off.

Alternatively, if no compensation is given, all income will go to consumer 1. This will reduce aggregate welfare, since by strict concavity of u,

$$\alpha\left[u\left(\frac{\omega_1}{2}\right) + u\left(\frac{\omega_1}{2}\right)\right] > \alpha[u(\omega_1) + u(0)];$$

thus a welfare loss occurs.

Note, however, that if we compute the new equilibrium without transfers, "aggregate" welfare at new weights will increase, since this is precisely the function being maximized at this equilibrium. This result is generalized in proposition 5.8.

PROPOSITION 5.8 (Welfare at the zero-tax, no-compensation equilibrium) Let $W(\alpha, \xi^c) = \Sigma_i \alpha_i u_i(x_i(\alpha, \xi^c))$, where the $x_i(.)$ are determined as in (5.16) and the weights are those prevailing at the zero-tax, no-compensation equilibrium. Then $W(\alpha, \xi^c) \leqslant W(\alpha, 0)$.

Proof The proof proceeds as in proposition 5.2. ■

5.3.6 Consequences of the Propositions for Policy Design

The propositions discussed above only provide sufficient conditions to avoid welfare losses. There are basically two options. In the first, one compensates losers and designs the (possibly) partial reform in a specific way. For consumption, say, one has to calculate $\xi^c = p^c - p$, reduce it by a factor s, and solve for the new equilibrium with compensation; new tax rates can then be computed (propositions 5.2 to 5.7).[23] Under the second option, one does not compensate and so eliminates all wedges at once (proposition 5.8).

In practice, policy experiments rarely fulfill the requirements imposed by

23. In the case of tariffs one could also introduce a premium on import licences and keep the wedges fixed.

the first option. Compensation is rarely considered, and when it is, it usually does not keep the relative welfare weights (marginal utilities of income) or the utilities at their pre-reform level. On the other hand, for obvious reasons, the experiments rarely focus on complete elimination of wedges, so they do not comply with the second option either. The practice is to run policy scenarios: a pre-reform equilibrium as hereby the so-called reference scenario is calculated,[24] tax rates are reduced on the basis of policy recommendations, and a new equilibrium is computed without compensating transfers. Then comparisons are made using some application-specific welfare criterion.[25]

Alternatively, one could use the general equilibrium model to compute allocations for various tax scenarios and select the one that maximizes the welfare criterion. One could also go a step further and try to optimize this criterion formally, but this would, in general, not be a convex program.[26]

5.4 Practices in Applied Modeling

5.4.1 Taxes and Tariffs in the CGE Model

The introduction of taxes and tariffs does not change the balances for goods, factors and imports of the CGE model (4.8) of section 4.4.1. It only affects the budgets and the price formation equations.

The treatment of consumer taxes is identical to the one given in definition 5.1 and leads to tax-ridden prices:

$$\begin{aligned} p_k^c &= (1 + \tau_k^c)p_k^q \quad \text{for goods,} \\ p_k^d &= (1 + \tau_k^d)p_k^f \quad \text{for factors.} \end{aligned} \tag{5.24a}$$

For goods, producer taxes on outputs (τ^q) and on inputs (τ^v) create a wedge as in definition 5.2. This generates

$$\begin{aligned} p_k^q &= (1 - \tau_k^q)p_k^q \quad \text{for ex-factory prices,} \\ p_k^v &= (1 + \tau_k^v)p_k^q \quad \text{for inputs of goods.} \end{aligned} \tag{5.24b}$$

Since factors are not produced, only inputs are taxed at rates τ^ω, and

24. If there are multiple equilibria, a "small" reform without compensation can lead to large changes in the equilibrium solution. This cannot happen when the reform is as specified in propositions 5.2 to 5.7, since it works through perturbations of convex programs.

25. Criteria for such comparisons are treated in section 2.4.2.

26. Some studies therefore are satisfied with a search for optimal directions of policy changes, locally, in the prevailing equilibrium; see Guesnerie (1977).

tax-ridden prices are

$$p_k^{\omega} = (1 + \tau_k^{\omega})p_k^f \qquad \text{for inputs of factors.} \tag{5.24c}$$

Recall that the CGE model assumes that exports are perfect substitutes for domestic goods, but imports are modeled like intermediate demand for factors, with given prices $\bar{p}^m$ and perfectly elastic supply. The exported goods do not have perfect substitutes in the rest of the world, so their prices p^e are endogenous. To represent tariffs on imports τ^m, we define tariff-ridden prices as

$$\hat{p}_k^m = (1 + \tau_k^m)\bar{p}_k^m \qquad \text{for imports of factors.} \tag{5.24d}$$

The price of the exported good k adjusts according to

$$p_k^e = \frac{p_k^g}{1 - \tau_k^e} \qquad \text{for exports of goods.} \tag{5.24e}$$

This price enters the ad hoc export function $e_k(\,.\,)$ and the budget constraint of the rest of the world, that is, the balance of trade of the country, $p^e e^g = \bar{p}^m m - \Sigma_i \Pi_i^w$.

Since we treat government as one of the consumers, it belongs to the index set of consumers $I = \{1, 2, \ldots, m\}$, and public consumption is allocated to it. The balance equations become

$$\Sigma_i x_i^g(p^c, p^d, h_i) + A^g(p^v, p^{\omega}, \hat{p}^m)q^g + e^g = q^g \qquad \text{for goods,} \tag{5.25a}$$

$$\Sigma_i x_i^f(p^c, p^d, h_i) + A^f(p^v, p^{\omega}, \hat{p}^m)q^g = \Sigma_i \omega_i^f \qquad \text{for factors,} \tag{5.25b}$$

$$A^m(p^v, p^{\omega}, \hat{p}^m)q^g = m \qquad \text{for imports,} \tag{5.25c}$$

where A^g, A^f, and A^m are matrices of intermediate demand coefficients and are thus functions of tax and tariff-ridden prices.

Price formation of goods (5.25d), exports (5.25e), income formation (5.25f–h), and budget equilibrium (5.25i) of the government are now

$$p^g = p^v A^g(p^v, p^{\omega}, \hat{p}^m) + p^{\omega} A^f(p^v, p^{\omega}, \hat{p}^m) + \hat{p}^m A^m(p^v, p^{\omega}, \hat{p}^m), \tag{5.25d}$$

$$e^g = \operatorname{argmax}_{e^g \geqslant 0}\{u^w(e^g) \mid p^e e^g = \bar{p}^m m - \Sigma_i \Pi_i^w\}, \tag{5.25e}$$

$$h_i^0 = p^f \omega_i^f + \Pi_i^w, \tag{5.25f}$$

$$T_i = T_i((p^g, p^f, \hat{p}^m), T, h_1^0, h_2^0, \ldots, h_m^0), \tag{5.25g}$$

$$h_i = h_i^0 + T_i, \tag{5.25h}$$

$$\begin{aligned} T = {} & (p^c - p^g)\Sigma_i x_i^g(\,.\,) + (p^d - p^f)\Sigma_i x_i^f(\,.\,) \\ & + (p^g - p^q)q^g + (p^v - p^g)A^g(\,.\,)q^g + (p^{\omega} - p^f)A^f(\,.\,)q^g \\ & + (\hat{p}^m - \bar{p}^m)m + (p^g - p^e)e^g. \end{aligned} \tag{5.25i}$$

To solve this model, we can start from given levels of prices p^f and total taxes T (and of course exogenous values for $\bar{p}^m$ and Π_i^w) and compute the prices of goods by substituting (5.24a)–(5.24d) in (5.25d) and solving (5.25d). Prices of exported goods are derived from (5.24e). Once all prices have been obtained, we can compute consumer demand and the exports and inputs coefficients. Outputs for goods (5.25a) and the imbalance on the factor markets (5.25b) as well as on the government budget (5.25i) will follow. The given values of p^f and T can be adjusted until equilibrium is reached.

5.4.2 Issues in Model Implementation

The Specification of Wedges

Tax rates are obtained in many applications as ratios calculated on the basis of a single-year observation. They are then treated as exogenous parameters, some of which can be varied in reform scenarios. This is the simplest approach. At the other extreme, optimal taxation provides a way to endogenize these wedges, but the approach is often found to be unrealistic or numerically cumbersome.

A third (intermediate) possibility is to specify rates as functions of other endogenous variables and to obtain the parameters of these functions by econometric estimation.[27] To maintain continuity and homogeneity of supplies and demands, these tax functions must themselves be continuous and homogeneous of degree zero in prices. Whereas continuity is a technical requirement, homogeneity is imposed to ensure that only relative prices can have an effect on allocations.

Indirect taxes are often used to maintain certain relative prices within a target range. Assume, for instance, that the international price of imported oil rises significantly and that the policy is to avoid the impact of such a shock on domestic relative prices. We can then introduce a so-called variable levy ξ_k^m whereby the import tariff adjusts so as to keep the tariff-ridden import price p_k^m fixed relative to some domestic price p_s $(s \neq k)$, that is, assign

$$\xi_k^m = (1 + \eta_k)p_s - p_k^m.$$

In this expression, $1 + \eta_k$ is the target ratio. Similarly, if we want to keep two consumption prices in a fixed ratio, this can be achieved by setting

$$\xi_k^c = (1 + \beta_k)p_s^c - p_k,$$

$$\xi_s^c = \tau_k^c p_s,$$

27. See among others Burniaux and van der Mensbrugghe (1990) or Fischer et al. (1988).

where $(1 + \beta_k)$ is the target ratio. However, such tax functions create relative price rigidities, and closed economy equilibrium solutions may fail to exist unless the reduced flexibility of prices is compensated by a perfectly elastic adjustment of quantities. This notion will be considered in chapter 6.

The two previous examples were concerned with maintaining relative prices, but tax functions can also be used to stabilize quantities. An import tariff, for instance, may be designed to rise with the import level in order to avoid excessive imports: There is a low (possibly zero) tariff that rises beyond a specified threshold level, sometimes called tariff quota, since it gives the level of imports that the country is allowing to enter without protection.

Under certain circumstances an exporter commits himself to restrain exports. These are voluntary export restraints (VERs), usually modeled as export quotas whose rent accrues to the exporter. As a reward the importing country will often offer a tariff quota.

Another trade-limiting practice is the imposition of a nontariff barrier (NTB). Imperfect substitution is artifically introduced by legislation that imposes norms that the imported commodity can only satisfy to a limited extent. The severity of the inspection by customs determines the degree of product heterogeneity. Modeling can use quotas or the "Armington" approach.[28] Another approach is to treat every commodity as a bundle of characteristics,[29] some of which the imported commodity misses (e.g., an ecologically friendly mode of production expressed through an eco-label). Nontariff barriers exist for commodities (meat, pharmaceuticals), services (insurance), and factors (labor). Labor immigration, for example, can be restricted via nationality quotas, nonrecognition of education certificates, and age limits.

The Specification of Direct Tax Functions

Direct tax functions are used to redistribute income (including the net proceeds of indirect taxation) to consumers. These have to satisfy assumption T of section 5.1.2. A possible theoretical way to derive such functions was presented in section 2.2.5. In practice, this approach is seldom implemented, and tax functions are constructed in a more ad hoc way. Positiveness of post-tax income is then not guaranteed, as the following frequently used linear expenditure system shows. Let

$$T_i(p, T, h_1^0, h_2^0, \ldots, h_m^0) = -\tau_i h_i^0 + p\delta_i + \gamma_i(T + \Sigma_i \tau_i h_i^0 - \Sigma_i p\delta_i),$$

where τ_i is the rate at which income of consumer i is taxed and, as before,

28. See the discussion of heterogeneous commodities in section 4.4.2.

29. See Lancaster (1966, 1971) and section 3.6.2.

T represents the total proceeds of indirect taxation, tariffs, and so on; γ_i, which satisfies $\Sigma_i \gamma_i = 1$, distributes the total net tax proceeds (direct and indirect) and $p\delta_i$ is an indexed lump-sum transfer (e.g. allowance for children). Here adding-up is ensured by construction, but $h_i^0 + T_i(.)$ may become negative. An accounting discrepancy may also arise if one accidentally chooses functions $T_i(.)$ that violate adding-up.

The Specification of Public Demand

So far we have considered government as a single utility-maximizing consumer. In applications, there is often a need to distinguish among several government programs (e.g., health care, environment, defense) and levels of government (central, regional). Budget allocation decisions can be seen as resulting from a single government utility maximization. Alternatively, separate utilities and budgets can be specified for each; an intermediate setup is to introduce functional separability among government programs and apply multi-stage budgeting.[30]

In many applications government consumption is given priority over private consumption and is set exogenously. Then taxation or adjustment of foreign savings Π_0 has to be used to balance the budget.[31] In the Negishi format the government utility function must be dropped from the objective and government consumption must be fixed exogenously in the commodity balances.

5.4.3 Empirical Elaboration: The SAM with Taxes and Tariffs

We consider the equilibrium solution of a model with consumer and producer taxes and with tariffs (combining definitions 5.1, 5.2, and 5.3) and follow the same steps as in section 3.6.1. The relevant equations are as follows:

1. Market equilibrium conditions

$$\Sigma_i x_{ik}^* - \Sigma_j(y_{jk}^{+*} - y_{jk}^{-*}) - m_k^* + e_k^* - \Sigma_i \omega_{ik} \leqslant 0 \perp p_k^* \geqslant 0, \qquad \text{all } k. \qquad (5.26a)$$

2. Producers' profits

$$\Pi_j^* = (p^{+*} y_j^{+*} - p^{-*} y_j^{-*}), \qquad \text{all } j. \qquad (5.26b)$$

3. Consumers' budget equations

$$p^{c*} x_i^* = p^* \omega_i + \Sigma_j \theta_{ij} \Pi_j^* + \Pi_i^w + T_i^*, \qquad \text{all } i. \qquad (5.26c)$$

4. The budget equation of the rest of the world (foreign sector)

$$\bar{p}^e e^* = \bar{p}^m m^* - \Sigma_i \Pi_i^w. \qquad (5.26d)$$

30. See section 2.2.5.

31. See section 3.4.2 on closure rules.

5. The government's budget balance

$$\Sigma_i T_i^* = (p^{c*} - p^*)\Sigma_i x_i^* + (p^* - p^{+*})\Sigma_j y_j^{+*} + (p^{-*} - p^*)\Sigma_j y_j^{-*}$$
$$+ (\hat{p}^m - \bar{p}^m)m^* + (\bar{p}^e - \hat{p}^e)e^*. \quad (5.26e)$$

The complementarity conditions (5.26a) imply that

$$p_k^* \Sigma_i x_{ik}^* + p_k^* e_k^* = p_k^* \Sigma_j y_{jk}^* + p_k^* \Sigma_i \omega_{ik} + p_k^* m_k^*, \quad (5.26f)$$

and it can be verified, as in section 4.4.3, that the balance of payments equilibrium (5.26d) is implied by the sum of consumers' budgets (including government) and (5.26f).

We represent in table 5.1 the SAM at clearing prices p^*, which include tariffs on imports and indirect taxes on outputs but exclude consumer taxes and taxes on inputs and on exports.[32] To obtain the SAM in tabular form, we add to table 4.1 a column and a row representing collection and redistribution of taxes (government is one of the consumers and its budget as a consumer is not represented separately) in accordance with equation (5.26e).

The "indirect taxes" (which includes tariffs) row is disaggregated by commodity to make it possible to recover the value of each entry at the tax-ridden prices agents actually face. For example, for consumer i there will be under the column heading "consumers" entries $p_k^* x_{ik}^*$ for "goods" and "factors" and entries $(p_k^{c*} - p_k^*)x_{ik}^*$ for "indirect taxes." For production taxes, firms may be further distinguished as buyers and sellers in order to keep track of taxes on inputs and outputs separately. For foreign trade separate records could be kept for import and export tariffs. Aggregate indirect taxes balance with transfers.

For any given commodity, either inputs or outputs (resp. imports or exports) will be zero in the model, but the data may contain positive values for both. This conflict has to be resolved by a decision that involves netting out or reclassification of commodities. Other conflicts may exist in the data; for example, the tariff-ridden price may conflict with the observed direction of the trade flow, and the like, and the earlier discussion in section 4.4.3 applies here.

5.4.4 Simulations

Solving the Model

Solving a national model with taxes and tariffs poses no new difficulty other than introducing total tax receipts T as an adjusting variable and ensuring

32. Obviously one could also construct SAMs using another convention, say, evaluating at consumer prices.

Table 5.1
Structure of the SAM with taxes and tariffs

	Goods	Factors	Firms	Consumers	Foreign sector	Government budget	Indirect taxes
Goods			Input	Consumption	Exports		
Factors			Input	Consumption	(Exports)		
Firms	Production	(Production)					
Consumers	(Endowments)	Endowments	Profits		Foreign profits	Transfers	
Foreign sector	(Imports)	Imports					
Government budget							Net receipts
Indirect taxes			Production taxes	Consumption taxes	Tariffs		

Note: In the SAM for the CGE model, the terms in parentheses are zero.

that consumers' incomes remain positive out of equilibrium. The international model can be solved as a single model with all the countries included. One has to determine international prices p^w (including the prices of nontraded commodities) and one T variable for each country. Alternatively, one can opt for a decomposition and compute a solution for every country model at given prices p^w, setting the T variable so as to satisfy balance of payments (and hence government budget) equilibrium. Under mild conditions[33] on monotonicity of consumer demand functions, this solution will be unique for each country and the net import correspondence will be well-behaved, so prices p^w can be adjusted in the linkage part.

Recall from section 4.4.4 that if there is one consumer per country and if indirect taxes and tariffs are absent, decomposition can be pursued by restricting the price adjustment in the linkage to internationally traded commodities. However, if indirect taxes or tariffs are present, such a national model may, for given prices of internationally traded commodities, have several equilibria. In this case the net import function will not be well-behaved, and this decompositon scheme may fail to work.

Tax and Tariff Reforms at National Level

It was recognized early in the public finance literature that the study of tax reform calls for numerical simulation because tax changes typically involve opposing effects, and it is difficult to predict the final result on theoretical grounds only. This intuition was corroborated later on, in particular, in the theoretical literature on transfer paradoxes,[34] which shows that paradoxical outcomes are crucially dependent on functional specifications and numerical values of parameters. This dependence makes it important to develop models with a sound empirical basis.

National tax reforms (sometimes including social security) were the first modeling applications for developed countries. The main issues are avoiding double taxation by replacing taxes on intermediate inputs by a value-added tax, namely a tax on consumption, making tax rates more uniform across social groups and commodities, replacing indirect taxes by direct ones, and reducing social security charges to increase employment. Some authors have also looked for optimal second-best tax structures when there is no possibility of giving lump-sum transfers or taxing consumption at a uniform rate (e.g., see Mitra 1992). The number of such applications is extremely large (for surveys, see Shoven and Whalley 1984; Powell and Lawson 1990; see also Bergman et al. 1990; Don et al. 1991; Scarf and Shoven 1984; Shoven and Whalley 1992).

33. See Keyzer (1981).

34. See, for example, Bhagwati et al. (1983). The paradoxes are in fact implied by the lack of generic properties of the excess demand function. See sections 1.2.4 and 2.2.3.

Many studies of international trade liberalization find that welfare gains are modest. One reason for this is that the important channel through which liberalization leads to growth (dynamics) is absent from the standard model. First, trade liberalization improves access to technology, either because it is incorporated in imported goods or because it is tied to foreign investment. Second, it improves competition, but this effect is not well captured in standard models, which assume perfect competition, ignore rent-seeking and keep trade margins fixed. Hence in such models the initial situation may be insufficiently distorted so that the potential gains are underestimated. Later applications have been able to show larger effects in various ways: by introducing rent-seeking and imperfect competition or by considering capitalization effects through an intertemporal model. De Melo (1988) and Shoven and Whalley (1984) give surveys of trade policy applications; see also Srinivasan and Whalley (1986).

Tariff Reforms at International Level

International trade reform AGE models have been constructed to study multilateral trade reforms by linking national AGE models. Whalley (1985) has used various multiregional models to assess Tokyo Round Proposals. One of his results is that regions often gain less from their own proposal than from the proposals of other negotiating partners. The reason for this is that the own proposal often maintains more internal protection. He also shows that a single developing country is likely to gain from unilateral trade liberalization (partly because consumers will gain from lower domestic prices of imports). The country will also gain when trade liberalization is global (so that developing countries have better access to markets in developed countries). However, when a group of developing countries liberalizes (without reciprocation by developed countries), they may lose because of an induced terms of trade loss. Similar results are obtained by Deardorff and Stern (1986).

The emphasis placed in the Uruguay Round on reform of agricultural policies has led to construct models tailored to this issue. Examples are the Basic Linked System models described in Parikh et al. (1987), Fischer et al. (1988), and the RUNS model used by the OECD (Gunning et al. 1982; Burniaux and van der Mensbrugghe 1990). A comparison of results of agricultural trade liberalization scenarios in various (general equilibrium and other) models is given in Goldin and Knudsen (1990) and Goldin and Winters (1992). One of the results of many of these models is that poor groups in developing countries may lose from trade liberalization because they are net buyers of food, and world food prices rise as a result of reform (see also Srinivasan 1989).

Welfare gains are often limited because factor mobility is restricted in the short run. Experiments for multilateral liberalization show much larger gains than for unilateral liberalization and losers can easily be compensated. Hamilton and Whalley (1984) and Fischer et al. (1991) emphasize that the barriers usually ignored in trade negotiations, namely constraints on international migration, are very important: The welfare gains obtained by allowing migration are significantly larger than those obtained through trade liberalization.

The consequences of regional integration resulting from the Single European Market of 1992 and of the North American Free Trade Agreement have been investigated by Burniaux and Waelbroeck (1992), Gasiorek et al. (1992a, 1992b), and Mercenier (1994) for Europe and by Cox and Harris (1985), de Melo and Tarr (1992), and Harris (1984) for North America. These studies focus on welfare gains from increasing returns (in fact setup costs) and improved competition. Such models tend to generate larger welfare gains than those that assume convex technologies and perfect competition.[35]

35. This issue will be taken up in more detail in section 11.5.2.

6 Price Rigidities

Price rigidities have played an important role in theoretical debates and macroeconomic modeling, especially in relation to labor and unemployment in market economies and rationing of consumer goods in centrally planned economies. Though the current literature places less emphasis on the topic, exogenous price rigidities that are implemented through rationing, buffer stock schemes, and a wide array of other forms of rationing (e.g., to help the poor with ration cards, or to support prices for domestic producers) are still part of current policies in many countries and have to be accounted for in applied modeling.

6.1 Price Rigidities and Supporting Mechanisms

In chapters 4 and 5 we saw that in an economy with international trade, export and import prices set bounds on domestic prices. Under the small-country assumption, these bounds are rigid, and if the domestic price reaches the import price, imports will adjust and clear the domestic market, while price remains fixed. This illustrates that quantity adjustments can be used to support price rigidities. The question naturally arises whether such a rigidity can also be enforced for a commodity that is not traded on international markets. We will see that this is the case if commodities can be purchased and sold from and to a buffer stock. This is in fact how we will introduce price rigidities in the model, distinguishing the various schemes according to the rules for the holding of the stocks and their financing.

The following examples illustrate the mechanism. In the European Union the buffer stock of milk powder, which accumulates as a consequence of milk price support policies, is centrally financed and controlled. On the labor market, wage agreements often contain provisions on minimum wages. These agreements can be sustained by an unemployment benefit fund that "buys" the labor supplied by the unemployed at the minimum wage; "unemployed" labor is then paid out of workers' contributions to the fund (or out of public funds). If the benefits fall short of the minimum wage, the difference can be interpreted as the unemployed's own contribution to the fund. Alternatively, this contribution can be looked at as a "virtual" payment that the unemployed pays to himself for hiring his own labor. A wage rigidity can also be implemented by rationing some workers in their supply of labor. The textbook case for rationing occurs if the unemployed have to carry the full cost of the ration.

In this chapter we treat these rigidities as exogenous bounds on prices, though the literature has emphasized the need for explaining them in an endogenous way. Price rigidities are commonly seen as resulting from short-run sluggishness of the market-clearing mechanism or from government (or private) intervention. In the first interpretation, all agents face the same prices as in the competitive model, but trading opportunities are restricted to account for the fact that competitive equilibrium is not achieved instantaneously. Such an approach is a compromise between the representations of the market-clearing process through perfect price flexibility and those in which agents are trading out of equilibrium.[1] Price adjustments are not represented explicitly, but this was already so in the competitive case.

The second interpretation considers explicit government (or private) intervention as the main cause of price rigidities. In the earlier examples of the milk powder market and unemployment, the price rigidity and the associated buffer stock adjustment are indeed the consequence of a government or of a collective arrangement, and the exogeneity of the bound is natural. In chapters 7 and 8 we will consider dynamic models in which stock accumulation (and depletion) over time and the apparent price rigidities are the consequences of optimizing intertemporal behavior; the price rigidity will then become endogenous through its relation to future prices. In chapters 9 and 10 (sections 9.1.5 and 10.4.2) wage rigidities are endogenized also: The productivity of the employed improves if they earn more. Finally, in chapter 11 (section 11.3), the rigidity is due to the price-setting behavior of producers who operate under imperfect competition.

6.2 Price Rigidities in General Equilibrium

In this section we consider a setting in which the commodities subject to price rigidities are stored centrally and consumers finance the scheme (section 6.2.1). We then turn to the setup used by Drèze and Benassy, among others, in which consumers are rationed in their supplies or their demands; the "stockholding" is decentralized over consumers through rations that are not tradable among agents (sections 6.2.2 to 6.2.4). If the opportunity of trading rations is available, the price for rations will be the same for all agents, and it will act like an indirect tax (section 6.2.5). These forms of rationing will implement price rigidities. In practice, rationing is often used in a different way, for example, as an entitlement to buy or sell a given quantity at a privileged price (section 6.2.6). Finally, the schemes of consumer rationing are extended to include producers as well (section 6.2.7).

We start with the open economy model of definition 3.5 and consider

1. Trading out of equilibrium means that exchanges occur already at prices that differ across exchange contracts. See Arrow and Hahn (1971, chs. 12, 13). See also section 1.2.5.

enforcing lower and upper bounds p^l and p^u on domestic prices such that[2]

$0 < \bar{p}^e < p^l \leqslant p^u < \bar{p}^m$.

Let p^* be an equilibrium solution of this model prior to any enforcement. If by coincidence, this solution happens to satisfy $p^l \leqslant p^* \leqslant p^u$, then no further action is required, since the bounds on prices do not matter. When both imports and exports happen to be zero, then, if for some positive ρ the constraint $p^l \leqslant \rho p^* \leqslant p^u$ can be satisfied, the bounds on domestic prices can be achieved by simple renormalization of prices. In the sequel, we will, however, be interested in solutions that need additional policy intervention. This intervention will take three forms: centrally held stocks, rations, and indirect taxes.

6.2.1 Centrally Held Buffer Stocks

Price constraints p^l and p^u are implemented through constrained purchases s^l and sales s^u by a buffer stock agency, in such a way that $s^l \leqslant \bar{s}^l$, $s^u \leqslant \bar{s}^u$, where $\bar{s}^l$ is the maximum stock increase allowed while $\bar{s}^u$ is the opening stock (an endowment). The model in open economy format is then written

$$\max \bar{p}^e e - \bar{p}^m m + p^l s^l - p^u s^u,$$

$$e, m \geqslant 0,\ s^l, s^u \geqslant 0,\ y_j, \text{ all } j,$$

$$\text{subject to} \qquad (6.1)$$

$$\Sigma_j y_j + m - e + s^u - s^l \geqslant \Sigma_i \hat{x}_i - \Sigma_i \omega_i \qquad (p),$$

$$s^l \leqslant \bar{s}^l \qquad (\mu^l),$$

$$s^u \leqslant \bar{s}^u \qquad (\mu^u),$$

$$y_j \in Y_j,$$

for given $\hat{x}_i$ and with feedback

$$\hat{x}_i = \operatorname{argmax}\{u_i(x_i) \text{ subject to } px_i \leqslant h_i,\ x_i \geqslant 0\}. \qquad (6.2)$$

The income h_i of consumer i contains a lump-sum transfer T_i and can be written

$$h_i = h_i^0 + T_i(\,.\,), \qquad (6.3)$$

where $h_i^0 = p\omega_i + \Sigma_j \theta_{ij} \Pi_j(p)$ is income before taxation; $T_i(\,.\,) = T_i(p, T, h_1^0,$

2. Here we consider bounds on nominal prices only. In section 6.4 we consider bounds on relative prices and cases in which some prices are pegged or related to other prices. Domestic prices then follow the normalization imposed by international prices, but this will have no effect on equilibrium allocations as long as demand and supply are homogeneous of degree zero in prices.

$h_2^0, \ldots, h_m^0$), for all i, satisfies assumption T of section 5.1.2 and

$$T = ps^u - ps^l. \tag{6.4}$$

Note that if an upper bound on the stock activities s^l and s^u is reached for some commodity k, the Lagrange multiplier p_k may fall outside the range $[p_k^l, p_k^u]$, in which case the price rigidity is not maintained. We can now define an equilibrium with buffer stocks as:

DEFINITION 6.1 (Equilibrium with buffer stocks) The allocation y_j^*, all j, x_i^*, all i, s^{l*}, s^{u*}, m^*, and e^* supported by the price vector $p^* \geqslant 0$ and a level of transfers T^* is an equilibrium with buffer stocks for the open economy if

1. for every producer j, $y_j^* \in \operatorname{argmax}\{p^* y_j$ subject to $y_j \in Y_j\}$; Π_j^* is his optimal profit;

2. for every consumer i, $x_i^* = \operatorname{argmax}\{u_i(x_i)$ subject to $p^* x_i \leqslant h_i^*$, $x_i \geqslant 0\}$, where $h_i^* = h_i^{0*} + T_i^*$ is after tax income and $h_i^{0*} = p^*\omega_i + \Sigma_j \theta_{ij} \Pi_j^*$ is pre-tax income;

3. proceeds from consumer taxes are fully redistributed: $T_i^* = T_i(p^*, T^*, h_1^{0*}, h_2^{0*}, \ldots, h_m^{0*})$, where $T^* = p^*(s^{u*} - s^{l*})$;

4. all markets are in equilibrium:

$$\Sigma_i x_i^* - \Sigma_j y_j^* - \Sigma_i \omega_i + e^* - m^* - s^{u*} + s^{l*} \leqslant 0 \perp p^* \geqslant 0,$$

$$e^* \geqslant 0 \perp p^* \geqslant \bar{p}^e \quad \text{and} \quad m^* \geqslant 0 \perp p^* \leqslant \bar{p}^m,$$

$$s^{u*} \leqslant \bar{s}^u \perp \mu^{u*} \geqslant 0 \quad \text{and} \quad s^{l*} \leqslant \bar{s}^l \perp \mu^{l*} \geqslant 0,$$

$$p^* \leqslant p^u + \mu^{u*} \perp s^{u*} \geqslant 0 \quad \text{and} \quad p^* \geqslant p^l - \mu^{l*} \perp s^{l*} \geqslant 0.$$

This equilibrium reduces to an equilibrium of a closed economy if $e^* - m^* = 0$.

Existence of equilibrium follows from proposition 6.1.

PROPOSITION 6.1 (Existence of equilibrium with buffer stocks) Under assumptions S and T:

1. If $0 < \bar{p}^e < p^l \leqslant p^u < \bar{p}^m$, then there exists an allocation x_i^*, all i, y_j^*, all j, s^{l*}, s^{u*}, e^*, m^*, a price vector p^*, and transfers T_i^*, which define an open economy equilibrium with buffer stocks as in definition 6.1.

2. All solutions in part 1 will satisfy the condition of trade balance equilibrium: $\bar{p}^m m^* - \bar{p}^e e^* = 0$.

3. If $0 < \bar{p}^e < p^l \leqslant p^u < \bar{p}^m$ and if $\bar{p}^e$ is sufficiently close to 0, the solution will be a closed economy equilibrium with buffer stocks.

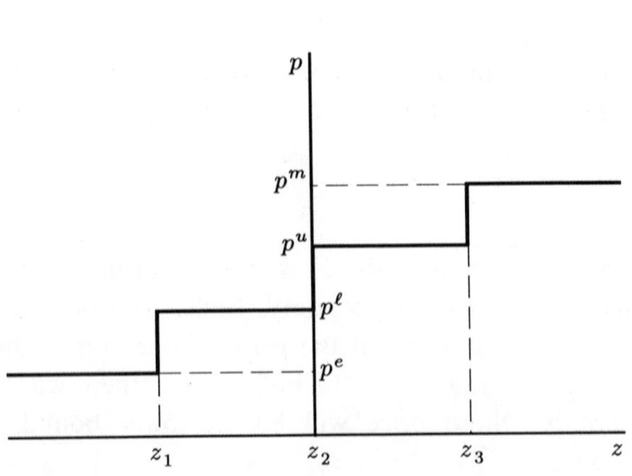

Figure 6.1
Alternative regimes under price rigidities.

Proof *Part 1:* The proof is analogous to that of proposition 5.1; the main difference is that when one proves that transfers T^* exceed T_0, the complementarity conditions 4 of definition 6.1 have to be used.

Part 2: This follows from the Kuhn-Tucker conditions and satisfaction of budget constraints.

Part 3: One can prove this as in part 2 of proposition 5.1. ■

We can further describe the properties of the equilibrium as follows: First, due to the price rigidities, the equilibrium solution can lie in one of several regimes. This is illustrated in figure 6.1 where $z = (m - e) + (s^u - s^l)$, the net deliveries from trade and buffer stock are shown in relation to prices. Consider, for example, the case $z = z_1$. At that point, $s^l = \bar{s}^l$, $s^u = m = e = 0$, and $p^e \leqslant p \leqslant p^l$. For $z < z_1$, $p = p^e$ and $e > 0$, while for $z_1 < z < z_2$, $0 < s^l < \bar{s}^l$, $e = 0$, and $p = p^l$.

Rigidities can no longer be maintained when constraints on stocks become binding. Prices then become flexible until they reach either the import or the export price; from that point onward exports or imports adjust. For nontradable commodities, such as labor,[3] the price range between $\bar{p}^e$ and $\bar{p}^m$ can be set sufficiently wide to avoid international trade altogether.

Second, note that $T = ps^u - ps^l$ represents the budget constraint of the buffer stock agency, whose operations are financed from direct taxes on consumers. In particular, if $\bar{s}^u$ is zero, stock operations, if any, will have to be fully financed by consumers. Alternatively, we could assume that domes-

3. In that case $\bar{s}^u = 0$, since labor cannot be stored; $\bar{s}^l$ will stand for the total labor endowment.

tic consumers do not contribute to the financing of the buffer agency so that $T_i = 0$ for all i. Then, in equilibrium, the trade balance will accommodate for the buffer stock deficit and

$$\bar{p}^e e - \bar{p}^m m = p^u s^u - p^l s^l.$$

It is now clear that if the trade balance is in equilibrium, $p^u s^u - p^l s^l = 0$, which shows that in the closed economy, buffer stocks can be made self-financing. It also indicates that if the price of one commodity hits its upper bound and requires sales from the buffer stock, there will be at least one other commodity whose price will hit its lower bound, requiring purchase by the agency.

Finally, the closed model may also be seen as a straight reinterpretation of the small economy model of proposition 4.2, where the import and export prices are p^u and p^l, respectively. The bounds on stocks are then quotas and an equilibrium will exist with $T^* = 0$ so that $T_i^* = 0$ also.

We have discussed two financing rules of the buffer agency; one where the consumer does not contribute at all ($T = T_i = 0$), and one where the consumer contributes through direct taxes.[4] We now show that decentralization of the stockholding activity through rationing can also be used to enforce price rigidities and makes it possible to reduce or eliminate buffer stock operations. We subsequently study how the same can be achieved through indirect taxation. These types of rationing have been studied extensively in the theoretical literature but are not often found in applied work. Applications use buffer stocks to implement price rigidities for nontradables (including labor), and foreign trade for tradables; rations are in general treated as fixed and implemented in the way discussed in section 6.2.6. The reader who is mainly interested in applications may immediately proceed to this section.

6.2.2 Shifting the Adjustment to Consumers

The Concepts of Drèze, Benassy, Virtual Tax Rationing, and Soft Rationing

We will consider three standard forms of decentralizing buffer stock activity through rationing; these are known as Drèze rationing (Drèze 1975), Benassy rationing (Benassy 1975), and virtual taxation (Cornielje and van der Laan 1986, building on ideas by Kurz 1982). In each scheme consumers receive vouchers to buy or sell rations; these vouchers are nontradable (Drèze and Benassy) or tradable (virtual taxation). In the Drèze scheme prices are constrained to lie within a given range, whereas Benassy assumes that the two bounds coincide so that all prices are fixed. A less important

4. Indirect taxes on goods for which no rigidity is effective could also be a source of funding; see chapter 5.

distinction between the two models concerns the way in which consumers are rationed: In Drèze's model every consumer is allowed to purchase or sell the same quantity (uniform rationing), while Benassy computes the rations taking into account the consumers' preferences.

We fully develop Drèze's model and then indicate how it can be modified in order to account for Benassy rationing and for virtual taxation. Our approach will deviate from the presentation that is standard in the literature because we want to present the various rationing schemes in a unified way, relating them to buffer stock operations. Unlike many theoretical models with rationing, our formulation will include producers, which is clearly an important requirement for applied work. We finally present a model with soft rationing, in which agents will be given the opportunity to buy or sell limited amounts at a privileged price. This form of rationing is often used in actual policy regulations and is relatively easy to model.

In terms of buffer stock operations, the idea of rationing schemes is to constrain purchases (resp. sales) when the price hits its upper (resp. lower) bound. In Drèze's model, the bounds do not coincide. Then, in equilibrium, there may be a surplus if the upper bound is reached and rationing of (net) sellers becomes effective. Likewise there may be a shortage if the lower bound is reached, and this will imply a rationing of (net) buyers. If, as in Benassy's model, the upper bound is equal to the lower bound, then the equilibrium price is obviously known, but it does not convey any information as to whether there is a shortage or a surplus. To determine the nature of the imbalance, Benassy introduces the concept of "effective" demand: Each consumer is "asked" to reveal whether he would like to buy more or sell less. Alternatively, mutations in central stocks can also provide the information on the imbalance: There is a shortage if stocks are sold and a surplus if they accumulate.

Once the side of the market to be rationed is known, the intensity of the rationing has to be determined. For this, let r_i^l and r_i^u be the lower and the upper bound on consumer i's net demand ($r_i^l \leqslant x_i - \omega_i \leqslant r_i^u$). Assume now that for some commodity k, a lower bound on prices is reached and $r_{ik}^l = 0$ for all i. Then no further reduction in sales is possible, and a surplus could persist, which necessitates a stock purchase s_k^l. Note that this can only happen if producers supply the commodity. There are basically two ways to avoid such a surplus. The first is to impose a zero upper bound on the stock so that the price constraint will not be maintained. The second is to involve producers in the rationing.[5]

5. This problem does not arise in the standard models with rationing, since these do not consider producers. Moreover in many applications unemployment is the main concern, and there is usually no need to ration producer demand for labor. Since producers cannot supply labor, there is no possibility of rationing their supply either.

To introduce virtual taxes, we write the model of the rationed consumer as

$$\max u_i(x_i),$$
$$x_i \geqslant 0,$$
$$\text{subject to} \tag{6.5}$$
$$px_i \leqslant h_i^0 \qquad (\lambda_i),$$
$$r_i^l \leqslant x_i - \omega_i \qquad (\psi_i^l),$$
$$x_i - \omega_i \leqslant r_i^u \qquad (\psi_i^u),$$

where $h_i^0 = p\omega_i + \Sigma_j \theta_{ij} \Pi_j(p)$.

Assuming that Slater's condition holds, the Lagrangean of the program can be written

$$L(.) = u_i(x_i) - \lambda_i(px_i - h_i^0) + \psi_i^l(x_i - \omega_i - r_i^l) - \psi_i^u(x_i - \omega_i - r_i^u),$$

or, alternatively, after rearrangement of terms

$$L(.) = u_i(x_i) - \lambda_i[(p - \xi_i^l + \xi_i^u)x_i - (h_i^0 - \xi_i^l(\omega_i + r_i^l) + \xi_i^u(\omega_i + r_i^u))],$$

where $\xi_i^l = \psi_i^l/\lambda_i$ and $\xi_i^u = \psi_i^u/\lambda_i$, assuming nonsatiation of utility, so that $\lambda_i > 0$.

Virtual indirect taxes are now $\xi_i^u - \xi_i^l$, and we can define a consumer price $p_i^c = p - \xi_i^l + \xi_i^u$ and a virtual direct tax $T_i = -\xi_i^l(\omega_i + r_i^l) + \xi_i^u(\omega_i + r_i^u)$ such that the solution of the original consumer problem coincides with the solution of the following nonrationed consumer problem:

$$\max u_i(x_i)$$
$$\text{subject to}$$
$$p_i^c x_i \leqslant h_i^0 + T_i.$$

Note that if consumer i sells commodity k, then $x_{ik} < \omega_{ik}$, and he cannot be constrained on his purchase, so $\xi_{ik}^u = 0$ and $p_{ik}^c = p_{ik}^l = p_k - \xi_{ik}^l$. Likewise, if $x_{ik} > \omega_{ik}$, then $p_{ik}^c = p_{ik}^u = p_k + \xi_{ik}^u$. We thus arrive at a model with two consumer prices p_{ik}^l and p_{ik}^u. Since $p_{ik}^l \leqslant p_{ik}^u$, utility maximization subject to a budget constraint will lead the consumer to first buy at the low price p_{ik}^l and only consider purchasing at the high price when he has exhausted the first possibility, that is his endowment ω_{ik}.

Under the virtual tax scheme the ξ's will be equal for all sellers and for all buyers, so the agent-specific prices will be set to $p_i^l = p - \xi^l$ for sellers and to $p_i^u = p + \xi^u$ for buyers.

Pure Equilibrium with Rationing of Consumers: Assumptions and Definitions

In the Drèze model we will assume that price bounds do not coincide:

ASSUMPTION **R1** (Positive price interval) $p^l < p^u$.

If all commodities are subject to rationing, it may happen that the budget constraint cannot be met for some consumer, since he cannot spend his income fully. For reasons that will become clear later on, we require consumers to have the possibility of spending all their income. Therefore we exclude one commodity from the rationing scheme:

ASSUMPTION **R2** (No rationing for one commodity) For at least one commodity, say, the first, $p_1^l < \bar{p}_1^e$, $p_1^u > \bar{p}_1^m$ and $\bar{p}_1^m = \bar{p}_1^e > 0$ so that price rigidities are always ineffective.

This nonrationed commodity will serve as numéraire. We will see that assumption R2 is not restrictive if only relative prices matter. To allow for positive consumption in case rationing does not permit a consumer to buy on the market, we will use assumption C2 instead of the standard assumption C2′. Finally, the proof will make use of single-valuedness of the Lagrange multipliers of the consumer problem; for this, we need to modify assumption C1″ (section 2.3) into CR1:

ASSUMPTION **CR1** The utility function u_i: $R_+^r \to R_+$, $u_i(x_i)$ is continuous, strictly concave, and satisfies $u_i(0) = 0$ for all i. Moreover it is continuously differentiable on R_{++}^r, satisfies the boundary condition $\lim_{x \to x^0} \| \partial u_i(x_i)/\partial x_i \| = \infty$ whenever $x_k^0 = 0$, and is uniformly bounded away from zero for the numéraire $k = 1$, that is, $\partial u_i(x_i)/\partial x_{i1} \geqslant \varepsilon > 0$.

As shown in the proof of proposition 6.2, this assumption ensures, in conjunction with assumptions R2 and C2, that in the rationed problem (6.5), consumption is positive, λ_i is positive, and Lagrange multipliers λ_i, ψ_i^l, and ψ_i^u are unique.

We can now define an equilibrium with pure consumer rationing as follows:

DEFINITION **6.2** (Equilibrium with pure consumer rationing) The allocation y_j^*, all j, x_i^*, all i, supported by the price vector $p^l \leqslant p^* \leqslant p^u$ and rations r_i^{l*}, r_i^{u*}, is an equilibrium with pure consumer rationing if the following conditions are satisfied:

1. For every producer j, $y_j^* \in \operatorname{argmax}\{p^* y_j$ subject to $y_j \in Y_j\}$; Π_j^* is his optimal profit.
2. For every consumer i, $x_i^* = \operatorname{argmax}\{u_i(x_i)$ subject to $p^* x_i \leqslant h_i^*$, $r_i^{l*} \leqslant x_i - \omega_i \leqslant r_i^{u*}$, $x_i \geqslant 0\}$ where $h_i^* = p^* \omega_i + \Sigma_j \theta_{ij} \Pi_j^*$.

3. All markets are in equilibrium, $\Sigma x_i^* = \Sigma_j y_j^* + \Sigma_i \omega_i$.

4. The rationing scheme satisfies the properties of pure rationing:

a. There is no forced trade. Every agent can choose not to trade, $r_i^l \leqslant 0$ and $r_i^u \geqslant 0$.

b. Only one side of the market is rationed. If for given k, $r_{ik}^{l*} = x_{ik}^* - \omega_{ik}$ for any i, then $x_{ik}^* - \omega_{ik} < r_{ik}^{u*}$ for all i, and if $r_{ik}^{u*} = x_{ik}^* - \omega_{ik}$ for any i, then $r_{ik}^{l*} < x_{ik}^* - \omega_{ik}$ for all i.

c. Rationing occurs only when the price is at bound. If for given k, $p_k^* > p_k^l$, then $r_{ik}^{l*} < x_{ik}^* - \omega_{ik}$ for all i, and if $p_k^* < p_k^u$, then $r_{ik}^{u*} > x_{ik}^* - \omega_{ik}$.

For models without production, price rigidities can be enforced by rationing consumers only. However, to make the model relevant for applications, we have to consider production. Then it may not be feasible to enforce price rigidities with consumer rationing alone, and buffer stock operations may be necessary. However, even this will be insufficient if the buffer stock is depleted, in which case imports are needed. We will show that if stock activities are zero in equilibrium, we obtain the equilibrium with pure consumer rationing of definition 6.2. Moreover, in such an equilibrium, exports and imports will also be zero. Later, in section 6.2.7, we will see that when producers can be rationed, no buffer stocks will be needed either.

6.2.3 Drèze Rationing

The Drèze model basically follows definition 6.2, but it assumes that every consumer faces the same exogenous rations $r_i^l = r^l \leqslant 0$, $r_i^u = r^u \geqslant 0$ on his net trades, and this is known as uniform rationing. We now introduce the existence proof for this model, but we first discuss in some detail the components of the fixed point mapping used in the proof because this will make the relation between Drèze rationing and buffer stocks more explicit. The fixed point mapping will, as in Proposition 3.4, consist of a mathematical program (in the open economy format) with feedbacks. This program determines the levels of buffer stocks and rations as primal variables and prices as dual variables. The feedback relations primarily consist of the rationed consumer problems (6.5) with uniform rations that generate consumptions and virtual tax rates. However, there are also two linking functions that ensure that the equilibrium will possess the properties of pure rationing given in part 4 of definition 6.2; these functions adjust virtual tax rates and rations differently according to whether there is a surplus or a shortage on the market.

The mathematical program is a modification of program (6.1), where we add indirect tax terms ξ_i^l and ξ_i^u in the objective, as was done in chapter 5 for indirect taxes.

We adjust the constraints of this program, introducing endogenous bounds $z^l \leqslant 0$ and $z^u \geqslant 0$ on net trades of every consumer, which will be set equal to the smallest (resp. largest) net demand whenever the tax ξ_i^l (resp. ξ_i^u) is positive for consumer i. This program is

$$\max \bar{p}^e e - \bar{p}^m m + p^l s^l - p^u s^u + (\Sigma_i \xi_i^l) z^l - (\Sigma_i \xi_i^u) z^u,$$

$$e, m \geqslant 0,\ s^l, s^u \geqslant 0,\ z^l \leqslant 0,\ z^u \geqslant 0,\ y_j \text{ all } j,$$

$$\text{subject to} \qquad (6.6)$$

$$\Sigma_j y_j + m - e + s^u - s^l \geqslant \Sigma_i \hat{x}_i - \Sigma_i \omega_i \qquad (p),$$

$$z^l \leqslant \hat{x}_i - \omega_i \leqslant z^u, \text{ all } i,$$

$$z^l \geqslant -\eta,$$

$$z^u \leqslant \eta,$$

$$y_j \in Y_j,$$

for given $\hat{x}_i$ and ξ_i^l, ξ_i^u and fixed parameter η, the upper bound on z^u and lower bound on z^l; for given $(\bar{p}^m, \bar{p}^e, p^u, p^l)$, the value of η is chosen sufficiently large to exceed any possible consumption that satisfies the budget constraint for $p \in [p^l, p^u]$. Since p^l is positive and the production sets are compact, η is finite. To make $-\eta$ suitable as a lower bound for z^l, it should also satisfy $\eta > \Sigma_i \omega_i$. This construction is only introduced to keep z^l and z^u compact-valued.

The consumer problems are specified according to (6.5), but with uniform rations; they generate demands $\hat{x}_i$ and Lagrange multipliers λ_i, ψ_i^l, and ψ_i^u, which make it possible to compute rates of (virtual) indirect taxes ψ_i^l/λ_i and ψ_i^u/λ_i; these rates can, in principle, be entered in the objective of program (6.6). To ensure that all the adjustment is shifted to consumers and that buffer stock activities are zero in equilibrium, the adjustments are also specified to make rations binding whenever s^l or s^u are positive:

$$\xi_{ik}^l = \max\left(\frac{\psi_i^l}{\lambda_i}, s_k^l\right),$$

$$\xi_{ik}^u = \max\left(\frac{\psi_i^u}{\lambda_i}, s_k^u\right), \qquad (6.7)$$

where s_k^l and s_k^u are given levels of stock activities. In the fixed point these have to be equal to the values obtained in the program (6.6). Note that if s_k^l (or s_k^u) is positive, so will be ξ_{ik}^l (or ξ_{ik}^u) and that if in (6.6), ξ_{ik}^l (or ξ_{ik}^u) is positive, then z_k^l (or z_k^u) will either be at a bound or zero.

Likewise the rations z^l and z^u generated by (6.6) are adjusted in order to make rations more binding if there is stock activity and less binding if price

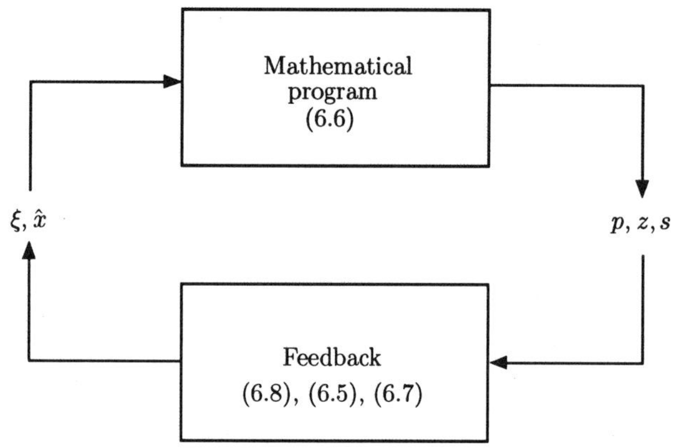

Figure 6.2
Fixed point mapping for a Drèze rationing equilibrium.

bounds are ineffective:

$$
\begin{aligned}
r_k^l &= \min(z_k^l + s_k^l - \gamma(p_k - p_k^l), 0), \\
r_k^u &= \max(z_k^u - s_k^u + \gamma(p_k^u - p_k), 0);
\end{aligned}
\tag{6.8}
$$

γ is a positive scalar that converts prices to quantities. In the fixed point we must have $r_k^l = z_k^l$ or $r_k^u = z_k^u$ whenever the price bound is reached. Note that by (6.8), we have $r_k^l \leqslant 0$ and $r_k^u \geqslant 0$, so the "no forced trade" requirement of definition 6.2 is satisfied.

The mathematical program (6.6) with the feedbacks (6.8), (6.5), and (6.7) defines a fixed point mapping that starts from given prices p, rations z, and stock activities s, updates these through the feedback, and feeds the resulting values of taxes ξ and consumptions $\hat{x}$ into the mathematical program which in turn generates new values for p, z, and s. This mapping is illustrated in figure 6.2.

We can now state and prove the following existence result:

PROPOSITION 6.2 (Existence of equilibrium with Drèze rationing) Under assumptions CR1, C2, P1′, R1, and R2:

1. There exists an allocation x_i^*, all i, y_j^*, all j, stock activities s^{l*}, s^{u*}, a price vector p^*, and rations r^{l*}, r^{u*}, defining an equilibrium with rationing and stocks that is a fixed point of the mapping (6.5)–(6.8). Moreover, for any given commodity k, the following properties hold at the fixed point:

a. *There is no forced trade.* Every agent can choose not to trade, $r_i^{l*} \leqslant 0$ and $r_i^{u*} \geqslant 0$.

b. *Only one side of the market is rationed.* If $r_{ik}^{l*} = x_{ik}^* - \omega_{ik}$ for any i,

$x_{ik}^* - \omega_{ik} < r_{ik}^{u*}$ for all i, and if $r_{ik}^{u*} = x_{ik}^* - \omega_{ik}$ for any i, then $r_{ik}^{l*} < x_{ik}^* - \omega_{ik}$ for all i.

c. *Rationing occurs only when the price is at bound.* If for given k, $p_k^* > p_k^l$, then $r_{ik}^{l*} < x_{ik}^* - \omega_{ik}$ for all i, and if $p_k^* < p_k^u$, then $r_{ik}^{u*} > x_{ik}^* - \omega_{ik}$.

d. *Stockholding.* No buying occurs if there is shortage, $r_k^{l*} < 0$ or $\Sigma_j y_{jk}^* \leqslant 0$ implies that $s_k^{l*} = 0$, and no selling occurs if there is a surplus, $r_k^{u*} > 0$ or $\Sigma_j y_{jk}^* \geqslant 0$ implies that $s_k^{u*} = 0$.

2. If stock activities are zero in equilibrium, then 1.a–d defines a pure rationing equilibrium as in definition 6.2, with uniform rations.

Proof *Part 1:* We start this proof with a review of the properties of the programs (6.5) and (6.6).

In the rationed consumer program (6.5), feasibility is ensured by requiring that $r^l \leqslant 0$ and $r^u \geqslant 0$, as expressed in (6.8) and by assumption R2 (which ensures that the budget equation can be satisfied also with equality). Boundedness follows from the fact that $p \geqslant \bar{p}^e$. To verify existence of Lagrange multipliers, note that by assumption C2, income will be positive; then by assumption CR1, consumption will be positive, and marginal utility of consumption will be unique and bounded. By assumption R2, there is no rationing on commodity $k = 1$, so $\partial u_i/\partial x_{i1} = \lambda_i p_1$. Hence, by assumption CR1, for given positive p_1 (by assumption R2), λ_i will be unique, positive, and greater than ε/p_1. Moreover, for the other commodities, the ψ's will also be uniquely determined, via the first-order conditions, by the marginal utility and the λ_i which is uniformly bounded away from zero. Therefore (6.5) defines continuous functions $\hat{x}_i$ and Lagrange multipliers λ_i, ψ_i^l, and ψ_i^u that are continuous functions of prices p and rations r^l and r^u. Since $\lambda_i > \varepsilon/p_1$, the virtual tax functions ψ_i^l/λ_i and ψ_i^u/λ_i will also be continuous.

In program (6.6), feasibility, boundedness, and satisfaction of Slater's constraint qualification are obvious.

Step 1. Construction of the Fixed Point Correspondence

The construction has been described above and illustrated in figure 6.2. Since the feedback functions (6.5), (6.7), and (6.8) are continuous and program (6.6) generates positive prices p, stocks s^l, s^u, and rations z^l, z^u through a convex- and compact-valued uppersemicontinuous correspondence, we can, as usual, invoke the Kakutani fixed point theorem A.4.2 and conclude that a fixed point exists.

Step 2. The Fixed Point Is an Equilibrium

This can be verified as follows:

a. Producers maximize profits. This follows from the Lagrange inequality of program (6.6).

b. Consumers maximize utility subject to their budget constraint and rationing constraints. This is ensured by (6.5).

c. Markets are in equilibrium. This is so by program (6.6).

d. In (6.6), the η-constraints are nonbinding. Since in the fixed point all $\hat{x}_i$'s satisfy the budget constraints, the upper bound η on z^u will not be reached, and since the consumer cannot sell more than his endowment, the lower bound $-\eta$ on z^l cannot be reached either.

The remainder of the proof consists of showing that in the fixed point, the rationing and the stock activities satisfy parts 1a through 1d.

1a. *No forced trade.* This is ensured by (6.8).

1d. *Stockholding.* We give the proof for lower bounds only. Consider commodity k, and assume that $s_k^{l*} > 0$. Now, by (6.7), $\xi_{ik}^{l*} > 0$, and by program (6.6), $p_k^* = p_k^{l*}$. Then (6.8) gives $r_k^l = \min(z_k^{l*} + s_k^{l*}, 0)$. Here we have two cases: (1) $r_k^{l*} < 0$ so that $r_k^{l*} = z_k^{l*} + s_k^{l*}$ or (2) $r_k^{l*} = 0$ so that $z_k^{l*} + s_k^{l*} \geqslant 0$. For the first case, since $\xi_{ik}^{l*} > 0$, program (6.6) will set $z_k^{l*} = \min_i(x_{ik}^* - \omega_{ik})$, and the consumer problem (6.5) ensures that $r_k^{l*} \leqslant \min_i(x_{ik}^* - \omega_{ik})$. Thus $z_k^{l*} \geqslant r_k^{l*} = z_k^{l*} + s_k^{l*}$, which can only be true for $s_k^{l*} = 0$, a contradiction. For case 2, we have $r_k^{l*} = 0$ and $\Sigma_j y_{jk}^* \leqslant 0$. Now $s_k^{l*} > 0$ requires either that $\Sigma_j y_{jk}^* > 0$, which is excluded, or that there are imports, but if $s_k^{l*} > 0$, program (6.6) cannot generate imports. We thus have a contradiction here also, and s_k^{l*} must be zero.

1b. *Only one side of the market rationed.* We first prove that $\xi_{ik}^{l*} > 0$ for any i implies that $\xi_{ik}^{u*} = 0$ for all i, and $\xi_{ik}^{u*} > 0$ for any i implies that $\xi_{ik}^{l*} = 0$ for all i. Assume that, on the contrary, for given k, $\xi_{ik}^{l*} > 0$ and $\xi_{ik}^{u*} > 0$. This means that $z_k^{l*} = \min_i(x_k^* - \omega_{ik})$ and $z_k^{u*} = \max_i(x_k^* - \omega_{ik})$. At the same time $r_k^{l*} \leqslant z_k^{l*}$ and $r_k^{u*} \geqslant z_k^{u*}$ by (6.5). Four possibilities on the equilibrium values of rations must now be distinguished.

1. If $r_k^{u*} > 0$ and $r_k^{l*} < 0$, then in the fixed point it must be that in (6.8) we have $s_k^{l*} = s_k^{u*} = 0$ so that

$$r_k^{l*} = z_k^{l*} - \gamma(p_k^* - p_k^l), \tag{6.9}$$

$$r_k^{u*} = z_k^{u*} + \gamma(p_k^u - p_k^*). \tag{6.10}$$

Since, by assumption R1, $p_k^l < p_k^u$, either of the price terms in (6.9) and (6.10) must be nonzero, and therefore either $r_k^{l*} < z_k^{l*}$ or $r_k^{u*} > z_k^{u*}$. If strict inequality holds, the associated ration will be nonbinding. Since stock was already seen to be zero, the corresponding ξ will be zero as well (see equation 6.7), a contradiction.

2. Since $\xi_k^{u*} > 0$ was just shown to imply that $p_k^u = p_k^*$, it follows that if $r_k^{u*} > 0$ and $r_k^{l*} = 0$, then $p_k^* > p_k^l$. Hence, by program (6.6), $s_k^{l*} = 0$. Now we

are left in (6.8) with $r_k^{l*} = 0 = z_k^{l*} - \gamma(p_k^* - p_k^l)$, which again implies that $z_k^{l*} > 0$, and this contradicts nonpositivity of z_k^{l*}.

3. The case where $r_k^{u*} = 0$ and $r_k^{l*} < 0$ leads to the same contradiction.

4. Finally, if $r_k^{u*} = r_k^{l*} = 0$, it must be that either s_k^{l*} or $s_k^{u*} = 0$, since (6.6) will not set both nonzero at the same time. A contradiction will again arise via (6.9) and (6.10).

This proves that for a given commodity k, there can be no binding constraint on purchase by any consumer if there is a binding constraint on sale for some consumer, and similarly there can be no constraint on sale if there is some on purchase. Hence the purchase (sale) rations can be set at nonbinding levels (i.e., with strict inequality) without affecting the allocation.

1c. *Rationing only when the price is at a bound.* We have just seen that $p_k^* > p_k^l$ contradicts $\xi_{ik}^{l*} > 0$ and similarly that $p_k^* < p_k^u$ contradicts $\xi_{ik}^{u*} > 0$. Therefore rations can only be binding when a price is at a bound.

Part 2: We first prove that if all stocks are zero, there is a closed economy equilibrium with $e^* = m^* = 0$. Note that in equilibrium, commodity balance implies that $p^*e^* + p^*\Sigma_i x_i^* + p^*s^{l*} = p^*m^* + p^*\Sigma_j y_j^* + p^*\Sigma_i\omega_i + p^*s^{u*}$, while budget equilibrium implies that $p^*\Sigma_i x_i^* = p^*\Sigma_j y_j^* + p^*\Sigma_i\omega_i$. Now, if $s^{l*} = s^{u*} = 0$, this reduces to $p^*e^* = p^*m^*$. At this point the nonrationed commodity $k = 1$ needs a special treatment. Since, for $k \neq 1$, $p_k^e < p_k^* < p_k^m$, it follows that $e_k^* = m_k^* = 0$. Thus the balance of trade condition implies that $e_1^* - m_1^*$ is also equal to 0, and there is a solution $e_1^* = m_1^* = 0$.

All the conditions of definition 6.2 have now been verified. The fixed point is an equilibrium with pure consumer rationing and uniform rations. ■

The properties of the equilibrium deserve some comments. First, if $r_k^{l*} = 0$, then $x_{ik}^* - \omega_{ik} \geqslant 0$, and no consumer can supply commodity k. If there is no production of that commodity (i.e., if $\Sigma_j y_{jk}^* \leqslant 0$), there will be no central stock activity. Production is not contributing to the surplus, so rationing of consumers can fully implement the bound. This will be the case for, say, labor and is the common form of a Drèze equilibrium. However, if the commodity can be produced and is supplied by firms in equilibrium, rationing of consumers only may be insufficient to cope with excess supply, so a central stock will be necessary. Similarly, if $r_k^{u*} = 0$, then $x_{ik}^* - \omega_{ik} \leqslant 0$ and no consumer is allowed to buy. If some producers need the commodity, they can obtain it from a sale s_k^u by the central stock, which was assumed to be unlimited. If an upper bound is imposed on this sale and is effective, then the upper bound p_k^u can no longer be enforced and will be exceeded, possibly up to the level at which imports become profitable. Second, if in equilibrium central stocks exist, they must be financed. This can be achieved by levying taxes on consumers or, as the model of proposition 6.2 suggests, purchases can be covered by sales in lieu of taxes. Third, one-sided rationing is meant to avoid situations where two consumers, one rationed on his sale and the other on his purchase, come together and make an exchange "outside the market." Fourth, the choice of the numéraire commodity in assumption R2 is arbitrary. The proposition actually shows that it is possible to constrain r prices with rationing on at most $r - 1$ markets.

Finally, we mention that existence proofs for a Drèze equilibrium are often formulated by introducing an artificial adjustment variable g_k for every commodity, which controls all price and quantity adjustments.[6] This makes it possible to reduce the dimension of the fixed point problem and to avoid equations like (6.7)–(6.8), but the approach will work only if buffer stock operations and international trade can be disregarded. The idea is illustrated in Figure 6.3.[7] As long as the price is within its bounds, rations are kept at nonbinding levels; as soon as a bound becomes effective, one-sided rationing comes into play. Note, however, that the figure does not show which variable has to adjust if a ration reaches a zero level.

6.2.4 Benassy Rationing: Fix-Price Exchange Equilibria

Recall that when $p_k^l = p_k^u$ and central stocks are zero in equilibrium, the Drèze scheme does not provide any information on the nature of the imbalance: One does not know whether it is purchasing or selling that should be constrained. In the original model of Benassy, prices are fixed, $p = p^l = p^u$, and he describes this as a representation of the very short term only, before prices have had time to adjust. Since prices are fixed, one cannot determine from the price level which side of the market is to be rationed. Therefore Benassy endows consumers with the capability to pass on the information of the net trade they would have liked to exercise had they not been constrained. This is implemented by defining so-called effective demands,[8] according to which each consumer calculates his target demand for commodity k by solving a rationed utility maximizing problem as in (6.5), in which the constraints on commodity k are lifted. Consumer i's problem is solved for each commodity s separately:

$$\max\ u_i(x_i^s),$$

$$x_i^s \geqslant 0,$$

6. See also section A.9 of the mathematical appendix for a discussion of the conversion of a complementarity problem into an equality-constrained system.

7. Here the value of the upper bound on the ration is, for simplicity, denoted as η. In terms of (6.8), it corresponds to $\eta + \gamma(p_k^u - p_k^*)$.

8. This concept builds on the work by Keynes (1936), Clower (1960), and Leijonhufvud (1968).

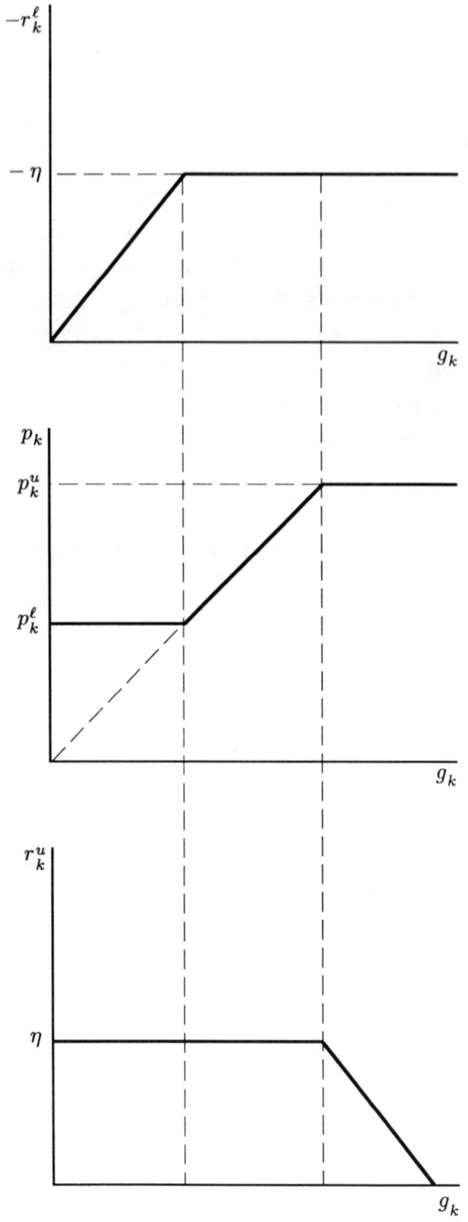

Figure 6.3
Adjustment regimes for rationing.

subject to (6.11)

$$px_i^s \leqslant h_i$$

$$r_{ik}^l \leqslant x_{ik}^s - \omega_{ik} \leqslant r_{ik}^u,\ k \neq s,$$

where $h_i = p\omega_i + \Sigma_j \theta_{ij} \Pi_j(p)$.

Thus x_i^s is consumer i's demand vector when he faces constraints on all markets except market s. Only element s is of interest here: It is called the effective demand for commodity s. These elements are stored in the net effective demand vector d_i, the elements of which are d_{is}:

$$d_{is} = x_{is}^s - \omega_{is}. \tag{6.12}$$

At fixed positive prices and under the usual strict concavity assumptions, d_{is} is a continuous function of the rations (r_i^l, r_i^u):

$$d_{is} = x_{is}^s(r_i^l, r_i^u). \tag{6.13}$$

Rations are distributed among consumers through (continuous) ration distribution functions F_{ik}^l, F_{ik}^u (defined for $k \neq 1$ only, since commodity 1 is not rationed):

$$\begin{aligned} r_{ik}^l &= F_{ik}^l(d_1, d_2, \ldots, d_m), \\ r_{ik}^u &= F_{ik}^u(d_1, d_2, \ldots, d_m). \end{aligned} \tag{6.14}$$

A Benassy equilibrium is a fixed point of the system (6.13)–(6.14). For such a solution to be a pure rationing equilibrium, the rationing functions have to possess some properties:

1. No forced trade. $r_{ik}^l \leqslant 0$ and $r_{ik}^u \geqslant 0$.
2. One-sided rationing. For given k, if $r_{ik}^l < d_{ik}$ for some i, then $r_{hk}^l \leqslant d_{hk}$ for all h, and vice versa for upper bounds.
3. Commodity 1 never rationed. $r_{i1}^l = -\infty$ and $r_{i1}^l = +\infty$.
4. Consistency. Rationed net trades[9] $z_{ik} = \min\{\max(r_{ik}^l, d_{ik}),\ r_{ik}^u\}$ should satisfy the condition $\Sigma_i z_{ik} = 0$ for all d.

These properties are also required to hold out of equilibrium. But this is unnecessarily restrictive; below we will require them to hold only in equilibrium.

In Drèze's scheme rationing was uniform. Benassy considers agents that are allowed to convey information on their preferences. The rationing scheme can reflect this and generate agent-specific rations. Whenever d_{is} does not lie within the range of the rations, the consumer problem (6.5) will

9. The same rationed net trades are obtained if one solves the consumer problem (6.5).

attribute a positive value to the corresponding Lagrange multiplier. Hence it is also possible to formulate ration distribution functions F that adjust rations away from uniformity, while remaining within the Drèze framework. For this, we generalize model (6.5)–(6.8), replacing uniform rationing by an alternative scheme. For $k \neq 1$ the rationing constraints of consumer i's problem (6.5) will now be written

$$\beta_{ik}^{l} r_k^{l} \leqslant x_{ik} - \omega_{ik} \leqslant \beta_{ik}^{u} r_k^{u}. \tag{6.15}$$

The β coefficients are agent-specific and satisfy β_{ik}^{l}, $\beta_{ik}^{u} \in [0, 1]$ and $\max_i \beta_{ik}^{l} = \max_i \beta_{ik}^{u} = 1$. In the Drèze scheme they are all equal to 1. In the Benassy scheme they are endogenized as continuous functions of effective demands d_i and possibly other variables:

$$\begin{aligned} \beta_{ik}^{l} &= G_{ik}^{l}(d_1, d_2, \ldots, d_m, r^{l}, \ldots), \\ \beta_{ik}^{u} &= G_{ik}^{u}(d_1, d_2, \ldots, d_m, r^{u}, \ldots). \end{aligned} \tag{6.16}$$

In the mathematical program (6.6) this leads to constraints

$$\beta_{ik}^{l} z_k^{l} \leqslant \hat{x}_{ik} - \omega_{ik} \leqslant \beta_{ik}^{u} z_k^{u}. \tag{6.17}$$

The objective function of the program has to be modified accordingly and is written

$$\max \bar{p}^{e} e - \bar{p}^{m} m + p^{l} s^{l} - p^{u} s^{u} + \Sigma_k \Sigma_j \xi_{ik}^{l} \beta_{ik}^{l} z_k^{l} - \Sigma_k \Sigma_i \xi_{ik}^{u} \beta_{ik}^{u} z_k^{u}. \tag{6.18}$$

The model now consists of rationed demand calculations (6.5) amended by (6.15), effective demand calculations (6.11) amended by (6.15) and (6.12) with an income that includes profits, the mathematical program (6.6) amended by (6.17) and (6.18), and the setting of rationing (6.7), (6.16), and (6.8).[10] This model will have a solution with rations r^{l*} and r^{u*}, stock activities s^{l*}, s^{u*}, consumptions x_i^*, and supplies y_j^*. Under the assumptions of proposition 6.2, this solution defines an equilibrium with rationing and buffer stocks, where consumer i faces rations $\beta_{ik}^{l*} r_k^{l*}$, $\beta_{ik}^{u*} r_k^{u*}$. An equilibrium with pure consumer rationing occurs if stock activities are zero.

6.2.5 Virtual Tax Rationing

Whereas in the Benassy framework the rationing function is specified in general terms only, Cornielje and van der Laan (1986), building on ideas of Kurz (1982), implement a special scheme through vouchers, traded among consumers at a given common price. In our framework every consumer i

10. Obviously one may be willing to leave specific consumers out of the scheme. Then the constraints (6.15) and (6.17) should be left out. This may lead to equilibria with zero rations on all other consumers and with positive stock operations.

will again face specific exogenous bounds on his net trades in (6.5):

$$\beta_{ik}^{l} r_k^{l} \leqslant x_{ik} - \omega_{ik} \leqslant \beta_{ik}^{u} r_{ik}^{u}, \tag{6.19}$$

which in the mathematical program (6.6) lead to

$$\beta_{ik}^{l} z_k^{l} \leqslant \hat{x}_{ik} - \omega_{ik} \leqslant \beta_{ik}^{u} z_k^{u}. \tag{6.20}$$

Here the β are not specified through functions but adjust until the $\zeta_{ik}^{l} = \psi_i^{l}/\lambda_i$ are equal among all sellers and the $\zeta_{ik}^{u} = \psi_i^{u}/\lambda_i$ are equal among all buyers. One could thus say that there is one market for buying and one market for selling vouchers, and a single voucher price will prevail on each market. One-sided rationing will ensure that one of the voucher prices will be zero in equilibrium.

The model now consists of rationed demand calculations (6.5) amended by (6.19), the mathematical program (6.6) amended by (6.20), the setting of rationing (6.7) and (6.8), and an adjustment rule for the β so as to equalize the price of vouchers across buyers and sellers. As in the Benassy model there will exist an equilibrium with rations r^{l*} and r^{u*} stock activities s^{l*}, s^{u*}, consumptions x_i^*, supplies y_j^* that, by the assumptions of proposition 6.2, define an equilibrium with rationing and buffer stocks. Consumer i faces rations $\beta_{ik}^{l*} r_k^{l*}$, $\beta_{ik}^{u*} r_k^{u*}$ that equalize voucher prices. If stock activities are zero, this is an equilibrium with pure consumer rationing.

6.2.6 Soft Rationing

In the fix-price literature, rationing serves as an instrument to maintain prices within bounds. In practice, there are many instances in which a government explicitly wants to ration. It will often keep these rations fixed in the short run and only adjust these with a lag so as to avoid excessive accumulation (or depletion) of stocks and budgetary costs. This government will also have to accept the fact that rations may be exceeded, and tolerate the existence of free markets with flexible prices. We refer to such rations as being "soft," since they can be exceeded.

Under soft rationing consumers may be allowed to buy limited rations at low price in a "fair price shop" or sell limited rations to a "procurement agency" at high price. Similarly producers may be allowed to buy limited quantities of inputs up to a quota or to sell limited quantities of their output at privileged prices. All this amounts to confronting consumers (resp. producers) with an upward-sloping net supply curve (resp. downward-sloping net demand curve). To illustrate this, we consider a model in which consumers are given the opportunity to buy at a fair price shop and write consumer i's problem as follows:[11]

$$\max u_i(x_i^l + x_i^r + x_i^u),$$

$$x_i^l \geqslant 0,\ x_i^r \geqslant 0,\ x_i^u \geqslant 0,$$

$$\text{subject to} \tag{6.21}$$

$$p^l x_i^l + p^r x_i^r + p^u x_i^u \leqslant p^l \omega_i,$$

$$x_i^l \leqslant \omega_i,$$

$$x_i^r \leqslant r_i,$$

for given prices $p^l \leqslant p^r \leqslant p^u$.

If the inequalities on prices hold strictly, and consumer i wants to consume more than his endowment ω_i, he will not sell and first set $x_i^l = \omega_i$. He will rather fill up the ration at price p^r and only turn to the "free" market where he has to pay a higher price, if he still needs more. In (6.21) it is assumed that selling takes place at the low price p^l only.

In this model the right to consume ω_i at prices p^l and buy a ration at price p^r is nontransferable among consumers: Any excess of ω_i over x_i^l and of r_i over x_i^r has a zero value. Alternatively, one can make these rights transferable so that they become entitlements that add to income. The consumer problem is then

$$\max u_i(x_i),$$

$$x_i \geqslant 0,$$

$$\text{subject to} \tag{6.22}$$

$$p^u x_i \leqslant p^l \omega_i + T_i^0,$$

where $T_i^0 = (p^u - p^l)\omega_i + (p^u - p^r)r_i$ is the imputed income from the rations.

In the competitive model there is only one price, say, $p^u(=p^l)$, and the r_i's are zero. When rations are transferable, (6.22) applies, and there is an income transfer $T_i^0 = (p^u - p^r)r_i$ which must be funded out of other taxes. Finally, if the rations are nontransferable, model (6.21) applies and again the implicit subsidy $(p^u - p^r)x_i^r$ has to be funded out of other taxes. In both cases we are led back to the competitive model with taxes.

6.2.7 Shifting the Adjustments Also to Producers

So far producers are not rationed. Buffer stock operations may be required in equilibrium if consumer rationing happens to have reached its limit, while producers are still demanding or supplying more than can be met at that

11. Here we disregard income from profits, only to simplify notation.

price. To avoid this, producers have to be rationed as well. The rationed producer's problem can be stated as

$$\Pi^r_j(p, r^l_j, r^u_j) = \max_{y_j,} py_j,$$

subject to (6.23)

$$y_j \in Y_j,$$

$$r^l_j \leqslant -y_j \qquad (\Psi^l_j),$$

$$-y_j \leqslant r^u_j \qquad (\Psi^u_j).$$

Feasibility is ensured whenever $r^l_j \leqslant 0$ and $r^u_j \geqslant 0$ (and $0 \in Y_j$), which amounts to requiring that the producer cannot be forced to trade.[12] We assume that Slater's constraint qualification is satisfied and that the multipliers are unique.[13] As in (6.7) these values can be used to construct the indirect taxes that will enter the objective of the mathematical program

$$\xi^l_{jk} = \max(\Psi^l_j, s^l_k),$$
$$\xi^u_{jk} = \max(\Psi^u_j, s^u_k), \qquad (6.24)$$

and program (6.6) is written

$$\max \bar{p}^e e - \bar{p}^m m + p^l s^l - p^u s^u + \Sigma_k(\Sigma_i \xi^l_{ik}\beta^l_{ik} + \Sigma_j \xi^l_{jk}\beta^l_{jk})z^l_k$$
$$- \Sigma_k(\Sigma_i \xi^u_{ik}\beta^u_{ik} + \Sigma_j \xi^u_{jk}\beta^u_{jk})z^u_k,$$

$$e, m \geqslant 0,\ s^l, s^u \geqslant 0,\ z^l \leqslant 0,\ z^u \geqslant 0,\ y_j, \text{ all } j,$$

subject to (6.25)

$$\Sigma_j y_j + m - e + s^u - s^l \geqslant \Sigma_i \hat{x}_i - \Sigma_i \omega_i,$$

$$\beta^l_{ik} z^l_k \leqslant \hat{x}_{ik} - \omega_{ik} \leqslant \beta^u_{ik} z^u_k,$$

$$\beta^l_{jk} z^l_k \leqslant -y_{jk} \leqslant \beta^u_{jk} z^u_k,$$

$$y_j \in Y_j,$$

for given $\hat{x}_i$ and ξ^l_i, ξ^u_i, ξ^l_j, ξ^u_j, set by (6.5), (6.7), (6.8), (6.24) and given values of β^l_{ik}, β^l_{jk}, β^u_{ik}, $\beta^u_{jk} \in [0, 1]$ such that for each k one of the β^u and one of the β^l must be equal to unity.

12. Rationing is done on net demands rather than on net supplies to maintain consistency in notation with the consumer problem.

13. Note that because of rationing, the solution set may fail to have a strict interior, so existence of Lagrange multipliers is not obvious. Even if Lagrange multipliers exist, they may be set valued, and the existence proof would have to be adjusted.

In the Drèze equilibrium the β are all equal to one (uniform rationing) or may be taken as exogenous. Alternatively, one may adapt the concept of effective demand to producers and use Benassy rationing or virtual taxation.[14] It can be shown that the central stock s^l (resp. s^u) cannot be positive when rations r^l (resp. r^u) are all zero, so a pure rationing equilibrium can be implemented.

6.3 Welfare Implications

In this section we interpret price rigidities as being generated by government policy (and not by institutional sluggishness) and proceed along the same lines as for taxes in section 5.3. We show that removing price rigidities will not reduce welfare if compensations are allowed for.

6.3.1 Centrally Held Stocks

As in (5.21) we write the problem in the welfare-maximizing format for the small open economy:

$$\max \Sigma_i \alpha_i u_i(x_i) + p^l s^l - p^u s^u,$$

$$e, m \geqslant 0,\ s^l,\ s^u \geqslant 0,\ x_i \geqslant 0, \text{ all } i,\ y_j, \text{ all } j,$$

subject to

$$\Sigma_i x_i - \Sigma_j y_j - m + e - s^u + s^l \leqslant \Sigma_i \omega_i \qquad (p),$$

$$s^u \leqslant \bar{s}^u,$$

$$s^l \leqslant \bar{s}^l,$$

$$\bar{p}^m m - \bar{p}^e e \leqslant \Pi_0,$$

$$y_j \in Y_j.$$

In this program the α_i are the Negishi equilibrium welfare weights; likewise Π_0 is the (possibly nonzero) given trade deficit.

We will now consider a removal of the price rigidities by lowering p^u and p^l until they fall below $\bar{p}^e$. Since p^u will fall below $\bar{p}^e$, the multiplier p will become larger than p^u and s^l will fall to zero as s^u reaches its upper bound $\bar{s}^u$. At this point a competitive equilibrium (with transfers among consumers) of the open economy will be reached, and consumers will not only share the production $\Sigma_j y_j$ and the endowments $\Sigma_i \omega_i$ but also the initial stock $\bar{s}^u$. This situation cannot be welfare deteriorating. It is the equivalent of proposition

14. It is simpler to impose virtual taxation on producers by rationing total $\Sigma_j y_j$ instead of rationing individual producers.

5.2, so that we can state (without proof):

PROPOSITION 6.3 (Welfare loss from price constraints supported by central stocks) Dismantling the buffer stock agency and releasing initial stocks cannot decrease welfare.

If in equilibrium there is no stock activity ($s^{u*} = s^{l*} = 0$), then there is no distortion and the equilibrium is Pareto efficient. Alternatively, if $s^{l*} \neq 0$ and there is no initial stock ($\bar{s}^u = 0$), stock operations must have been financed out of taxes. Bringing p^l below $\bar{p}^e$ will not only bring s^l to zero, it will also eliminate these taxes.

One can also prove the equivalent of proposition 5.6 on gradual dismantling of the scheme, through (proportional) reductions on p^l and p^u. The following point must, however, be stressed: If rather than reducing both p^l and p^u, one chooses to reduce p^l and increase p^u, welfare may decrease. In particular, if p^u is shifted beyond $\bar{p}^m$, then s^u will be zero and consumers will be unable to buy from the stock, so total demand is reduced by $\bar{s}^u$.

6.3.2 Consumer (and Producer) Rationing

Under consumer rationing, the welfare program initialized at equilibrium values is

$$\max \Sigma_i \alpha_i u_i(x_i) + \alpha_0(p^l s^l - p^u s^u + \Sigma_k \Sigma_i \xi^l_{ik} \beta^l_{ik} z^l_k - \Sigma_k \Sigma_i \xi^u_{ik} \beta^u_{ik} z^u_k)$$

$$e, m \geqslant 0,\ s^l,\ s^u \geqslant 0,\ z^l \leqslant 0,\ z^u \geqslant 0,\ x_i \geqslant 0, \text{ all } i,\ y_j, \text{ all } j,$$

$$\text{subject to} \qquad (6.26)$$

$$\Sigma_i x_i - \Sigma_j y_j - m + e - s^u + s^l \leqslant \Sigma_i \omega_i,$$

$$s^u \leqslant \bar{s}^u,$$

$$s^l \leqslant \bar{s}^l,$$

$$\beta^l_{ik} z^l_k \leqslant x_{ik} - \omega_{ik} \leqslant \beta^u_{ik} z^u_k,$$

$$\bar{p}^m m - \bar{p}^e e \leqslant \Pi_0,$$

$$y_j \in Y_j.$$

In this program the values of α_i, ξ^l_{ik}, β^l_{ik}, ξ^u_{jk}, β^u_{jk} are set at their equilibrium values, the trade deficit Π_0 is given, and $\alpha_0 = 1$. Note that there are now bounds $\bar{s}^u$ and $\bar{s}^l$ on stock operations, that were absent from (6.6). If we were to allow unbounded s^u, then under a reduction of p^u the welfare program would also become unbounded. For symmetry reasons we introduce the bound $\bar{s}^l$, but s^l is already bounded by total supply and the cost of imports.

Recall that Drèze uniform rationing amounts to set all the β equal to one. In Benassy's scheme the β are determined by the rationing functions, while virtual taxation sets them so as to equalize the values ξ of the vouchers. The following remarks are now in order:

1. If p^l and p^u were to lie below $\bar{p}^e$, then clearly there would be no rationing in the economy and the equilibrium would be Pareto efficient.

2. If in (6.26) α_0 is reduced as part of a gradualist reform like in proposition 5.6, then by proposition 1.11, welfare cannot decrease. This requires the ξ to be interpreted as virtual taxes that are reduced when α_0 is reduced while the β-coefficients are left unchanged. Along the path of such a reduction, bounds on stocks may become effective and price constraints may be violated; this will eventually occur when p^u falls below $\bar{p}^e$.

3. If the reform keeps the rations $\beta_{ik}^l z_k^l$ and $\beta_{ik}^u z_k^u$ at their equilibrium values, then the two last terms in the objective function of (6.26) are constants, so they can be dropped. Government can now implement a welfare-improving reform in two stages. First, it should reduce the price bounds. This cannot decrease welfare (see proposition 6.3). Eventually p^l and p^u will drop below $\bar{p}^e$. Now government can move to the second stage and relax the remaining rations, increasing welfare if some rations were still binding. Note that as long as p^l or p^u lie above $\bar{p}^e$, the mere relaxation of rations does not guarantee welfare improvements, since central stocks operations may have to be intensified. However, if the values of s^l and s^u are kept constant, relaxing rations cannot reduce welfare.

4. Producer rationing can also be incorporated in the previous program, and the above remarks will apply. If p^l and p^u are below $\bar{p}^e$, moving from individual rationing constraints to collective rationing (i.e., rationing $\Sigma_j y_j$), will not reduce welfare.

5. As long as price rigidities are not abolished, welfare may be reduced if rations become tradable. To see this, consider the example of producer rationing under a Drèze or a Benassy scheme, where virtual taxes differ between producers. Producers can increase their profits and should therefore be interested in trading rations. However, so long as p^l and p^u are larger than $\bar{p}^e$, this will not necessarily increase welfare, since it could shift more resources to overvalued commodities.

6. All of the several forms of soft rationing discussed in section 6.2.6 generate different prices for the same commodity and are therefore distortionary unless vouchers can be traded. They then become lump-sum transfers, for the vouchers will carry a market value equal to the price difference $\max[p - p^l, 0]$ (resp. $\max[p^u - p, 0]$) and consumer i will receive a transfer equal to the value of his vouchers.

7. If a commodity cannot be transfered over time (e.g., labor or perishable goods), the initial stock will be zero, and the commodity purchased by the

central stock wasted. For that reason it may be more efficient to ration and spend on lump-sum transfers than to spend on purchasing. For labor, consumers could then enjoy leisure instead of supplying their labor time to the buffer agency.[15] Perishable commodities can be consumed rather than rot in a warehouse. But, if producers are rationed in the supply of a commodity that is in surplus, they will save the inputs required to produce it.

8. If the commodity is nonperishable and can be stored, the price rigidity p^l may be interpreted as representing the (discounted) value of a future sale, serving as an end-valuation of the stock. Then its storage is not necessarily distortionary, as will be seen in chapter 7.

6.4 Practices in Applied Modeling

6.4.1 Fixed Prices in the CGE Model

In the CGE model of section 3.4.1, fixing prices amounts to bringing the number of endogenous variables below the number of equations. Since this could make the system of equations inconsistent, additional variables, which can be interpreted as closure variables, have to be introduced. In the present context these are either stocks or rations.

Recall that in the CGE model discussed so far, production of goods is performed under constant returns to scale. This sets the price of a good equal to its average cost, so that sectors make no profits. If we impose a fixed price for such a good, the sector producing it will either make a loss or a profit. In the first case production will be zero. If there are profits, the producer will try to increase his supply until the input prices, which are flexible, rise so much that his profit becomes zero. This does not ensure that these prices can make all markets clear. If moreover all input prices of that sector are fixed, profits will never become zero, and the producer will try to produce infinite amounts unless rationing signals are given to stop him. Therefore it is not practical to fix prices for goods in the producer model without rationing or taxation. Price rigidities in the CGE model are usually implemented on the factor market and restricted to consumers only.

To represent a surplus s_k^l of factor k when there is a price rigidity, we relax the kth commodity balance and determine the income from that factor as $p_k(\omega_{ik} - \sigma_{ik}s_k^l)$, where $\omega_{ik} - \sigma_{ik}s_k^l$ represents the part of the initial endowment consumer i can sell ($\Sigma_i\sigma_{ik} = 1$). Alternatively, we can interpret $p_k\sigma_{ik}s_k^l$ as a tax consumer i pays to finance the purchase s_k^l by the buffer stock. This

15. Here we assume that the conditions of the unemployment scheme completely exclude the possibility to exercise other activities for the time paid under the scheme. In practice, moonlighting, studying, recreation, are bound to occur.

is equivalent to a rationing constraint so long as the factor does not enter consumer i's utility function; otherwise, the ration would pick a shadow price that would affect the consumer choice through spillovers. Here we have implicitly assumed that $s_k^l \geqslant 0$, but this is not guaranteed by the procedure outlined above and shortages may arise.

Factor shortages and rigidities on the goods markets can be dealt with by rationing consumers. To avoid the lack of realism of uniform rationing and also the difficulty of specifying Benassy rationing functions, applied modelers tend to use virtual taxation or soft rationing. However, this implies leaving the CGE format, since the model will involve inequalities and complementary slackness conditions.[16]

The Single Consumer–Single Producer Case with Three Commodities

A simplified version of the Benassy model has often been used, particularly in applied macroeconomics. Though it is an equality constrained model, it does not follow the CGE format. It distinguishes three commodities—labor, a consumption good, and a numéraire good—one consumer and one producer.

We have seen that pure rationing schemes can be classified according to the way in which they share rations among agents. Empirically it is difficult to indicate who gets rationed and by how much if the scheme is not specified explicitly by government policy. This difficulty is obviously avoided in the single consumer–single producer model when either the consumer or the producer exclusively demands (supplies) a given commodity so that one immediately knows whom to ration. Usually labor will be supplied by the consumer and the consumer good by the producer. Since in this case it is not necessary to distribute the rations over agents, a Benassy equilibrium will also be a Drèze and a virtual tax equilibrium, though prices are fixed.[17]

To describe this setting, we return to the Benassy model in which prices are fixed. The single consumer computes his effective demands as follows:

$$\max\{u(x^s)\,|\,px^s \leqslant p\omega + \Pi,\ x^s \geqslant 0,\ r_k^l \leqslant x_k^s - \omega_k \leqslant r_k^u,\ k \neq s\} \quad \text{for } s = 1, 2, 3.$$

As before, x_s^s is the effective demand for commodity s, and the net effective demand vector d contains the elements

$$d_s = x_s^s(p, r^l, r^u) - \omega_s.$$

16. See section A.9 of the mathematical appendix. Though it is possible to maintain equalities if one uses the artificial variables discussed in section 6.2, figure 6.3, this approach has serious practical disadvantages because the artificial variable triggers other variables expressed in different units, so its nondifferentiability can create problems in algorithms.

17. For this equivalence to hold, the assumptions on concavity and convexity should be made strict so as to keep all decisions single-valued.

The rationed net demand of the consumer is now

$$z_k = \min\{\max(r_k^l, d_k), r_k^u\}, \tag{6.27}$$

which yields the same net demands as model (6.5) of the rationed consumer.

Similarly the producer solves

$$\max\{py^s \mid q_k^l \leqslant -y_k^s \leqslant q_k^u, k \neq s, y^s \in Y\} \qquad \text{for } s = 1, 2, 3.$$

His net effective demand is

$$d_s' = -y_s^s(p, q^l, q^u),$$

and his rationed net demand is

$$z_k' = \min\{\max(q_k^l, d_k'), q_k^u\}. \tag{6.28}$$

Market equilibrium at fixed prices is the solution (r^l, r^u, q^l, q^u) of

$$z_k + z_k' = 0. \tag{6.29}$$

There are thus as many equations as commodities minus one (the non-rationed commodity) and four times as many unknowns (r^l, r^u, q^l, q^u). To make the model determinate without having to specify rationing functions, we could limit the number of possible regimes by exogenously setting some of the rations at levels where they cannot be binding and by specifying rationing functions for others.

In a three commodity setting, labor, a consumption good, and a numéraire that cannot be rationed, one specification suggested by Benassy (1982, ch. 11) is given in table 6.1.

The consumer faces no constraint on his demand for labor, since he cannot be offered any labor,[18] but he may be rationed on his supply of labor by employment opportunities. He faces no constraint on his supply of the consumer goods, since he does not possess them, but he may be unable to buy as much as he wants. Similarly the producer can perceive constraints on his demand for labor and on his sales of consumer goods. Out of the twelve variables, we have eliminated four, by assuming that the last commodity is the numéraire and four others by setting them at plus or minus infinity. The four remaining are functions of rationed net demands. This gives us as many unknowns as equations.

When these rations are replaced in the two (rationed) net demand

18. Of course the consumer cannot have more leisure than he has labor endowment. Thus there is the constraint $x_1 \leqslant 0$, which is not a constraint imposed by rationing.

Table 6.1
Rationing rules in the three-commodity case

	r^l	r^u	q^l	q^u
Labor	$-z_l'$	$+\infty$	$-\infty$	$-z_l$
Consumption	$-\infty$	$-z_2'$	$-z_2$	$+\infty$
Numéraire	$-\infty$	$+\infty$	$-\infty$	$+\infty$

functions (6.27) and (6.28), one obtains

$$
\begin{aligned}
z_1 &= \min\{\max(d_1, -z_1'), +\infty\} = \max(d_1, -z_1'),\\
z_2 &= \min(d_2, -z_2'),\\
z_3 &= d_3,\\
z_1' &= \min(d_1', -z_1),\\
z_2' &= \max(d_2', -z_2),\\
z_3 &= d_3',
\end{aligned}
\tag{6.30}
$$

which are rationing conditions of the type discussed, for example, by Malinvaud (1977).

In equilibrium, conditions (6.29) have to hold, so $z_k = -z_k'$. The solutions can be classified according to whether $z_k = d_k$ or $z_k' = d_k'$ is satisfied. Now four regimes are possible depending on which rations are strictly binding:

1. *Keynesian unemployment.* The consumer is rationed on his supply of labor, $z_1 > d_1$; the producer on his supply of goods, $z_2' > d_2'$.[19]

2. *Classical unemployment.* The consumer is rationed on his supply of labor, $z_1 > d_1$, and on his demand for goods, $z_2 < d_2$.

3. *Repressed inflation.* The consumer is rationed on his demand for goods, $z_2 < d_2$; the producer on his demand for labor, $z_1' < d_1'$.

4. The consumer is not rationed, and the producer is rationed on both his sales and his demand for labor. It is unlikely that a profit-maximizing producer will be constrained on both markets, and this fourth case is not interesting.

Clearly there can be intermediate cases where, at equilibrium, some of the constraints are not strictly binding, namely where some $z_k = d_k$ (or $z_k' < d_k'$). In particular, when no constraint is binding, the fixed prices are those of the

19. It is questionable whether "Keynesian" is an appropriate label. In this model unemployment will be avoided if only prices are set right, which is not a Keynesian view point.

Walrasian equilibrium, and given our assumptions of strict concavity (and convexity), there is only one equilibrium.

This type of model has been used to determine whether an adjustment Δp_2 of the consumer price (or Δp_1 of the wage) or an exogenous new demand Δg_2 for the consumer good (i.e., public consumption) would make constraints less binding and improve welfare. Multiplier analysis is performed to study the consequences of specific perturbations of the commodity balances at one of the three equilibria listed above. Consider the following equilibrium conditions:

$$z(p + \Delta p, z') + z'(p + \Delta p, z) + \Delta g = 0,$$

which may be rewritten as

$$z(p + \Delta p, z'(p + \Delta p, z)) + z'(p + \Delta p, z) + \Delta g = 0.$$

At given Δp and Δg, this is an implicit function in z, and it is possible to compute analytical derivatives of z with respect to g, for fixed p:

$$\frac{\partial z}{\partial z'} \frac{\partial z'}{\partial z} dz + \frac{\partial z'}{\partial z} dz + dg = 0.$$

If then the inverse is well-defined,

$$\frac{dz}{dg} = -\left[\left(\frac{\partial z}{\partial z'} + I\right) \frac{\partial z'}{\partial z}\right]^{-1},$$

where only the derivative dz/dg_2 is of interest here. The same calculation can be done for the partial derivative with respect to p (with g_2 fixed).[20]

The signs of the multipliers dz/dg_2, dz/dp_1, and dz/dp_2 will depend on the specification of the utility and production functions. In general, there will be effects across markets, the so-called spillover effects, that may lead to unexpected signs. For more complete discussions on policy issues, see Benassy (1982, 1986), Böhm (1978), and Malinvaud (1977).

The results given by the multiplier equations only provide information for small changes. For larger changes, regimes may change, and new constraints may become binding. This has been extensively studied in the so-called two-gap literature, in a slightly different context, and without explicit reference to rationing as a supporting mechanism of price rigidities.[21]

In the single consumer–single producer case, when more than two commodities can be rationed, the number of possible regimes may become

20. For discussions on the nonsingularity of this matrix and, associated to it, on the uniqueness of the fix-price equilibrium, see among others Gourieroux et al. (1980) and Ito (1980).

21. See Chenery and Strout (1966), Blitzer, Clark, and Taylor (1976), and Taylor (1979). Gunning (1983) does make the link explicit.

large, but otherwise, the earlier discussion applies and multipliers can be calculated.

6.4.2 Issues in Model Implementation

The Specification of Bounds on Prices

Until now we have kept p^l and p^u fixed; therefore they define bounds on absolute as well as on relative prices and hence determine the price normalization. We will now assume that these bounds are functions P^l and P^u of p (and possibly other variables):

$$p^l = P^l(p, \ldots) \quad \text{and} \quad p^u = P^u(p, \ldots).$$

To ensure existence of equilibrium, we want these functions to satisfy the following properties:

1. P^l and P^u are continuous.
2. $0 < P^l(p, \ldots) \leqslant P^u(p, \ldots)$ for all p.
3. $P^u(p, \ldots) \leqslant \theta$ for all p.

Condition 1 is technical, condition 2 avoids any inconsistent price indexation,[22] and condition 3 ensures that the feedback relation will be bounded.

To avoid money illusion, the additional requirement that functions P^l and P^u are homogeneous of degree zero (or, more generally, homothetic) should be imposed. The rigidity is then said to be real; if money illusion is not ruled out, it is called nominal.

Stocks versus Rationing

We have discussed three ways of implementing price rigidities: buffer stocks, quantity rationing (and virtual taxes), and soft rationing. We will now discuss their relative merits in applications.

Initial stocks $\bar{s}^u$ can only exist for commodities that are storable and have been carried over from a previous period. Factor services, like labor or housing services, can obviously not be stored. Additions to stocks s^l can, however, take place even for such services. This can be thought of as a rental of the service by the central stock.

Drèze or Benassy pure rationing operates like an invisible hand mechanism: Only the equilibrium matters. These schemes can hardly be interpreted as implementing some government policy. Governments do ration, but they cannot adjust the rations with sufficient accuracy to maintain the price

22. As an example one might think of a wage indexed on a price (e.g., $w^l = w^u = w = \gamma p$), while the price is indexed on the wage ($p^l = p^u = p = \delta w$). If $\gamma \neq 1/\delta$, there will be no solution other than $w = p = 0$.

rigidity, without recurring to buffer stock operations, multiple pricing with soft rationing, and adjustment of foreign trade.

Price support policies for dairy in the European Union illustrate the simultaneous use of the various instruments. Originally there was a scheme of high intra-EU prices to support farmers' income. Eventually, however, a surplus occurred that had to be disposed of with export subsidies. When international demand could no longer absorb the surplus, public buffer stocks started to build up. At first, the EU tried artificially to raise demand through food subsidy programs such as school milk. When this no longer worked, farmers were asked to contribute through a soft rationing scheme that allowed them to sell a quota at high price supported by the buffer stock, and the remainder at very low (international) price. Allowing buffer stock purchases to coexist with rationing leads to production inefficiencies among farmers and to consumer losses due to high prices, so reforms had been adopted in order to lower the buffer stock price and provide a compensation by direct income support.

In many developing countries there are schemes to ensure satisfaction of basic needs. Consumers receive vouchers to buy rations at a state-shop price and can supplement their rations on the free market. The state shops get their supply from a central stock which buys at high (procurement) prices from producers (or forces state firms to deliver) or imports what they cannot buy on the domestic market. The production in excess of the procurement quota is sold at a low price on the free market. Unfortunately, when the procurement price falls below the free market price, as has often occurred in the past, the system breaks down.[23] Similar schemes exist in many other countries for various basic consumer goods—housing, transportation, fuel, and health care. These are the very systems that can lead to abuse. For example, producers have been known to sell their worst quality to the buffer stock, so elaborate grading systems are necessary.

The labor market usually has some special characteristics. Labor may be viewed as a heterogeneous commodity differentiated by consumer-specific characteristics such as skill or age. Some types of labor such as the highly skilled are demanded and have a positive competitive equilibrium price, while others such as workers with multiple disabilities are in excess supply and have a zero competitive price. Unemployment can therefore be represented in the competitive model without the need for any downward price rigidity or rationing. If, however, labor is considered in a more aggregate way, and there is only one wage rate, some mechanism has to be introduced that allows for unemployment at a positive wage. Moreover, if minimum wage or unemployment benefit schemes exist, their effects have to be represented also. A lower price bound can be enforced through buffer stocks;

this presupposes that leisure time is either not valued by the consumer or used up by the buffer stock. Alternatively, one could shift the leisure time to the consumer through a rationing scheme. These representations ignore indivisibilities of jobs and increasing returns, as well as the fact that labor has to be produced through consumption. These are issues to which we return in sections 9.1.5 and 10.4.2, in our discussion of efficiency wage relations.

Vouchers as Fiat Money

In most of the literature on fix-price models, the numéraire commodity is referred to as money. This commodity is treated as an ordinary consumer good that is not rationed and whose supply is fixed as an initial endowment. Increases in the prices of other goods, with respect to this numéraire, are qualified as "inflationary." This assumes that money can be treated as a commodity, like gold. In section 12.1 money will not be an ordinary commodity; it will be treated as fiat (or paper) money or coupons. The tradable vouchers in the virtual tax and the soft rationing schemes may already be interpreted as fiat money: They are pieces of paper that carry a value in exchange. They bring no utility and can therefore have a zero price.

6.4.3 Empirical Elaboration

Construction of the SAMs

As usual, we start from the equilibrium conditions of the model to derive the equations that underlie the SAM (social accounting matrix). For buffer stocks (definition 6.1) we have the following conditions:

1. The market equilibrium conditions

$$\Sigma_i x_{ik}^* - \Sigma_j y_{jk}^* - \Sigma_i \omega_{ik} + e_k^* - m_k^* - s_k^{u*} + s_k^{l*} \leqslant 0 \perp p_k^* \geqslant 0. \tag{6.31a}$$

2. The profits of producers

$$\Pi_j^* = p^* y_j^*, \text{ all } j. \tag{6.31b}$$

3. The budget equations of consumers

$$p^* x_i^* = p^* \omega_i + \Sigma_j \theta_{ij} \Pi_j^* + T_i^*, \text{ all } i. \tag{6.31c}$$

4. The budget equation of the foreign sector

$$\bar{p}^e e^* = \bar{p}^m m^*. \tag{6.31d}$$

5. The budget balance of the government

$$\Sigma_i T_i^* = p^*(s^{u*} - s^{l*}). \tag{6.31e}$$

As usual, the complementarity conditions (6.31a) as well as (6.31b), (6.31c) and (6.31e) imply (6.31d). The components of the SAM are given in table 6.2.

This will lead to a SAM in value terms; the commodity balances are in quantity terms so that unit values p^* can be computed. However, we must still account for the complementarity condition 5 in definition 6.1, which may lead to conflicts with the data used. The resolution of these conflicts has to proceed in the same way as in section 4.4.3 for international trade prices.

To represent pure rationing within the SAM, we turn to definition 6.2 and see that the SAM has exactly the same components as those in table 6.2, though in case of pure rationing the buffer stock operations, and consequently the government and the foreign sector, will vanish so that the structure of the SAM reduces to the one of the competitive model in table 3.1.

Calibration and Estimation of the Model for the Rationed Consumer

The main problems in the empirical implementation of rationing schemes is that the parameters of model (6.5) of the rationed consumer have to be calibrated or estimated, and the data only report on transacted quantities and market prices, not on rations or on the Lagrange multipliers, and hence not on the implicit virtual taxes. These are latent variables that can be determined once the parameters are known. Calibration and estimation procedures have been designed that estimate the parameters of the model and the rations so as to fit the observed data. Usually the number of parameters is restricted by limiting the number of commodities and agents and estimating models such as (6.30)[24] or by making exogenous assumptions on the side of the market that is rationed.[25]

6.4.4 Simulations

The CGE model by Adelman and Robinson (1978) already considers price rigidities on factor markets, and many subsequent models include some form of wage rigidity. Models with fixed rations and buffer stocks have been applied to India by Narayana et al. (1990) and to Bangladesh by Salauddin Ahmad et al. (1986). These rations appear to be an effective tool for income redistribution to the poor (particularly as part of food for work programs). Erlich et al. (1987) treat the excess supply of labor as a buffer stock; it appears that a reduction in the wage rigidity is insufficient to eliminate

24. The min or max forms in these models require using censored regression methods. See, for example, Amemiya (1985) or Greene (1991).

25. See, for example, Cornielje (1990).

Table 6.2
Structure of the SAM with buffer stocks

	Goods	Factors	Firms	Consumers	Foreign sector	Government budget	Buffer stock
Goods			Input	Consumption	Exports		
Factors			Input	Consumption	(Exports)		
Firms	Production	(Production)					
Consumers	(Endowments)	Endowments	(Profits)			Transfers	
Foreign sector	(Imports)	Imports					
Government budget							Net receipts
Buffer stock	Net sales	Net sales					

Note: In the SAM for the CGE model, the terms in parentheses are zero.

unemployment in the short run because of limited capital labor substitution and insufficient capital. Nguyen and Whalley (1985) and Devarajan et al. (1989) have developed a variant in which a market with price controls coexists with a black market. The actual clearing takes place on the black market, but there are fines and other risks that introduce special transaction costs. They find, not unexpectedly, that higher fines reduce welfare.

7 Finite-Horizon Dynamics

So far, commodities were differentiated by date of delivery: oranges for today, tomorrow and the day after. Since r, the number of commodities, was finite and given, the number of periods, say, T, had to remain finite as well. Consumers and producers were assumed to be present throughout. Thus the models discussed in earlier chapters can be interpreted in dynamic terms, but all this has remained implicit. We now make the time dependence explicit and specify finite-horizon dynamics.

7.1 Elements of Dynamics

In dynamics the aim is to show how variables of interest change over time, from past to present and possibly to future. Changes may result from allocations over time, as endogenously generated by producer and consumer behavior as well as from shifts in exogenous variables or parameters, such as initial endowments, tax rates, or coefficients of functional forms. In this chapter we will focus on endogenous shifts. Exogenous shifts can be represented by allowing for time-dependent functions in the model, and will not be dealt with explicitly.

Dynamics may also be concerned with changes in the number and composition of commodities and agents. For commodities, one can maintain a fixed number of varieties, old and new, that coexist in the model; shifts in allocations then determine the change in composition. Alternatively, one may represent changes in the characteristics of varieties as a continuous exogenous or endogenous process.

In the case of consumers, similar distinctions apply: A finite number of social classes, occupational groups, age cohorts, for instance, might be represented. For consumers who live shorter than the horizon of the model and thus only coexist in some periods, one speaks of overlapping generations. While shifts between groups, due to social transition, migration, and demography can reflect changes in composition, shifts in functional forms will reflect changes in the characteristics of the agents themselves. All this should be accounted for in welfare evaluations, as will be seen in section 7.3. On the production side there may be entry and exit of firms. Given the possibility of inaction in assumption P1′, this can simply be considered as transitions from action to inaction, and vice versa, while the total number of producers remains constant. Again, here we only cover shifts in composition; changes in characteristics are reflected through possible time dependence of the functions.

In this chapter the time horizon is finite. This creates the well-known problem of valuating terminal stocks. If stocks are given zero value, there is no incentive to leave anything for use after T; this will affect even first-period decisions. Moreover, since age cohorts of consumers overlap, some consumers will still be young in period T and lack the possibility of planning their future. These problems will be dealt with in chapter 8, where the analysis extends to an infinite horizon, for models with infinitely lived consumers and models with overlapping generations.

7.2 Finite-Horizon Dynamics

We begin by rewriting the competitive equilibrium in dynamic terms. We assume that each agent draws up a plan for the current and future periods, up to T, but does not value what may happen afterwards. Plans are made compatible on future markets, which exist for each period, and plans can and will be carried out. This setup, to be developed in section 7.2.1, will be referred to as the T-period competitive equilibrium. The model is evidently unrealistic in several respects: (1) the world will not end after T, (2) it is not possible to organize all future markets, (3) unforeseen events will occur, and (4) agents may change their minds when the time has come to carry out their plans. Some of these shortcomings will be addressed in section 7.2.2 using the temporary equilibrium framework. There agents will plan for two periods, the present and the future; only present-period markets will be assumed to exist, while, for the future, agents form expectations of prices and profits.

The model will be restricted further in section 7.2.3, where we introduce a single-period model in which agents take all future variables as exogenously given. This model is of special relevance, since it provides the basis for much of the applied work in dynamics. In section 7.2.4 we introduce international trade, taxes, and price rigidities within the dynamic framework.

7.2.1 *T*-Period Competitive Equilibrium

Definition and Existence

We introduce the time dimension in definition 1.2 of a competitive equilibrium, simply by partitioning the relevant vectors but without articulating the recursive nature of physical processes caused by production lags, accumulation of stocks, and so on. The time subscript t will run from 1 to T. This leads to the following definition:

DEFINITION 7.1 (T-period competitive equilibrium) The allocation y_{jt}^*, all j, x_{it}^*, all i, supported by the price vectors $p_t^* \geqslant 0$ is a T-period competitive equilibrium if the following conditions are satisfied:

1. For every producer j $(y_{j1}^*, \ldots, y_{jT}^*)$ is an optimal choice in

$\max\{\Sigma_t p_t^* y_{jt}$ subject to $(y_{j1}, \ldots, y_{jT}) \in Y_j\}$; $\Pi_j(p_1^*, \ldots, p_T^*)$ is the optimal profit.

2. For every consumer i $(x_{i1}^*, \ldots, x_{iT}^*)$ is the optimal choice in

$\max\{u_i(x_{i1}, \ldots, x_{iT})$ subject to $\Sigma_t p_t^* x_{it} \leqslant h_i^*,\ x_{it} \geqslant 0\}$, where $h_i^* = \Sigma_t p_t^* \omega_{it} + \Sigma_j \theta_{ij} \Pi_j(p_1^*, \ldots, p_T^*)$.

3. Markets are in equilibrium for all t,

$\Sigma_i x_{it}^* - \Sigma_j y_{jt}^* - \Sigma_i \omega_{it} \leqslant 0 \perp p_t^* \geqslant 0.$

In definition 7.1 the vectors x_{it}, y_{jt}, ω_{it}, and p_t have dimension a_t, the number of commodities in period t and $\Sigma_t a_t = r$. Unless specified otherwise, the number of commodities will be the same in every period. Then the subscript t will be dropped so that $T \cdot a = r$. Agents are assumed to set up their plans at period $t = 0$ for all coming periods.

Suppose that after one period, when all plans pertaining to that period have been executed and come out as anticipated, an agent decides to plan again. Then his new plan for the coming periods will coincide with the earlier one. Indeed, when the agent sets up his first plan, he had to know all future prices and take them as given. Each agent's "new" plan will fit within the earlier one as long as the utility functions, production sets, and endowments shift in a way that agents can anticipate and the end period remains fixed.[1,2] A sequence of plans drawn at $t = 2$, $t = 3$, etc., with a common end-period T (and thus of shortening length) will be consistent and require no replanning.[3] Such an economy is then said to move along a perfect-foresight trajectory.

Since the model merely recasts, in dynamic terms, the competitive equilibrium model of definition 3.1, for which various proofs of existence of equilibrium have been given in chapter 3, we can now state, without proof:

PROPOSITION 7.1 (Existence of a T-period competitive equilibrium) Under assumption S,[4] there exist for every t an allocation x_{it}^*, y_{jt}^* and a price vector p_t^* in accordance with the T-period competitive equilibrium of definition 7.1.

Yet the dynamics are only implicit in the agents' behavior of definition 7.1, where (1) each producer maximizes the sum of his profits over time subject to a production set Y_j, for which the time dependence still has to be spelled out, (2) each consumer maximizes an intertemporal utility function subject to a budget (wealth) constraint over the whole period, (3) market

1. The discussion implies that at given prices, every agent's plan is single-valued and that those prices are unique. But this is only a matter of exposition, and the principle that no replanning is needed still holds when set-valuedness and/or multiplicity occur.

2. This can be viewed as a consequence of the Bellman principle.

3. Clearly, if T were shifted so as to keep the length of the plan fixed, the consecutive plans could be inconsistent.

4. See section 2.3.

equilibrium conditions stipulate that for every commodity, markets exist in all periods; they only show the balances at each point in time and not carryovers from one period to the next. These carryovers are also "hidden" in the production sets Y_j.

Prices belonging to different periods measure value in a common unit so that the calculation of profits and consumer budgets can proceed by straight summation. For each of two consecutive periods, the ratio $p_{kt}/p_{k,t-1}$ embodies changes of relative scarcities as well as discounting between these two periods, and there is no meaningful way to separate the two effects.

Profits may become negative in any given period, but profit maximization with possibility of inaction will keep the sum of discounted profits nonnegative. Similarly the consumer is not required to balance his consumption against his income in each period separately.[5] He plans over the full T periods and can borrow (or lend) as long as accumulated debt is nonpositive after T periods. This setup is referred to as the life-cycle hypothesis.

We will now make the time dependence more explicit.

Dynamics in Producer Decisions

In the equilibrium of definition 7.1, the producer maximizes the sum of (discounted) profits over time. In applications his production set Y_j must be given an explicit dynamic structure. This will limit the number of parameters that describe the set and is also a natural way to define production and capital accumulation over time. An endowment or a commodity produced in $t+1$ can hardly serve as an input in t. Moreover some separability over time must apply, since production is seen as a physical process that can be repeated in every period provided that inputs are available.

This structure can be represented by transformation functions. We now describe the technology in each period $t = 1, 2, \dots, T$ by means of two constraints:

$$F_{jt}(y_{jt}, \tilde{k}_{j,t+1}, k_{jt}) \leqslant 0$$

$$k_{jt} \leqslant \tilde{k}_{jt}.$$

The first contains the transformation function $F_{jt}(y_{jt}, \tilde{k}_{j,t+1}, k_{jt})$, as introduced in section 2.1.3. y_{jt} is the net supply of commodities to the market, k_{jt} is the stock of capital goods[6] used as inputs, and $\tilde{k}_{j,t+1}$ stands for stocks of capital goods that, given y_{jt} and k_{jt}, will be made available as initial capital

5. Since profits in a given period can become negative, consumer income can be negative as well.

6. Here we use the term "capital good" for every commodity that is carried over from one period to the next. Since these capital goods relate to the internal structure of the production process, their number can be different from r, the number of commodities.

stocks in period $t + 1$. Some of these capital stocks will be produced and use inputs of commodities; others may grow at a natural rate or not change at all. The second constraint links consecutive periods by making capital goods $\tilde{k}_{jt}$ produced in period $t - 1$ available (with free disposal) as inputs in t. At the outset of period 1, the capital stock $\tilde{k}_{j1}$ is given. The distinction between k_{jt} and $\tilde{k}_{jt}$ is introduced to make Lagrange multipliers ψ_{jt} explicit in the following problem solved by producer j:

$$\max \Sigma_t p_t y_{jt},$$

$$k_{jt}, \tilde{k}_{j,t+1} \geqslant 0, \; y_{jt}, \text{ all } t,$$

subject to (7.1)

$$F_{jt}(y_{jt}, \tilde{k}_{j,t+1}, k_{jt}) \leqslant 0 \qquad (\rho_{jt}),$$

$$k_{jt} \leqslant \tilde{k}_{jt} \qquad (\psi_{jt}),$$

where $\tilde{k}_{j1}$ is given.

We make the following assumption on the transformation function for all j and t:

ASSUMPTION **PT1** (The transformation function) The transformation function $F_{jt}: R^a \times R^a_+ \times R^a_+ \to R$ is continuous, convex, nondecreasing in $(y_{jt}, \tilde{k}_{j,t+1}, -k_{jt})$. The constraint $F_{jt}(y_{jt}, \tilde{k}_{j,t+1}, k_{jt}) \leqslant 0$ satisfies (1) the possibility of inaction: $F_{jt}(0, 0, k_{jt}) \leqslant 0$ for all $k_{jt} \geqslant 0$; (2) boundedness: for any given $k_{jt} \geqslant 0$, $p_t \geqslant 0$, $\psi_{jt} \geqslant 0$, the set $\{y_{jt}, \tilde{k}_{j,t+1} | p_t y_{jt} + \psi_{j,t+1}\tilde{k}_{j,t+1} \geqslant 0, F_{jt}(y_{jt}, \tilde{k}_{j,t+1}, k_{jt}) \leqslant 0\}$ is bounded; (3) the property that $F_{jt}(y_{jt}, \tilde{k}_{j,t+1}, k_{jt})$ is continuously differentiable.[7]

Continuity of F_{jt}, together with possibility of inaction, allows the firm to switch without setup costs between production and inaction. Storing is seen as a part of the production process. It "transforms" a capital good at t into one at $t + 1$. In every period t, k_{jt} enters as an input, and the firm has joint outputs of commodities and capital goods y_{jt} and $\tilde{k}_{j,t+1}$, whose proportions depend on F_{jt}.[8] A firm can also purchase and sell the capital goods it uses in production (these are then included in y_{jt}). The purchases and sales may involve installation or dismantling costs, often called adjustment costs. In the model such costs can be reflected through various degrees of substitution between y and $\tilde{k}$ (or y and k).

7. More formally, F_{jt} is continuously differentiable on an open set that contains $R^a \times R^a_+ \times R^a_+$.

8. As we will see in section 7.4.4, purchases are often assumed to be perfect substitutes for installed capacities; then adjustment costs are zero.

The first-order conditions of (7.1) imply for every t,

$$p_t = \rho_{jt}\frac{\partial F_{jt}}{\partial y_{jt}},$$

$$\rho_{jt}\frac{\partial F_{jt}}{\partial \tilde{k}_{j,t+1}} \geqslant \psi_{j,t+1} \perp \tilde{k}_{j,t+1} \geqslant 0,$$

$$-\rho_{jt}\frac{\partial F_{jt}}{\partial k_{jt}} \leqslant \psi_{jt} \perp k_{jt} \geqslant 0.$$

For known initial value $\tilde{k}_{j1}$, given p_1 and optimal ψ_{j1}, one can solve a single-period problem. p_1 is the price of marketed commodities, and ψ_{j1} is the price of firm-specific capital goods. If the firm is given $\tilde{k}_{j1}$, prices p_t, and its own optimal ψ_{jt}, it can solve a sequençe of single-period profit maximization problems:

$$\max_{k_{jt},\, \tilde{k}_{j,t+1} \geqslant 0,\, y_{jt},} p_t y_{jt} + \psi_{j,t+1}\tilde{k}_{j,t+1},$$

subject to (7.1′)

$$F_{jt}(y_{jt}, \tilde{k}_{j,t+1}, k_{jt}) \leqslant 0,$$

$$k_{jt} \leqslant \tilde{k}_{jt},$$

for given $\tilde{k}_{jt}$ optimal in $t-1$.

This decomposition into single-period problems shows that the critical issue in t is the valuation of the capital goods in $t+1$. In particular, since $\psi_{j,T+1} = 0$, the producer has no incentive to carry over any capital goods to period $T+1$. By contrast, setting $\psi_{j,T+1}$ to any positive value may greatly affect allocations and to obtain appropriate valuations $\psi_{j,T+1}$, one needs to study the economy after T, which requires analyzing infinite horizon problems, as will be done in chapter 8.

Though $\Sigma_t p_t y_{jt}$ will be nonnegative, producers can plan $p_t y_{jt}$ to be negative in certain periods; $p_t y_{jt}$ is the value of outputs minus the value of inputs which, in an intertemporal context, can easily become negative, since investment is part of the inputs. However, negative $p_t y_{jt}$ would, in the current setting, lead to negative "dividends," which may be unrealistic. This situation is avoided by allowing firms to borrow and lend, like consumers, thus enabling them to smoothen their dividend payments over time. Here we make the simple assumption that period's t dividends constitute a fixed proportion of total profits and that firms adjust borrowing accordingly. We can then write the producer problem as

$$\max_{k_{jt},\, \tilde{k}_{j,t+1} \geqslant 0,\, y_{jt},\, \Pi_{jt},\, S_{jt},\, V_{j,t+1},\ \text{all } t,} \Pi_j,$$

subject to

$$F_{jt}(y_{jt}, \tilde{k}_{j,t+1}, k_{jt}) \leqslant 0,$$

$$k_{jt} \leqslant \tilde{k}_{jt}; \tag{7.2a}$$

$$\Pi_j = \Sigma_t p_t y_{jt},$$

$$\Pi_{jt} = \sigma_{jt} \Pi_j,$$

$$S_{jt} = p_t y_{jt} - \Pi_{jt}, \tag{7.2b}$$

$$V_{j,t+1} = V_{jt} + S_{jt},$$

$$V_{j,T+1} \geqslant 0,$$

for given $\tilde{k}_{j1} \geqslant 0$ and $V_{j1} = 0$.

In (7.2a), the constraints of (7.1) are repeated. In (7.2b), Π_j is the intertemporal profit; Π_{jt} is the nonnegative dividend distributed in period t, as a fixed share $\sigma_{jt} \geqslant 0$ ($\Sigma_t \sigma_{jt} = 1$) of total profit; S_{jt} denotes net savings; terminal net worth $V_{j,T+1}$ is constrained to be nonnegative. We have also assumed that $V_{j1} = 0$: The producer starts with zero net claims. Note, however, that the producer plan will not change if $V_{j1} = \delta$ and $V_{j,T+1} \geqslant \delta$ for given δ. Whereas the constraints in (7.2b) can be viewed as postoptimal accounting and do not change the optimal plans y_{jt} in the T-period equilibrium, they will become relevant in the temporary equilibrium model.

Dynamics in Consumer Decisions

The consumer maximizes his intertemporal utility function, subject to a budget constraint which is intertemporal also. Here recursivity does not impose itself as naturally as in production processes. By condition 2 of definition 7.1, it is apparent that each period's consumption can contribute to utility in its own way, and there is no immediate necessity for introducing functional separability over time. Separability is nonetheless commonly found in theoretical and applied work. It reflects the idea that past consumptions influence future ones only through the budget constraint and not through the utility function: This is often referred to as "Let bygones be bygones" or "No regrets."

A common form for the intertemporal utility function is[9]

$$u_i(x_{i1}, x_{i2}, \ldots, x_{iT}) = \Sigma_t (\beta_i)^t u_i(x_{it}), \tag{7.3}$$

which assumes that the discount factor β_i is constant over time for consumer i. It also has the property that once consumption levels have been set for,

9. In the objective function, $(\beta_i)^t$ denotes β_i to the power t.

say, $t = 1$ and $t = 2$, the subsequent choices for, say, $t = 3$ until 4 are only affected by past decisions through the excess of past expenditure over past revenue. This is relevant in relation to replanning, when the consumer solves his problem in $t = 3$, given the realizations of earlier periods. He will now solve

$$\max \beta_i u_i(x_{i1}) + (\beta_i)^2 u_i(x_{i2}) + (\beta_i)^3 u_i(x_{i3}) + (\beta_i)^4 u_i(x_{i4}),$$

$$x_{i3}, x_{i4} \geqslant 0,$$

subject to

$$p_3 x_{i3} + p_4 x_{i4} \leqslant h_i - p_1 x_{i1} - p_2 x_{i2},$$

for given x_{i1}, x_{i2}. Then the first two terms in the objective have no effect on the allocations. However, x_{i1} and x_{i2} would be choice variables and allocations would be affected if his utility function were, say, $\beta_i u_{i1}(x_{i1}) + (\beta_i)^2 u_{i2}(x_{i1}, x_{i2}) + (\beta_i)^3 u_{i3}(x_{i1}, x_{i2}, x_{i3}) + (\beta_i)^4 u_{i4}(x_{i1}, x_{i2}, x_{i3}, x_{i4})$ for nonlinear functions $u_{it}(.)$, $t = 1, 2, \ldots, 4$.

The intertemporal utility function defined in (7.3) can also be obtained from the recurrence relation

$$u_{it} = u_i(x_{it}) + \beta_i u_{i,t+1} \qquad \text{for } t = 1, 2, \ldots, T, \tag{7.4}$$

$$u_{i,T+1} = 0.$$

As can be seen, the form (7.4) can be generalized to nonlinear specifications by adopting the formulation introduced in section 2.2.5, with recursive single-period utility function $W_i(x_{it}, u_{i,t+1})$:

$$u_i(x_{i1}, x_{i2}, \ldots, x_{iT}) = u_{i1},$$

where u_{i1} is obtained by recursion from[10]

$$u_{it} = W_i(x_{it}, u_{i,t+1}), \qquad t = 1, 2, \ldots, T, \tag{7.5}$$

$$u_{i,T+1} = 0.$$

For example, in a model with four periods, this leads to

$$u_i(x_{i1}, x_{i2}, x_{i3}, x_{i4}) = W_i(x_{i1}, W_i(x_{i2}, W_i(x_{i3}, W_i(x_{i4}, 0)))).$$

Choosing such a nonlinear recursive form has the advantage of allowing for a discount factor (the derivative of W_i with respect to $u_{i,t+1}$) that may vary over time while maintaining the "no regrets" property. The properties of the recursive utility function can be summarized as:

10. The bound on $u_{i,T+1}$ could be given any value. Zero is used here because consumption after time T is zero, and under our assumptions on the consumer, $u_i(0) = 0$. For specific functional forms (e.g., when $\ln W_i$ is linear in $\ln u_{i,t+1}$), a positive value will have to be used.

ASSUMPTION **CT1** (The recursive utility function) (1) $W_i: R^r_+ \times R \to R$, $W_i(x_i\ u_i)$ is continuous, strictly concave and continuously differentiable on $R^r_{++} \times R$; (2) $W_i(x_i, u_i)$ is nonsatiated in x_i and increasing in x_i for $i = 1$; (3) $W_i(0, 0) = 0$.[11]

Assuming that the utility function $u_i(x_{i1}, x_{i2}, \ldots, x_{iT})$ has the recursive structure (7.5), we can now turn to the time recursivity of the budget constraint and write the consumer problem as

$$
\begin{aligned}
&\max u_{i1}, \\
&x_{it} \geqslant 0,\ u_{it} \geqslant 0,\ S_{it},\ V_{i,t+1},\ \text{all } t, \\
&\text{subject to} \qquad\qquad (7.6) \\
&u_{it} = W_i(x_{it}, u_{i,t+1}), \\
&p_t x_{it} + S_{it} = h_{it}, \\
&V_{i,t+1} = V_{it} + S_{it}, \\
&V_{i,T+1} \geqslant 0,
\end{aligned}
$$

for given $h_{it} = p_t\omega_{it} + \Sigma_j\theta_{ij}\Pi_{jt}$ and $V_{i1} = 0$.

In (7.6) the usual budget constraint $\Sigma_t p_t x_{it} \leqslant h_i$, where $h_i = \Sigma_t p_t \omega_{it} + \Sigma_j \theta_{ij}\Pi_j$, is written in the form of a sequence of budgets, one for each period, by making explicit net savings or lending S_{it} and the accumulation of net worth V_{it}. The second equation is the period t budget constraint; the third accumulates net worth; interest receipts are covered via the discounted prices, and the constraint $V_{i,T+1} \geqslant 0$ forces the consumer to repay his debt before $T+1$. We have also assumed that $V_{i1} = 0$: The consumer starts with zero net worth.

For known initial value V_{i1}, given p_1 and h_{i1}, and an optimal pair (V_{i2}, u_{i2}), the consumer can solve for his first-period optimal plan. Then, when given a full sequence of (p_t, h_{it}) and optimal pairs ($V_{i,t+1}$, $u_{i,t+1}$), he can move forward in time and solve a sequence of single-period optimal plans.[12]

Recall that in the producer problem, $\tilde{k}_{j1}$ is given as an initial condition and there is accumulation of stocks as time goes by. In the consumer problem, V_{i1} is also given, but the recurrence relation $u_{i0} = W_i(x_{i0}, u_{i1})$ which, given the past (x_{i0}, u_{i0}), would uniquely set the value of u_{i1}, is being ignored by the consumer. This, together with the recursive structure of the utility function, generates the desired property of "no regrets."[13]

11. Differentiability is not needed for existence of equilibrium.

12. Alternatively, the same information can be conveyed to the consumer by shadow prices on V_{it} and u_{it}.

13. If regrets were allowed, one could also maximize $u_{i,T+1}$ given u_{i0} and thus have past utility affect current decisions.

In the optimum, consumer i's net worth $V_{i,T+1}$ will be zero, unless it receives a value in the objective of the problem. As was the case for producers, attributing such a value may generate misallocations whenever the consumer lives beyond T or wants to bequest part of his wealth.

We might conceive of consumers who live shorter than T as agents with zero marginal utility of consumption for the periods during which they are not alive. However, this formulation has several drawbacks. All consumers, even those born at T, are no longer alive at $T+1$. Moreover, agents not yet born at $t = 0$ participate in the planning decisions made before their birth. The first characteristic calls for the analysis of overlapping generations under infinite horizon (chapter 8); the second, for a relaxation of the assumption that at $t = 0$, all markets exist (this is done in section 7.2.2 for the temporary equilibrium model[14] and will be pursued further in chapter 8).

In the budget constraint, prices p_t are discounted at unobservable, commodity-specific rates. Discounting of wealth and savings is only implicit, so the values generated cannot readily be compared with statistical observations. For later reference, we rewrite model (7.6) in terms of current prices and discount rates r_t. In current prices, net worth will accumulate according to

$$\tilde{V}_{i,t+1} = (1 + r_t)(\tilde{V}_{it} + \tilde{S}_{it}), \tag{7.7}$$

where the tilde ($\sim$) distinguishes current from discounted values and prices. The program will now read

$$\begin{aligned}
&\max u_i(x_{i1}, x_{i2}, \ldots, x_{iT}),\\
&x_{it} \geqslant 0, \tilde{S}_{it}, \tilde{V}_{i,t+1}, \text{ all } t,\\
&\text{subject to} \qquad\qquad (7.8)\\
&\tilde{p}_t x_{it} + \tilde{S}_{it} = \tilde{h}_{it},\\
&\tilde{V}_{i,t+1} = (1 + r_t)(\tilde{V}_{it} + \tilde{S}_{it}),\\
&\tilde{V}_{i,T+1} \geqslant 0,
\end{aligned}$$

for given $h_{it} = \tilde{p}_t \omega_{it} + \Sigma_j \theta_{ij} \tilde{\Pi}_{jt}$ and $\tilde{V}_{i1} = 0$.

Solving for $\tilde{S}_{it}$ in (7.7) and substituting the result in period t's budget equation, makes it possible to write the budget equation in terms of $\tilde{V}_{i,t}$, $\tilde{V}_{i,t+1}$, and x_{it}. Let λ_{it} be the associated shadow price, that is, period t's marginal utility of income. Among the first-order conditions, we find

$$\frac{\partial u_i}{\partial x_{it}} \leqslant \lambda_{it} \tilde{p}_t \perp x_{it} \geqslant 0 \qquad \text{for all } t,$$

14. For a more thorough discussion, see Grandmont and Laroque (1975).

and

$$\frac{\lambda_{i,t-1}}{(1+r_t)} = \lambda_{it} \qquad \text{for } t = 2, 3, \ldots, T.$$

Thus we may write $\lambda_{it} = \lambda_{i1}/[(1+r_2)\ldots(1+r_t)]$ for all $t \geq 2$. From this it follows that the sequence of budget constraints can be collapsed into one intertemporal constraint (with shadow price λ_{i1}) and $p_t = \tilde{p}_t/[(1+r_2)\ldots(1+r_t)]$ as the discounted price. This program makes it possible to derive the discounting rule, rather than to postulate it beforehand. While in this simple case the discounting rule can be obtained from intuition, it becomes less obvious, as we will see, when taxes or constraints on savings are present.

Dynamic Implications of the Market Equilibrium Conditions

Recall that producers can make losses in specific periods and that consumers are not required to balance each period's expenditure with the income earned during that period. Complementarity condition 3 in definition 7.1 implies that

$$p_t \Sigma_i x_{it} = p_t \Sigma_i \omega_{it} + p_t \Sigma_j y_{jt} \tag{7.9}$$

in each period. Adding up the budget equations over consumers and substituting income for period t leads to

$$\Sigma_i p_t x_{it} + \Sigma_i S_{it} = \Sigma_i p_t \omega_{it} + \Sigma_i \Sigma_j \theta_{ij}(p_t y_{jt} - S_{jt}). \tag{7.10}$$

Since we have assumed that $\Sigma_i \theta_{ij} = 1$ for each j, (7.9) and (7.10) imply that $\Sigma_i S_{it} + \Sigma_j S_{jt} = 0$, so net savings balance over agents in each period. Within period t net savings operate like nonimposed transfers between agents. Though looking at the budget constraint of, say, a single consumer, could lead one to infer that savings transfer value over time, this is not so, since savings add up to zero in each period.[15] As a consequence we have

$$\Sigma_i V_{i,t+1} + \Sigma_j V_{j,t+1} = \Sigma_i V_{it} + \Sigma_j V_{jt} \; (=0), \tag{7.11}$$

and there is no increase in (discounted) total net worth.

Furthermore net savings do not just add up to zero as a residual variable. If all agents wanted to spend more than they collectively earned within the period, prices would adjust to restore savings balance: The "rate of interest" is implicitly accounted for by prices.

Hence in the finite-horizon competitive equilibrium, commodity prices convey all the intertemporal information and the unconstrained possibility for agents to borrow and lend creates a perfect, frictionless capital market

15. Commodities can of course be stocked by producers to transfer value over time.

and makes net worth a perfect financial asset. There is no need for any other asset.

7.2.2 Temporary Equilibrium

Definition and Existence

For simplicity, we distinguish between two phases: Phase 1 consisting of periods 1 to $t-1$, and phase 2 consisting of periods t to T. This simply amounts to treating commodities in different periods as separate commodities in each phase. Hence, there is no loss of generality in setting $T=2$ as long as one keeps in mind that the number of commodities does not have to be the same in both phases.

To define a temporary equilibrium, we take up[16] from definition 7.1 for $t=1, 2$, and replace the market equilibrium conditions for $t=2$ by producer and consumer-specific functions $\phi_j(p_1)$, $\phi_i(p_1)$ that generate expectations of future prices;[17] these functions can depend on past and current prices and other variables but not on future prices and allocations. Each consumer i is also assumed to form expectations on second-period dividends $\Pi_{i2}(p_1)$.

DEFINITION 7.2 (Temporary equilibrium) The allocation y^*_{j1}, all j, x^*_{i1}, all i, supported by the price vector $p^*_1 \geqslant 0$ and with expectations ϕ^*_j, ϕ^*_i, Π^*_{i2}, is a temporary equilibrium if:

1. for every producer j, (y^*_{i1}, y^*_{i2}) solves

$$\Pi_j(p^*_1, \phi^*_j) = \max\{p^*_1 y_{j1} + \phi^*_j y_{j2} \text{ subject to } (y_{j1}, y_{j2}) \in Y_j\};$$

2. for every consumer i, (x^*_{i1}, x^*_{i2}) solves

$$\max\{u_i(x_{i1}, x_{i2}) \text{ subject to } p^*_1 x_{i1} + \phi^*_i x_{i2} \leqslant h^*_i + V_{i1}, x_{i1}, x_{i2} \geqslant 0\},$$

for given $h^*_i = p^*_1 \omega_{i1} + \phi^*_i \omega_{i2} + \Sigma_j \theta_{ij} \Pi^*_{j1} + \Pi^*_{i2}$ and V_{i1}, where $\Pi^*_{j1} = \sigma_{j1} \Pi_j(p^*_1, \phi^*_j)$;

3. first-period markets are in equilibrium,

$$\Sigma_i x^*_{i1} - \Sigma_j y^*_{j1} - \Sigma_i \omega_{i1} \leqslant 0;$$

4. expectations on period 2 prices and profit income are

$$\phi^*_j = \phi_j(p^*_1) \quad \text{for all } j,$$

$$\phi^*_i = \phi_i(p^*_1) \quad \text{for all } i,$$

$$\Pi^*_{i2} = \Pi_{i2}(p^*_1) \quad \text{for all } i.$$

16. The model (without production) has been extensively studied by Grandmont (1974, 1977b, 1982) and Grandmont and Laroque (1975).

17. The future is represented through a single (average) expected value. One could also consider a distribution of states, with associated probabilities, maximize expected utilities and profits over these states, and treat outcomes of period 2 as results from these maximizations. See, for example, Radner (1972).

The main changes with respect to the two-period finite horizon model are that only first-period markets have to clear, the price and profit signals for the second period are generated through agent-specific expectations, and these expectations will, in general, not clear markets beyond the first period.[18]

It will be convenient to make producer j's borrowing explicit. Distribution of nonnegative dividends requires that[19]

$$\begin{aligned} \Pi_{j1} + V_{j2} &= p_1 y_{j1} + V_{j1}, \\ \Pi_{j2} &= \phi_j(p_1) y_{j2} + V_{j2}, \end{aligned} \qquad (7.12a)$$

where the rule for distributing dividends is

$$\begin{aligned} \Pi_{j1} &= \sigma_{j1}\Pi_j, \\ \Pi_{j2} &= \sigma_{j2}\Pi_j, \end{aligned} \qquad (7.12b)$$

with $\sigma_{j1} + \sigma_{j2} = 1$.

Likewise consumer i's intertemporal budget constraint can be split into

$$\begin{aligned} p_1 x_{i1} + V_{i2} &= h_{i1} + V_{i1}, \\ \phi_i(p_1) x_{i2} &= h_{i2} + V_{i2}, \\ x_{i1}, x_{i2} &\geqslant 0, \end{aligned} \qquad (7.13a)$$

with

$$\begin{aligned} h_{i1} &= p_1\omega_{i1} + \Sigma_j \theta_{ij}\Pi_{j1}(p_1, \phi_j(p_1)), \\ h_{i2} &= \phi_i(p_1)\omega_{i2} + \Pi_{i2}(p_1). \end{aligned} \qquad (7.13b)$$

Note that in (7.12a) and (7.13a), the (V_{i1}, V_{j1}) are given and that the (V_{i2}, V_{j2}) can be negative, zero, or positive. So borrowing and lending are possible, and all debt is expected to be repaid in period 2. The consumer, say, faces $p_1 x_{i1} + \phi_i(p_1) x_{i2} = h_{i1} + h_{i2} + V_{i1}$ as two-period budget constraint. He does not plan to leave any debt behind, even if he starts with negative V_{i1}.

Before turning to existence of equilibrium, we compare the excess demand function of the competitive and the temporary equilibrium models. Let excess demand for period 1 be written

$$z_1(p_1, \phi(p_1), \Pi_2(p_1)) = \Sigma_i x_{i1}(p_1, \phi_i(p_1), \Pi_{i2}(p_1)) - \Sigma_j y_{j1}(p_1, \phi_j(p_1)) - \Sigma_i \omega_{i1},$$

18. Additional mechanisms may be introduced to make these expectations compatible. This leads to a perfect-foresight equilibrium that coincides with the two-period equilibrium.

19. This leaves producer j's optimal plans unaffected.

where $\phi_i(p_1)$, $\Pi_{i2}(p_1)$, and $\phi_j(p_1)$ are agent-specific expectations for $t = 2$; the vector $\phi(p_1)$ denotes all expectations, and similarly $\Pi_2(p_1)$ denotes all the Π_{i2}. An equilibrium requires that

$$z_1(p_1^*, \phi(p_1^*), \Pi_2(p_1^*)) \leqslant 0,$$

which has the same form as the excess demand equilibrium of definition 1.1. Such an equilibrium was shown to exist in proposition 1.4.[20]

Despite this similarity there are two important differences: First, $z_1(.)$ may not be homogeneous of degree zero in p_1 because we have not introduced any requirement for $\phi(p_1)$ and $\Pi_2(p_1)$ to be homogeneous of degree one in p_1. Moreover, a given (nominal) value V_{i1} appears in the budget of consumer i, and prices will be expressed in the units of V_{i1}. Second, the condition $p_1 z_1(p_1, \phi(p_1), \Pi_2(p_1)) = 0$ will in general be violated. Walras's law holds in equilibrium ($p_1^* z_1(p_1^*, \phi(p_1^*), \Pi_2(p_1^*)) = 0$) but not necessarily out of equilibrium. Summation of the budget constraints over consumers gives the Walras law $p_1 z_1(.) + \eta(.) = 0$, where η stands for the terms beyond period 1, and nothing forces η to be zero. Therefore existence of equilibrium calls for the following additional assumptions:

ASSUMPTION E1 (Expectations) Expectations $\phi_j(p_1)$, $\phi_i(p_1)$, $\Pi_{i2}(p_1)$ are nonnegative, continuous, and uniformly bounded above. Moreover, for $i = 1$, $\phi_i(p_1) \neq 0$ whenever $p_1 \neq 0$.

ASSUMPTION E2 (Nonnegativity of initial net worth) $V_{i1} \geqslant 0$ for all i and $V_{i1} > 0$ for $i = 1$; $V_{j1} \geqslant 0$ for all j.

In principle, assumption E1 (uniform boundedness) rules out functions that are homogeneous of degree one for all $p_1 \geqslant 0$ (and in particular, the identity function). However, the function may be chosen to be homogeneous of degree one (and hence exclude money illusion) on a bounded subset of prices. The proof that follows will show that assumption E1 keeps p_1 bounded. Assumption E2 ensures that at zero prices, consumer 1 will have positive income. We will further illustrate the role of these assumptions in section 7.2.3 in the context of single-period models with given terminal conditions.

We will use our standard assumptions on production sets and utility functions, but for simplicity of exposition, we take production sets to be strictly convex. In the proof we will construct a function that in equilibrium possesses all the properties of the excess demand function used in proposition 1.1. We state and prove:

PROPOSITION 7.2 (Existence of temporary equilibrium) Under assumptions P1, C1′, C2′, E1, and E2 there exist allocations x_{i1}^*, y_{j1}^*, and a price

20. For simplicity of exposition, we take $z_1(\cdot)$ to be single-valued.

vector p_1^* that are consistent with the temporary equilibrium of definition 7.2.

Proof

Step 1. Construction of the Excess Demand Function

We need an excess demand function that satisfies assumptions Z1, Z3–Z5, of section 1.2.1: Walras's law, homogeneity of degree zero, single valuedness and continuity at positive prices, and desirability. To construct it, we extend the set (of period-1) markets to include one additional market for net savings, cleared by a price $\rho_1 = 1/(1 + r_1)$, where r_1 is the interest rate.

We discount prices p_1 accordingly, and denote the discounted prices by $\tilde{p}_1 = \rho_1 p_1$. We express (7.12)–(7.13) in discounted prices by multiplying all period 1 equations by ρ_1:

$$\rho_1 \Pi_{j1} + \rho_1 V_{j2} = \tilde{p}_1 y_{j1} + \rho_1 V_{j1}, \tag{7.14}$$

$$\Pi_{j2} = \phi_j\left(\frac{\tilde{p}_1}{\rho_1}\right) y_{j2} + V_{j2};$$

$$\rho_1 \Pi_{j1} = \sigma_{j1} \rho_1 \Pi_j, \tag{7.15}$$

$$\Pi_{j2} = \sigma_{j2} \Pi_j,$$

$$\rho_1 \Pi_j = \tilde{p}_1 y_{j1} + \rho_1 \phi_j\left(\frac{\tilde{p}_1}{\rho_1}\right) y_{j2};$$

$$\tilde{p}_1 x_{i1} + \rho_1 V_{i2} = \rho_1 h_{i1} + \rho_1 V_{i1}, \tag{7.16}$$

$$\phi_i\left(\frac{\tilde{p}_1}{\rho_1}\right) x_{i2} = h_{i2} + V_{i2};$$

$$\rho_1 h_{i1} = \tilde{p}_1 \omega_{i1} + \rho_1 \Sigma_j \theta_{ij} \Pi_{j1}, \tag{7.17}$$

$$h_{i2} = \phi_i\left(\frac{\tilde{p}_1}{\rho_1}\right) \omega_{i2} + \Pi_{i2}\left(\frac{\tilde{p}_1}{\rho_1}\right).$$

We can define excess demand as

$$\tilde{z}_{k1} = \Sigma_i x_{ik1} - \Sigma_j y_{jk1} - \Sigma_i \omega_{ik1} \qquad \text{for commodities } k = 1, 2, \ldots, r',$$

and

$$\tilde{z}_{k1} = \Sigma_i (V_{i2} - V_{i1}) + \Sigma_j (V_{j2} - V_{j1}) \qquad \text{for net savings } k = r' + 1.$$

Note that excess demands for commodities are bounded from below, since supply is bounded, but that for net savings boundedness from below is not obvious because consumers may "hope" to earn sufficiently in period 2 to be able to borrow infinite amounts in period 1; as we will see, assumptions E1 and E2 rule out this possibility.

Step 2. Properties of the Excess Demand Function

The properties of the excess demand function can be verified as follows:

1. *Walras's law.* By substituting (7.14), (7.15) and (7.17) in (7.16), Walras's law follows.

2. *Homogeneity of degree zero for* $(\tilde{p}_1, \rho_1) > 0$. Producer j's profits are $\tilde{p}_1 y_{j1} + \rho_1 \phi_j(\tilde{p}_1/\rho_1) y_{j2}$; consumer i's budget constraint is (7.16)–(7.17). Hence it can be seen that optimal decisions are unaffected by a rescaling of $(\tilde{p}_1, \rho_1)$, and we may normalize on the simplex so that $\Sigma_k \tilde{p}_{k1} + \rho_1 = 1$, $(\tilde{p}_1, \rho_1) \geqslant 0$.

3. *Single-valuedness and continuity for* $(\tilde{p}_1, \rho_1) > 0$. Since $V_{i1} \geqslant 0$ and $V_{j1} \geqslant 0$, and by assumptions P1, C1′, *C2′*, and E2, both the producer and the consumer programs are feasible, compact-valued, strictly convex; the solutions are unique (theorems A.1.1 and A.1.3). By continuity in assumption E1 and the maximum theorem A.3.1, they are continuous functions of $(\tilde{p}_1, \rho_1)$.

4. *Desirability.* This property will be used to avoid discontinuities in the individual net demand functions. We will show first that consumer 1's net demand for $k = 1, 2, \ldots, r' + 1$ will go to infinity as the associated price goes to zero. Then we will establish that all agents' net demands are bounded from below. Therefore, whenever consumer 1's demand goes to infinity, so will aggregate excess demand, and desirability will hold. We now demonstrate this in steps 4a to 4e.

4a. Individual net demands are bounded from below for commodities. For producers, this is an immediate consequence of the compactness of the production sets. For consumers, ω_{i1} is given and $x_{i1} \geqslant 0$, so $-\omega_{i1}$ is the lower bound.

4b. Individual net savings are bounded from below. All agents plan to repay first period's loans in period 2, since V_{j2} and V_{i2} appear in the right-hand sides of second period's budgets. For producers, uniform boundedness of ϕ_j and compactness of Y_j imply that they cannot expect infinite profits; therefore their loans will be bounded. For consumers, boundedness of ϕ_i and Π_{i2} and nonnegativity of x_{i2} impose a bound on net borrowing.

4c. Suppose that $\tilde{p}_{k1}$ goes to zero for one or several k. By assumption E2, $V_{i1} > 0$ for $i = 1$ and by assumption C2′, $\omega_{i1} > 0$ for $i = 1$. Thus consumer 1 will always have strictly positive income; now by increasingness of utility of 1, his net demand will go to infinity for these k. (See also proposition 2.11).

4d. Suppose that ρ_1 goes to zero. Then, since $(\tilde{p}_1, \rho_1)$ lies on the simplex, $\tilde{p}_1 \neq 0$ and $\phi_{i1}(\tilde{p}_1) \neq 0$ for $i = 1$ (assumption E1). This means that consumer 1 expects to have to pay a positive price for at least one commodity in period 2. By assumption C1′ this commodity is desirable to the consumer so that he will save infinite amounts when ρ_1 goes to zero (see equation 7.16).

4e. Given boundedness from below (steps 4a and 4b) and desirability for consumer 1 on all markets (steps 4c and 4d), the boundedness and desirability properties carry over to aggregate excess demand $\tilde{z}_1(\tilde{p}_1, \rho_1)$.

Step 3. Truncating Excess Demand

Since excess demand can only lack continuity when a price goes to zero, desirability makes it now possible to define a truncated excess demand function that is continuous on the $(\tilde{p}_1, \rho_1)$ simplex so that Brouwer's fixed-point theorem can be used. We define the truncated excess demand $\zeta_k(\tilde{p}_1, \rho_1) = \min\,[\tilde{z}_{k1}(\tilde{p}_1, \rho_1), \varepsilon]$ for any given $\varepsilon > 0$. It is single-valued, continuous on the simplex, homogeneous of degree zero. However, while Walras's law holds as an equality for the excess demand $\tilde{z}_1(\tilde{p}_1, \rho_1)$, it only holds as an inequality ($\leqslant$) for this truncated excess demand. Nonetheless, the proof of proposition 1.4 applies to $\zeta(\tilde{p}_1, \rho_1)$ so that there exists a value $(\tilde{p}_1, \rho_1)$ with $\zeta(\tilde{p}_1, \rho_1) \leqslant 0$. By construction of ζ, this implies that at this point $\tilde{z}_1(\tilde{p}_1, \rho_1) \leqslant 0$ as well, so the truncation is ineffective.

Step 4. Existence of a Temporary Equilibrium

By property 4e, aggregate excess demand $\tilde{z}_1(\tilde{p}_1, \rho_1)$ has the desirability property, so $\tilde{z}_1(\tilde{p}_1, \rho_1) \leqslant 0$ implies that $(\tilde{p}_1, \rho_1) > 0$. Then by Walras's law $\tilde{z}_1(\tilde{p}_1, \rho_1) = 0$. Finally, since $\rho_1 > 0$ at this point, we can recover $p_1 = \tilde{p}_1/\rho_1$ and obtain a temporary equilibrium where $z_1(p_1) = 0$. ■

Net Savings in Temporary Equilibrium

The proof extends the set of period 1 commodities by attributing a price ρ_1 to net worth, which clears the market for net savings. Commodity prices are then discounted as $\tilde{p}_1 = \rho_1 p_1$, and a normalization is performed on $(\tilde{p}_1, \rho_1)$ to have them add up to one. In step 4 of the proof, p_1 is recovered as $\tilde{p}_1/\rho_1$ so that the normalization does not affect equilibrium prices p_1 which therefore are not normalized; they cannot be normalized because the price expectation functions ϕ already do so, since they are not homogeneous of degree one.

The normalization used restricts ρ_1 to lie between 0 and 1. Then the rate of interest r, defined by $\rho_1 = 1/(1 + r)$, must be nonnegative. If we choose another normalization, we can obtain any value $r > -1$ so that the interest rate can be negative (implying a "cost" of holding the asset) or zero. If it is zero, $\rho_1 = 1$, and carryovers are costless.

If an institution (a bank) serves as an intermediary, there may be two interest rates, one for lending and one for borrowing (with the borrowing rate not exceeding the lending rate). This will add one more asset market. If the rate on the lending market is normalized to zero, this asset can be interpreted as money and the other asset as a bond. For the borrowing and the lending rates to differ, one needs to assume either that the intermediary charges a fee, or that access to borrowing and lending varies among agents.[21] One can pursue this disaggregation further into several assets, and this leads to modeling portfolio choices.[22]

The amounts that agents can borrow depend on their subjective expectations. To avoid infinite borrowing, price expectations should not become infinite and agents should be given the intention to redeem their debts. The model can be made more realistic by allowing for institutions that constrain the individuals' decisions to borrow. Assumption E1 would then be weakened, the intention to repay relaxed, and the existence proof adjusted accordingly.

Sequences of Temporary Equilibria

Once a two-period plan has been obtained, it is natural to let one period go by and have agents replan, with previous period's equilibrium allocations as initial conditions. One then obtains equilibrium prices and allocations for period 2, and so on and thus a sequence of temporary equilibria, which generates a so-called recursively dynamic trajectory.

If the terminal year is kept fixed at $T \geqslant 2$, and if agents have perfect foresight, in the sense that expectations are common to all agents and clear markets over the remaining planning horizon, then there is no need for replanning, and we obtain a T-period competitive equilibrium. Even if agents do not have perfect foresight throughout, there may be forces at work that bring the allocations close to a perfect foresight path.[23]

If the length of each agent's plan is kept fixed at, say, $T = 2$, so that the horizon moves forward, replanning is required. This is called a rolling-plan procedure. It will lead to a consistent sequence of plans if the end-valuation of each plan lies on the path of a longer, say, $5T$-period competitive equilibrium. When there is no such consistency, problems will arise from one plan to the next, especially if it proves impossible for some agents to repay accumulated debt.

21. In Grandmont and Laroque (1975), for instance, some agents are not allowed to borrow.

22. See chapter 12 for a further discussion.

23. Turnpike theory studies such dynamics, and in particular, convergence to a trajectory along which relative prices are constant so that in the neighborhood of this trajectory expectation functions can be made very simple. Another subject of study is the design of expectation functions that reflect the agents' capacity to learn from past events.

In considering a sequence of two-period plans, we would relax assumption E2, since agents would start with negative net worth V_{i1} or V_{j1} inherited from the past. For a redistribution of income to be feasible at all prices even when $p_1 = 0$, we need an assumption of positive aggregate net worth and a redistributing agency. Assumption E2 on net worth then has to be replaced by E2′:

ASSUMPTION E2′ (Positive aggregate net worth) $\Sigma_i V_{i1} + \Sigma_j V_{j1} > 0$ and $V_{i1} > 0$ for $i = 1$.

We have seen in (7.11) that $\Sigma_i V_{it} + \Sigma_j V_{jt}$ remains constant over time. The assumption thus has to be imposed only for the first plan, which results in a positive lower bound on V_{i2} for $i = 1$. To ensure that all agents have positive wealth, the redistributing agency must perform the same role as the tax-redistributing government in assumption T of chapter 5. The transfers may reflect an explicit policy of wealth redistribution or simply the fact that when a debt is not repaid, the debtor receives, by default, a wealth transfer from the creditor.[24] For convenience we rephrase the redistribution assumption in the new context, taking into account that government only redistributes and does not need taxes for other purposes. This assumption given assumption E2′, ensures that every agent has a positive wealth.

ASSUMPTION TD (Wealth distribution) Wealth is distributed according to $T_i(p_1, 0, h^0)$ for consumer i and to $T_j(p_1, 0, h^0)$ for producer j, where h^0 is the vector with the expected wealth of every agent as elements. This function satisfies (1) continuity, (2) homogeneity of degree 1 in $(p_1, 0, h^0)$, (3) adding-up whereby $\Sigma_i T_i(.) + \Sigma_j T_j(.) = 0$, and (4) positiveness whereby $h_i^0 + T_i > 0$ for consumer i and $h_j^0 + T_j > 0$ for producer j.

Under assumptions P1, C1′, C2′, E1, and E2′, a temporary equilibrium will exist. This can be proved as in proposition 7.2.

7.2.3 Single-Period Equilibrium

We have seen that the temporary equilibrium model can be considered a two-period model in which equilibrium is enforced only in the first period. The horizon can be contracted further to $T = 1$ so that the model reduces to the competitive model of chapter 1.[25] If the transformation and utility functions are taken to be recursive and the terminal conditions (k_{j2}, u_{i2}) are set exogenously, we can write the single-period model as follows:

24. Grandmont and Laroque (1975) perform the redistribution by having the bank issue new money.

25. At the beginning of this chapter, the T-period equilibrium was introduced as an application of the static model in which commodities were dated. Here the static model can also be considered as a special case of a single-period one, with V_{i1}, V_{j1}, and all terminal conditions set to zero.

For the behavior of producer j (see equation 7.2),

$$\max \Pi_{j1}$$

$$y_{j1}, \Pi_{j1}, \tilde{k}_{j2}, S_{j1}, V_{j2},$$

$$\text{subject to} \tag{7.18}$$

$$F_{j1}(y_{j1}, \tilde{k}_{j2}, k_{j1}) \leqslant 0,$$

$$\tilde{k}_{j2} \geqslant k_{j2},$$

$$\Pi_{j1} = p_1 y_{j1},$$

$$S_{j1} = p_1 y_{j1} - \Pi_{j1},$$

$$V_{j2} = V_{j1} + S_{j1},$$

for given k_{j1}, k_{j2}, V_{j1}.

Program (7.18) can be simplified, since y_{j1} is the only "true" choice variable: k_{j1} and $\tilde{k}_{j2} = k_{j2}$ are given, while Π_{j1}, V_{j2}, and S_{j1} can be computed ex post. The three last constraints only serves to calculate net savings, once the optimal allocation y_{j1} has been computed.

For the behavior of consumer i (see equations 7.6),

$$\max u_{i1},$$

$$x_{i1} \geqslant 0, S_{i1}, V_{i2}, u_{i1},$$

$$\text{subject to} \tag{7.19}$$

$$u_{i1} = W_i(x_{i1}, u_{i2}),$$

$$p_1 x_{i1} + S_{i1} = h_{i1},$$

$$h_{i1} = p_1 \omega_{i1} + \Sigma_j \theta_{ij} \Pi_{j1},$$

$$V_{i2} = V_{i1} + S_{i1},$$

for given u_{i2}, V_{i1}.

Since V_{i1} is given, S_{i1}, consumer i's net saving is given and can be determined before optimization takes place.

For the commodities market equilibrium,

$$\Sigma_i x_{i1} - \Sigma_j y_{j1} - \Sigma_i \omega_{i1} \leqslant 0 \perp p_1 \geqslant 0. \tag{7.20}$$

For the balance of exogenous net savings,

$$\Sigma_i (V_{i2} - V_{i1}) + \Sigma_j (V_{j2} - V_{j1}) = 0.$$

This last condition is a consistency requirement on initial and terminal values.

Clearly the values of (k_{j2}, u_{i2}) significantly affect first-period allocations and should therefore not be set arbitrarily. Four further aspects deserve attention. First, setting the V_{j1}, and V_{i1} at fixed levels will make consumer demands nonhomogeneous in prices. Existence of finite and positive equilibrium prices can be established nonetheless if $V_{i1} > 0$ (see assumption E2) and if assumptions CT1 and C2′ hold. This determines the absolute price level, but allocations are homogeneous of degree zero in the Vs. An alternative is to define "savings functions" by making net worth V a function of p_1 and normalizing on p_1.[26]

Second, the terminal conditions should not rule out existence of equilibrium. The setting of terminal capital stocks k_{j2} eliminates producer j's possibility of inaction, and profits as well as dividend payments may become negative. Since this feeds into the consumers' income, $h_{i1} - S_{i1}$ may become nonpositive. This income may also become negative if $S_{i1} = V_{i2} - V_{i1}$ is set too high. Consumers will then go bankrupt. In the temporary equilibrium model, the difficulty is avoided by assumption TD on redistribution of initial wealth and the redistribution functions T_i and T_j of that model can also be interpreted as "saving functions."

Third, as in temporary equilibrium with perfect foresight, a single-period equilibrium solution could be made to coincide with the first-period allocation of a T-period equilibrium provided that the terminal values are taken from there. As we will see in section 7.3.2, the welfare properties of the single-period equilibrium will depend on how the terminal values are chosen.

Finally, in the single-period model, savings S_{i1} are often represented through savings functions that operate like direct tax functions $T_i(.)$ and have the properties specified in assumption T: Within one period they redistribute income among agents to finance consumption in excess of that period's income and to finance investment.

7.2.4 Dynamics in the Presence of International Trade, Taxes, and Price Rigidities

Here we indicate how dynamics can be incorporated within the models of chapters 4, 5, and 6.

International Trade

In the static model the trade balance was, under the small-country assumption,

$$\bar{p}^m m - \bar{p}^e e = \Pi_0,$$

where Π_0 is the trade deficit of the country (denoted with subscript 0), which can also be considered the gross foreign savings generated during the period covered by the model. In the T-period model this constraint is intertemporal,

26. See section 7.4.1.

so it takes the form

$$\Sigma_t \bar{p}_t^m m_t - \Sigma_t \bar{p}_t^e e_t = \Pi_0,$$

where Π_0 is given. This budget constraint can, as in the case of the consumer, be rewritten as a sequence of trade balances:

$$\bar{p}_t^m m_t - \bar{p}_t^e e_t = \Pi_{0t},$$

where Π_{0t} is period's t deficit (or surplus) which is endogenous and obviously satisfies $\Sigma_t \Pi_{0t} = \Pi_0$. If we disregard possible exogenous grants and wage remittances, net worth accumulates according to

$$V_{0,t+1} = V_{0t} + S_{0t}.$$

$$S_{0t} = -\Pi_{0t},$$

where S_{0t} represent net foreign savings. If, as in (7.8), we make the discounting and interest payments explicit, the first equation becomes

$$\tilde{V}_{0,t+1} = (1 + r_{0t})(\tilde{V}_{0t} + \tilde{S}_{0t}),$$

where r_{0t} is the rate of interest to be paid on foreign debt.[27] This equation shows the accumulation and repayment of foreign debt. In a T-period equilibrium model we have to set exogenously the value of $\tilde{V}_{0,T+1}$, that is, the terminal debt. In the temporary model we can also incorporate trade balances and derive period 1's trade deficit from expectations on prices, including import and export prices and the interest rate on foreign debt r_{0t}.

Taxes

While commodity taxation (indirect taxes and tariffs) is dealt with through price wedges in the same way as in the static model, direct taxes require some elaboration. In chapter 5 we defined consumer i's budget as

$$px_i \leqslant h_i = p\omega_i + \Sigma_j \theta_{ij} \Pi_j(p) + T_i(\cdot),$$

where $T_i(\cdot)$ is a direct tax. When h_i is treated as intertemporal income, it has to be interpreted as the sum of discounted incomes (i.e., wealth), so T_i is a wealth tax (which is nondistortionary as in the static model). To represent income taxes, we need to consider the sequence of incomes

$$h_{it} = p_t \omega_{it} + \Sigma_j \theta_{ij} \Pi_{jt} + T_{it}.$$

In this formulation the tax T_{it} is levied on discounted income. In terms of current income this amounts to writing the budget constraint as in (7.8):

27. Under the small country assumption, foreign prices are exogenous, and therefore the interest rate r_{0t} should also be. To simplify, we have assumed that old and new debts carry the same interest rate.

$$\tilde{p}_t x_{it} + \tilde{S}_{it} = \tilde{h}_{it},$$

$$\tilde{V}_{i,t+1} = (1 + r_t)(\tilde{V}_{it} + \tilde{S}_{it}), \tag{7.21}$$

where $\tilde{h}_{it}$ is after-tax income of period t. Since this is equivalent to the earlier equation, the tax rule is also nondistortionary. However, in practice, interest receipts $r_t(\tilde{V}_{it} + \tilde{S}_{it})$ will also be taxed, say, at a rate τ, and this is as if the interest rate were reduced by τr_t, affecting discounting, savings, and consumption. Thus tax T_{it} is distortionary.

This example shows how discounting rules can be derived from the consumer model in a nontrivial way. Similarly bounds on borrowing (or a distinction between borrowing and lending rates) will make the discount rate endogenous (indeed, when the bound is effective, the discount rate will carry a premium). Such a representation of the sequence of undiscounted budgets helps in representing institutional characteristics that cannot be incorporated in the consumer model of definition 7.1.[28]

Price Rigidities

In the T-period competitive equilibrium, producers will carry stocks between periods, even in the absence of price rigidities. The transformation function allows for this. When stocks are carried forward from t to $t + 1$, undiscounted prices in t cannot be lower than the undiscounted prices in t, minus the storage cost and the interest charge. This generates price rigidities and buffer stock operations in an endogenous way: There is a "downward rigidity" when the stock accumulates and an "upward rigidity" when it is used.

Exogenous price rigidities can also be taken into account by direct reinterpretation of the static model. The buffer stocks of storable commodities that result from this policy are no longer "lost" for the economy; they have value in the future and therefore do not cause distortions. However, allowing the existence of rigidities in the long run may be questionable. Usually they are only imposed in early periods of the T-period model.[29]

In the temporary and the single-period equilibrium models, price expectations enter and markets beyond period 1 are not necessarily in equilibrium. For a commodity that can be stored, price expectations will induce a lower bound on the price in period 1, and operate as a downward price rigidity in period 1. The rigidity is thus not exogenous, since it is

28. See Pereira and Shoven (1988).

29. See, for example, Erlich et al. (1987).

generated by price expectations that have been assumed to depend on period 1 prices.

7.3 Welfare Implications

7.3.1 Intertemporal Pareto-Efficiency

Since the T-period competitive equilibrium is merely a reformulation of the static competitive equilibrium of chapters 1 and 3, the first and the second welfare theorems apply: Equilibrium allocations are intertemporally Pareto-efficient (proposition 1.5), and any intertemporally Pareto-efficient allocation is an equilibrium with transfers (proposition 1.10). Under a finite horizon, intertemporal Pareto-efficiency requires the conditions of the T-period equilibrium to be fulfilled. Consumers and producers must optimize over the full horizon and anticipate prices and incomes with perfect accuracy. For this to occur, all future markets must be in place. Anything else, such as attributing a value $\psi_{j,T+1}$ to terminal stocks $k_{j,T+1}$, will reduce welfare in periods 1 to T, since it will add the term $\Sigma_j \psi_{j,T+1} k_{j,T+1}$ to the welfare function $\Sigma_i \alpha_i u_i(\cdot)$; it could nonetheless promote welfare after T.

We have already noted that under perfect foresight the sequence of temporary equilibria over a finite horizon T will coincide with the T-period equilibrium. The welfare propositions will apply, but under imperfect foresight, the efficiency properties are unclear.

The intertemporal welfare maximum is obtained as

$$\begin{aligned}
&\max \Sigma_i \alpha_i u_i(x_{i1}, x_{i2}, \ldots, x_{iT}),\\
&x_{it} \geqslant 0, \text{ all } i, \; y_{jt}, \text{ all } j, \text{ all } t,\\
&\text{subject to} \qquad\qquad (7.22)\\
&\Sigma_i x_{it} - \Sigma_j y_{jt} - \Sigma_i \omega_{it} \leqslant 0,\\
&(y_{j1}, y_{j2}, \ldots, y_{jT}) \in Y_j.
\end{aligned}$$

For a recursive specification, this becomes

$$\begin{aligned}
&\max \Sigma_i \alpha_i u_{i1},\\
&x_{it} \geqslant 0 \text{ and } u_{it}, \text{ all } i, \; k_{jt}, \tilde{k}_{j,t+1} \geqslant 0 \text{ and } y_{jt}, \text{ all } j, \text{ all } t,\\
&\text{subject to}\\
&u_{it} = W_i(x_{it}, u_{i,t+1}),\\
&F_{jt}(y_{jt}, \tilde{k}_{j,t+1}, k_{jt}) \leqslant 0, \qquad\qquad (7.23)\\
&k_{jt} \leqslant \tilde{k}_{jt},\\
&\Sigma_i x_{it} - \Sigma_j y_{jt} - \Sigma_i \omega_{it} \leqslant 0,
\end{aligned}$$

Since the T-period model is only a rewriting of the competitive equilibrium of definition 1.2, all its elaborations considered in chapters 4 and 6 can be introduced and the welfare propositions presented there apply.

7.3.2 Static Pareto Efficiency

The recursive problem (7.23) makes it possible to consider the period 1 allocation obtained by assigning fixed and feasible values to u_{i2} for all i and k_{j2} for all j. Then the single-period problem with given initial and terminal values may be written

$$\max \Sigma_i \alpha_i u_{i1},$$

$$x_{i1} \geqslant 0,\ u_{i1},\ \text{all } i,\ k_{j1},\ \tilde{k}_{j2} \geqslant 0,\ y_{j1},\ \text{all } j,$$

$$\text{subject to} \tag{7.24}$$

$$u_{i1} = W_i(x_{i1}, u_{i2}),$$

$$F_{j1}(y_{j1}, \tilde{k}_{j2}, k_{j1}) \leqslant 0,$$

$$k_{j1} \leqslant \tilde{k}_{j1},$$

$$\Sigma_i x_{i1} - \Sigma_j y_{j1} - \Sigma_i \omega_{i1} \leqslant 0,$$

for given $\tilde{k}_{j1}$ and u_{i2}.

If in (7.23) and (7.24) the welfare weights α_i are set at their T-period equilibrium values, and in (7.24) the u_{i2} and the k_{j1} are set at the optimal (equilibrium) values of (7.23), then the solution of (7.24) will coincide with the first-period allocations of (7.23). Both welfare propositions would then also apply to the single-period equilibrium, conditional on initial and terminal conditions. This property also holds for a temporary equilibrium allocation, conditional on the terminal values it generates. Therefore the welfare propositions of chapters 5 and 6 apply provided that the terminal conditions are kept fixed when tax or other policy reforms are implemented.

7.4 Practices in Applied Modeling

Most of the applied models have been built as single-period models and used in comparative statics. As mentioned in section 3.6.3, in practice, a benchmark solution is computed as a reference. Then, some parameters are changed, and a new solution is computed and compared to the benchmark case. This solution is interpreted as a short-term equilibrium or as a long-term equilibrium if all price adjustments and all intersectoral allocations of capital goods have taken place.

An applied model should do more than replicate the data for a single period. Ideally it should track a historical period of ten to twenty years. Such a model would then be used to run simulation scenarios that extrapolate past developments into the future and to study the effects of policy changes (and of uncertain events). But this raises the question of whether general equilibrium models are suited to such purposes.

First, there is the matter of the "length" of the time period between t and $t + 1$ which, in principle, might be a day or a decade. However, the fact that markets clear in every period implies that all agents are faced with the same price and abide by trade in equilibrium. This is not realistic for the very short run. Therefore, even in the T-period model, the time path cannot be thought of as a sequence of very short periods and does not represent the transition from one equilibrium to another.[30] On the other hand, when the chosen length of the time period is too long, there is loss of information due to aggregation over time, and substitution possibilities may be exaggerated, leading to an overestimation of the opportunities for growth.[31]

Second, there are the assumptions of "perfect foresight" and of existence of all markets at the beginning of the first time period. Though this is interesting in order to generate a time path into the future (because it abstracts from unexpected shocks and information problems), it will not hold for the past. While temporary and single-period models may be easier to estimate or calibrate, their time path into the future depends on ad hoc assumptions with respect to saving and investment and, as we have seen, is not Pareto-efficient.

We now briefly review the procedure one takes in constructing a dynamic model, distinguishing the three types discussed so far: the T-period competitive model, the temporary equilibrium model, and the single-period model.

If one is willing to ignore end-period valuations and the difficulties in assuming perfect foresight and complete markets, the T-period model offers a simple framework. To implement it, one has only to quantify intertemporal transformation and utility functions. The results derived in chapter 2 apply, so dual representations can be obtained.

In temporary equilibrium the future can be restricted to a single period, which somewhat simplifies the T-period implementation. But expectation functions need to be quantified, and since loans and borrowings are driven by expectations, this will lead to intertemporal inefficiencies.

30. To make the general equilibrium represent some of the frictions on the transition, Kydland and Prescott (1982) have introduced adjustment costs. See also the discussion of adjustment costs in section 7.4.2.

31. It is an empirical and still open question what the ideal length of the time period should be. In practice, the length is dictated by the availability of data. Note that the length does not have to be the same for all time periods. One could, for example, choose the early ones to be shorter than the later ones. See Mercenier and Michel (1994a, 1994b).

In single-period equilibrium models, the debts, capital stocks, and utilities of period 2 are exogenous. Since realistic values are difficult to obtain for these variables, savings and investment functions are often used instead.[32]

7.4.1 Dynamics in the CGE Model

The CGE model for the T-period equilibrium has exactly the same structure as the one of section 3.4.1, except that commodities carry a time subscript. The temporary equilibrium CGE model is obtained by leaving out market-clearing conditions for period 2 and specifying instead expectation functions for prices and profits as in definition 7.2, property 4. Most of the applications deal with the single-period model, either in a static application or for use in recursive dynamics. This involves extending the version of section 3.4.1 with a demand b^g (b for building) for investment goods, specified as an ad hoc function of prices of goods p^g and factors p^f. Dropping the t subscript, the equations are

$$\Sigma_i x_i^g(p^g, p^f, h_i) + b^g(p^g, p^f) + A^g(p^g, p^f)q^g = q^g \text{ for goods,}$$

$$\Sigma_i x_i^f(p^g, p^f, h_i) + A^f(p^g, p^f)q^g = \Sigma_i \omega_i^f \qquad \text{for factors,}$$

$$h_i = p^f \omega_i^f.$$

For prices of goods the equations are

$$p^g = p^g A^g(p^g, p^f) + p^f A^f(p^g, p^f).$$

Capital stocks are included in ω^f. The corresponding entries in x^f will be zero, since capital goods are not consumed.

In the usual recursive dynamic setting of CGE models, all these variables will carry a time subscript t, and demands and input coefficients in t will only be affected by prices in t (or before t if there are rigidities) and not by prices after t. The dynamics only enter through investment b^g which generates a change in capital stocks of the following period, according to relations to be discussed in section 7.4.2.

If the capital stocks are distinguished by type of capital but not by sector of production, capital services will be allocated to sectors so as to equalize marginal productivity among users, through the coefficients A^f. If the

32. In principle, the T-period model also requires such exogenous values for what happens after T. Neglecting this simply means that these values are set to zero. For T sufficiently large, one could argue that this does not matter too much for consumers who discount the future. For production, the issue is far more intricate, and we will return to it in chapter 8.

capital stocks are sector specific, the allocation of capital to sectors is determined as a consequence of the investment decisions.[33]

The vector of endowments ω^f may also contain net supplies of firms that undergo a lag in production and have to base their production decisions on expected prices only. This will, for example, be the case for farmers who must decide in advance on the crops to plant.

Consumer i's demands $x_i^g(p^g, p^f)$ and $x_i^f(p^g, p^f)$ are derived from the usual utility maximization under a budget constraint that deducts savings S_i from period 1 income h_i as a direct tax:

$$\max u_i(x_i^g, x_i^f),$$

$$x_i^g, x_i^f \geqslant 0,$$

subject to

$$p^g x_i^g + p^f x_i^f = h_i - S_i(p^g, p^f, h_i),$$

where $S_i(p^g, p^f, h_i)$ is the savings function for which different types of specifications have been used, depending on the nature of the closure rule that is adopted:[34]

1. Savings can be specified in the same way as income taxes and can be assumed to accrue as income to an agent, the "investor." This agent satisfies a budget constraint that equates this income to his investment expenditure. Then total savings will be equal to total investment at all prices, so that Walras's law is maintained:

$$\Sigma_k p_k^g b_k^g(p^g, p^f) = \Sigma_i S_i(p^g, p^f, h_i(p^g, p^f)).$$

2. In many applied models investment is set exogenously as part of a public investment program. Then the savings generated will not be equal to investment unless an additional adjustment mechanism is specified. For this there are several choices: One could balance savings with total investment by dropping, say, consumer i's saving function (and for several consumers if this is not sufficient to ensure equilibrium),[35] one could also balance investment with savings. Alternatively, one could attempt to reach equilibrium by dropping some other function(s), such as government consumption

33. If capital stocks are not sector specific, the part of the A^f matrix corresponding to capital stocks will have as many rows as there are capital goods. When capital is sector specific, the number of such rows will be equal to the number of sectors times the number of capital goods. Accordingly, the prices p^f of these capital goods will either be equalized and be market prices for capital services or sector-specific internal accounting prices.

34. See also section 3.4.2.

35. The consumer's savings are then adjusted residually. Since his consumption cannot be negative, it may be necessary for more than one consumer to adjust.

making the corresponding variable a slack. In some cases the equilibrium condition may be left out, generating an imbalance in a particular market.

3. A natural step is to introduce an interest rate as an equilibrating variable ($S(p, r) = I(p, r)$ as in Adelman and Robinson 1978). This approach has been extended to segmented credit "markets" where one sector refers to formal, rationed credit, and the other to credit in the "curb" market.[36]

7.4.2 Issues in Model Implementation

Making Investment Explicit in Producer Decisions

We start from the producer model (7.1′), substituting $k_{j,t+1}$ for $\tilde{k}_{j,t+1}$:

$$\max p_t y_{jt} + \psi_{j,t+1} k_{j,t+1},$$

$$k_{j,t+1} \geqslant 0, \; y_{jt},$$

subject to

$$F_{jt}(y_{jt}, k_{j,t+1}, k_{jt}) \leqslant 0,$$

for given k_{jt} optimal in $t - 1$. We recall that $k_{j,t+1}$ is produced jointly with net outputs y_{jt}, while k_{jt} denotes capital inputs.

In practice, it is convenient to separate within-period production from accumulation, so that one can distinguish between two types of equations:

$$G_{jt}(y_{jt}, k_{jt}) \leqslant 0, \tag{7.25a}$$

$$b_{jt} = K_j(k_{j,t+1} - (I - \Delta_j)k_{jt}), \tag{7.25b}$$

where K_j is the matrix of investment requirements and Δ_j is the (usually diagonal) depreciation matrix; $k_{j,t+1} - k_{jt}$ represents net additions, $\Delta_j k_{jt}$ is the investment for replacement, and the sum of the two is gross investment.

The first (scalar) equation is the within-period transformation function, where y_{jt} represents net outputs, excluding demand for investment as an input.[37] The second set of relations generate these requirements so as to permit variation of capital stocks. The vector b_{jt} has as many elements as there are commodities in the period. Since capital is not traded, k_{jt} and $k_{j,t+1}$ can have any dimension. In this equation there only is a one-period gestation lag (time to build) between investment and changes in capital stocks. Some investments may take longer to mature and thus remain in a pipeline for more than one period. This can be tackled by considering these

36. See van Wijnbergen (1982) and Nsengiyumva and Decaluwé (1992).

37. Demand for investment goods b_{jt} will now enter the commodity balances, which will read $\Sigma_j b_{jt} + \Sigma_i x_{it} - \Sigma_j y_{jt} - \Sigma_i \omega_{it} \leqslant 0$. Here we do not constrain investment requirements to be nonnegative, so invested goods can be sold. One could also exclude sales of goods invested earlier or make the sale "costly."

investments as separate capital goods. Such a disaggregation also makes it possible to distinguish various lifetimes for different types of equipment and to keep track of the different vintages for every type of equipment.[38]

Equations (7.25) can replace the constraint on F_{jt} in the one-period program, which then becomes

$$\max p_t y_{jt} + \psi_{j,t+1} k_{j,t+1} - p_t b_{jt},$$
$$k_{j,t+1} \geqslant 0, b_{jt}, y_{jt},$$
$$\text{subject to} \qquad (7.26)$$
$$G_{jt}(y_{jt}, k_{jt}) \leqslant 0 \qquad (\rho_{jt}),$$
$$b_{jt} = K_j(k_{j,t+1} - k_{jt} + \Delta_j k_{jt}),$$
$$\text{for given } k_{jt} \text{ optimal in } t-1.$$

This formulation poses two difficulties. First, if $\psi_{j,t+1} > p_t K_j$, it may happen that investment becomes unbounded.[39] Second, even if all variables are bounded in the optimum, a sequence of one-period plans may lead to a situation in which all investment will take place in a single period (bang-bang behavior), as shown in the following first-order conditions:[40]

$$\rho_{jt} \frac{\partial G_{jt}(\cdot)}{\partial y_{jt}} = p_t, \qquad (7.27)$$
$$\psi_{j,t+1} \leqslant p_t K_j \perp k_{j,t+1} \geqslant 0.$$

For $k_{j,t+1} > 0$ equality holds in the second relation. Since all these are given data when the producer makes his decision, the condition will only be satisfied by coincidence. In the multiperiod program (7.1), ψ_{jt} will be endogenous, but the intertemporal conditions on the ψ_{jt} will see to it that investment takes place in a single period (or the producer will be indifferent as to the period in which he chooses to invest).

Such problems due to the linearity of the investment requirement function can be avoided by introducing decreasing returns to scale in installation of new capacities. Then K_j can be treated as a function of gross (or net)

38. Often there will only be one capital good in each firm. K_j is then a column vector and gross additions to capital can be written as $k_{j,t+1} = (1 - \Delta_j)k_{jt} + B_{jt}$, with corresponding demands for capital goods given by $b_{jt} = K_j B_{jt}$.

39. Such unboundedness does not endanger existence of equilibrium because an infinite level of $k_{j,t+1}$ necessarily implies an infinite value for $p_t b_{jt}$ so that the boundedness of the intertemporal production set is still ensured.

40. Note that these are the standard first-order conditions of profit maximization subject to a transformation function, so the dual approach from chapter 2, using a profit function, can be applied. It will involve valuating terminal stocks; this is avoided in the continuous-time approach. See Epstein and Denny (1983).

investment, and the allocation of investment overtime will be smoother.[41] The K_j functions can be considered the derivatives of the vector of investment-cost functions and be written as $K_j(p_t, k_{j,t+1} - k_{jt} + \Delta_j k_{jt})$. Such forms can also be derived from the q theory of investment.[42]

Representing Producer Decisions through Profit Functions

Condition (7.27) is the first-order condition of the restricted transformation function, $G(y_{jt}, k_{jt})$ for given k_{jt}, which is dual to the restricted profit function $\Pi_{jt}(p_t, k_{jt})$. We can now write the intertemporal problem of capital accumulation in two stages, starting with

$$\max \Sigma_t [\Pi_{jt}(p_t, k_{jt}) - p_t b_{jt}] + \psi_{j,T+1} k_{j,T+1},$$

$$k_{j,t+1} \geqslant 0,\ b_{jt},\ \text{all } t,$$

$$\text{subject to} \qquad (7.28)$$

$$b_{jt} = K_j(p_t, k_{j,t+1} - k_{jt} + \Delta_j k_{jt}),$$

for given k_{j1}.

In the second stage the net outputs y_{jt} are calculated using Hotelling's lemma of proposition 2.1.

Expectations in Temporary Equilibrium

In the temporary equilibrium model, future prices have to be generated as producer-specific expectations $\phi_j(\cdot)$. Recall that these functions must satisfy assumption E1 and be bounded from above. This requirement excludes, for example, naive expectations $\phi_j = p_1/(1 + r)$ commonly found in applied models where the future extends to one period only. Since ϕ_j is linear, one has to impose a bound on borrowing as an additional constraint in (7.26) or (7.28), in order to ensure existence.

Single-Period Equilibrium

The single-period model is the model most commonly used. Equation (7.26) can, in principle, be implemented by specifying ψ_{jt} as a function of other variables. However, this approach has not proved very successful, and many modelers prefer to fix net investment $(k_{j,t+1} - k_{jt})$ exogenously in the producer model. In the full equilibrium model, net investment can be endogenized through investment functions, with prices and possibly other

41. To ensure boundedness, one either bounds investment or makes it fully improductive beyond a certain level of capacity expansion.

42. See, for example, Hayashi (1982) and Abel and Blanchard (1986) for surveys. An explicit representation of gestation lags and lifetimes of equipments will also avoid bang-bang behavior.

variables as arguments. Apart from the requirement that these functions should be continuous and homogeneous of degree zero in prices, theory provides few indications on how to construct them outside of the optimization framework.

Distributing Profits

In (7.2) the profit distribution of model (7.1) is made explicit, and it is assumed that intertemporal profits are distributed in fixed shares σ_{jt}. Since by assumption P1′ producers have the possibility of inaction in every period t, the profit distributed in period t cannot become negative. In applications one rarely finds such a rule, and this profit may become negative. This does not create problems for existence of equilibrium as long as the consumer has income from other sources or opportunities to borrow. Usually these models distribute profits $p_t y_{jt}$ (rather than $p_t(y_{jt} - b_{jt})$), without deducting any depreciation allowance.

Non-Firm Specific Capital Stocks

So far we have considered that each producer makes his own investment plan and that capital goods are held within the firm. Most applied models follow a different, neoclassical approach of defining a total availability of each capital good (machines, buildings, etc.) to be allocated over sectors, firms, or producers so as to equalize the marginal productivity of capital among these. This marginal productivity is equal to $p\partial y_j/\partial k_j = \psi$, the rental cost of the capital goods concerned.[43] Now, capital goods are not held by firms but by capital holders who lease them at a price ψ and may also be given the role of investor. The capital goods used by one firm can freely be redirected to another one during the next period. To avoid this unrealistic feature, one could restrict gross investments by all firms to remain nonnegative and equalize marginal productivity of gross additions to capital rather than of capital stocks themselves. The investment levels can also be kept exogenous, calculated so as to reach sectoral growth targets. One could also opt for investment functions that have sectoral profits among the arguments. This reflects the notion that the average profit determines the liquidity position of the firm and as a result any self-financing. It amounts to formulating investment portfolios, to which we will return in chapter 12.

Consumer Decisions

In the consumer problem of definition 7.1, the value function is an ordinary indirect utility function $v_i(p_1/h_i, p_2/h_i, \ldots, p_T/h_i)$, from which we can derive

43. In the single-commodity economy, ψ/p is often referred to as the real rate of interest. We avoid this terminology, which may lead to confusion with the rate of interest on financial assets.

the demands for each period in accordance with Roy's identity in proposition 2.8. However, since only total wealth h_i is explicit, we cannot directly compute each period's saving, debt accumulation, and so on. To make savings more explicit, we replace the intertemporal wealth constraint by a sequence of budgets as in (7.6). Under assumption C1′ this program has a unique solution and implicitly defines a saving function $S_{it}(p_1, p_2, \ldots, p_T, h_{i1}, h_{i2}, \ldots, h_{iT}, V_{i1})$ for each period t.[44] This function will be continuous and homogeneous of degree one in its arguments.

In the temporary equilibrium model, the variables for $t \neq 1$ will be expectations, so savings will be a function of p_1, h_{i1}, and V_{i1}, and current income rather than wealth will drive savings. This result has been the subject of extensive debate: The notion that savings are better explained by current income than by wealth violates, under perfect foresight, the life-cycle hypothesis.[45]

Number of Agents

As discussed in chapter 2, changes in the number of producers via entry and exit can, by assumptions P1′, be assimilated with the decision of whether to produce or not to produce. All producers are potentially present, and therefore changes in their number does not require special treatment.

Consumers are characterized by endowments, profit shares, and utility functions. The sequence of consumers' endowments ω_{it} is exogenous, and the equilibrium allocations are Pareto-optimal given these endowments. Thus endowments do not shift endogenously across markets (geographically or in time) or across agents. They can be transported or stored by producers or exchanged among agents. Still the location of the source of the commodity remains fixed, and this may cause inefficiency, in comparison to a setup where the location of endowments if flexible. This naturally leads us to the notion of migration of factors, particularly of labor.

Apart from having their endowments move among locations or in time, consumers may undergo changes in lifestyle as well as in social and demographic status. This will also affect the "location" of their endowments and has to be taken into account in the formulation of the consumer problem.

Most applied models classify consumers into groups, whose population is adjusted over time through demography of birth and death as well as through transitions between groups. The adjustment is exogenous or a

44. This saving function can be given a more explicit form if one introduces recursivity in the utility function. One can then define for each period the indirect recursive utility function and use it in program (7.28), which then only needs to be solved for each period's total consumption expenditure.

45. For a discussion, see Deaton (1991).

function of relative wages, employment opportunities, and the like. Consumer behavior is then specified on a per capita basis. When the number of members of a group increases, the labor endowment rises and per capita income changes. As the consumption demand of the group changes so does the equilibrium allocation. Here we are confronted with two issues.

First, in any period $t \neq 1$, group i consists of consumers who may have belonged to other groups in $t = 1$. Maximizing utility of the group over time is not the same as maximizing utility of the consumers as defined in $t = 1$, which is the relevant concept for a consumer maximizing utility over his lifespan. Otherwise, it is the group that maximizes per capita utility: Groups of consumers operate like organizations with varying membership. Even if one accepts that it is the group that maximizes utility,[46] there is a second problem that concerns the specification of the intertemporal utility function in the presence of variations in the number of members over time. In section 2.2.1 the static case was considered. There the "group" was maximizing $n_i u_i(x_i/n_i)$, and the solution was identical to the one in which per caput utility was being maximized. In the dynamic context n_i will vary over time. The problem is now that specifying intertemporal utility on per caput basis as $u_i(x_{i1}/n_{i1}, x_{i2}/n_{i2}, \ldots, x_{iT}/n_{iT})$ does not lead to a measure of group utility.

Consider an agent i defined in $t = 1$ and planning for $t = 1$ to T, and denote his utility by $u_i(\cdot)$. To handle the two issues discussed above, we take his utility function to be separable over time and over the social groups to which he may belong in the future (to be called destinations $d = 1, 2, \ldots, D$):

$$u_i(\cdot) = W_i(\hat{u}_{11}, \ldots, \hat{u}_{dt}, \ldots, \hat{u}_{DT}), \tag{7.29}$$

where $\hat{u}_{dt}(x_{idt}, n_{idt}) = n_{idt} u_{dt}(x_{idt}/n_{idt})$ is the total utility obtained from members in destination d at period t and n_{idt} is the number of such members.[47] Alternatively, we could write utility on a per caput basis as

$$u_i(\cdot) = W_i(u_{11}, \ldots, u_{dt}, \ldots, u_{DT}). \tag{7.30}$$

In practice, expression (7.30) may be simplified by setting $u_{dt}(x_{idt}) = u_d(x_{idt})$ and by taking a linear form for W_i so that $u_i(\cdot) = \Sigma_d \Sigma_t (\beta_i)^t u_d(x_{idt}/n_{idt})$ or $u_i(\cdot) = \Sigma_d \Sigma_t (\beta_i)^t n_{idt} u_d(x_{idt}/n_{idt})$. This implies that the weights on the destinations are equal (independently of the number of members in every destination) or proportional to the number of members. One could also choose other weights.

When the n_{idt} are set exogenously with consumer behavior specified according to (7.29) or (7.30), the T-period competitive equilibrium will be

46. This is obviously correct if there is only a single consumer group.

47. The formulation has the limitation that $\hat{u}_{dt}$ is homogeneous of degree one in its arguments, but this can be generalized.

Pareto-efficient given the fixed n_{idt}. Temporary and single-period equilibria can also be obtained along these lines without affecting their (lack of) efficiency.[48]

The n_{idt} can also be endogenized. Harris and Todaro (1970) introduced the idea of making the n_{idt} functions of relative wages among destinations and the ω_{it} functions of n_{idt}: When a farmer moves, his labor endowment becomes urban.[49] The n_{idt} can also be treated as a choice variable by consumer i,[50] reflecting decisions on the number of children to rear as well as on the choice of lifestyle. The n_{idt} affect the endowments and will usually be bounded by demographic and social constraints on $\Sigma_d n_{idt}$ and $\Sigma_i n_{idt}$, respectively. Two types of efficiency gains will result: First, consumers may choose their destination, and therefore their lifestyle; this leads to a gain in consumer efficiency. Second, endowments may be shifted (possibly at some cost) between, say locations, leading to a gain in producer efficiency. These gains appear to be large.[51]

7.4.3 Empirical Implementation

Construction of the SAMs with Savings and Investment

SAMs should, in principle, be constructed for every period t, though modelers usually only construct one for $t = 1$. The SAMs for the T-period model, the temporary equilibrium model, and the single-period model have the same structure, since any nonoptimal savings and investments are only reflected in the numerical values and not in the structure.

In the text so far, investment was carried out by a single "investor." In a model with consumers, a foreign sector, and a government, this would mean that the savings of all three groups of agents accrue to the single investor. The usual approach is to channel all foreign savings via the government and to distinguish only private and public investors. There can also be a separate representation of investment for the three agents.

Dropping the period subscript t, we can write the equilibrium conditions that underlie the SAM (with a single investor) for any given period as follows:

48. Many applications of sequences of single-period equilibria use utility functions defined by group of destination: A farmer who moves to the city loses his membership of the group of farmers and becomes an urban worker. This makes it even more problematic to compare welfare of groups of agents over time and between alternative policy settings.

49. See also Kelley and Williamson (1984).

50. For this to work, the utility function $\hat{u}_{dt}$ must be concave, nonstationary, and nondecreasing. See theorem A.1.5.

51. See Fischer et al. (1991) and Hamilton and Whalley (1984).

For the market equilibrium conditions,

$$\Sigma_i x_{ik}^* + b_k^* - \Sigma_j y_{jk}^* - \Sigma_i \omega_{ik} \leqslant 0 \perp p_k^* \geqslant 0. \tag{7.31a}$$

For the profits of producers,

$$\Pi_j^* = p^* y_j^*, \qquad \text{all } j. \tag{7.31b}$$

For the budget equations of private consumers,

$$p^* x_i^* = p^* \omega_i + \Sigma_j \theta_{ij} \Pi_j^* - S_i^* + T_i^*, \qquad \text{all } i. \tag{7.31c}$$

For the budget equation of the government (consumer with subscript g),

$$p^* x_g^* = p^* \omega_g + \Sigma_j \theta_{gj} \Pi_j^* - S_g^* + T_g^*. \tag{7.31d}$$

For the budget equation of the foreign sector,

$$\bar{p}^e e^* = \bar{p}^m m^* - \Pi_0. \tag{7.31e}$$

For the budget equation of the investor,

$$\Sigma_k p_k^* b_k^* = \Sigma_i S_i^* + S_g^* + \Pi_0. \tag{7.31f}$$

For the budget balance of the government (in the absence of indirect taxes),

$$\Sigma_i T_i^* + T_g^* = 0. \tag{7.31g}$$

The SAM structure is given in Table 7.1.

Calibration and Estimation of the Model

SAMs can only be used to calibrate the base year rather than the time paths of variables. As was noted in section 3.6.1, the behavioral equations of the model cannot reproduce every SAM even when using a sequence of SAMs. Modelers who construct *T*-period models will usually calibrate a selected number of critical variables in benchmark years, such as aggregate savings and consumption. Single-period models use trends or shifts in parameters to fine-tune the model with respect to a large number of variables.

The components of the model can be estimated by more formal methods. For consumers, one could estimate the first-order conditions of life-cycle models by assuming some form of perfect foresight.[52] For producers, demand for investment could be estimated by applying duality theory. Simulation results obtained with such consumer and producer models are, however, not very satisfactory,[53] and in practice modelers fine-tune coefficients until the model reproduces selected benchmark values.

52. See, for example, Arellano and Meghir (1992), Deaton (1985), and Phimister (1993).

53. The main explanation is that the lifetime budget neglects asymmetries due to period-specific constraints on borrowing and accumulation effects of nontradable household capital. See Deaton (1995) and Kydland (1995).

Table 7.1
Structure of the SAM with savings and investment

	Goods and factor	Firms	Consumers	Government	Foreign sector	Investor	Government budget
Goods and factors		Input	Consumption	Consumption	Exports	Investment	
Firms	Production						
Consumers	Endowments	Profits					Transfers
Government	Endowments	Profits					Transfers
Foreign sector	Imports						
Investor			Savings	Savings	Savings		
Government budget							

7.4.4 Simulations

Solving the Model

For the *T*-period competitive equilibrium, the same computational methods can be used as for the static model, especially if the time horizon extends to a few periods. But these methods are impractical for longer time horizons because dimensionality poses a problem. In the dynamic model then, one proceeds by iterations that adjust price expectations. For given initial price expectations over the whole horizon, one moves forward in time, accumulating capital and computing for each period prices that (almost) clear that period's markets. The computed prices serve as the expectations of a new round of calculations. One concludes that a solution has been found when no further price adjustments are needed.[54]

Running Simulations

If perfect foresight and intertemporal efficiency are considered essential, a finite-horizon model with some end-stock valuation may be appropriate. Blitzer et al. (1986, 1994) solve a finite horizon model for Mexico and Egypt with particular reference to the energy sector. Manne and Richels (1990, 1992) study world energy supplies and prices. These models were originally developed to obtain efficient investment decisions in the energy sector, and they were well suited to analyze carbon dioxide emissions. Moreover *T*-period models have been implemented to incorporate short-run rigidities and lack of substitution in the first period, while the second period represents a long-term future with flexible prices and substitution. Finite-horizon models are also used as approximations for infinite-horizon models.[55]

If perfect foresight and intertemporal efficiency are not essential, recursively dynamic simulation can be used.[56] It may still be desirable to perform welfare comparisons over an infinite horizon. For this Levy and van Wijnbergen (1991) have introduced a valuation of end-stocks and debts in a post-T steady state. Although they treat investment and savings rates as exogenous variables, their approach makes it possible to compare scenarios in a better way than with a finite-horizon measure. Demographic aspects of fertility, mortality, and migration are covered by Kelley and Williamson (1984).

54. See, for example, Jorgenson and Slesnick (1985). Codsi et al. (1992) use a different method that approximates the model linearly and solves it simultaneously.

55. See section 8.4.

56. Since the wide majority of applications is of this nature, we refer to the survey papers mentioned in the introduction to this book.

8 Infinite-Horizon Dynamics

There are two types of infinite-horizon models: dynastic models, with a finite number of infinitely lived consumers (section 8.1) and overlapping generations models, with an infinite number of generations of finitely lived consumers whose generations overlap (sections 8.2 and 8.3). The two types are treated separately not only because their mathematical structures differ but also because they have different economic content. Dynastic models describe infinitely lived families whose generations anticipate each others' utility and leave bequests to (or borrow from) future generations: The wealth constraint of each family extends over the infinite horizon. By contrast, in overlapping generations models, each finitely lived consumer has a wealth constraint over his lifetime. Although he may leave bequests, he cannot leave any debt behind. Section 8.4 compares the various types of models and discusses how they have been applied.

8.1 Dynastic Models

It is convenient to start the discussion of the dynastic model with a single infinite-horizon (Negishi) program, since this enables us to use properties of dynamic programming as well as Bellman's principle, as is now common in growth and capital theory (section 8.1.1). We prove that welfare weights exist for the Negishi program such that each consumer satisfies his intertemporal budget constraint, and we prove that as in the finite-horizon model, this solution is a competitive equilibrium and therefore is Pareto-efficient (sections 8.1.2 and 8.1.3). We discuss multiplicity of equilibria in the presence of distorsions such as commodity taxes; we show that unlike in the finitehorizon model, the number of equilibria tends to become infinite (section 8.1.4). We also define steady states, study their number, and discuss their asymptotic stability properties (section 8.1.5). In section 8.1.6 we show how the infinite-horizon model may be brought to numerical implementation through a finite-horizon approximation. Finally, in section 8.1.7 we briefly cover real business cycle models, a field of application that has received much attention in the macroeconomic literature.

8.1.1 Specification of the Dynastic Negishi Program

We start from the intertemporal welfare program (7.23) with given welfare weights α_i. We subsequently introduce budget constraints and adjust the weights to satisfy these constraints. The analysis of infinite-horizon programs is easier when the number of parameters is finite. The recursive presentation of chapter 7 still contains time-dependent transformation functions $F_{jt}(.)$ and endowments ω_{it}. If these can be represented by a finite number of parameters, for instance, by using trends, all adjustments of the transformation functions and the endowments can be treated as accumulation of stocks, and the dimension of the vector k_{jt} can be extended accordingly. To focus on the essential properties of the model, and without loss of generality, we take aggregate net supply y_t to include the total endowment[1] $\Sigma_i \omega_{it}$, and we do not separate technology by firm. Thus program (7.23) becomes

$$\max \Sigma_i \alpha_i u_{i1},$$

$$x_{it} \geqslant 0,\ u_{it},\ \text{all } i,\ k_t,\ \tilde{k}_{t+1} \geqslant 0,\ y_t,\ t = 1, 2, \ldots,$$

subject to (8.1)

$$u_{it} = W_i(x_{it}, u_{i,t+1}),$$

$$F(y_t, \tilde{k}_{t+1}, k_t) \leqslant 0,$$

$$k_t \leqslant \tilde{k}_t,$$

$$\Sigma_i x_{it} - y_t \leqslant 0,$$

for given $\tilde{k}_1$.

Although program (8.1) seems close to (7.23), its mathematical properties are very different because it is defined in infinite-dimensional space, with an infinite number of commodities and balance constraints, and in this case standard Lagrange multiplier theory is not readily applicable.

Another difference is that the solution of (8.1) may be unbounded. Clearly y_t, $\tilde{k}_{t+1}$ (and hence all x_{it}) are bounded when k_t is given. However, over time some elements of k_t may become infinite through accumulation, and this can cause unboundedness of the objective,[2] which makes it impossible to compute the budget deficits needed in the Negishi approach. The following assumption, in which parts (1) to (3) are the same as in assumption CT1 on the recursive utility function of chapter 7, will prevent this. Part (4) introduces discounting, while unboundedness is avoided in part (5):[3]

ASSUMPTION CD1 (The dynastic utility function) (1) $W_i : R^r_+ \times R \to R$, $W_i(x_i, u_i)$ is continuous, strictly concave and continuously differentiable on $R^r_{++} \times R$; (2) $W_i(x_i, u_i)$ is nonsatiated in x_i and increasing in x_i for $i = 1$; (3) $W_i(0, 0) = 0$; (4) discounting: $0 < \partial W_i(.)/\partial u_i \leqslant \gamma$ for some given $\gamma \in (0, 1)$ common to all i, and (5) $W_i(x_i, u_i)$ is uniformly bounded above with upper bound $\bar{v}$.

1. In section 8.4 we will see that if a dynasty is interpreted as consisting of a succession of generations, then the distinction between endowments and capital stocks matters. While endowments are obtained freely, capital stocks have to be purchased by every generation.

2. Unboundedness of the objective makes it impossible to compare welfare between two allocations; more refined optimality concepts are then necessary. See, for example, Carlson and Haurie (1987).

3. Uniform boundedness will be seen to facilitate finite-horizon approximations.

A special but commonly adopted case is

ASSUMPTION CD1′ (Dynastic utility function with fixed discount factor) $W_i(x_i, u_i)$ satisfies parts 1 to 5 of assumption CD1 with the additional assumption that the discount factor is constant and common to all $i: \partial W_i/\partial u_i = \beta$ so $W_i(x_{it}, u_{i,t+1})$ is linear in $u_{i,t+1}$ and can be written as $u_i(x_{it}) + \beta u_{i,t+1}$; then $u_{i1} = \Sigma_{t=1}^{\infty} (\beta)^{t-1} u_i(x_{it})$.[4]

Assumption CD1′ is often made for mathematical convenience. Its drawback is that it requires all consumers to have the same discount factor β. Otherwise, as we will see, the consumer whose discount factor is highest will eventually be the only one to consume.

We also need an assumption on initial capital stocks, that is the equivalent of Assumption C2 on endowments. It will ensure (in conjunction with assumption PD1, part 1) that all consumers have positive income:

ASSUMPTION K (Initial stocks) In period 1 each consumer i owns nonnegative and nonzero initial stocks ($\tilde{k}_{i1} \geqslant 0, \tilde{k}_{i1} \neq 0$) and a positive stock of the first element of the vector ($\tilde{k}_{i1,1} > 0$), say, labor. We denote $\Sigma_i \tilde{k}_{i1}$ by $\tilde{k}_1$.

Finally, assumption PT1 of chapter 7 needs to be strengthened in two respects: Constant returns to scale are imposed, and it is assumed that starting from initial stocks, one can attain a level of stocks such that positive consumption is possible and no stock is depleted (sustainability):

ASSUMPTION PD1 (Transformation function) (1) $F(y_t, \tilde{k}_{t+1}, k_t)$ is convex, nondecreasing in $(y_t, \tilde{k}_{t+1}, -k_t)$, and increasing in $-k_{1t}$, where k_{1t} denotes the aggregate stock of capital good 1 in period t. The constraint $F(y_t, \tilde{k}_{t+1}, k_t) \leqslant 0$ satisfies (2) homogeneity of degree one: If $F(y_t, \tilde{k}_{t+1}, k_t) \leqslant 0$, then $F(\lambda y_t, \lambda \tilde{k}_{t+1}, \lambda k_t) \leqslant 0$ also for any $\lambda > 0$; and (3) boundedness: y_t and $\tilde{k}_{t+1}$ are bounded above for any given bounded k_t. (4) The set $M = \{k > 0 \mid F(y, k, k) < 0, y > 0\}$ is nonempty; (5) there exists a value $k \in M \cap (0, \tilde{k}_1)$, where $\tilde{k}_1$ is the initial stock.

In assumption PD1, by part 1 some output may be increased if k_{1t} increases; hence the initial stock $\tilde{k}_{11} = \Sigma_i \tilde{k}_{i1,1}$ (labor) will have positive shadow price, and (by assumption K) all consumers will have positive income. Part 2 amounts to requiring that multiplication of stocks k_t by a positive factor raises net outputs y_t and stocks $\tilde{k}_{t+1}$ by the same factor; it will ensure that the value of outputs is equal to the value of stock inputs for each t so that the reward of factors exhausts all profits.[5] Part 3 implies that the value of outputs is bounded for all t as long as k_t is bounded. Part 4

4. To avoid confusion later on (when the superscript t will denote a consumer born in period t), we write $(\beta)^{t-1}$ to denote exponentiation of β.

5. See the producer program (7.1′).

ensures that the technology makes it possible indefinitely to produce all commodities in constant amounts, and part 5 ensures that the available initial stock $\tilde{k}_1$ is sufficient to reach this level (so that the constraint set of program 8.1 has a strict interior).

8.1.2 Negishi Equilibrium: Fixed Discount Factor β

When the discount factor is constant and identical for all consumers, it is possible to circumvent the difficulties caused by the infinite-dimensional problem by changing the format of presentation in three ways: (1) extending the utility function, (2) applying dynamic programming, and (3) recovering shadow prices.

First, we note that expenditure is now the sum of an infinite number of terms. Nonethless, we can compute it as the difference between full expenditure and consumer surplus, using the extended (homogeneous of degree one) utility function $\tilde{u}_i(x_i, n_i) = n_i u_i(x_i/n_i)$, where $n_i = 1$, as introduced in section 2.4.2. The intertemporal utility for consumer i may now be obtained as

$$u_{i1} = \sum_{t=1}^{\infty} (\beta)^{t-1} \tilde{u}_i(x_{it}, n_i), \tag{8.2}$$

which leads to the following welfare program where we substitute $\Sigma_i x_{it}$ for y_t and k_{t+1} for $\tilde{k}_{t+1}$:[6]

$$V(\alpha, k_1, n) = \max \Sigma_i \alpha_i \sum_{t=1}^{\infty} (\beta)^{t-1} \tilde{u}_i(x_{it}, n_i),$$

$$x_{it} \geqslant 0, \text{ all } i, k_{t+1} \geqslant 0, t = 1, 2, \ldots,$$

subject to (8.3)

$$F(\Sigma_i x_{it}, k_{t+1}, k_t) \leqslant 0,$$

for given k_1 and $n_i = 1$ for all i.

In this program $V(.)$ is the value function which is well defined, since, by assumption CD1, the objective is bounded. It depends on the parameters α, k_1, and n, where n denotes the vector of n_i.

Second, we write program (8.3) as a dynamic program,[7] namely as a sequence of single-period programs for $t = 1, 2, \ldots$, starting from a given initial stock k_1 and with $n_i = 1$ for all i:

6. These substitutions are made to bring the model in the form that is commonly used to derive the dynamic programming format.

7. For a proof, see Stokey and Lucas (1989, ch. 4).

$$V(\alpha, k_t, n) = \max \Sigma_i \alpha_i \tilde{u}_i(x_{it}, n_i) + \beta V(\alpha, k_{t+1}, n),$$

$$x_{it} \geqslant 0, \text{ all } i, k_{t+1} \geqslant 0,$$

subject to (8.4)

$$F(\Sigma_i x_{it}, k_{t+1}, k_t) \leqslant 0.$$

By the perturbation theorem A.3.2, $V(\alpha, k, n)$ is convex increasing in α and concave monotonic in k; by assumption PD1, part 1, it is also increasing in the first element of k (labor). It should be noted that (8.4) is a single-period welfare program with a term $\beta V(\alpha, k_{t+1}, n)$ added in the objective for final stock valuation: The infinite-horizon program thus reduces to an infinite sequence of single-period programs. This is analytically convenient, since, for a given function $V(\alpha, k, n)$, program (8.4) is a standard convex program, and the usual tools of convex programming can be used. Note that by recurrently substituting the value function T times, one obtains a T-period program:

$$V(\alpha, k_1, n) = \max \Sigma_i \alpha_i \sum_{t=1}^{T} (\beta)^{t-1} \tilde{u}_i(x_{it}, n_i) + (\beta)^T V(\alpha, k_{T+1}, n),$$

$$x_{it} \geqslant 0, \text{ all } i, k_{t+1} \geqslant 0, t = 1, 2, \ldots, T,$$

subject to (8.5)

$$F(\Sigma_i x_{it}, k_{t+1}, k_t) \leqslant 0.$$

For T sufficiently large, the second term of the objective function will vanish, since, by assumption CD1, part 4, $\beta < 1$, and the first term will approximate the value $V(\alpha, k_1, n)$ of (8.3).

Third, it is convenient to return to the formulation with explicit commodity balances and restrictions on capital stocks, rewriting (8.4) and introducing multipliers p_t and ψ_t:

$$V(\alpha, \tilde{k}_t, n) = \max \Sigma_i \alpha_i \tilde{u}_i(x_{it}, n_i) + \beta V(\alpha, \tilde{k}_{t+1}, n),$$

$$x_{it} \geqslant 0, \text{ all } i, k_t, \tilde{k}_{t+1} \geqslant 0, y_t,$$

subject to (8.6)

$$F(y_t, \tilde{k}_{t+1}, k_t) \leqslant 0,$$

$$k_t \leqslant \tilde{k}_t \qquad (\psi_t),$$

$$\Sigma_i x_{it} - y_t \leqslant 0 \qquad (p_t),$$

where for the program solved at $t = 1$, $\tilde{k}_1 = \Sigma_i \tilde{k}_{i1}$, which is the aggregate initial endowment, and for subsequent programs, $\tilde{k}_{t+1}$ is the optimal value of capital stocks from the program of period t.

Finally, from the solutions of program (8.6) solved for $t = 1, 2, \ldots$, we can compute budget deficits as

$$b_i(\alpha, \bar{k}_1, n) = \sum_{t=1}^{\infty} p_t x_{it} - \psi_1 \tilde{k}_{i1}, \tag{8.7}$$

where ψ_1 is associated to the constraint $k_1 \leqslant \tilde{k}_1$ and $\bar{k}_1$ denotes the vector $(\tilde{k}_{11}, \tilde{k}_{21}, \ldots, \tilde{k}_{i1}, \ldots, \tilde{k}_{m1})$ of initial stocks owned by the consumers. Because of constant returns to scale, profits of the firm are zero, and $\psi_1 \tilde{k}_{i1}$ is the income accruing to consumer i from his given initial resources only. The expenditure in (8.7) is still expressed as a sum with an infinite number of terms, but as mentioned earlier, we will compute it as the difference between two variables, full income and consumer surplus.

We now state in proposition 8.1 an existence result:

PROPOSITION 8.1 (Existence of a dynastic Negishi equilibrium with fixed discount factor) Let assumptions CD1′, PD1, and K hold. Then there exist positive welfare weights α_i^* such that the solution of (8.6) solved for $t = 1, 2, \ldots$, yields zero budget deficits (8.7).

Proof This proposition is a special case of proposition 8.2, for which a proof will be given later. ∎

8.1.3 Existence and Efficiency of Equilibrium with Recursive Utility Functions

In this section we return to the more general form of the dynastic utility function of assumption CD1 and prove first that we can find Negishi weights such that each consumer satisfies his budget constraint (proposition 8.2). We call this a dynastic Negishi equilibrium. Second, evoking the second welfare theorem for infinite-horizon models, we show that this dynastic Negishi equilibrium is a competitive (dynastic) equilibrium (proposition 8.3).

The Negishi Equilibrium

When one uses the more general form of the dynastic utility function of assumption CD1, where discount rates differ among consumers, it is no longer possible to write the infinite-horizon program in the dynamic programming format (8.6). Yet, following Lucas and Stokey (1984), one may define a sequence of single-period programs. In each of these programs all effects of decisions made in t on future utilities and capital stocks will be subsumed in the requirement that these variables remain within a set G_{t+1}, that is, $(u_{i,t+1}, \text{all } i, \tilde{k}_{t+1}) \in G_{t+1}$. This formulation reduces the infinite-dimensional problem to a sequence of finite-dimensional ones. But one still has to show that the set G_{t+1} has the properties needed in convex programming.

The program is

$$V(\alpha_t, \tilde{k}_t, n) = \max \Sigma_i \alpha_{it} n_i W_i\left(\frac{x_{it}}{n_i}, \frac{u_{i,t+1}}{n_i}\right),$$

$$x_{it} \geqslant 0, u_{i,t+1}, \text{all } i, k_t, \tilde{k}_{t+1} \geqslant 0,$$

subject to (8.8)

$$F(y_t, \tilde{k}_{t+1}, k_t) \leqslant 0,$$

$$k_t \leqslant \tilde{k}_t \qquad (\psi_t),$$

$$\Sigma_i x_{it} - y_t \leqslant 0 \qquad (p_t),$$

$$(u_{i,t+1}, \text{all } i, \tilde{k}_{t+1}) \in G_{t+1}.$$

In program (8.8) the set G_{t+1} is defined as

$$G_{t+1} = \{(u_{i,t+1}, \text{all } i, k_{t+1}) \mid u_{i,t+\tau} \leqslant W_i(x_{i,t+\tau}, u_{i,t+\tau+1}),$$
$$F(\Sigma_i x_{i,t+\tau}, k_{t+\tau+1}, k_{t+\tau}) \leqslant 0, x_{i,t+\tau} \geqslant 0, k_{t+\tau+1} \geqslant 0,$$
$$u_{i,t+\tau+1} \geqslant 0 \text{ for } \tau = 1, 2, \ldots\},$$

and $n_i = 1$, all i. The α_{it} are the given positive welfare weights of period t.[8,9]

Program (8.8) thus defines a sequence of single-period programs. We will see that every consumer's expenditure can be recovered from the first-period program so that we may restrict attention to this program only. We state and prove:

PROPOSITION 8.2 (Existence of a dynastic Negishi equilibrium) Let assumptions CD1, PD1, and K hold. Then there exist positive welfare weights α^*_{i1} such that the solution of (8.8) yields zero budget deficits.

Proof The proof will proceed in three parts. In part 1, we show that the set G_2 in (8.8) is nonempty, compact, and convex so that (8.8) is a standard convex program. We then use this program in part 2 to derive budget deficit functions required in part 3 to define the mapping for adjusting the welfare weights like in proposition 3.1 (Negishi's theorem).

Part 1: G_2 is a nonempty, compact, and convex set. Nonemptiness is immediate, since zero values for all $x_{i,t+\tau}$, $u_{i,t+\tau}$, and $k_{t+\tau}$ are feasible for $t = 2$. Boundedness of x_{i1} follows from assumption PD1, part 3;

8. Unlike in the case with fixed discount factor, welfare weights will be seen to be time dependent, and therefore they must carry a time subscript. We will use α_t as the vector of welfare weights for a program solved in t, with elements α_{it}.

9. In the set G_{t+1} we substitute $\Sigma_i x_{it}$ for y_t and k_{t+1} for $\tilde{k}_{t+1}$. We also write $u_{i,t+\tau} \leqslant W_i(x_{i,t+\tau}, u_{i,t+\tau+1})$ with inequality, but in any optimum, equality will hold.

boundedness of u_{i2} from assumption CD1, part 5. To prove convexity, consider two different sequences $\{x_{i,t+\tau}, u_{i,t+\tau}, k_{t+\tau}\}$ for $t = 1$ and $\tau = 1, 2, \ldots,$ which belong to G_2. Any convex combination also belongs to G_2 as can be checked for every constraint separately in every τ, since every constraint involves a finite number of variables and is convex. Hence the set is convex. The proof of closedness is technical and will be skipped. It can be found in Lucas and Stokey (1984, lemma 1).

Therefore (8.8) is a convex program with a strictly concave objective and a nonempty, compact, and convex constraint set G_2; hence a solution exists. The value function is bounded and continuous. By the perturbation theorem A.3.2, it is convex, increasing in α, and concave in k_1 and n; by assumption PD1, parts 4 and 5, the feasible set has a strict interior. So Lagrange multipliers $\psi_1 \in \Psi_1(\alpha_1, \tilde{k}_1, n)$ are well defined ($\Psi_1(.)$ is the set of Lagrange multipliers associated with the constraints $k_1 \leqslant \tilde{k}_1$).

Part 2: To derive the budget deficit, we first compute expenditure following Kehoe (1991) and replace consumer expenditure in (8.7) by an expression that is a continuous function in a finite number of variables. Recall that in the static model (section 2.4.2, equation 2.44), consumer expenditure was obtained, by Euler's rule, as full expenditure minus consumer surplus:

$$px_i = \alpha_i \tilde{u}_i(x_i, n_i) - \alpha_i \frac{\partial \tilde{u}_i(x_i, n_i)}{\partial n_i}.$$

Here this may be written

$$\sum_{t=1}^{\infty} p_t x_{it} = \alpha_{i1} u_{i1} - \alpha_{i1} \frac{\partial(n_i W_i(x_{i1}/n_i, u_{i,2}/n_i))}{\partial n_i}.$$

By assumption CD1, part 1, W_i is strictly concave (so that the optimal x_{i1} and u_{i2} are unique) and continuously differentiable. By the envelope theorem A.3.3, the value function $V(.)$ is continuously differentiable in α and n, and this yields

$$\frac{\partial V(\alpha_1, \tilde{k}_1, n)}{\partial \alpha_{i1}} = u_{i1},$$

$$\frac{\partial V(\alpha_1, \tilde{k}_1, n)}{\partial n_i} = \alpha_{i1} \frac{\partial(n_i W_i(x_{i1}/n_i, u_{i,2}/n_i))}{\partial n_i}.$$

The budget deficit function of consumer i is therefore

$$b_i(\alpha_1, \bar{k}_1, n, \psi_1) = \frac{\alpha_{i1} \partial V(\alpha_1, \tilde{k}_1, n)}{\partial \alpha_i} - \frac{\partial V(\alpha_1, \tilde{k}_1, n)}{\partial n_i} - \psi_1 \tilde{k}_{i1} \tag{8.9}$$

for given $\psi_1 \in \Psi_1(\alpha_1, \tilde{k}_1, n)$, where we recall that $\bar{k}_1$ denotes the vector $(\tilde{k}_{11}, \tilde{k}_{21}, \ldots, \tilde{k}_{i1}, \ldots, \tilde{k}_{m1})$ of initial stocks owned by the consumers.

By the Envelope Theorem again, the first two terms on the right-hand side of (8.9) are continuous in $(\alpha_1, \bar{k}_1, n)$; the third term is obviously continuous in ψ_1. Hence b_i is continuous.

Note also that $V(.)$ is homogeneous of degree one in α_1 and homogeneous of degree one in $(\tilde{k}_1, n)$. Euler's rule and the Kuhn-Tucker conditions (using homogeneity in assumption PD1) therefore imply that

$$V(\alpha_1, \tilde{k}_1, n) = \frac{\Sigma_i \alpha_i \partial V(\alpha, \tilde{k}_1, n)}{\partial \alpha_i}$$

and

$$V(\alpha_1, \tilde{k}_1, n) = \frac{\Sigma_i n_i \partial V(\alpha, \tilde{k}_1, n)}{\partial n_i} + \psi_1 \tilde{k}_1,$$

for all $\psi_1 \in \Psi_1(\alpha_1, \tilde{k}_1, n)$. Since $n_i = 1$ for all i and $\psi_1 \tilde{k}_1 = \psi_1 \Sigma_i \tilde{k}_{i1}$, it follows that $\Sigma_i b_i(\alpha_1, \bar{k}_1, n) = 0$ for all positive α_{i1}.[10] Thus the budget deficits $b_i(\alpha_1, \bar{k}_1, n, \psi_1)$ are continuous functions, homogeneous of degree one in α_1 and $\Sigma_i b_i = 0$ for all $\psi_1 \in \Psi_1(\alpha_1, \tilde{k}_1, n)$.

Part 3: We can now proceed along the lines of the proof of proposition 3.1 and show that there exist α_{i1}'s such that the program generates a Negishi equilibrium, that is, a solution with zero budget deficits. (Here the prices ψ_1 play the role of prices p in proposition 3.1 and the budget deficits actually are functions of α_1, ψ_1, since $\bar{k}_1$ and n do not vary.) Assumptions K and PD1, part 1, ensure that the equilibrium α_{i1}'s are positive. Since consumer 1's utility is increasing in all commodities, all prices are positive as well. ■

There is an important difference between the case with discount factor β fixed and the more general recursive formulation. Recall that in the objective of (8.8), the welfare weights carry a time index. The reason for this time dependency is the following: Suppose that we have found a Negishi equilibrium with weights α_{i1}^* and capital stocks k_2^* for period 2. To restart the program at $t = 2$ with given k_2^*, we need welfare weights $\alpha_{i2}^* = (\partial W_i(x_{i1}, u_{i2})/\partial u_{i2})\alpha_{i1}^*$ which will lead to the same allocation as the program started in $t = 1$ (time consistency).[11] More generally,

$$\alpha_{i,t+1}^* = \frac{\partial W_i(x_{it}, u_{i,t+1})}{\partial u_{i,t+1}} \alpha_{it}^*, \qquad t = 1, 2, \ldots. \tag{8.10}$$

10. Here we use a generalized form of Euler's rule because we did not impose the assumption of $V(\alpha_1, \tilde{k}_1, n)$ being differentiable with respect to $\tilde{k}_1$. If this condition holds, $\psi_1 = \partial V/\partial \tilde{k}_1$ so that the standard form of Euler's rule applies.

11. For any $t > 1$ and any finite time horizon $T \geqslant t$, the first-order conditions in an optimum will satisfy $\alpha_{i1} = p_t$ (if $x_{it} > 0$). This may also be written as $\alpha_{it} = p_t$, which leads to the recurrence relation for the α's. A similar reasoning applies when the horizon is infinite.

This updating also applies when the discount factor β is constant and common to all consumers, in which case $\partial W_i(x_{i1}, u_{i2})/\partial u_{i2} = \beta$. As a result we can write the objective in $t = 1$ as $\Sigma_i \alpha_i u_i(x_{i1}) + \beta \Sigma_i \alpha_i u_{i2}$. Then restarting in $t = 2$ will lead to the same allocation if welfare weights $\alpha_{i2} = \beta \alpha_i$ are chosen. However, since only relative welfare weights matter, such an updating of weights is unnecessary. Taking the value of β to be common to all consumers amounts to assuming that they have the same time preference, which is obviously restrictive. A natural alternative is to choose discount factors that differ across consumers, but this leads to another difficulty: If, for instance, consumer 2 has the highest discount factor (i.e., the lowest discount rate), his relative welfare weight will rise over time, and eventually he will be the only person to consume.

Proposition 8.2 shows that Negishi equilibrium weights can be found. At those weights, program (8.8) for $t = 1$ makes it possible to compute the equilibrium x_{i1} and u_{i2} for all i, as well as $\tilde{k}_2, y_1$, and prices p_1 and ψ_1 as the associated Lagrange multipliers. To compute the allocations and prices for $t > 1$, we need to solve the sequence of programs (8.8) for $t = 2, 3, \ldots$. However, we still need to prove that the sequence so obtained is a competitive dynastic equilibrium.

The Dynastic Competitive Equilibrium

We define a dynastic equilibrium as follows:[12]

DEFINITION 8.1 (Dynastic competitive equilibrium) The allocation k_1^*, y_t^*, x_{it}^*, all i, supported by the price vectors ψ_1^* and $p_t^* \geqslant 0$, $t = 1, 2, \ldots$, such that $\Sigma_t p_t^* \zeta_t < \infty$ for any bounded sequence of ζ_t is a dynastic competitive equilibrium if the following conditions are satisfied:

1. $(k_1^*, y_1^*, y_2^*, \ldots)$ is an optimal solution of

$$\max \sum_{t=1}^{\infty} p_t^* y_t - \psi_1^* k_1,$$

$$k_t, \tilde{k}_{t+1} \geqslant 0, y_t, t = 1, 2, \ldots,$$

subject to

$$F(y_t, \tilde{k}_{t+1}, k_t) \leqslant 0,$$

$$k_{t+1} \leqslant \tilde{k}_{t+1}.$$

12. In this definition we write infinite-horizon programs with a "max" operator, assuming that the optimal value is bounded and attained. This condition will be seen to hold under assumptions CD1, K, and PD1. Since the producer operates under constant returns, in equilibrium, he is indifferent with respect to scale of production, and the scale is determined from consumer demand and initial stocks.

2. For every consumer i, $(x_{i1}^*, x_{i2}^*, \dots)$ is the optimal solution of

$$\max u_{i1},$$

$$x_{it} \geqslant 0, u_{it}, \text{ all } i \text{ and } t = 1, 2, \dots,$$

subject to

$$u_{it} = W_i(x_{it}, u_{i,t+1}),$$

$$\sum_{t=1}^{\infty} p_t^* x_{it} \leqslant \psi_1^* \tilde{k}_{i1}.$$

3. Markets are in equilibrium for $t = 1, 2, \dots,$

$$\Sigma_i x_{it}^* - y_t^* \leqslant 0 \perp p_t^* \geqslant 0,$$

$$k_1^* \leqslant \tilde{k}_1 \perp \psi_1^* \geqslant 0.$$

Part 1 of this definition tells us that the producer has to rent his initial capital stocks k_1 at prices ψ_1 from the consumers who owns $\tilde{k}_1 = \Sigma_i \tilde{k}_{i1}$ as endowments; from period 2 onward all stocks are variables within the producer problem. Since there is constant returns to scale (assumption PD1, part 2), there are no profits to be distributed to consumers. In part 2, consumer i faces an intertemporal budget constraint, with total wealth equal to $\psi_1 \tilde{k}_{i1}$. Market equilibrium is specified in part 3; the constraint on capital stocks in period 1 has been taken out from the producer problem (compare with equation 8.8) and treated as a market equilibrium condition; the price ψ_1 will, in equilibrium, be such that capital demand k_1 computed by the producer in part 1 will not exceed $\tilde{k}_1$.

Proposition 8.3 relates the allocation obtained from the dynastic Negishi equilibrium to the dynastic competitive equilibrium of definition 8.1.

PROPOSITION 8.3 (Existence and efficiency of a dynastic competitive equilibrium) Let assumptions CD1, PD1, and K hold. Then a dynastic Negishi equilibrium is a dynastic competitive equilibrium and is Paretoefficient.

Proof

Existence

We may proceed as in proposition 1.9 for the static case, where we proved that a welfare optimum is a competitive equilibrium (but transfers will now be zero by proposition 8.2). We will not prove this but rely on the second welfare theorem as stated by Debreu (1954) for an infinite-dimensional commodity space (here the infinite dimension is due to the infinite horizon). Assumption PD1, parts 4 and 5, ensures that the feasible set of the program has a strict interior. This means that like Slater's constraint qualification, "Lagrange multipliers" (here linear functionals $\phi(.)$) exist such that we can compute the intertemporal expenditure of consumer i as $\phi(x_{i1}, x_{i2}, \ldots)$ and profits as $\phi(y_1, y_2, \ldots)$. Prescott and Lucas (1972) show that the prices p_t generated by the sequence of programs (8.8) can be used to construct this linear functional.[13]

Efficiency

Pareto-efficiency is obvious, since the equilibrium is obtained from a welfare program with positive weights for all consumers. ■

8.1.4 Multiplicity and Indeterminacy of Equilibrium

Since we use the Negishi format to prove existence of equilibrium, it is natural to discuss multiplicity in this framework and work with the budget deficits (8.9) as depending on welfare weights α_{i1}. Note that if ψ_1 is unique,[14] it will be a continuous function of the α_{i1}, so the budget deficits will also be continuous functions of the α_{i1},[15] and the discussion in section 3.1.2 on local uniqueness of regular equilibria will apply. There may then exist several equilibria, but the number of regular equilibria will be finite.

These budget deficit functions constitute a system of $m - 1$ equations, where m is the number of consumers. Since their number is equal to the number of unknowns (number of consumers minus one), any equilibrium in which the Jacobian of the budget deficit functions is regular will be locally unique.[16]

When, however, indirect taxes (e.g., on consumptions x_{it} for $t = 1, 2, \ldots$) or other period-specific distortions (commodity taxes, tariffs, price rigidities, markups) are introduced, the number of equations and unknowns will become infinite. Then the approach outlined above becomes pointless because there is an infinite number of equations and unknowns.

Recall that in a static model such as model (5.16), introducing taxes adds as many parameters ξ_h^c to the Negishi welfare problem as there are taxed commodities (e.g., r_1). Assuming that shadow prices are unique, the system of equalities to solve is

$$\tilde{b}(\tilde{\alpha}, \xi^c) = 0 \tag{8.11a}$$

$$\xi_h^c - p_h(\tilde{\alpha}, \xi^c)(1 + \tau_h^c) = 0 \qquad \text{for } h = 1, 2, \ldots, r_1, \tag{8.11b}$$

13. Lucas and Stokey (1984) use the same result in their existence proof.

14. This will be satisfied if $F(.)$ is locally strongly convex in $(y_t, \tilde{k}_{t+1}, -k_t)$, and satisfies a boundary condition w.r.t. k_t so that k_1 is positive in any optimum.

15. There are several ways to ensure uniqueness. One of them is to assume that all stocks are owned in the same proportions σ_i by consumer i (so $k_{i1} = \sigma_i k_1$, where σ_i is a scalar). Another possibility is to assume that $F(.)$ is continuously differentiable with respect to k_1 and strictly convex in $(y_t, \tilde{k}_{t+1})$. Then $(y_t, \tilde{k}_{t+1})$ will be uniquely determined, and therefore $\partial F(y_t, \tilde{k}_{t+1}, k_t)/\partial k_t$ will be unique also. Hence the b_i are continuous functions.

16. See also theorem A.7.6.

where $\tilde{b}$ is the vector of budget deficits of the first $m-1$ consumers, $\tilde{\alpha}$ is the vector of the $m-1$ welfare weights, while the τ_h^c are given tax rates. When the horizon is infinite, r_1 will be infinite as well, so there is an infinite number of equations in (8.11), though the number of consumers is finite. Now the standard theorems (theorems A.7.1, A.7.3, and A.7.6) on local uniqueness no longer apply, and it may happen that there is a continuum of equilibria that cannot be eliminated by a small perturbation of the parameters of the functions. This is often referred to as indeterminacy.[17] Note, however, that if beyond some horizon T the distorsions ξ^c are dropped,[18] there will be no indeterminacy.

Geanakoplos and Polemarchakis (1991) suggest that indeterminacy leaves "degrees of freedom" that can be used to incorporate exogenous rigidities, inherited from the past or generated by common beliefs about the future (like "sunspots"). Indeterminacy also leaves scope for policy intervention, particularly with respect to the distribution of income. However, exogenous expectations and interventions add equations or constraints to an already infinite number of equations, which may then become inconsistent or be indeterminate as before.

At any rate the notion of an infinite horizon is introduced as a way of avoiding the arbitrary valuation of final stocks at some finite horizon T. Extending the horizon to infinity is thought to determine this value. But the indeterminacy problem itself makes clear that this is not always possible.

8.1.5 Dynamics and Steady States

Theory has been concerned with the path followed by the variables of a dynamic model in order to find out whether it converges, diverges, cycles, or is chaotic. Traditionally the emphasis has been on convergence to a steady state, a condition in which all consumptions and stocks are constant over time. Such a steady state can be defined as follows: We start from program (8.8). Dropping the extended utility function notation with n_i, we have

$$\tilde{V}(\alpha_t, \tilde{k}_t) = \max \Sigma_i \alpha_{it} W_i(x_{it}, u_{i,t+1}),$$

$$x_{it} \geqslant 0, u_{i,t+1}, \text{ all } i, k_t, \tilde{k}_{t+1} \geqslant 0,$$

17. For further discussion, see Kehoe (1991) and Kehoe, Levine, and Romer (1989). Although, in the absence of distortions, our version of the dynastic model does not generate indeterminacy, Ryan and Bean (1989) give examples of models in which indeterminacy arises within the infinite-horizon program itself. This is referred to as degeneracy, which disappears for sufficiently high discount rates.

18. More generally, it is sufficient for the tax rates ξ to remain independent from allocations and prices, in which case the corresponding equations (8.11b) can be dropped.

subject to

$$F(y_t, \tilde{k}_{t+1}, k_t) \leqslant 0,$$

$$k_t \leqslant \tilde{k}_t,$$

$$\Sigma_i x_{it} - y_t \leqslant 0,$$

$$(u_{i,t+1}, \text{all } i, \tilde{k}_{t+1}) \in G_{t+1}.$$

Suppose that in solving this program, we find that (1) there is a resulting optimal value $\tilde{k}_{t+1}$ equal to $\tilde{k}_t$ and (2) all optimal $\partial W_i/\partial u_{i,t+1} = \beta$ for all i so that by (8.10) $\alpha_{i,t+1} = \beta\alpha_{it}$. Then we have a steady state, since solving the problem again at $t + 1$ with the new values of $\tilde{k}$ and α will reproduce the same $\tilde{k}$ and a new α, proportional to the previous one by the same factor β.[19] Obviously the x_{it} and the $u_{i,t+1}$ will also remain constant.[20,21]

Conditions 1 and 2 above characterize a steady state. In practice, the following characterization is more operational because it does not involve the set G_{t+1}: Define the profit function

$$\Pi(p, \tilde{\psi}, k) = \max py + \tilde{\psi}\tilde{k},$$

$$\tilde{k} \geqslant 0, y,$$

subject to

$$F(y, \tilde{k}, k) \leqslant 0.$$

Assume as before that in addition to assumption PD1, the transformation function $F(.)$ is strictly convex with respect to $(y, \tilde{k})$. Then $\Pi(.)$ will be continuously differentiable with respect to $(p, \tilde{\psi})$ so that by Hotelling's lemma, $y = \partial\Pi(.)/\partial p$, $\tilde{k} = \partial\Pi(.)/\partial\tilde{\psi}$. Assume also that $F(.)$ is continuously

19. The factor β does not affect real allocations; it merely scales the objective function, and a renormalization can be performed to keep the α constant over time.

20. When some stocks follow exogenous trends, there will be no steady state. If there is only one such exogenous growth rate, then a steady state may exist in which the variables are expressed relatively to the exogenous trend variable, say, population.

21. The steady state is obtained as a fixed point of the mapping of (normalized) welfare weights and capital stocks into itself. Existence of such a fixed point may be verified as follows: By the maximum theorem A.3.1, the mapping is convex valued and uppersemicontinuous. The difficulty that makes it impossible to readily apply Kakutani's theorem is that unbounded capital stocks will lead to unbounded output due to constant returns to scale. However, recall that the constant-returns-to-scale formulation is obtained by treating all resources as part of the accumulation process. If one of the stocks is fixed (e.g., land), then its value will already be at steady-state level, and decreasing returns will prevail with respect to the other stocks. Then if one adds the rather mild assumption that this imposes an upper bound on the level of the other stocks, the mapping will have a uniformly bounded range that is a subset of its domain, and Kakutani's theorem can be applied (using this range as the domain of the fixed point mapping). Other assumptions will also ensure existence of a steady state. For example, $\tilde{k}_{t+1}$ may be bounded not because of restrictions on the transformation function but because of impatience of consumers. The (uniform) boundedness assumption can also be relaxed somewhat, but then other fixed point theorems have to be used.

differentiable and has a negative derivative with respect to k. Then $\Pi(.)$ is also differentiable with respect to k so that $\psi = \partial\Pi(.)/\partial k$. The steady state can now be defined by the following equations:

Conditions for consumer i,

$$\frac{\partial W_i(x_i, u_i)}{\partial u_i} = \beta,$$

$$u_i = W_i(x_i, u_i), \tag{8.12a}$$

$$\frac{\alpha_i \partial W_i(x_i, u_i)}{\partial x_i} \leqslant p \perp x_i \geqslant 0,$$

$$\Sigma_i \alpha_i = 1, \alpha_i \geqslant 0.$$

Conditions for the producer,

$$y = \frac{\partial \Pi(p, \tilde{\psi}, k)}{\partial p},$$

$$k = \frac{\partial \Pi(p, \tilde{\psi}, k)}{\partial \tilde{\psi}}, \tag{8.12b}$$

$$\tilde{\psi} = \frac{\beta \partial \Pi(p, \tilde{\psi}, k)}{\partial k}.$$

Commodity balances,

$$\Sigma_i x_i = y. \tag{8.12c}$$

In (8.12b), we use the steady-state conditions $\tilde{k} = k$ and $\tilde{\psi} = \beta\psi$. In equations (8.12) all variables are endogenous, including β and the α_i.

If all the x_i are positive, then (8.12) defines a system of equations $H(z) = 0$, where H is continuously differentiable. The number of regular steady states (i.e., solutions z such that $H'(z)$ is nonsingular) will be finite, and algorithms can be designed to compute such a solution.[22]

In the dynastic competitive equilibrium, welfare weights are such that budget constraints are satisfied over the infinite horizon. In a steady state there are also welfare weights,[23] but they do not refer to budgets. The steady state defines a level of aggregate capital stocks k from which, as in the second welfare theorem, one can compute a distribution of stock values or of transfers that supports the welfare weights. Any other given distribution of stocks (possibly the one linked to k_{i1}, the initial stocks) will require some

22. See sections A.7 to A.9 of the mathematical appendix.

23. In the case with a fixed discount rate, there will exist a steady state for any choice of α. One could of course choose the α of the dynastic Negishi equilibrium, but this implies that one has to solve an infinite-horizon problem first.

consumers indefinitely to transfer income to others. This may be interpreted as servicing of a debt that was incurred before the steady state was reached.

We now return to the sequence of supplies, demands, and capital stocks generated by an optimal solution of the dynastic program and examine whether it converges to such a steady state. Recall that by assumption PD1, part 5, it is technologically feasible to attain the steady state, even within the first period; that is, $F(y, k, k) \leqslant 0$ and $k \in [0, \tilde{k}_1]$. However, global asymptotic convergence is not a natural property of the trajectory. Indeed Boldrin and Montrucchio (1986) show that any compact-valued mapping from k_t to k_{t+1} can be generated through a dynastic model like (8.3) with fixed discount factors and assumptions comparable to CD1′ and PD1 (there is no uniform boundedness or differentiability on the utility function, and the production technology is represented by a compact convex set). This finding shows that the path generated by a dynastic model can follow any pattern. It may converge to a steady state, but it may also diverge, cycle, or be chaotic. On the other hand, for economies with a single consumer good (and two consumers), Lucas and Stokey (1984) prove, under somewhat more restrictive assumptions than CD1 and PD1, that there is a unique steady state, to which the trajectory converges.[24] Obviously global asymptotic convergence requires the steady state to be unique.[25] Convergence results for the multicommodity case are summarized in Epstein (1987); they depend on the discount rate (which should not be too high) and on the curvature of utility and transformation functions. A high discount rate requires the curvature to be higher as well.

In summary, the result by Boldrin and Montrucchio shows that convergence to a steady state is problematic in the dynastic model. The reason is that consumers do not care much about the distant future and therefore about sustainability as implied by the steady state. At the same time such discounting limits the importance of the convergence issue because early allocations will hardly be affected by what happens in the long run.

24. For continuous-time models, properties that are similar to the Lucas and Stokey result have been established by Arrow and Kurz (1980, ch. 3) for the single-consumer–one-sector model. Extensions can be found in Carlson and Haurie (1987) and Epstein (1987).

25. To study local convergence properties of such a steady state, one considers a quadratic value function evaluated around the steady state and the associated linear Euler equations, which may be turned into a system of first-order linear difference equations. Evaluating the eigenvalues of this system makes it possible to characterize stability properties of the steady state. A sufficient condition for local asymptotic stability is that in this system the number of roots with modulus smaller than one be equal to the number of initial conditions. (In the fixed discount rate case, this is the dimension of k_1; in the recursive case, it is the dimension of k_1 plus the dimension of α, the number of consumers, minus one.) Since the path is determinate, the number of such roots cannot be larger than the number of initial conditions, but it can be smaller. In that case the trajectory does generally not converge, although there exist initial values from which it may converge. When there are several steady states one can only study local asymptotic convergence (i.e., from $\tilde{k}_1$ and α_1 close to a steady state). For further details, see Kehoe (1991).

8.1.6 Finite-Horizon Approximations

Obviously one cannot solve numerically a system with an infinite number of equations and unknowns. Therefore some approximation procedure is needed. There are basically two ways of doing this. The first is to approximate the model itself by a more tractable form (e.g., a quadratic objective with linear constraints). This is only possible if the model is small or assumed to remain close to its given steady state and discussed in connection with real business cycle models in the next section. For larger models not close to their steady state, one solves a finite-horizon approximation, where the horizon is sufficiently long so as not to affect significantly the allocations.[26]

We start from the T-period version of (8.1) with terminal utilities in $T + 1$ equal to $v_i \in [0, \bar{v}]$, where $\bar{v}$ is the upper bound on W_i in assumption CD1, part 5:[27]

$$
\begin{aligned}
& V_T(\alpha_1, \tilde{k}_1, v) = \max \Sigma_i \alpha_{i1} u_{i1}, \\
& x_{it} \geqslant 0, u_{it}, \text{ all } i, k_t, \tilde{k}_{t+1} \geqslant 0, y_t, t = 1, 2, \ldots, T, u_{i,T+1}, \text{ all } i, \\
& \text{subject to} \\
& u_{it} = W_i(x_{it}, u_{i,t+1}), \\
& F(y_t, \tilde{k}_{t+1}, k_t) \leqslant 0, \\
& k_t \leqslant \tilde{k}_t, \\
& \Sigma_i x_{it} - y_t \leqslant 0, \\
& u_{i,T+1} = v.
\end{aligned}
\tag{8.13}
$$

We now use the uniform boundedness assumption made in assumption CD1, part 5, to derive upper and lower bounds that bracket the value function $V(\alpha_1, \tilde{k}_1)$ of the infinite-horizon model. These bounds can be computed from the T-period version of the model, and the interval between the bounds vanishes as T increases.

PROPOSITION 8.4 (Approximation of the value function) Under assumptions CD1 and PD1, the following inequalities hold:

$$
\begin{aligned}
V_T(\alpha_1, \tilde{k}_1, 0) &\leqslant V_{T+1}(\alpha_1, \tilde{k}_1, 0) \\
&\leqslant V(\alpha_1, \tilde{k}_1) \leqslant V_{T+1}(\alpha_1, \tilde{k}_1, \bar{v}) \\
&\leqslant V_T(\alpha_1, \tilde{k}_1, \bar{v})
\end{aligned}
\tag{8.14}
$$

26. See Keyzer (1991).

27. To simplify, we drop the n_i, and the last argument in $V_T(.)$ is the upper bound on W_i.

and

$$V_T(\alpha_1, \tilde{k}_1, \bar{v}) - V_T(\alpha_1, \tilde{k}_1, 0) \leqslant \gamma^T \bar{v}. \tag{8.15}$$

Proof Let $P_{T,v}$ denote program (8.13) with end period T and $u_{i,T+1} = v$. We first prove (8.14). An allocation that is optimal for program $P_{T,0}$ is feasible for $P_{T+1,0}$, and the first inequality follows from this, since consumption is allowed for one more time period. Similarly an allocation optimal for $P_{T+1,0}$ is feasible in the infinite-horizon problem with value function $V(\alpha_1, \tilde{k}_1)$, which proves the second inequality in (8.14). To prove the third inequality in (8.14), note that in $P_{T+1,\bar{v}}$, the maximal utility $\bar{v}$ is provided in $T + 2$, without a need to reserve capital for $T + 1$. The last inequality holds because the maximal utility is given one period earlier. This proves (8.14).

For (8.15) we consider an optimal allocation in $P_{T,v}$ with $v = \bar{v}$, and we evaluate the reduction in welfare when v is brought to zero, keeping the previous allocation fixed. It follows from the discounting property of W_i that the resulting reduction in welfare cannot exceed $\gamma^T \bar{v}$, where $\gamma \in (0, 1)$ is such that $\partial W_i / \partial u_i \leqslant \gamma$ (we also use the fact that $\Sigma_i \alpha_{i1} = 1$). Since the inequality holds when the allocation is kept fixed, it must also hold when the allocation can adjust to the new value of v.[28] ■

This proposition not only provides a bound to the value function $V(\alpha_1, \tilde{k}_1)$; it also yields an approximation that, as T increases, converges to the true value. This can be seen from (8.14) and (8.15) combined. Furthermore the proposition makes it possible to establish that early allocations converge as the value function converges. This is most easily shown for the case with fixed discount factor using a dynamic programming argument, but the property holds for the case of recursive utility as well.[29] To see this, it suffices to consider the dynamic program (8.4) solved for period 1, with the value function for end stock valuations approximated by V_T:

$$\max \Sigma_i \alpha_i \tilde{u}_i(x_{i1}, n_i) + \beta V_T(\alpha, \tilde{k}_2, v),$$

$$x_{i1} \geqslant 0, \text{ all } i, \tilde{k}_2 \geqslant 0, y_1,$$

subject to

$$F(y_1, \tilde{k}_2, k_1) \leqslant 0,$$

$$\Sigma_i x_{i1} - y_1 \leqslant 0.$$

28. Note also that (8.14) implies that $V_T(\alpha_1, \tilde{k}_1, \bar{v}) - V_T(\alpha_1, \tilde{k}_1, 0) \geqslant 0$.

29. We reason in terms of the model with fixed discount factor, for which we have given the dynamic programming formulation. For the model with recursive utility, the dynamic program is given in Lucas and Stokey (1984).

Since, as shown in proposition 8.4, V_T converges to V as T increases, the period 1 allocation will also converge. Hence model (8.5) can also be applied to show convergence of the allocations taken for a fixed number of periods τ, and $\beta^\tau V_T(\alpha, \tilde{k}_{\tau+1}, v)$ can be used as the value function for final stocks. An approximation of utility and expenditure of consumer i over the infinite horizon can be derived by reintroducing the n_i variables and computing derivatives of $V_T(\alpha, \tilde{k}_2, v, n)$ with respect to α_i and n_i as in (8.9).

8.1.7 Real Business Cycles

Research in real business cycles (RBC) is concerned with reproducing the moments of observed business fluctuations in macroeconomic variables such as output, employment, and asset prices by generating them through simple dynastic models in which agents have rational expectations,[30] and in which markets clear. Work in this field started with the papers by Kydland and Prescott (1982) and Long and Plosser (1983) and was soon followed by an expanding literature.[31] In this section we only sketch the general economic orientation: AGE modelers can learn from this literature how to solve and simulate infinite-horizon models with or without stochastic elements, while RBC modelers can look at the AGE literature for possible extensions of the competitive model. RBC research also pays a great deal of attention to the design of elaborate solution schemes, whose details are beyond the scope of this book. Since in this field the specification of the model, its estimation or calibration, the solution techniques, and the stochastic simulations are intimately linked, we discuss them here, deviating from our practice of postponing the discussion of numerical implementation until the last section of the chapter.

Research on real business cycles is concerned with the stochastic version of the dynastic model (8.3). In the simplest version the economy consists of a unique representative dynastic consumer and one producer. The basic framework of the RBC model is a stochastic extension of the following dynastic program:

$$V(k_1) = \max \sum_{t=1}^{\infty} (\beta)^{t-1} u(x_t),$$

$$x_t \geqslant 0, k_{t+1} \geqslant 0, t = 1, 2, \ldots,$$

30. Since these models allow for random shocks in output, future prices are not known with certainty. Thus, unlike in the dynastic model treated so far, agents lack perfect foresight. Under the rational expectations hypothesis, they are assumed to have the capacity of computing their expected utility in an unbiased way.

31. See the surveys by Danthine and Donaldson (1993) and Stadler (1994) as well as the book by Cooley (1995).

subject to

$F(x_t, k_{t+1}, k_t) \leqslant 0$

for given k_1.

The stochastic aspect is introduced by a scalar error term in the transformation function and is usually specified according to the autocorrelated process $\zeta_t = \rho\zeta_{t-1} + \varepsilon_t$, where ε_t follows some stochastic distribution that is independent across time periods and ρ is a nonnegative autocorrelation term. The model is then

$$V(k_1, \zeta_1) = \max E\left[\sum_{t=1}^{\infty} (\beta)^{t-1} u(x_t)\right],$$

$$x_t \geqslant 0, k_{t+1} \geqslant 0, t = 1, 2, \ldots,$$

subject to (8.16)

$$F(x_t, k_{t+1}, k_t, \zeta_t) \leqslant 0$$

for given k_1 and the distribution of ζ_t, where $E[.]$ denotes the expectation operator.

Obviously this program can be decentralized into a consumer and a producer problem as in definition 8.1. In this model the agent plans at $t = 1$, knowing k_1 and ζ_1 and the distribution of future disturbances. If he replans at $t = 2$, he will take as given the value of k_2 generated in $t = 1$ as well as the value ζ_2 revealed at the beginning of period 2. Under appropriate assumptions on the way the disturbance enters the transformation function, it is possible to represent the infinite-horizon program through a single-period dynamic programming formulation, as in (8.4):

$$V(k_t, \zeta_t) = \max u(x_t) + \beta E[V(k_{t+1}, \zeta_{t+1})],$$

$$x_t \geqslant 0, k_{t+1} \geqslant 0,$$

subject to (8.17)

$$F(x_t, k_{t+1}, k_t, \zeta_t) \leqslant 0$$

for given k_t, ζ_t.

Solving the Competitive Model

If $E[V(k_{t+1}, \zeta_{t+1})]$ were a known function with given parameters, there would be no difficulty in solving the RBC model. Program (8.17) would yield so-called policy functions $X(k_t, \zeta_t)$ and $K(k_t, \zeta_t)$ for consumption and capital stocks, respectively. One could then run simulations, starting with given k_1 at $t = 1$, generating ζ_1 from a random drawing of ε_1 and thus obtaining k_2, and so on. The main problem is, however, that from a finite data set, only an approximation of $E[V(k_{t+1}, \zeta_{t+1})]$ can be obtained. Alternatively one could work directly with the first-order (Euler) conditions of program (8.16), but then one would be faced with an infinite number of equations.

There are thus two difficulties: the infinite number of time periods and the (possibly) infinite number of uncertain states. This suggests four basic strategies to solve the model, though within these four many different approaches may be followed[32] depending on the length of the time horizon (finite approximation or infinite) and on the nature of the disturbance (finite number of states or continuum of states).

Finite Horizon, Finite Number of States Consider the finite-horizon version of program (8.16). Suppose that over the T-period horizon there are S states, indexed s, which describe one possible future "history" such as weather (good weather in $t = 1$, bad weather in $t = 2$, etc.). Uncertainty can be represented by a given vector ε_s, with elements ε_{st}, and associated probability P_s. One then solves a standard convex program:

$$\max u(x_1) + \sum_{s=1}^{S} P_s \sum_{t=2}^{T} (\beta)^{t-1} u(x_{st}),$$

$$x_1, x_{st} \geqslant 0, k_2, k_{s,t+1} \geqslant 0, s = 1, 2, \ldots, S, t = 2, 3, \ldots, T,$$

subject to

$$F(x_1, k_2, k_1, \zeta_1) \leqslant 0,$$

$$F(x_{st}, k_{s,t+1}, k_{st}, \zeta_{st}) \leqslant 0,$$

for given k_1, ζ_1 and ζ_{st}.

This program will, in general, be large, but its recursive nature makes it suitable for application of decomposition techniques that use deterministic algorithms. When the number of states is very large, stochastic programming techniques can be used.[33]

Finite Horizon, Continuum of States Under a continuum of states the expectation operator is an integral instead of a finite sum. Then, for computing solutions, at least three possibilities can be distinguished. First, one could once again use stochastic optimization by sampling from this continuum. Second, one could approximate the objective function of the model in the neighborhood of the steady state, via the expectation of a quadratic function which can be replaced by a deterministic quadratic

32. See Danthine and Donaldson (1995), Hansen and Prescott (1995), Marcet and Marshall (1994), Rios-Rull (1995), Taylor and Uhlig (1990), and Uhlig (1995).

33. See, for example, Ermoliev and Wets (1988).

function where the covariance matrix appears in the quadratic term. Third, one could approximate the distribution by a piecewise linear function so that it becomes possible to perform analytically the integration of the expectation in the objective function.

Infinite Horizon, Finite Number of States Treating the time horizon as infinite introduces the additional difficulty that one has an infinite number of equations. This can be avoided in roughly three different ways; in practical terms, the first two are only applicable if the number of stochastic shocks and state variables (capital stocks) is small say, less than 3 or 4. The first way works with the dynamic programming format (8.17), and the issue is then to approximate the value function or the associated policy functions without knowing their functional form. This is done either by assuming a quadratic form, for which coefficients are estimated, or by computing the functions on a discrete grid of (k, ζ) points[34] and approximating them by interpolation between these points.[35] Second, one could deal with the infinite number of first-order condition (Euler equations) and derive from there the policy functions (again, via a quadratic approximation or via a grid). Finally, one could work with a tractable (linear-quadratic[36]) approximation of the model itself or of its Euler conditions, and this also leads to policy functions.

Infinite Horizon, Infinite Number of States Either the model itself or its Euler conditions are approximated, in which case continuous distributions can be accommodated[37] or one considers the discretization on a grid discussed above as a representation of a continuous distribution.

Specifying an Application of the Competitive Model

Applications of RBC models in macroeconomics usually distinguish as commodities a good that is consumed (c_t) and invested (b_t) and labor used as input in production (e_t) or as leisure (a_t). Capital accumulates according to $k_{t+1} = (1 - \delta)k_t + b_t$, where δ is the rate of depreciation. The good is generated through a production function of the firm which uses capital k_t and labor e_t as inputs and produces the output y_t. Production is subject to stochastic technological shocks ζ_t, a so-called productivity disturbance, that shifts the production function and that is often assumed to follow a first-order autocorrelated process with $0 < \rho < 1$. Program (8.16) may now

34. Clearly, if there are several capital stocks and/or random variables, the dimension of the grid and thus the number of points to be visited can become intractably large.

35. See Marcet and Marshall (1994) for an alternative to the grid method.

36. See Wonham (1979).

37. See Uhlig (1995).

be written

$$\max E \sum_{t=1}^{\infty} (\beta)^{t-1} u(c_t, a_t),$$

$$a_t, b_t, c_t, e_t, k_t \geqslant 0, \; t = 1, 2, \ldots,$$

$$\text{subject to} \tag{8.18}$$

$$c_t + b_t \leqslant y_t,$$

$$a_t + e_t \leqslant 1,$$

$$k_{t+1} = (1 - \delta)k_t + b_t,$$

$$y_t = G(k_t, e_t) \exp(\zeta_t),$$

for given k_1 and ζ_t, where $\zeta_t = \rho\zeta_{t-1} + \varepsilon_t$, ε_t follows some stochastic distribution and ρ is an autocorrelation parameter.

Estimating and Calibrating

To begin computation, one has to obtain (1) the various coefficients of the utility and the production function, as well as the discount rate β and the depreciation rate δ, and (2) the distribution of the ε_t and the value of ρ. In the more recent literature the coefficients are chosen by calibration assuming that observed values from national statistics over a particular range of years define steady-state values of the deterministic version of model (8.18), namely with $\zeta_t = 0$ for all t. Coefficients are identified by feeding empirical data into the steady-state equations. The probability distribution of ε_t is usually derived from a detrended[38] time series of output y_t and characterized by its first and second moments, and the autocorrelation is also estimated from this series.

Deriving Policy Functions

The policy functions associated with (8.18) can be derived through one of the methods outlined above. These lead to either an analytical form, say, $k_{t+1} = K(k_t, \zeta_t)$, for capital or to a table where for pairs of capital and shock values, specified on a given grid, one can directly read off the corresponding new value of capital, while for intermediate values an interpolation can be performed.

Simulating

Once the policy function has been obtained, one can run Monte Carlo simulations, starting with given k_1 at $t = 1$, generating ζ_1 from a random

38. This is often done by filtering the data using the Hodrick and Prescott (1980) filter. See also Harvey and Jaeger (1993) for more flexible methods.

drawing of ε_1 and obtaining k_2 using the policy function, and so on. This generates a trajectory characterized by fluctuations due to the shocks and the various lags (e.g., time to build). The time series of this trajectory must be sufficiently long (e.g., 5,000 periods) to serve as a basis for statistical estimation of the moments, possibly after detrending. It is possible to improve statistical reliability by repeating this exercise for several drawings of ε.

The results can be used to check whether the model is able to reproduce stylized facts about the economy under consideration. This is usually done by comparing the moments of the distributions with those of the observed time series; correlations between simulated series are also checked against those of observed data.[39] Though RBC modelers realize that actual economies contain many imperfections and government interventions, initially they have used these models in order to show that by a very simple competitive model with rational expectations, persistent, nonexplosive cycles can be generated in a robust manner without invoking institutional rigidities or interventions of any sort. It would then be clear that such interventions have little stabilizing effect.

Noncompetitive Extensions

The success of this first line of research was, however, modest. Nowadays imperfections and interventions of many types are being incorporated to improve the explanatory power of the model. Taxes have been accounted for,[40] the labor market has been modeled more carefully using efficiency wage relations[41] and wage rigidities,[42] and imperfect competition in commodity markets has been dealt with.[43] Multiple agents (countries) can and have been introduced in the welfare program, either with fixed welfare weights[44] or with agent-specific budget equations.[45] Virtually every extension covered in this book would qualify for incorporation in a business cycle model.

It may be added that introduction of government actions implies a significant departure from the original RBC research program. Rather than showing the ineffectiveness of government policy in models where agents have rational expectations, one is led back to a framework in which policies are effective and can thus be used for controlling the economy.

39. An overview of results obtained for a certain number of countries is given in Backus and Kehoe (1989). See also Danthine and Donaldson (1993) and Stadler (1994).

40. See Chang (1995) and Danthine and Donaldson (1995).

41. See Danthine and Donaldson (1992).

42. See Danthine and Donaldson (1989).

43. See Rotemberg and Woodford (1989).

44. See Backus and Kehoe (1989).

45. See Rioss-Rull (1995) and Garcia-Mila et al. (1995).

Finally, as discussed in section 8.1.4, taxes and other distortions may undermine the single-agent construct that uses dynamic programming. One then has to cope with the Euler equations of every consumer separately and link these through capital accumulation and commodity balance equations. For this procedure one can use a grid or an approximation. Recall also that such imperfections can lead to indeterminacy of solutions, though this problem is neglected in the RBC literature.

8.2 Overlapping Generations Models of Pure Exchange

We introduce the overlapping generations model (OLG) in the version with pure exchange, as is commonly done in the literature. We first mention the main distinctions between various OLG models (section 8.2.1) and then, as in dynastic models, we analyze the existence of equilibrium (section 8.2.2), possible lack of efficiency of this equilibrium (section 8.2.3), indeterminacy (section 8.2.4), dynamics and steady states (section 8.2.5), and finite-horizon approximations (section 8.2.6). In the pure exchange model savings cannot be invested into physical capital stocks that can be transferred from one period to the next. Therefore they must be expressed as financial claims, which are treated in the present section. In section 8.3 a parallel treatment will be given for the OLG model in which all savings are invested in capital goods.

8.2.1 Alternative OLG Models

Overlapping generations models describe consumers who live two (or G) periods, so that two (or G) generations coexist in every period t. Each consumer faces a lifetime budget constraint. He works while he is "young" and saves for "old" age. There is an infinite succession of such consumers.

OLG models have two interesting characteristics: They allow for a more natural treatment of savings by taking into account the life cycle of consumers, and they can easily accommodate exogenous demographic variables. In the competitive model of definition 1.2, consumers derive utility from their own consumption only, and they do not leave any bequest. This is in contrast to dynastic consumers who may be thought of as consisting of a succession of individuals who do not overlap with their parents and anticipate utility of their children with perfect accuracy. The dynastic and the pure exchange OLG model represent two extremes; many intermediate cases can be envisaged, such as the OLG model with capital accumulation to be studied in section 8.3 or the model with altruism briefly discussed in section 8.4.1.

Consumption by overlapping generations can be represented in a stair-like configuration, where generations appear in "rows" and time periods (which we also refer to as "years") in "columns" (figure 8.1). Consumptions are denoted by x_t^t and x_{t+1}^t, for $t = 0, 1, \ldots$, where the superscript denotes the year of birth of the generation and the subscript the date of the consumption.

Let us suppose that the model starts at $t = 1$, at the top of the staircase, with the old age consumption of the generation born in $t = 0$. There are markets corresponding to each column so that in each period, two consumers of different generations compete, and this extends until infinity. This scheme is a simplification in two respects at least: There is only one consumer in each generation, and each generation (consumer) lives only two periods.

The scheme in figure 8.2 does not have these limitations and will be used to explain how the treatment of the two-period model can be adapted to more complicated cases. The rows represent generations of five consumers who live for three periods, while columns denote the date of consumption. Consumers in every generation can have heterogeneous preferences, and their endowments may differ. Likewise the number of consumers can vary between generations, representing exogenous demographic changes. At the top of the staircase, there are now two generations of consumers who already have a past: Those born in $t = -1$ and those born in $t = 0$; they have (possibly zero or negative) accumulated savings.

OLG models can be distinguished further according to how these accumulated savings are transferred over years. We discuss this in the two-period scheme.

The OLG model is usually started in year $t = 1$, with the consumer born in $t = 0$ coming in with a claim M^0 possibly equal to the savings he made in $t = 0$ and with a commodity endowment ω_1^0. He has to compute his optimal period 1 consumption x_1^0, and his consumption x_0^0 is a given

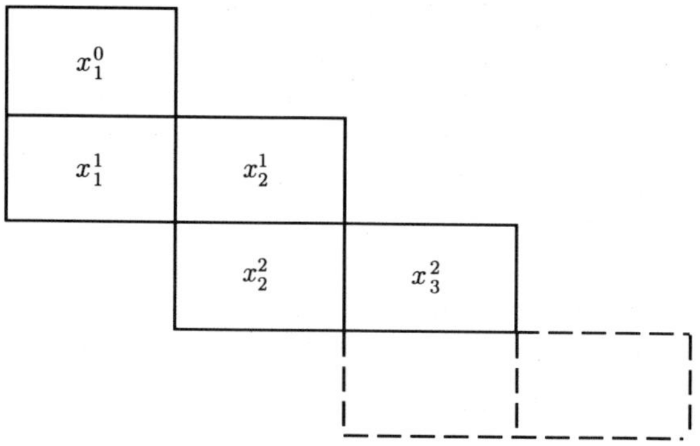

Figure 8.1
Overlapping generations of consumers who live two periods.

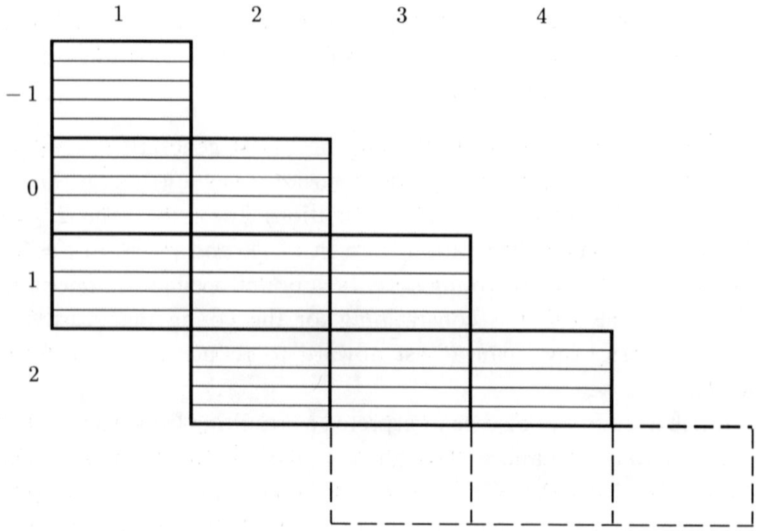

Figure 8.2
Overlapping generations of five consumers who live three periods.

parameter inherited from the past. He solves

$$\begin{aligned}&\max u(x_0^0, x_1^0),\\&x_1^0 \geqslant 0,\\&\text{subject to}\\&p_1 x_1^0 \leqslant M^0 + p_1\omega_1^0,\end{aligned} \tag{8.19}$$

where the superscript denotes the year of birth and the subscript the year in which the endowment is available or the consumption takes place.

The claim M^0 has important implications for the structure of the model. It is a financial claim not backed by any commodities or capital goods that can be sold to the new generation, or by any explicit contract with that generation. This claim will impair the satisfaction of Walras's law over any finite number of periods and will require artifacts in existence proofs. It can be expressed in different ways, which distinguish various kinds of pure exchange OLG models:

1. M^0 is equal to $p_1 m^0$, where m^0 is a given vector expressed in terms of commodities (a "real" claim).
2. M^0 is a fixed amount (a nominal claim).[46]

46. Recall that in the T-period model, nominal claims V_{i0} impair on homogeneity of excess demand functions and may create problems of existence of equilibrium.

3. M^0 is enforced by a government through a direct tax levied on the generation born in $t = 1$, which in turn will have a (possibly different) claim on the next generation, etc.

Case 3 is common when the claim of the old generation is safeguarded by government through a tax-funded pension scheme (though this may eventually lead to infeasible levels of taxation). In cases 1 and 2 there are no taxes levied on the generation born in $t = 1$, and prices in $t = 1, 2 \dots$ have to adjust to make the young save the amount needed to honor the claim of the old. Case 2 is less constraining for the young than case 1, since the absolute price level can adjust upward to reduce (or even eliminate) the claim.

Such a claim can also be interpreted as resulting from public consumption in $t = 1$ that is financed through, say, public loans from the savings of the generation born in $t = 1$. It may then be possible to increase consumption of the old in $t = 1$ (or in any later period), compensating the young when they are old themselves, and so on, indefinitely, so that an increase in consumption can be achieved without harming anyone. Obviously there can be no such gain if the equilibrium is Pareto-efficient before the increase. This is why it is especially important to study Pareto-efficiency in OLG models.

If there is production with accumulation of physical capital stocks, the claim may consist of the value of these stocks, which are owned by the consumer born in $t = 0$. Generation 1 will buy them, and sell them one period later to the newly born generation, and so on. If the claim is expressed in terms of physical capital only, it also is less constraining for the young than a real claim, since the old sell a stock of capital goods which the young have to reward according to its productivity only.

Although in reality all these claims coexist, and the way of honoring claims may change over time, the theoretical models that will be discussed only contain one type of claim. In this section we focus on case 1 and only briefly cover cases 2 and 3.

8.2.2 Specification of the Model with Financial Claims and Existence of Equilibrium

Equilibrium with Real Claims

We now specify the pure exchange OLG model for case 1 with real claims and define its equilibrium under the following assumptions, which replace our standard assumptions C1′ and C2′:

ASSUMPTION CO1 The consumer born in t has utility function $u^t: R^r_+ \times R^r_+ \to R_+$; $u(x^t_t, x^t_{t+1})$ is strictly concave, uniformly bounded above, nondecreasing for all t. It is continuously differentiable for positive consumption. Consumption is necessary in both periods: $\lim_{x \to 0} \|\partial u(x, y)/\partial x\| = \infty$ for all $y \geqslant 0$ and $\lim_{y \to 0} \|\partial u(x, y)/\partial y\| = \infty$ for all $x \geqslant 0$. Finally, the utility function is increasing for at least one commodity k in both periods t and $t+1$, and we denote by K the set of such commodities.

The assumption that $u(.)$ is uniformly bounded from above is not restrictive, since one can always find a monotonic transformation of utility that ensures this without affecting consumer demands. The limit conditions operate as boundary conditions and ensure that under utility maximization, consumption will be positive in both periods of a consumer's life.

ASSUMPTION **CO2** (Nonzero endowments) Endowments are nonnegative and stationary (and thus bounded), so $\omega_t^{t-1} = \omega_1^0$ and $\omega_t^t = \omega_1^1$, and $\omega_1^0 + \omega_1^1 > 0$.[47] Each consumer has a positive endowment of at least one commodity k in which his utility is increasing in t or in $t+1$.

Invariance of the endowments and the utility functions is assumed for convenience. Invariance beyond some year T is needed only for existence of a steady state. Obviously, for the claim to be honored, we need the requirement:

ASSUMPTION **CO3** (Feasibility of the real claim) $\omega_{k,1}^0 + m_k^0 \geqslant 0$ and strictly positive for some commodity $k \in K$ and $\omega_{k,1}^1 - m_k^0 \geqslant 0$ for all k.

The following technical assumption on the boundedness of the slopes of the indifference curves is also imposed:

ASSUMPTION **CO4** (Slopes of the indifference curves) There exists a positive constant δ such that $(\partial u(.)/\partial x_{h,t+1}^t)/(\partial u(.)/\partial x_{k,t}^t) \leqslant \delta$ for all commodities h and k and all nonzero consumptions.

Now the consumer born in $t = 0$ solves

$$\max u(x_0^0, x_1^0),$$

$$x_1^0 \geqslant 0,$$

$$\text{subject to} \tag{8.20a}$$

$$p_1 x_1^0 \leqslant p_1 m^0 + p_1 \omega_1^0.$$

The consumer born in $t = 1, 2 \ldots$ (denoted by the superscript t) solves the two-period consumer problem (the period is denoted by the subscripts t and $t+1$):

$$\max u(x_t^t, x_{t+1}^t),$$

$$x_t^t, x_{t+1}^t \geqslant 0,$$

47. There is positive supply of all commodities in each period t; an alternative is to drop the commodity from the list in the period where it is not available.

subject to (8.20b)

$$p_t x_t^t + p_{t+1} x_{t+1}^t \leqslant p_t \omega_t^t + p_{t+1} \omega_{t+1}^t.$$

Markets clear in every period $t = 1, 2, \ldots$:

$$x_t^{t-1} + x_t^t \leqslant \omega_t^{t-1} + \omega_t^t \perp p_t \geqslant 0. \tag{8.20c}$$

In analogy to (7.11), in equilibrium, savings of consumers will be constant over time and equal to the initial claim. This property can be verified as follows: The budget constraint of consumer born in $t = 0$ is

$$p_1 x_1^0 = p_1 \omega_1^0 + p_1 m^0. \tag{8.21}$$

We split the intertemporal budget constraint of consumer 1 to obtain

$$\begin{aligned} p_1 x_1^1 + S^1 &= p_1 \omega_1^1, \\ p_2 x_2^1 &= p_2 \omega_2^1 + S^1. \end{aligned} \tag{8.22}$$

Commodity balances imply that

$$p_1(x_1^0 + x_1^1) = p_1(\omega_1^0 + \omega_1^1). \tag{8.23}$$

Adding (8.21) and (8.22) and subtracting (8.23) yields $S^1 = p_1 m^0$. By recurrently applying this rule, we obtain

$$S^t = p_1 m^0,$$

which shows that intertemporal savings are constant, as in (7.11),[48] and equal to the initial claim.[49] Hence, if m^0 is positive, every young generation will spend less than the value of its endowment $p_t \omega_t^t$.

We can now define:

DEFINITION 8.2 (OLG pure exchange equilibrium with real claims) The allocation x_t^{t*}, x_{t+1}^{t*}, supported by the price vectors $p_t^* \geqslant 0, t = 1, 2, \ldots$, which are bounded for any finite t, is an OLG pure exchange equilibrium if it solves (8.20a) and (8.20b), and all markets clear as in (8.20c).

To prove that such an equilibrium exists, we will use a T-period OLG model (to be referred to as the truncated T-period model) in which there is a last consumer, born in T, who pays a transfer $p_1 m^0$ to the consumer born

48. In (7.11) the initial claim is assumed to be zero, but the reasoning given there also applies if it is not.

49. When there are several consumers in each generation, the property that S^t is constant only applies in the aggregate, for the whole generation. Consumers within a generation may borrow from or lend to other consumers in the same generation.

in 0. This is the artifact used to maintain Walras's law in the model.[50] The value x_{T+1}^T, consumer T's consumption in $T+1$, is exogenous and does not appear in any commodity balance, just as x_0^0 for consumer 0.

In this truncated model consumer T solves

$$\max u^T(x_T^T, x_{T+1}^T),$$

$$x_T^T \text{ unconstrained,}$$

$$\text{subject to} \tag{8.24}$$

$$p_T x_T^T = -p_1 m^0 + p_T \omega_T^T.$$

For this consumer, both x_{T+1}^T, his consumption after T and m^0, the initial claim of consumer 0, are given. The claim may exceed consumer T's income, and therefore x_T^T has to be unconstrained in sign so that the domain of his utility function is redefined accordingly: $u^T(.)$ is an (artificial) utility function that has the same properties as $u(.)$ on the nonnegative orthant, but it is also defined for values outside the nonnegative orthant. However, the specification of $u^T(.)$ will not play any role in the existence proof. Consumer T has also forced savings equal to $p_1 m^0$ (which, in equilibrium, will give the same savings S^T to him as to any other generation).

The resulting model is an Arrow-Debreu model with imposed transfers for the last agent. This will be shown in the first part of the existence proof. In the second part we consider the solutions obtained from the sequence of models with T increasing to infinity. The sequence of consumptions for consumers 0 to $T-1$ will then also be of infinite length, and we will show that it converges to the equilibrium of definition 8.2, so it is as if consumer T were to disappear into the horizon.

PROPOSITION 8.5 (Existence of equilibrium for the pure exchange OLG model with real claims) Let assumptions CO1, CO2, CO3, and CO4 be satisfied. Then there exists an OLG equilibrium that satisfies definition 8.2 with nonzero consumption for each consumer in every period of his life.

Proof

Part 1: Equilibrium in the truncated T-period model. We follow the approach of section 3.2 and define the T-period full-format version of (8.20), with welfare weights α^t (where the superscript denotes the generation)

50. This construction replaces the infinite sequence of budgets of definition 8.2 by a "circle" of payments. Note that if the consumer lives more than two periods, the same construct can be used, and a sharing rule among all consumers who live beyond T will have to be specified to honor the claim.

incorporating the consumer born in T only via his budget equation:[51]

$$\max \sum_{t=0}^{T-1} \alpha^t u(x_t^t, x_{t+1}^t),$$

$$x_t^{t-1}, x_t^t \geqslant 0, t = 1, 2, \ldots, T-1, x_T^{T-1} \geqslant 0, x_T^T \text{ unconstrained,}$$

subject to (8.25)

$$x_t^{t-1} + x_t^t \leqslant \omega_t^{t-1} + \omega_t^t, \quad t = 1, 2, \ldots, T \qquad (p_t),$$

$$\bar{p}_t x_t^t + \bar{p}_{t+1} x_{t+1}^t \leqslant \bar{p}_t \omega_t^t + \bar{p}_{t+1} \omega_{t+1}^t + \varepsilon, \quad t = 1, 2, \ldots, T-1 \quad (\rho^t),$$

$$\bar{p}_1 x_1^0 \leqslant \bar{p}_1 m^0 + \bar{p}_1 \omega_1^0 + \varepsilon \qquad (\rho^0),$$

$$\bar{\bar{p}}_T x_T^T = -\bar{p}_1 m^0 + \bar{\bar{p}}_T \omega_T^T \qquad (\rho^T),$$

where m^0, x_0^0, and x_{T+1}^T are given, the vectors $\alpha = (\alpha^0, \alpha^1, \ldots, \alpha^{T-1})$, and $\bar{p} = (\bar{p}_1, \bar{p}_2, \ldots, \bar{p}_{T-1})$ lie on simplices S^T and $S^{r(T-1)}$, respectively, ε is a positive constant and $\bar{\bar{p}}_T$ is positive and given.

For every such $(\alpha, \bar{p})$, program (8.25) is feasible, compact valued (because $\bar{\bar{p}}_T$ is positive, x_T^T and hence x_T^{T-1} must be bounded) and satisfies Slater's constraint qualification (here we make use of $\varepsilon > 0$ for all consumers except T, for whom this is not needed because his T-period consumption can be made negative and he is automatically on his budget constraint). Therefore it will always have a solution as well as Lagrange multipliers p_t and ρ^t. Moreover first-order conditions with respect to x_T^T ensure that $p_T = \bar{\bar{p}}_T$.

We now follow the two steps of the proof of proposition 3.3, using assumptions CO1 and CO2 instead of C1′ and C2′.

Step 1 Construction of the Fixed Point Correspondence

The construction of the fixed point correspondence proceeds exactly as in step 1 of proposition 3.3, defining one correspondence and two functions. The correspondence generates Lagrange multipliers from (8.25), and the functions update welfare weights and prices. For welfare weights the function is

$$\alpha^t = \mu(\hat{\alpha}^t + \delta s^t) \qquad \text{for } t = 0, 1, \ldots, T-1, \tag{8.26}$$

where δ is a positive constant, μ is a scalar such that $\Sigma_{t=0}^{T-1} \alpha^t = 1$, s^t is the excess budget of consumer t, and $\hat{\alpha}^t$ is the previous value of α^t.

For prices the updating function is

$$\begin{aligned} \bar{p}_t &= \lambda(p_t + \rho \hat{p}_t) \qquad \text{for } t = 0, 1, \ldots, T-1, \\ \bar{\bar{p}}_T &= \lambda \pi \qquad \text{for fixed } \pi \in R_{++}^r, \end{aligned} \tag{8.27}$$

51. Here we are only interested in the allocation to consumers $0, 1, \ldots, T-1$, and consumer T merely serves as a residual.

where $\rho = \max_t \rho^t$, λ is such that $\bar{p}$ lies on the simplex $S^{r(T-1)}$, and $\hat{p}_t$ is the previous value of $\bar{p}_t$. It follows from step 1 of proposition 3.3 that the correspondence defined in this way has a fixed point.

Step 2 The Fixed Point Is an Equilibrium

Here we cannot fully rely on step 2 of proposition 3.3, since the assumptions used are somewhat different. In the fixed point we have[52]

$$\alpha^t = \mu(\alpha^t + \delta s^t) \qquad \text{for } t = 0, 1, \ldots, T-1. \tag{8.28}$$

We show that α^t is positive for all t. It follows from normalization of welfare weights that α^t must be positive for some t, say, t'. Then, by increasingness of $u(.)$ in the commodities $k \in K$ (assumption CO1), we have for $k \in K$ and $t = t'$, $t' \neq 0$, and $t' \neq T$, $\bar{p}_{kt} > 0$ and $\bar{p}_{k,t+1} > 0$, which gives positive income to consumers $t'-1$, t', and $t'+1$. For consumer 0 only prices $\bar{p}_{k,1}$ are needed. Suppose now that consumer $t'-1$ has a zero welfare weight. This would leave him with zero expenditure, positive income, and thus a budget surplus. Then by (8.28) his welfare weight is positive, a contradiction. By the same reasoning the welfare weight of $t'+1$ is positive, and hence all welfare weights are positive. It follows that $\bar{p}_t \neq 0$, and by the same reasoning as in proposition 3.3, all budget surpluses are zero.

Hence, due to the introduction of ε, all budget constraints in (8.25) will be nonbinding and ρ^t will be equal to zero for all t in any equilibrium. Obviously this ε may be reduced to zero without affecting the optimal allocation. Therefore $\rho = \max_t \rho^t = 0$. Now (8.27) reduces to

$$\bar{p}_t = \lambda p_t \qquad \text{for } t = 1, 2, \ldots, T-1,$$

which shows that $\bar{p}_t$ is proportional to p_t. For period T we can also define $\bar{p}_T = \lambda p_T$ since $p_T = \pi$ in equilibrium. Since $\rho^t = 0$ in the fixed point, proposition 3.3 applies, and the solution of (8.25) is a competitive equilibrium for the truncated T-period model.

Part 2: Extending T to infinity.

a. *Reformulating the truncated model.* Note that in part 1 prices are normalized over the time horizon T. When we extend T to infinity, we cannot use this normalization. Therefore we define, for further use, consumer-specific prices with consumer-specific normalizations in the truncated T-period model:

$$\bar{p}_t^t = \lambda^t \bar{p}_t \quad \text{and} \quad \bar{p}_{t+1}^t = \lambda^t \bar{p}_{t+1}, \tag{8.29}$$

52. For notational convenience, we leave out the * superscript for denoting equilibrium values.

where λ^t is a positive scalar such that $\Sigma_k(\bar{p}^t_{kt} + \bar{p}^t_{k,t+1}) = 1$ for $t = 1, 2, \ldots, T-1$;

$$\bar{p}^0_1 = \lambda^0 \bar{p}_1, \tag{8.30}$$

where $\Sigma_k \bar{p}^0_{k1} = 1$ for $t = 0$.

These normalizations are feasible (because, as we have seen, $\bar{p}_t$ is nonzero for every t) and do not change the allocations (they simply multiply prices in the budget constraint of consumer t by a constant in the full-format program 8.25).

From part 1 follow equilibrium values for the α^t, for x^T_T, and for the newly normalized prices $\bar{p}^t_t, \bar{p}^t_{t+1}$. The associated allocations for consumers 0 to $T-1$ may now also be obtained from the following program where we treat consumption x^T_T as a given parameter in the commodity balance of period T, so we can drop consumer T's budget constraint:

$$\max \sum_{t=0}^{T-1} \alpha^t u(x^t_t, x^t_{t+1}),$$

$$x^{t-1}_t, x^t_t \geqslant 0, t = 1, 2, \ldots, T-1, x^{T-1}_T \geqslant 0,$$

subject to (8.31)

$$x^{t-1}_t + x^t_t \leqslant \omega^{t-1}_t + \omega^t_t, \quad t = 1, 2, \ldots, T \qquad (p_t),$$

$$\bar{p}^t_t x^t_t + \bar{p}^t_{t+1} x^t_{t+1} \leqslant \bar{p}^t_t \omega^t_t + \bar{p}^t_{t+1} \omega^t_{t+1}, \quad t = 1, 2, \ldots, T-1 \quad (\rho^t),$$

$$\bar{p}^0_1 x^0_1 \leqslant \bar{p}^0_1 m^0 + \bar{p}^0_1 \omega^0_1 \qquad (\rho^0).$$

We note that if for some finite T the x^T_T found as an equilibrium value of the truncated model in step 1, happens to be an equilibrium value of the infinite-horizon OLG model in definition 8.2, then for consumers $t = 0, 1, \ldots, T-1$, the solution to (8.31) is also a solution of the infinite-horizon OLG equilibrium. To find such a value, we proceed as follows:

In part 1, we have shown that we can find at least one value x^T_T for which a truncated T-period equilibrium exists, but there may be many such x^T_T and to each of them corresponds at least one solution $\bar{p}^t_t$ for $t = 1, 2, \ldots, T-1$. Below we show that as T increases, the set of such price trajectories shrinks and eventually yields the desired x^T_T value.

b. *Convergence for consumer-specific prices.* We now follow the arguments of Balasko, Cass, and Shell (1980). We first define a two-period price vector q^t, partitioned as $q^t = (q^t_0, q^t_1)$, according to the following rules:

- For $t = 0$: $q^t = (0, \bar{p}^0_1)$.
- For $t = 1, 2, \ldots, T-1$: $q^t = (\bar{p}^t_t, \bar{p}^t_{t+1})$.
- For $t = T, T+1, \ldots$: $q^t \in S^{2r}$.

Thus, for $t = 0, 1, \ldots, T-1$, the $\bar{p}_t$ are equilibrium values of the T-period truncation, while for $t = T+1, \ldots$, we set all prices arbitrarily on the simplex. This way we construct a sequence $\{q^t\}_{t=0}^{\infty}$.

For any x_T^T for which a truncated equilibrium exists, we can define such a sequence, and there may exist many other x_T^T values and corresponding price sequences. Define $Q(T)$ as the set of all such possible sequences obtained for the time-horizon T. Since a truncated equilibrium exists, the set $Q(T)$ is nonempty and by construction, it belongs to the infinite-dimensional set $B = S^{2r} \times S^{2r} \times \ldots$.

Now the main step in the proof is to note that since the set of possible x_T^T values cannot increase as the horizon is being extended from T to $T+1$, it must be that the set of possible price sequences cannot become larger, $Q(T) \supseteq Q(T+1)$.

Since B is an infinite-dimensional product of compact sets (simplices S^{2r}), it is compact as well in the product topology (see theorem A.6.1). Hence $Q(T)$ defines a nonincreasing sequence $\{Q(T)\}_{T=1}^{\infty}$ of compact nonempty sets, so $Q(\infty)$, the intersection of a nonincreasing sequence of nonempty compact sets is nonempty. This proves that there exists a sequence $\{q^t\}_{t=0}^{\infty}$ for T extended indefinitely.

c. *Recovering prices.* We have to recover bounded equilibrium prices p_t from the values of q^t, which are consumer specific. Define first $g^t = \Sigma_s \bar{p}_{s,t+1}^t / \Sigma_s \bar{p}_{s,t}^t$. Express all prices in terms of consumer 0's normalization $\Sigma_s \bar{p}_{s,1}^0 = 1$, and use (8.29) and (8.30) to return from consumer-specific prices to market prices p_t as

$$p_t = (\Pi_{\tau=1}^t g^\tau)\bar{p}_t^t.$$

Hence the sequence of p_t is bounded for finite t provided that the ratio g^τ is bounded for $\tau = 1, 2, \ldots, t$. To ensure this, we use assumption CO4. Since income is positive, there will be positive consumption of some commodity in both periods of each consumer's life. Then, using the first-order conditions, we obtain the ratio $[\partial u(.)/\partial x_{h,t+1}^t]/[\partial u(.)/\partial x_{k,t}^t] = \bar{p}_{h,t+1}^t/\bar{p}_{k,t}^t$ for some pair of commodities h, k, and this ratio is bounded above by assumption CO4. Hence

$$\frac{\Sigma_s \bar{p}_{s,t+1}^t}{\Sigma_s \bar{p}_{s,t}^t} \leqslant \frac{\Sigma_s \partial u(x_t^t, x_t^t)/\partial x_{s,t+1}^t}{\partial u(x_t^t, x_t^t)/\partial x_{kt}^t} \leqslant r\delta,$$

where k is chosen so that $x_{kt}^t > 0$. Therefore the ratios g^t are bounded as well, and for any finite t, p_t is bounded (but this does not prevent p_t from becoming unbounded when t goes to infinity, since it then is the infinite product of bounded values).

d. *Convergence of the sequence of consumer demands.* Finally, we can show that the associated equilibrium consumer t's demand (obtained by solving his utility maximization problem at given normalized prices q^t) converges also. Indeed, for such q^t, demand is continuous in q^t, which is of finite dimension $2r$ and clears the markets. Then, by theorem A.6.2, it is also continuous as T goes to infinity. Therefore consumer demand converges as $Q(T)$ converges to $Q(\infty)$.

Now all the conditions of definition 8.2 have been verified and an OLG equilibrium exists. ∎

Recall that in the finite-horizon truncation, consumer T has to transfer to consumer 0. Therefore he may have negative income, and has no choice as to his consumption in $T + 1$ (which is exogenously given). Hence there are prices that clear markets in period T but not in a competitive way. All consumers $t \neq T$ rely on competitive prices to make their intertemporal consumption, and hence their savings decision. For instance, consumer 1 voluntarily saves sufficiently to honor the claim $p_1 m^0$ of consumer 0, since he knows that his claim will be honored by consumer 2, and so on. It is only in year T that problems arise. If T can be extended to infinity, the system will work "forever" under competitive conditions, except possibly at infinity.

Finally, it is important to note that, though all prices p_t are bounded for finite t, they may become unbounded at infinity, in which case the associated welfare weight will also be unbounded. Even if each welfare weight is bounded, their sum may not be. Then social welfare will be infinite and, as we will see, the equilibrium may fail to be Pareto-efficient. However, if $\alpha^t \leqslant (\gamma)^{t-1}$ for some discount factor $\gamma \in (0, 1)$, welfare will also be, and $\lim_{t\to\infty} p_t = 0$, ensuring efficiency as in the dynastic model.[53]

Equilibrium with Nominal Claims

Recall that if claims are real, $M^0 = p_1 m^0$ and that, in equilibrium, the budget equations for each period of consumer t's life can be written

$$p_t x_t^t = p_t \omega_t^t - p_1 m^0, \tag{8.32a}$$

$$p_{t+1} x_{t+1}^t = p_{t+1} \omega_{t+1}^t + p_1 m^0. \tag{8.32b}$$

If claims are nominal, these budget equations read

$$p_t x_t^t = p_t \omega_t^t - M^0, \tag{8.33a}$$

$$p_{t+1} x_{t+1}^t = p_{t+1} \omega_{t+1}^t + M^0. \tag{8.33b}$$

This leads to the following equations for the second period of consumer 0's

53. See section 8.2.3 below.

life and the first period of consumer T's life:

$$p_1 x_1^0 \leqslant p_1 \omega_1^0 + M^0,$$

$$p_T x_T^T \leqslant p_T \omega_T^T - M^0.$$

Since M^0 is given, it determines the price normalization. To prove existence of equilibrium for this model, it is convenient to keep prices on the simplex and to introduce the price μ_1 of M^0. Obviously prices p_t which use M^0 as normalization will then be equal to $\tilde{p}_t/\mu_1$, and the above budget equations will read as

$$\tilde{p}_1 x_1^0 \leqslant \tilde{p}_1 \omega_1^0 + \mu_1 M^0, \tag{8.34}$$

$$\tilde{p}_T x_T^T \leqslant \tilde{p}_T \omega_T^T - \mu_1 M^0. \tag{8.35}$$

By setting $\mu_1 = 0$, we can eliminate the claim of consumer 0 and return to the model with zero real claims. This amounts to ignoring the claim or accepting infinite prices p_t and defines one possible equilibrium. We now investigate the possibility for other solutions to exist.

The value μ_1 can be interpreted as the price of "voluntary savings by consumer T," whose choice problem (8.24) becomes

$$\max u^T(x_T^T, x_{T+1}^T),$$

$$x_T^T, x_{T+1}^T \geqslant 0,$$

subject to (8.36)

$$\tilde{p}_T x_T^T \leqslant -\mu_1 \tilde{p}_{T+1}(x_{T+1}^T - \omega_{T+1}^T) + \tilde{p}_T \omega_T^T,$$

where $\tilde{p}_{T+1} > 0$ is exogenously given, the budget constraint in the program replaces (8.35), consumer T can choose his consumption x_{T+1}^T at prices $\mu_1 \tilde{p}_{T+1}$, and he has an ordinary utility function that satisfies assumption CO1. There also is a market-clearing condition for these voluntary savings:

$$\tilde{p}_{T+1}(x_{T+1}^T - \omega_{T+1}^T) \leqslant M^0 \perp \mu_1 \geqslant 0. \tag{8.37}$$

We can now bring consumer 0 and consumer T with budget constraints (8.34) and (8.36) as well as the new market clearing condition (8.37) within the full format (8.25), and the proof can proceed in terms of prices $\tilde{p}_t$ and the additional price μ_1. An equilibrium will exist. If, in equilibrium, the value μ_1 is positive, it may be used as normalization, and one has a solution of the original model in prices p_t. In this case money has a scarcity value, which it does not have in the finite-horizon model.[54]

54. See, for example, Hahn (1984).

The T-period equilibrium solution depends on the restrictions imposed on consumer T. Other such restrictions are possible, and one cannot, with the arguments given so far, exclude that this would lead to another equilibrium of the OLG model.

Equilibrium with Taxes

Imposing restrictions on the last consumer does not require "government intervention" before period T, which means never, since T goes to infinity. One can also conceive of interventions in earlier periods: This leads to a model in which consumer t pays a tax $p_t m^0$ and receives a pension $p_{t+1}m^0$. His intertemporal budget constraint is then

$$p_t x_t^t + p_{t+1}x_{t+1}^t = p_t\omega_t^t + p_{t+1}\omega_{t+1}^t - p_t m^0 + p_{t+1}m^0.$$

One has to ensure that the income of this consumer is positive at all prices. A sufficient condition is to take m^0 as a fraction of ω_t^t $(=\omega_1^1)$. This budget constraint can be split as

$$p_t x_t^t + S^t = p_t\omega_t^t - p_t m^0, \tag{8.38a}$$

$$p_{t+1}x_{t+1}^t = p_{t+1}\omega_{t+1}^t + S^t + p_{t+1}m^0, \tag{8.38b}$$

showing that, in equilibrium it is optimal for every consumer t to set $S^t = 0$. However, consumer T does not have this choice. He is forced to set $S^T = 0$ and to pay taxes $p_T m^0$. His problem is

$$\max u(x_T^T, x_{T+1}^T),$$

$$x_T^T \geqslant 0,$$

subject to (8.39)

$$p_T x_T^T = p_T\omega_T^T - p_T m^0$$

with x_{T+1}^T given.

The existence proof follows the same line as for proposition 8.5.

8.2.3 Inefficiency of Equilibrium

As mentioned earlier, for an OLG equilibrium that is inefficient, it is feasible to increase the consumption of some generation when it is old (i.e., the consumption of generation 0) without having to reduce the utility of any other generation because the young are compensated when they are old. An OLG equilibrium with real claims resulting from the finite-horizon truncation will obviously be inefficient, since consumer T is forced to save and in $T+1$ he is forced to consume a given amount x_{T+1}^T. In part 2 of the existence proof, the OLG equilibrium is obtained as the limit of a sequence of equilibria in which the value x_T^T is restricted to the set of values that allow an equilibrium to exist for the periods 1 to T. Though the set of possible x_T^T values shrinks when the horizon is extended, there may still be values that lead to inefficient allocations; the literature contains many examples of such inefficient equilibria. However, sufficient conditions for efficiency have been formulated, and we discuss two of them. We first consider the case in which the value of endowments becomes zero in the long run.

PROPOSITION 8.6 (Efficiency of OLG equilibria when values converge to zero[55]) If in the OLG equilibrium of definition 8.2, $\lim_{t\to\infty} p_t(\omega_t^{t-1}+\omega_t^t) = 0$, then the equilibrium is Pareto-efficient.

Proof The proof proceeds by contradiction. If the equilibrium is inefficient, there exists some alternative allocation that improves the welfare of, say, consumer 1. To represent such an allocation, we consider the sequence of prices p_t^* associated with an equilibrium and the following T-period program in which utility of consumer 1 is maximized, subject to feasibility constraints and the restriction that the other consumers may not lose:

$$U(\alpha^{1*}, u^*, x_T^{T*}) = \max \alpha^{1*} u(x_1^1, x_2^1),$$

$$x_t^{t-1}, x_t^t \geqslant 0, t = 1, 2, \dots, T,$$

subject to (8.40)

$$x_t^{t-1} + x_t^t \leqslant \omega_t^{t-1} + \omega_t^t, \qquad t = 1, 2, \dots, T,$$

$$u(x_t^t, x_{t+1}^t) \geqslant u^{t*}, \qquad t = 0, 2, \dots, T-1,$$

$$x_T^T = x_T^{T*} \qquad (p_T^*),$$

for given x_0^0, and where x_T^{T*} and the u^{t*} are equilibrium values, α^{1*} is the inverse of the marginal utility of income of consumer 1 at equilibrium prices (p_1^*, p_2^*), and p_T^* is the equilibrium price in T.

Obviously the allocation generated by (8.40) is efficient for given x_T^{T*} (and given u^{t*}) and is an OLG equilibrium. Therefore the OLG equilibrium can only be inefficient if some feasible variation Δ of x_T^{T*} yields a higher utility for, say, consumer 1 without hurting consumers $0, 2, \dots, T-1$ who are within the program nor any consumer beyond $T-1$. This means that

$$U(\alpha^{1*}, u^*, x_T^{T*} + \Delta) > U(\alpha^{1*}, u^*, x_T^{T*}). \tag{8.41}$$

By the perturbation theorem A.3.2, $U(.)$ is concave in x_T^{T*}. Since p_T^* is a

55. See Brown and Geanakoplos (1982). The proof is also given in Geanakoplos and Polemarchakis (1991).

subgradient of $U(.)$ with respect to x_T^{T*} (see theorem A.3.4),

$$U(\alpha^{1*}, u^*, x_T^{T*} + \Delta) \leqslant U(\alpha^{1*}, u^*, x_T^{T*}) + p_T^*\Delta. \tag{8.42}$$

Since the variation Δ is feasible,

$$p_T^*(x_T^{T*} + \Delta) \leqslant p_T^*(\omega_T^{T-1} + \omega_T^T).$$

Therefore

$$p_T^*\Delta \leqslant p_T^*(\omega_T^{T-1} + \omega_T^T).$$

By assumption, $\lim_{t\to\infty} p_t(\omega_t^{t-1} + \omega_t^t) = 0$, so the right-hand side will converge to zero as T increases. As a result $p_T^*\Delta$ cannot be positive, yielding a contradiction between (8.41) and (8.42). ■

We now consider the implications of this proposition for different types of claims. Recall from equations (8.32a), (8.33a), and (8.38a), with $S^t = 0$, that in equilibrium the first-period budget constraint of consumer t reads

$$p_t x_t^t + M^t = p_t \omega_t^t,$$

where M^t is either a real claim $(p_1 m^0)$, a given constant nominal claim (M^0), or a tax $(p_t m^0)$. If $\lim_{t\to\infty} p_t(\omega_t^{t-1} + \omega_t^t) = 0$, then $p_t\omega_t^t$, and therefore all M^t, go to zero as well.

For real claims this limit condition implies that $p_1 m^0$ will become zero. Since $\Sigma_k p_{1k}$ can be normalized to 1 (see part 2 in the proof of proposition 8.5), $p_1 \neq 0$. Then for $p_1 m^0$ to be zero, it must be that the positive elements of m^0 have a zero price. Thus, if m^0 has positive elements for some commodity $k \in K$ defined in assumption CO1 (i.e., a commodity for which the price will be positive), the equilibrium cannot have the property $p_1 m^0 = 0$, and the limit condition cannot hold. For nominal claims the limit condition cannot be satisfied in terms of (8.33a). However, in terms of prices of (8.34), which do not use the claim as numéraire, the condition can be fulfilled if $\mu_1 = 0$, which means that the claim disappears through inflation. Finally, in the case of taxes the limit condition can be satisfied because the claim is to be paid in current prices p_t.

An alternative sufficient condition for efficiency involves an assumption of nonvanishing curvature for the expenditure function (finite intertemporal substitutability) and the assumption that prices cannot be infinite for an infinite number of periods ($\lim_{T\to\infty} \Sigma_{t=t_0}^T 1/(c_t\|p_t\|) = \infty$, where c_t is some constant greater than $\varepsilon > 0$). Note that if prices go to zero or become constant, the latter condition will be satisfied. By contrast, a sufficient condition for inefficiency is that the utility satisfies a bounded curvature between periods (nonzero intertemporal substitutability) and that prices go to infinity ($\lim_{T \to \infty} \Sigma_{t=t_0}^{T} 1/\|p_t\| < \infty$).[56]

These conditions only make it possible to prove that given paths of prices are efficient. Unfortunately, in practice, one cannot extend the path to infinity to check whether the condition holds. We will see that when production (and capital accumulation) is introduced, restrictions on technology can be formulated that ensure efficiency.

8.2.4 Indeterminacy of Equilibrium

There are basically two ways of understanding why OLG equilibria may be indeterminate. The first is to note that there is an infinite number of prices and also of consumers so that we have an equilibrium problem in infinite dimensions for which theorems A.7.5 and A.7.6 do not apply.[57] The second is to recall that in the existence proof, x_T^T is fixed arbitrarily and that although the set of possible equilibria shrinks, it may not reduce to a single trajectory.

Even if the efficiency conditions of proposition 8.6 hold, indeterminacy may arise. But, since late allocations do not affect the earlier ones because of discounting, the equilibria that correspond to different x_T^T will hardly differ for sufficiently large T.

However, in a one-commodity economy in which consumers live two periods, there is no indeterminacy. To see this, one sets $p_1 = 1$ and adjusts p_2 to make consumer 1 voluntarily save $p_1 m^0$. This process is repeated for consumer 2, and so on, and generates a unique infinite sequence of prices and allocations that is an equilibrium of the OLG model.

8.2.5 Dynamics and Steady States

Since the full format representation of the OLG model is similar to the representation of the dynastic model, we can expect the path followed by the variables to converge, diverge, cycle, or be chaotic.

The dynamics of the pure exchange model are described by a time-invariant excess demand function $z(p_{t-1}, p_t, p_{t+1}) = a(p_{t-1}, p_t) + c(p_t, p_{t+1})$, where $a(p_{t-1}, p_t)$ is the net demand by the old (a for adult), while $c(p_t, p_{t+1})$ is the net demand by the young (c for child). A steady state of this economy

56. See Geanakoplos and Polemarchakis (1991). The proofs proceed by checking whether it is possible to tax the young and compensate them when they are old. When the curvature is vanishing, intertemporal substitution is high, so compensation is made "easy" and therefore inefficiency is more likely to appear. Conversely, when the curvature is infinite, there is no intertemporal substitutability, so there is little room for compensation. The result on inefficiency is proved for an exchange economy with given endowments. If it is possible to provide compensation while the endowments are fixed, this will a fortiori be the case when they are variable, that is, when there is production.

57. See also Kehoe et al. (1989) and Mas-Collel and Zame (1991).

is a price vector $p \geqslant 0$, and a discount factor $\beta \in (0, 1]$ such that[58]

$$a\left(\frac{p}{\beta}, p\right) + c(p, \beta p) \leqslant 0 \perp p \geqslant 0,$$

which, by homogeneity of $a(.)$, is equivalent to

$$a(p, \beta p) + c(p, \beta p) \leqslant 0 \perp p \geqslant 0. \tag{8.43}$$

The steady states can be further characterized as follows: Premultiplying (8.43) by p yields

$$pa(p, \beta p) + pc(p, \beta p) = 0. \tag{8.44}$$

The budget constraint of each consumer is

$$pc(p, \beta p) + \beta pa(p, \beta p) = 0. \tag{8.45}$$

Subtracting (8.45) from (8.44) yields

$$(1 - \beta)pa(p, \beta p) = 0.$$

The last equation shows that steady states can be of two types:

1. If $\beta \neq 1$, $pa(p, \beta p) = 0$. The value of the consumption of the old, and hence also the young, is equal to the value of their endowment. The young do not save for old age, and since savings are equal to the claim, this claim is zero. This is referred to as a real steady state.
2. If $\beta = 1$, there is no restriction on $pa(p, \beta p)$, and there may be savings. This is a monetary steady state.

In case 1 homogeneity is satisfied, and one can for every p find a unique $\beta(p)$ such that $pa(p, \beta p) = 0$. Under appropriate continuity assumption,[59] a steady state can be computed in which β may nevertheless turn out to be 1. In case 2 the excess demand function of the monetary steady state is homogeneous of degree zero in prices and satisfies Walras's law within a single period. Again, under appropriate continuity assumptions, the steady state will exist.

To compute these steady states, note that in case 1 there are r equations[60] (one for each commodity and not $r - 1$, since Walras's law does not hold within a single period) and r unknowns ($r - 1$ prices, due to homogeneity,

58. Note that while allocations are constant in this steady state, prices are discounted by the factor β, but relative prices do not change over time.

59. Continuity is not ensured under assumptions CO1 and CO2 because some prices may be zero.

60. These are in fact complementarity conditions. See section A.9 of the mathematical appendix.

and β). Prices p adjust to clear markets, and the discount factor β to ensure $pa(p, \beta p) = 0$ at equilibrium prices. In case 2 there are $r - 1$ equations (here Walras's law holds within a single period) in $r - 1$ prices.

If $\beta < 1$, the steady state describes a path with prices converging to zero, and by proposition 8.6, the steady state is efficient. Conversely, if $\beta > 1$, prices increase, and under the bounded curvature assumption, the steady state may be inefficient.[61] When $\beta = 1$, prices are constant in the steady state, and under the nonvanishing curvature assumption, the steady state will be efficient.

In a monetary steady state, $\beta = 1$, and the savings of the young, $pc(p, \beta p)$, can be of any sign. If this sign is the same as the sign of the initial claim $p^1 m^0$, prices p can be renormalized to make the value of steady-state savings equal to the initial claim. However, the way in which steady states are defined does not ensure the matching of signs. Thus, if the signs do not agree, the path generated by the sequence constructed in proposition 8.5 cannot converge to a steady state.[62] Even if the signs do agree, the path may fail to converge. Finally, if initial claims have a nonzero value, the trajectory cannot converge to a real steady state in finite time, since the claim is transmitted, and there is no claim in this steady state.

In practice, convergence is studied locally around the steady state. One considers the Jacobian of the excess demand function and assumes regularity of this steady state so that for prices a first-order difference equation can be derived. The eigenvalues of this system characterize its stability properties. For local stability the number of roots with modulus smaller than β (stable roots) should be equal to the number of initial conditions ($r - 1$ when there are real claims). However, if the number of such roots is larger than $r - 1$, there will be indeterminacy, for the path will converge starting from many, in fact an infinity of initial prices. Finally, when the number of stable roots is smaller than $r - 1$, the steady state is locally unstable. For further details, see Kehoe (1991).

8.2.6 Finite-Horizon Approximations

The finite-horizon approximations can be characterized in terms of the specification of the model for consumer T (see equations 8.24, 8.36, and 8.39 for real claims, nominal claims and taxes, respectively). Each of these specifications can be incorporated in a full format specification similar to

61. Inefficiency can be verified by using the alternative allocation defined in the proof of proposition 8.6. Efficiency is an immediate consequence of proposition 8.6.

62. Recall that in the dynastic model the steady-state welfare weights are independent of the distribution of initial stocks, and therefore they may involve transfers among consumers. In the steady state of the OLG model, budget constraints are satisfied. There are no such transfers, but as we have seen, the initial claims may not be respected.

(8.25). In applied models it is often assumed that the path converges to a steady state, reached with a sufficient degree of precision, within a finite-time horizon. This means that the steady-state values are used as end conditions. We return to this in section 8.4.

8.3 Overlapping Generations Models with Production

This section is organized like section 8.2. In section 8.3.1 we introduce the model, and then we study existence of equilibrium (section 8.3.2), efficiency properties (section 8.3.3), indeterminacy (section 8.3.4), dynamics and steady states (section 8.3.5), and finite-horizon approximations (section 8.3.6). In the model with production, we only consider savings that are invested into physical capital stocks, thus excluding financial claims.

8.3.1 Specification of the Model

Whereas in the pure exchange OLG model, the young consumer saves for old age by buying a "paper" claim that will be honored when he becomes old, the introduction of capital goods makes it possible to introduce "classical" savings that are used for capital stock accumulation. If the young save for old age by buying the physical capital stock, and all savings are invested into that stock, then there is no need for a paper claim, for the young will simply sell the capital stock when they are old.

As in the dynastic model we assume, without loss of generality, that there is only one firm that lives forever (or a sequence of firms) with a technology satisfying assumption PD1 of section 8.1.1.[63] The firm is owned by the young generation; at the beginning of every period t, it buys capital stocks k_t at market prices ψ_t from the old generation. During the period it produces net outputs y_t and capital stocks $\tilde{k}_{t+1}$, which it sells at the end of the period at prices p_t and ψ_{t+1}, respectively. Due to constant returns to scale, there will be no profits (hence the young only pay for the capital stock, and not for the firm which has no value when they acquire it). For period t the producer solves

$$\max_{k_t, \tilde{k}_{t+1} \geqslant 0,\, y_t} p_t y_t + \psi_{t+1}\tilde{k}_{t+1} - \psi_t k_t,$$

subject to (8.46a)

$$F(y_t, \tilde{k}_{t+1}, k_t) \leqslant 0.$$

Consumer 0 comes in at old age with the capital stock $\tilde{k}_1$, sells it to the

63. Capital stock 1 in assumption PD1 may be interpreted as education owned by every consumer in positive amounts; labor itself is part of the endowment ω.

firm, and chooses his consumption in $t = 1$ according to

$$\max u(x_0^0, x_1^0),$$

$$x_1^0 \geqslant 0,$$

subject to (8.46b)

$$p_1 x_1^0 \leqslant \psi_1 \tilde{k}_1 + p_1 \omega_1^0.$$

When he is young, consumer $t \neq 0$ spends $p_t x_t^t$ on consumption, buys the capital stock at value $\psi_t \tilde{k}_t$, and has financial savings S^t. He receives net profit $p_t y_t$ and income from his endowment $p_t \omega_t^t$. When old, he consumes $p_{t+1} x_{t+1}^t$. His income consists of the value of the capital stock $\psi_{t+1} \tilde{k}_{t+1}$, which he sells, of his endowment $p_{t+1} \omega_{t+1}^t$ and of his financial savings S^t. Hence his two budget constraints are

$$p_t x_t^t + \psi_t \tilde{k}_t + S^t = p_t y_t + p_t \omega_t^t, \tag{8.47}$$

$$p_{t+1} x_{t+1}^t = \psi_{t+1} \tilde{k}_{t+1} + p_{t+1} \omega_{t+1}^t + S^t. \tag{8.48}$$

These can be aggregated into a wealth constraint:

$$p_t x_t^t + p_{t+1} x_{t+1}^t + \psi_t \tilde{k}_t = p_t y_t + \psi_{t+1} \tilde{k}_{t+1} + p_t \omega_t^t + p_{t+1} \omega_{t+1}^t. \tag{8.49}$$

Due to constant returns to scale in assumption PD1, the full value of the firm accrues to inputs, so $\psi_t \tilde{k}_t = p_t y_t + \psi_{t+1} \tilde{k}_{t+1}$. Thus the wealth constraint of consumer t is the same as in the pure exchange model. He solves

$$\max u(x_t^t, x_{t+1}^t),$$

$$x_t^t, x_{t+1}^t \geqslant 0,$$

subject to (8.46c)

$$p_t x_t^t + p_{t+1} x_{t+1}^t \leqslant p_t \omega_t^t + p_{t+1} \omega_{t+1}^t.$$

Markets for commodities and capital stocks clear in every period $t = 1, 2, \ldots$:

$$x_t^{t-1} + x_t^t \leqslant y_t + \omega_t^{t-1} + \omega_t^t \perp p_t \geqslant 0, \tag{8.46d}$$

$$k_t \leqslant \tilde{k}_t \perp \psi_t \geqslant 0. \tag{8.46e}$$

Since there are no initial financial claims, S^1 will be zero in equilibrium, and so will be all S^t, as in the finite horizon model (see equation 7.11) and in the pure exchange OLG model. We now define:

DEFINITION 8.3 (OLG equilibrium with production and no financial claims) The allocation $x_t^{t*}, x_{t+1}^{t*}, y_t^*, k_t^*, \tilde{k}_{t+1}^*$ supported by the price vector $p_t^* \geqslant 0$, $\psi_t^* \geqslant 0$, $t = 1, 2, \ldots$, which are bounded for any finite t, is an OLG equilibrium with production and no financial claims if it solves (8.46a)–(8.46c) and markets clear as in (8.46d)–(8.46e).

8.3.2 Existence of Equilibrium

As for the pure exchange OLG model, the existence proof will use a T-period full format. To maintain Walras's law in the truncated T-period model, the consumer born in T will buy at the going market price the capital stock from the consumer who is old in T. This last consumer solves

$$\max V(x_T^T, \tilde{k}_{T+1}),$$

$$x_T^T \geqslant 0, k_T, \tilde{k}_{T+1} \geqslant 0,$$

$$\text{subject to} \tag{8.50}$$

$$p_T x_T^T + \psi_T k_T \leqslant p_T \omega_T^T + p_T y_T,$$

$$F(y_T, \tilde{k}_{T+1}, k_T) \leqslant 0,$$

where $V(.)$ can be any strictly concave increasing utility function.[64] This "consumer T problem" performs the same role as (8.24), but there are three differences: (1) Since there is no financial claim to pay to generation 0, x_T^T can be constrained to nonnegative values. (2) Since consumer T has no successor, he cannot sell the capital stocks to anyone; he only has a first period budget constraint (8.47) but has no possibility to borrow and no incentive to save, thus $S_T = 0$. (3) He has a subjective valuation for final capital stocks $\tilde{k}_{T+1}$.

We can now state and prove:

PROPOSITION 8.7 (Existence of equilibrium for the OLG model with production) Let assumptions CO1, CO2, CO3, and PD1 be satisfied. Then there exists an OLG equilibrium that satisfies definition 8.3, with nonzero consumption for all consumers.

Proof The proof consists of two parts. It follows the steps of proposition 8.5, but we skip those that are virtually identical. In part 1 we show that the truncated T-period model, with the last consumer behaving according to (8.50), has an equilibrium. In part 2 we establish that as T increases, the solution converges to the equilibrium of definition 8.3.

Part 1: Equilibrium in the truncated T-period model. We define the T-period full-format version of (8.46), with welfare weights α^t (the superscript denotes the generation), adding the consumer born in T:

$$\max \sum_{t=0}^{T-1} \alpha^t u(x_t^t, x_{t+1}^t) + \alpha^T V(x_T^T, \tilde{k}_{T+1}),$$

$$x_t^{t-1}, x_t^t \geqslant 0, k_t, \tilde{k}_{t+1} \geqslant 0, y_t, t = 1, 2, \ldots, T,$$

64. For approximations it will be desirable to specify $V(.)$ as the value function over the post-T horizon as in the dynastic model. We return to this in section 8.3.6.

subject to (8.51)

$$x_t^{t-1} + x_t^t \leqslant y_t + \omega_t^{t-1} + \omega_t^t \qquad (p_t),$$

$$k_t \leqslant \tilde{k}_t \qquad (\psi_t),$$

$$F(y_t, \tilde{k}_{t+1}, k_t) \leqslant 0,$$

$$\bar{p}_t x_t^t + \bar{p}_{t+1} x_{t+1}^t \leqslant \bar{p}_t \omega_t^t + \bar{p}_{t+1} \omega_{t+1}^t + \varepsilon, \quad t = 1, 2, \ldots, T-1 \quad (\rho^t),$$

$$\bar{p}_1 x_1^0 \leqslant \bar{\psi}_1 \tilde{k}_1 + \bar{p}_1 \omega_1^0 + \varepsilon \qquad (\rho^0),$$

$$\bar{p}_T x_T^T + \bar{\psi}_T k_T \leqslant \bar{p}_T \omega_T^T + \bar{p}_T y_T + \varepsilon \qquad (\rho^T),$$

where $x_0^0, \tilde{k}_1$ are given, the vectors $\alpha = (\alpha^0, \alpha^1, \ldots, \alpha^T)$ and the extended price vector $(\bar{p}, \bar{\psi}) = (\bar{p}_1, \bar{p}_2, \ldots, \bar{p}_T, \bar{\psi}_1, \bar{\psi}_2, \ldots, \bar{\psi}_T)$ lie on simplices S^{T+1} and $S^{(r+r')T}$ (r and r' are the number of commodities and capital stocks, respectively), and ε is a positive constant. (Note that only $\bar{\psi}_1$ and $\bar{\psi}_T$ enter as parameters; the other prices ψ_t will be needed later on.)

For every such $(\alpha, \bar{p}, \bar{\psi})$, program (8.51) is feasible, compact valued. By assumption PD1, parts 4 and 5, and since $\varepsilon > 0$, it satisfies Slater's constraint qualification. Therefore it will always have a solution, as well as Lagrange multipliers p_t, ψ_t, and ρ^t.

Step 1 Construction of the Fixed Point Correspondence

To obtain an equilibrium, we can define, like in the proof of proposition 8.5, a correspondence and adjustment functions. The correspondence generates Lagrange multipliers and the functions update the $T+1$ welfare weights and prices $(\bar{p}, \bar{\psi})$. We have thus defined a correspondence from $(\alpha, \bar{p}, \bar{\psi}, p, \psi, \rho)$ into itself, and as in proposition 8.5, a fixed point will exist.

Step 2 The Fixed Point Is an Equilibrium

The same arguments as in step 2 of proposition 8.5 can be used to prove that in the fixed point, welfare weights will be such that all consumers satisfy their budget constraints so that $\rho^t = 0$ and that $\bar{p}_t$, proportional to p_t, will be nonzero for every t. The solution coincides with that of a Negishi program and hence is a competitive equilibrium for the truncated T-period model.

Part 2: Extending T to infinity

a. *Reformulating the truncated model.* Like in step 2a of the proof of proposition 8.5, we define consumer-specific prices with consumer-specific normalizations for commodities

$$\bar{p}_t^t = \lambda^t \bar{p}_t \quad \text{and} \quad \bar{p}_{t+1}^t = \lambda^t \bar{p}_{t+1},$$

where λ^t is a positive scalar such that

$\Sigma_k(\bar{p}_{kt}^t + \bar{p}_{k,t+1}^t) = 1$ for generations $t = 1, 2, \ldots, T$,

$\bar{p}_1^0 = \lambda^0 \bar{p}_1$ for generation 0, where $\Sigma_k \bar{p}_{k1}^0 = 1$.

The normalizations for prices of capital stocks follow as

$\bar{\psi}_t^t = \lambda^t \bar{\psi}_t$ for $t = 1, 2, \ldots, T$,

$\bar{\psi}_1^0 = \lambda^0 \bar{\psi}_1$ for generation 0.

As we have seen, these normalizations are feasible because $\bar{p}_t$ is nonzero for every t. From part 1 follow equilibrium values for $\alpha^t, \bar{p}_t^t, \bar{p}_{t+1}^t, \bar{\psi}_1^0, \bar{\psi}_T^T$, and $x_T^T, \tilde{k}_{T+1}$. The associated allocations for periods 0 to $T-1$ may now also be obtained from the following program where we drop consumer T, while keeping his consumption x_T^T and his choice of $\tilde{k}_{T+1}$ as given parameters in the commodity balance and the transformation function for period t, respectively:

$$\max \sum_{t=0}^{T-1} \alpha^t u(x_t^t, x_{t+1}^t),$$

$$x_t^{t-1}, x_t^t \geqslant 0,\ k_t, \tilde{k}_{t+1} \geqslant 0,\ y_t,\ t = 1, 2, \ldots, T-1,\ x_T^{T-1} \geqslant 0,\ k_T \geqslant 0,\ y_T,$$

subject to (8.52)

$$x_t^{t-1} + x_t^t \leqslant y_t + \omega_t^{t-1} + \omega_t^t \qquad (p_t),$$

$$k_t \leqslant \tilde{k}_t \qquad (\psi_t),$$

$$F(y_t, \tilde{k}_{t+1}, k_t) \leqslant 0,$$

$$\bar{p}_t^t x_t^t + \bar{p}_{t+1}^t x_{t+1}^t \leqslant \bar{p}_t^t \omega_t^t + \bar{p}_{t+1}^t \omega_{t+1}^t, \quad t = 1, 2, \ldots, T-1 \quad (\rho^t),$$

$$\bar{p}_1^0 x_1^0 \leqslant \bar{\psi}_1^0 \tilde{k}_1 + \bar{p}_1^0 \omega_1^0 \qquad (\rho^0),$$

$$\bar{p}_T^T x_T^T + \bar{\psi}_T^T k_T \leqslant \bar{p}_T^T \omega_T^T + \bar{p}_T^T y_T \qquad (\rho^T).$$

If, for some finite T, the $x_T^T, \tilde{k}_{T+1}$ found as equilibrium values of the truncated model in part 1, happen to be equilibrium values of the infinite-horizon OLG model in definition 8.3, then for periods $t = 0, 1, \ldots, T-1$, the solution to (8.52) is also a solution of the infinite-horizon OLG equilibrium. To find such values, we proceed as follows.

In part 1 we have shown that we can find at least one pair $x_T^T, \tilde{k}_{T+1}$ for which a truncated T-period equilibrium exists, but there may be many such pairs and to each of them corresponds at least one solution $(\bar{p}_t^t, \bar{\psi}_t^t)$ for $t = 1, 2, \ldots, T-1$. In part 2b we show that as T increases, the set of such price trajectories shrinks and eventually yields the desired $x_T^T, \tilde{k}_{T+1}$ values.

b. *Convergence for consumer specific prices.* As in proposition 8.5 we can define a two-period price vector q^t for the consumer specific extended price vector $(\bar{p}_t^t, \bar{p}_{t+1}^t, \bar{\psi}_t^t)$ and associated to it a set of possible equilibrium sequences $Q(T)$, which can be shown to be nonempty and to converge for T going to infinity.

c. *Recovering prices.* We recover the prices p_t, using consumer 0's normalization as in part 2c of proposition 8.5. We also use this normalization to recover the ψ_t. If T is finite, the value of the initial capital, $\psi_1\tilde{k}_1$ will be finite and so will be all stock values at later dates. However, when the horizon is infinite, it must be shown that $\psi_1\tilde{k}_1$ is bounded because welfare (the weighted sum of utilities) is not necessarily bounded. Suppose that $\psi_1\tilde{k}_1$ is not bounded; then consumer 0 has infinite income. His prices are normalized to add up to unity (see above), and by assumption CO1, his utility function is nonsatiated (he leaves no bequest). Therefore he must have infinite demand, and this cannot be an equilibrium. Hence $\psi_1\tilde{k}_1$ is bounded.

Moreover, in equilibrium, y_1 is bounded from below, and therefore $p_1 y_1$ is bounded from below as well, since prices for period 1 add up to unity. By constant returns to scale, $\psi_2\tilde{k}_2 = \psi_1\tilde{k}_1 - p_1 y_1$. Hence $\psi_2\tilde{k}_2$ is bounded also. In the capital stocks balances in (8.52) for $t = 1$, ψ_1 appears as the shadow price and is therefore bounded (since constraint qualification holds). Because p_2 is bounded, this step can be repeated for $t = 2$, and so on, and thus for finite t, ψ_t is bounded. Hence for any finite t, p_t and ψ_t are bounded.

d. *Convergence of the sequence of consumer demands and of net supply by the firm.* For consumer demands, see part 2d of proposition 8.5. Supply by the firm in period t is then determined by demand, so demand and net supply converge.

We have now verified all the conditions of definition 8.3, and an OLG equilibrium with production exists. ■

As in the pure exchange model, prices (p_t, ψ_t) may become unbounded at infinity, but restrictions can be imposed on the transformation function of the firm, which avoid unboundedness. For example:

ASSUMPTION **PD1′** The number of capital stocks is equal to the number of commodities. The transformation function satisfies assumption PD1 and can be written[65]

$$F(y_t, k_{t+1}, k_t) = \max_h (y_{ht} + G_h(k_{t+1}) - k_{ht}),$$

where subscript h refers to a commodity.

This assumption merely amounts to treating capital stocks as ordinary commodities, assuming that a given capital stock can be used either for consumption (y_{ht}) or for expansion of future capacities. The assumption leads to the commodity balances

$$y_{ht} + G_h(k_{t+1}) \leqslant k_{ht}, \qquad h = 1, 2, \ldots, r.$$

65. See the derivation of the transformation function in section 2.1.3, where we have written the transformation function $T(y)$ as $\max_h g_h(y)$.

PROPOSITION 8.8 Under assumption PD1′ and if $\|\partial G(k_{t+1})/\partial k_{t+1}\| \leqslant \delta$, where δ is a positive constant smaller than unity and where $\| \ \|$ is any matrix norm,[66] then in the OLG model with production, commodity prices satisfy $\delta\|p_t\| \geqslant \|p_{t+1}\|$ for $t = 1, 2 \ldots$.

Proof Consider the following Negishi program:

$$\max \sum_{t=0}^{T-1} \alpha^t u(x_t^t, x_{t+1}^t),$$

$$x_t^{t-1}, x_t^t \geqslant 0, k_{t+1} \geqslant 0, t = 1, 2, \ldots, T-1, x_T^{T-1} \geqslant 0,$$

subject to

$$x_t^{t-1} + x_t^t + G(k_{t+1}) \leqslant k_t + \omega_t^{t-1} + \omega_t^t \qquad (p_t),$$

for given equilibrium values of α^t, x_T^T, and $\tilde{k}_{T+1}$, where we have substituted out y_t and $\tilde{k}_{t+1}$. The first-order condition with respect to k_{t+1} imply that

$$p_t G' \geqslant p_{t+1} \perp k_{t+1} \geqslant 0, \tag{8.53}$$

where $G' = \partial G(k_{t+1})/\partial k_{t+1}$. If $\|G'\| \leqslant \delta$, then by definition of the matrix norm, $\delta\|p_t\| \geqslant \|p_t G'\|$. By (8.53), $\|p_t G'\| \geqslant \|p_{t+1}\|$. Hence $\delta\|p_t\| \geqslant \|p_{t+1}\|$, and for T sufficiently large, this sequence can be extended indefinitely. Thus for the normalization $\|p_1\| = 1$, we have $\lim_{t\to\infty} p_t = 0$. ■

The condition $\|\partial G(k_{t+1})/\partial k_{t+1}\| \leqslant \delta$ means that the production technology allows all commodity supplies to grow at a positive rate. If this norm is equal to unity, then $\|p_t\| \geqslant \|p_{t+1}\|$, so stationary prices that will not go to infinity are possible. If the norm exceeds unity, prices will go to infinity unless, as can be seen from (8.53), $k_{h,t+1}$ becomes 0 for some h.

8.3.3 Inefficiency of Equilibrium

The discussion on inefficiency in section 8.2.3 can be extended to cover the model with production. In particular, it can be shown that allocations are efficient if prices go to zero. Therefore the allocation obtained in proposition 8.8 will be efficient.

If prices do not go to zero over time, the sufficient condition for inefficiency given for the pure exchange OLG model (bounded curvature of the consumer's expenditure function and $\lim_{T\to\infty} \Sigma_{t=t_0}^{T} 1/\|p_t\| < \infty$) also applies here.

If $\|\partial G(k_{t+1})/\partial k_{t+1}\| \leqslant \delta$, but $\delta = 1$ cannot be ruled out, prices cannot go to infinity, so the limit condition above cannot hold, and inefficiency cannot be established. However, since prices do not go to zero either, efficiency cannot be guaranteed.

66. For a definition, see the notation list in the front of the book.

8.3.4 Indeterminacy of Equilibrium

When efficiency can be established, it is possible to prove determinacy as well, but in the general case the discussion of section 8.2.4 applies.

8.3.5 Dynamics and Steady States

For the OLG model with production, the dynamics are described by the excess demand correspondence $z = a(p_{t-1}, p_t) + c(p_t, p_{t+1}) - y_t$. Here $a(p_{t-1}, p_t)$ and $c(p_t, p_{t+1})$ are the net demand by the old and the young, respectively, and y_t is the net supply, which may be set-valued. If differentiability[67] of the profit function is assumed, y_t is a function and it is set according to (8.12b):

$$y = \frac{\partial \Pi(p, \tilde{\psi}, k)}{\partial p},$$

$$k = \frac{\partial \Pi(p, \tilde{\psi}, k)}{\partial \tilde{\psi}}, \tag{8.54a}$$

$$\tilde{\psi} = \frac{\beta \partial \Pi(p, \tilde{\psi}, k)}{\partial k},$$

where we use the steady-state conditions $\tilde{k} = k$ and $\tilde{\psi} = \beta\psi$.

Commodity balances are

$$a\left(\frac{p}{\beta}, p\right) + c(p, \beta p) \leqslant y \perp p \geqslant 0. \tag{8.54b}$$

Equations (8.54a) and (8.54b) define the steady state of the OLG model with production and differentiable profit function. By homogeneity of $a(.)$, (8.54b) is equivalent to

$$a(p, \beta p) + c(p, \beta p) \leqslant y \perp p \geqslant 0, \tag{8.55}$$

and a steady state can be further characterized as follows: Premultiplying (8.55) by p yields

$$pa(p, \beta p) + pc(p, \beta p) = py. \tag{8.56}$$

The budget constraint of each consumer is

$$pc(p, \beta p) + \beta pa(p, \beta p) = 0. \tag{8.57}$$

Subtracting (8.57) from (8.56) gives

$$(1 - \beta)pa(p, \beta p) = py,$$

67. See section 8.1.5 for sufficient conditions.

and due to the constant returns to scale,

$$py = (1 - \beta)\psi k. \tag{8.58}$$

The previous equality then becomes

$$(1 - \beta)[pa(p, \beta p) - \psi k] = 0.$$

Hence $\beta \neq 1$ does not add any restriction, since in the model with production considered in this chapter, nominal claims are zero. Thus no distinction is needed between real and monetary steady states. Nonetheless, the constant returns condition (8.58) has important implications:

1. If $\beta < 1$, prices go to zero, and the steady state is efficient. Then from (8.58), we have $py \geqslant 0$. For $py = 0$, it follows that $\psi k = 0$, so there is no value in production and capital has no value. For $py > 0$, we have $\psi k > 0$.
2. If $\beta > 1$, then under the bounded curvature condition, the steady state is inefficient, since prices rise over time.[68] Now $py \leqslant 0$ and $\psi k \geqslant 0$. For $py = 0$, $\psi k = 0$ also. Capital has no value, but if $\psi k > 0$, capital has value, though production makes a loss. Capital is then merely a "device" to carry savings from one period to the next. The young buy it at a value ψk, and sell it when they are old at a value $\beta\psi k$, having invested py into it.[69]
3. Finally, if $\beta = 1$, then necessarily, $py = 0$; there is no value in production. As in the previous case, ψk may be positive but efficiency is not excluded.

The issues of computation and convergence to the steady state are as discussed in section 8.2.5. The convergence properties have been fully characterized for models with very few commodities.[70]

8.3.6 Finite-Horizon Approximations

In program (8.52) we have a representation of consumer T with utility function $V(x_T^T, \tilde{k}_{T+1})$ that follows an essentially arbitrary specification. This did not matter in the existence proof, where T is extended indefinitely. However, in practice, a finite T has to be specified. When, as in the dynastic model, there is a discounting property for welfare weights α^t so that beyond some year T_1 there exists a value $\gamma \in (0, 1)$ with $\alpha^t \leqslant (\gamma)^{t-1}$, then, as in proposition 8.4, bounds can be computed for the error made by approximating the infinite horizon by a finite one.

68. A Pareto-superior allocation can be constructed as in the proof of proposition 8.6.

69. In this model the firm's objective is to maximize the value of net supply plus final stocks $py + \beta\psi k$ at given availability of stocks. If the objective were $\Sigma_t(\beta)^{t-1}py$, implying that it only attributes value to stocks that serve production, then the condition $py < 0$ would lead to inaction and to a zero value for stocks. If prices would go to zero, the two objectives would be equivalent as there would be no value for stocks at infinity.

70. See Galor (1992).

When the condition $\alpha^t \leqslant (\gamma)^{t-1}$ holds (as for the OLG model with production of proposition 8.8), zero will be the lower bound, while the upper bound on the value function can be obtained as

$$V(x_T^T, \tilde{k}_{T+1}) = \max \sum_{t=T}^{\infty} (\gamma)^{t-1} u(x^t{}_t, x^t_{t+1}),$$

$$x_t^{t-1}, x_t^t \geqslant 0, k_t, \tilde{k}_{t+1} \geqslant 0, y_t, t = T+1, T+2, \ldots,$$

subject to (8.59)

$$x_t^{t-1} + x_t^t \leqslant y_t + \omega_t^{t-1} + \omega_t^t$$

$$F(y_t, \tilde{k}_{t+1}, k_t) \leqslant 0,$$

$$k_t \leqslant \tilde{k}_t,$$

where $x_T^T, \tilde{k}_{T+1}$ are given.

The value function of this program can be approximated, using dynamic programming arguments.[71] However, in the general case where welfare weights do not converge to zero, no such approximation can be obtained, since the welfare program will have an objective with infinite value.

8.4 Practices in Applied Modeling

8.4.1 Issues in Model Implementation

So far dynastic and OLG models have been presented separately. We now show that the dividing line is not so sharp and that intermediate formulations are possible. Moreover there are also many similarities from the computational point of view.

Choosing between Dynastic and OLG Models

Every consumer i in the dynastic model of definition 8.1, part 2, can be considered as consisting of a succession of consumers who live for one period only, and who care for the utility $u_{i,t+1}$ of their immediate descendant. In period t the utility function of this "single-period" consumer is $W_i(x_{it}, u_{i,t+1})$, and he solves

$$u_{it} = \max W_i(x_{it}, v_{i,t+1}),$$

$$x_{it} \geqslant 0, v_{i,t+1} \geqslant 0,$$

subject to

$$p_t x_{it} + \mu_{it} v_{i,t+1} = \psi_t \tilde{k}_{it},$$

71. See section 8.1.6.

where μ_{it} is the perceived cost of the utility level $v_{i,t+1}$ desired for the descendant,[72] ψ_t is the value of capital, and $\tilde{k}_{it}$ is the capital stock he "owns" at the beginning of period t and which he has inherited from his predecessor who lived in $t - 1$. It may be verified that

$$\psi_k \tilde{k}_{it} = \mu_{i,t-1} u_{it} \qquad \text{for } t = 2, 3, \ldots,$$

which implies that the consumer living in t inherits exactly the value that his predecessor chooses to bequest.

The generation that is "born" in t is thus fully dependent on the altruism of the generation born in $t - 1$, since it has no other source of revenue (e.g., from initial endowments) besides the capital it receives. If there are endowments ω_{it} (natural resources not available before t, e.g., sunshine or labor), the commodity balances and the budget equations become

$$\Sigma x_{it} + y_t \leqslant \Sigma_i \omega_{it},$$

$$p_t x_{it} + \mu_{it} v_{i,t+1} = \psi_t \tilde{k}_{it} + p_t \omega_{it}.$$

Thus in period t consumer i will always be able to spend $p_t \omega_{it}$. As long as $\mu_{it} v_{i,t+1}$ is positive (i.e., if the utility $v_{i,t+1}$ desired by the predecessor does not fall below the level realized by the successor), the predecessor determines the marginal utility of income of his successor and, if this holds for every consumer i and all periods t, we remain within the dynastic setting. Consequently indeterminacy and inefficiency of equilibrium are ruled out.

However, if $\mu_{it} v_{i,t+1}$ becomes zero for some (i, t) the marginal utility of income of consumer in period $t + 1$ is no longer determined by the predecessor. The previous dynasty ends, and a new one is created. Now the wealth constraint of former dynasty i has to be split (in a part before t and a part from t onward), and, as in any Negishi format, a welfare weight has to be associated to every dynasty so created. If this happens only a finite number of times, we can still cast the model in the dynastic framework (and the equilibrium will be determinate and efficient). If it occurs an infinite number of times, the model becomes similar to the OLG model of earlier sections, except that the young cannot buy the capital from the old, since they do not coexist with them and have no endowment to offer in exchange.

In the OLG models of sections 8.2 and 8.3, all dynasties i are split in every period t because all generations were assumed to be selfish. In this context, introducing altruism amounts to specifying the utility function of generation (i, t) as

$$u_i^t = W_i(x_{it}^t, x_{i,t+1}^t, v_i^{t+1}),$$

72. The price μ_{it} is derived as the shadow price of the constraint $v_{i,t+1} \leqslant u_{i,t+1} = W_i(x_{i,t+1}, v_{i,t+2})$, where $u_{i,t+1}$ is the utility level of the descendent. In an equilibrium of the dynastic model, $v_{i,t+1} = u_{i,t+1}$.

where x_{it}^t and $x_{i,t+1}^t$ are the consumptions of the members of "dynasty" i born in t and v_i^{t+1} is the utility of the successor. Generation (i, t) will solve

$$u_i^t = \max W_i(x_{it}^t, x_{i,t+1}^t, v_i^{t+1}),$$

$$x_{it}^t, x_{i,t+1}^t \geqslant 0, v_i^{t+1} \geqslant 0,$$

subject to

$$p_t x_{it}^t + p_{t+1} x_{i,t+1}^t + \mu_{i,t+1}^t v_i^{t+1} \leqslant p_t \omega_{it}^t + p_{t+1} \omega_{i,t+1}^t + \mu_{it}^{t-1} v_i^t.$$

When young, this generation receives a bequest $\mu_{it}^{t-1} v_i^t$, and when old, it gives $\mu_{i,t+1}^t v_i^{t+1}$.[73] Note that as in the OLG model with capital, capital stocks do not appear in the budget constraint. If $\mu_{i,t+1}^t v_i^{t+1}$ is zero for all (i, t) we are back in the standard OLG model. This is also the case if this term is zero for an infinite number of (i, t). However, if it is zero for a finite number of (i, t)—and hence positive for all periods beyond T_1—then the model can, as above, be cast in the dynastic framework. Thus inefficiency and indeterminacy are ruled out.

Note, finally, that if bequests are determined by the donor, independently of the utility level achieved by his descendent (e.g., according to some savings function), then the OLG model cannot be expressed in a dynastic setting.

There are thus four important differences between dynastic and OLG models: (1) in the dynastic model, generations do not overlap; (2) each generation is altruistic; (3) in the OLG model, generations have to buy capital but receive endowments freely, whereas in the dynastic model, consumers receive capital (and possibly, endowments) as an explicit gift; (4) in the dynastic model, the individual consumer can borrow from every other agent in the economy (although collectively, this is not possible as can be shown like in chapter 7, equation 7.11); in the OLG model, the young cannot borrow from the old, since they cannot repay them later.

Finite-Horizon Approximations

In practice, it is obviously impossible to simulate over a horizon that is truly infinite. All applied versions of infinite horizon models are therefore finite-horizon truncations that distinguish themselves from those of chapter 7, in two respects only. First, the reference run is usually a steady state. The policy exercise begins with a shock imposed on this steady state and proceeds computing the transition path to a new steady state. This new steady state is supposedly reached within finite time, but the length of the transition period is endogenous. By contrast, in finite-horizon models the

73. In equilibrium the μ_{it} ensure consistency in such a way that $v_i^t = u_i^t$, all i and t.

number of periods is given, and the trajectory is not required to end in a steady state. Second, since the time horizon has to be taken very long, dimensions often become excessively large unless the recursivity of the structure is exploited, but then problems of numerical error accumulation have to be tackled.

Thus the equations of applied infinite-horizon models are almost identical to their finite-horizon counterparts. The reference steady state is almost a static model and can be cast in the CGE format of section 3.4 and initialized through an idealized steady-state SAM. The discussion of sections 3.4, 3.6, 7.4.2, and 7.4.3 applies here, and we only need to be concerned with issues related to solving the model and running simulations. However, if the dimensions of the problem are small or if an approximation in the neighborhood of a steady state is deemed appropriate, one can use the methods discussed in section 8.1.7 in the context of real business cycle models.

8.4.2 Simulations

Solving the Model

Applied models are often written in an excess demand format with individual agents optimizing under perfect foresight. Distortions and exogenous variables are kept constant beyond a given year. A solution consists of prices and allocations such that all markets clear. Usually three solutions are computed: an initial steady state prior to some shift in exogenous variables, a steady state after this shift, and a transition path from one steady state to the other. The main purpose is to determine the transition path which is commonly obtained as follows: For given initial prices over the whole (infinite) period, one moves forward in time, accumulating capital and computing in each period the prices that clear the market of that period. Calculation proceeds over a T-period horizon using the steady-state prices as end valuation for the expectations beyond T. The computed prices serve as the "expected" prices for a new round of calculations.[74] It is concluded that a solution has been found when no further price adjustments are needed, and there is a convergence to a steady state toward the end of the simulation period.[75] This approach takes into account the recursive structure and appears to work well in practice, but a theoretically more formal

74. The calculation of the market-clearing prices for every period can make use of any formal algorithm, but there are two problems: First, there may be no such equilibrium at the prevailing expected prices, and second, one can only compute an approximate solution, so the adjustment rule for price expectations cannot rely on formal methods, since these require accurate function evaluations.

75. See, for example, Fair and Taylor (1983).

method is to iterate over all prices and compute excess demands until all markets clear for all t.[76]

The rationale for the approach is that if one succeeds in obtaining a path that converges to a steady state, one has a solution to the original model. Hence, while accepting that the solution may not converge to a steady state,[77] one only looks for convergent paths and is satisfied if one finds one, but there is no formal criterion to check whether the finite-horizon solution in fact lies on the infinite-horizon path.

In dynastic models it may happen that shifting the end date leads to different solutions, but due to discounting, this is not likely to affect early allocations too much. Most authors (e.g., Jorgenson and Wilcoxen 1993; Goulder 1991; Mercenier and Sampaio de Souza 1994; Bovenberg and Goulder 1991) suggest that indeed, truncation has little effect. For a more formal assessment of the quality of the approximation one may use measures based on the maximum post-T welfare (e.g., see Flam and Wets 1987; Keyzer 1991). Since serious unresolved problems remain in assessing whether a computed transition path is a good approximation of the infinite-horizon solution, there still is scope for using simpler tools, and in many applications this is also the practice. If perfect foresight and intertemporal efficiency are considered essential a finite-horizon model with some end-stock valuation may be appropriate. For example, Codsi et al. (1992) use a finite-horizon model with explicit terminal conditions, but without any requirement of convergence to a steady state. The model is linearly approximated, and linear methods are applied to solve for a finite-horizon equilibrium.

Though convergence to a steady state is not ensured for the OLG model either, it is often claimed to hold in applications. Kehoe (1991) shows how sensitive the outcomes in early periods may be for small variations in the neighborhood of the same steady state. He therefore suggests to linearize the model or to use a finite-horizon version. Alternatively, one could stop imposing budget constraints after T, assuming that transfers may be nonzero thereafter or introduce altruism as in the OLG model of section 8.4.1. This leads back to the dynastic model which, in the absence of distortions, will be determinate and efficient.

Running Simulations

Dynastic models have been used mainly to study investment in the energy and industry sectors, particularly in relation to environmental taxes and to capital taxation. These are typically areas where dynamics of gestation lags

76. See, for example, Codsi et al. (1992).

77. See section 8.1.5.

in investment and intertemporal efficiency of investment decisions matter. Indeed, in recursively dynamic simulations one can only run scenarios so that investment is either exogenous or specified according to a rule that is not derived from intertemporal optimization. In dynastic models intertemporal time-consistency is maintained.

For the reasons mentioned above, in applications, the line that separates finite from infinite-horizon models is thin. Dynastic infinite-horizon models have been developed for Brazil by Mercenier and Sampaio de Souza (1994) to study unemployment and by Goulder (1991) and Jorgenson and Wilcoxen (1993) to study energy policies and taxes on carbon dioxide emissions. One of the first applications of OLG models is the model by Auerbach and Kotlikoff (1987) that analyzes fiscal policy for the United States. It describes a single-commodity economy but distinguishes as many as fifty-five age cohorts, all with the same homothetic utility function. Because of this homotheticity, distributional changes do not affect aggregate demand and prices. As a result Auerbach and Kotlikoff can analytically solve many subproblems, such as the computation of a steady-state solution. In addition uniqueness of the steady state can be ensured. The simulation experiments compare transition paths from an initial reference steady state to a new one over a finite horizon. Our earlier remarks on convergence to a steady state and determinacy apply. Davies et al. (1989) have constructed a two-commodity version of the model and find that such disaggregation significantly affects the results. Finally, Ginsburgh and Sneessens (1992) also consider a two-sector version and study transition paths resulting from technological shocks or changing tastes.

9 Externalities

The competitive model of definition 1.4 is cast in terms of commodities that directly give utility to the consumer or enter as an input in the production process. It is a natural extension to consider individual utilities and technology sets as depending on, say, demand by other agents, who are thus the cause of externalities (or external effects). Such externalities are said to be positive if they are beneficial (section 9.1), but they can also be negative, for example in the case of pollution (section 9.2).

For each type of externality the discussion starts with a welfare program, which is subsequently decentralized through commodity and agent-specific signals (e.g., prices, taxes, transfers, and rations). These are provided either by markets or by institutional arrangements when markets cannot perform this function. Since these signals will be derived from a welfare optimum, the resulting equilibrium is necessarily Pareto-efficient and describes first-best policies. The equilibrium is then said to provide signals that internalize the externalities within the decisions of the individual agent.

However, in practice, it is difficult to generate such signals, and individuals may have the power to influence them or try to evade them. Therefore the efficient solution is rarely a realistic option. Policies will at best be inspired by theory, but they will rarely be able to achieve efficiency. It is the role of applied general equilibrium models to study and compare the implications of rules that are more easily implemented (section 9.3).

9.1 Positive Externalities

In the model of chapter 5, public consumption is set by a separate agent (the government) without any welfare justification. This consumption is financed through taxes imposed on consumers via a tax distribution function. In section 9.1.1 consumers will decide by themselves on the volume of public consumption and on the level of their contribution to finance it. The equilibrium solution considered in the second welfare theorem involves transfers among consumers. Selfish consumers are thus forced to accept a redistribution of income. In section 9.1.2 we abandon the selfishness hypothesis and allow agents to decide on the income transfers they want to give to others. Indirect taxes and subsidies, which were also set exogenously in chapter 5, will in section 9.1.3 result from externalities in consumption. Most of the discussion relates to consumption, but externalities can also arise in production when firms share common inputs; this is covered in section 9.1.4. Finally, in section 9.1.5 endowments are made dependent on consumption. This enables us to study the so-called efficiency wage argument.

9.1.1 Nonrival Commodities in Consumption

In section 5.1.1 public consumption is decided upon by a government that is treated like a consumer, maximizing its own utility under a budget constraint,[1] although it has the capacity to redistribute income among consumers. This consumption does not provide utility to consumers and should therefore be equal to zero in any Pareto-efficient allocation. Here we will assume that public consumption directly affects utility of consumers, and we specify a model that does lead to a Pareto-efficient allocation, distinguishing between rival and nonrival public commodities.

Rival public commodities, such as health care or education, only differ from other commodities because they are provided by the government, possibly at a subsidized price. Consumer i's utility function then can be written as $u_i(x_i, x_{ig})$, where x_{ig} is consumer i's consumption of these commodities. These consumptions add up to x_g, a quantity supplied by government production activities or purchased on the market. Alternatively, the provision of these commodities can be treated without government intermediation, except possibly for subsidies. The fact that the rival commodity is called public is not due to its physical characteristics. It either reflects an institutional setting (public education) or terminological practice (a highly subsidized consumption is often called public).

We will focus on nonrival commodities and consider a public consumption x_{gk} of commodity k that provides utility to all consumers at the same time. This is a physical characteristic of the commodity that distinguishes it from usual consumer goods appearing in x_i. Such a commodity is said to be nonrival, since, when one agent consumes more of it, this does not go at the expense of any other agent. Each agent can consume the full supply.[2] Although we have, for notational convenience, assumed that both vectors x_i and x_g refer to the same commodities, they will usually not have positive entries for the same k.

It may be noted that the distinction between rival and nonrival commodities is more of a conceptual than of an operational nature. Public services will always have an element of both. A service to a particular consumer, like issuing a passport, is rival. A TV broadcast is for the public at large and thus nonrival. A painting in a museum possesses both characteristics. It renders a private service for those who visit the museum; it is a nonrival good for all those who feel pride in having it in their museum. These examples should make it clear that nonrival goods are not necessarily shared by all but may be restricted to smaller circles (club goods). They are not necessarily supplied by any public authority or publicly owned firm. In the model there is no need for (and there will be no) public firms, and

1. In this chapter we will not consider government as a consumer. Government will not own any endowments or shares in profits; it will merely act as a planner who computes signals.

2. This notion was introduced by Lindahl in the early 1920s and formalized by Samuelson (1954).

nonrival commodities are supplied by private firms or are available from consumer endowments.

Since the utility function now has nonrival commodities as arguments, its properties have to be redefined:

ASSUMPTION **CG1** (Utility functions with nonrival commodities) The utility function u_i: $R^r_+ \times R^r_+ \to R_+$, $u_i(x_i, x_g)$ is continuous, concave, and satisfies $u_i(0, x_g) = 0$ for all i. For $i = 1$, it is strictly concave, increasing in (x_i, x_g), and for $i = 2, 3, \ldots, m$, it is strictly concave, nonsatiated in x_i, and nondecreasing in x_g.

Unlike in assumption C1′, we do not require the utility function to be strictly concave in all arguments for $i \neq 1$ because, except for $i = 1$, we do not want consumers to use the same commodity both in a rival and a nonrival way. The requirement $u_i(0, x_g) = 0$ means that the consumer has no utility when consuming nonrival goods only, and it already implies that his utility cannot be strictly concave in all arguments. We also allow all consumers except $i = 1$ to be satiated in x_g.

Clearly, for exogenously given x_g, consumer i can decide as before on x_i, given an income that may have been taxed to pay for x_g. Since it is the same x_g that enters all utility functions, a consumer cannot decide on his own about the provision of x_g, and some coordination mechanism is needed. In line with the approach of proposition 1.9 (a welfare optimum is an equilibrium with transfers) and proposition 3.1 (welfare weights can be found such that no transfers are needed), we proceed by (1) setting the level of x_g "optimally" through a welfare program with given welfare weights, (2) decentralizing the decisions to individuals through adequate "price" signals, and (3) adjusting the welfare weights to obtain an equilibrium without transfers in which the nonrival goods are financed.

The Welfare Problem

The optimal level of x_g can be set by solving the following welfare problem for given α_i:

$$\max \Sigma_i \alpha_i u_i(x_i, x_g),$$

$$x_i \geqslant 0, \text{ all } i, \ x_g \geqslant 0, \ y_j \text{ all } j,$$

$$\text{subject to} \qquad (9.1)$$

$$\Sigma_i x_i + x_g - \Sigma_j y_j \leqslant \Sigma_i \omega_i \qquad (p),$$

$$y_j \in Y_j.$$

Note that all consumers have to consume the same x_g, though they might be satiated in some nonrival good. A consumer may, for example, be willing to settle for less national security, yet he will be faced with the level that is actually supplied. Nonetheless, this will not decrease his utility, since his utility function is nondecreasing.

Therefore the allocations will not be different if we distinguish consumer specific nonrival consumptions $g_i \leqslant x_g$. This will also make it possible to obtain explicit Lagrange mutipliers ϕ_i for the value that each consumer attributes to x_g. We rewrite (9.1) as

$$\max \Sigma_i \alpha_i u_i(x_i, g_i)$$

$$x_i, g_i \geqslant 0, \text{ all } i, x_g \geqslant 0, y_j, \text{ all } j,$$

$$\text{subject to} \tag{9.2}$$

$$\Sigma_i x_i + x_g - \Sigma_j y_j \leqslant \Sigma_i \omega_i \qquad (p),$$

$$g_i \leqslant x_g \qquad (\phi_i),$$

$$y_j \in Y_j.$$

"Price" Signals for Nonrival Goods under Fixed Welfare Weights

To simplify the presentation, suppose that $u_i(x_i, x_g)$ is differentiable.[3] This leads to the following first-order conditions of (9.2):

$$\alpha_i \frac{\partial u_i}{\partial g_{ik}} = \phi_{ik} \qquad \text{if } g_{ik} > 0. \tag{9.3}$$

If consumer i has to pay the price ϕ_{ik} and has marginal utility of income $\lambda_i = 1/\alpha_i$, then (9.3) is also a first-order condition of the following consumer problem:

$$\max u_i(x_i, g_i),$$

$$x_i, g_i \geqslant 0,$$

$$\text{subject to} \tag{9.4}$$

$$px_i + \Sigma_k \phi_{ik} g_{ik} = h_i \qquad (\lambda_i),$$

for some given income h_i which may include lump-sum transfers. Moreover at these prices the value $g_i = x_g$ is optimal in (9.4).

The first-order conditions of (9.2) also imply that

$$\Sigma_i \phi_{ik} = p_k \qquad \text{if } x_{gk} > 0, \tag{9.5}$$

so if consumer i contributes ϕ_{ik}, often called consumer i's willingness to pay

3. Differentiability of the utility function will not be used to prove existence of equilibrium.

for the use of nonrival good k, then p_k, its value on the market, will be covered by the individual consumers' contributions.

Note that if the collective of consumers with utility function $\Sigma_i \alpha_i u_i(x_i, x_g)$ is considered a single consumer with budget constraint $p\Sigma_i x_i + px_g = \Sigma_i h_i$, then the demand for the nonrival goods is in no way different from the demand for a rival good. Thus decentralization between this collective and individual producers is possible, and yields an efficient equilibrium. Below we will discuss the case in which several collectives (e.g., villages or clubs) can be distinguished.

Price Signals and Financing for Nonrival Goods: The Lindahl Equilibrium

We now look for an equilibrium without transfers, by changing the welfare weights like in proposition 3.1 so that each consumer will satisfy his budget constraint:

$$px_i + \phi_i g_i = p\omega_i + \Sigma_i \theta_{ij} \Pi_j(p).$$

The resulting allocation is a Lindahl equilibrium, defined as[4]

DEFINITION 9.1 (Lindahl equilibrium) The allocation y_j^*, all j, x_i^*, g_i^*, all i, x_g^*, supported by the price vectors ϕ_i^*, $p^* \geqslant 0$, $p^* \neq 0$, is a Lindahl equilibrium if the following conditions are satisfied:

1. For every producer j, y_j^* solves $\max_{y_j}\{p^* y_j$ subject to $y_j \in Y_j\}$; Π_j^* denotes the optimal profit.

2. For every consumer i (x_i^*, g_i^*) solves $\max_{x_i, g_i \geq 0}\{u_i(x_i, g_i)$ subject to $p^* x_i + \phi_i^* g_i \leqslant h_i^*$, $x_i, g_i \geqslant 0\}$, where $h_i^* = p^*\omega_i + \Sigma_j \theta_{ij} \Pi_j^*$.

3. All markets are in equilibrium:

$$\Sigma_i x_i^* + x_g^* - \Sigma_j y_j^* - \Sigma_i \omega_i \leqslant 0 \perp p^* \geqslant 0.$$

4. Consumers agree on the provision of nonrival commodities and jointly finance them:

$$g_i^* = x_g^*,$$

$$\Sigma_i \phi_i^* \leqslant p^* \perp x_g^* \geqslant 0.$$

We state and prove:

PROPOSITION 9.1 (Existence of a Lindahl equilibrium) Let assumptions P1′, CG1, and C2′ be satisfied. Then there exists a Lindahl equilibrium as specified in definition 9.1.

4. In the definition we write $g_i^* = x_g^*$, while in program (9.2) we have an inequality. We will, however, see that the two formulations are equivalent.

Proof We note that program (9.2) is a welfare program that possesses the properties specified in proposition 2.14, except that in the price vector $(p, \phi_1, \phi_2, \ldots, \phi_m)$, the pair (p, ϕ_1) is positive whenever α_1 is positive and expenditures include the nonrival good.

The proof of proposition 3.1 can be followed, and there exist welfare weights such that each consumer is on his budget. We have seen that this implies that each consumer should solve condition 2 of definition 9.1. Profit maximization by producers and market equilibrium (conditions 1 and 3 of definition 9.1) follow as in proposition 1.9.

We now verify condition 4 of definition 9.1. Suppose that in equilibrium $g_{ik}^* < x_{gk}^*$ for some i and k. This implies that $\phi_{ik}^* = 0$. Hence increasing the consumption of the nonrival good until x_{gk}^* will not affect consumer i's budget, and given the concavity of utility and the nondecreasingness in g_{ik}, we can raise g_{ik} indefinitely without decreasing utility. Therefore $g_{ik} = x_{gk}^*$ is also an equilibrium. ■

What makes the proof almost identical to the proof of proposition 3.1 is that nonrival goods simply add linear constraints. These are similar to the balance constraints (but without free disposal), and they reduce the nonrival property to a technological restriction.

Interpreting the Lindahl Equilibrium

Though the proof is straightforward and illustrative of the similarity between the mathematical structure of the Lindahl and the competitive equilibrium, the consumer-specific prices for the nonrival commodities raise some issues in interpretation.

First, we consider the distribution of the financial burden imposed by the nonrival commodity under decentralization. Consumer i will be faced with a share $\sigma_{ik} = \phi_{ik}/p_k$ in the cost of commodity k ($\Sigma_i \sigma_{ik} = 1$ whenever $x_{gk} > 0$). This does not only provide him with the information on his own contribution but also on how much more he would have to pay for an additional unit, could he affect the decision to provide it. Nonusers and satiated users do not pay. If case $u_i(\,.\,)$ is differentiable, it follows from (9.3) that

$$\sigma_{ik} = \frac{\phi_{ik}}{p_k} = \frac{\alpha_i}{p_k} \frac{\partial u_i}{\partial g_{ik}},$$

which shows that σ_{ik} falls with marginal utility of income ($\lambda_i = 1/\alpha_i$) (the rich pay more), rises as marginal utility of the nonrival good goes up (those who like the nonrival good more pay more), and falls as the good becomes more expensive.

A nonrival good that is not desired by high-income groups will have to be financed by low-income groups, who may lack the means to secure the commodity. The Lindahl equilibrium fails to ensure equity and is, in this respect, not different from the competitive equilibrium. However, the second welfare theorem carries through: To ensure a more equitable outcome, transfers can be used and should be the only redistributive instrument. Tampering with the distribution of costs σ_{ik} is just as distortive as manipulating prices in the competitive equilibrium.

Though we have not discussed how prices are arrived at in the competitive equilibrium, at least some intuition exists about the auctioneer who sets prices to clear markets.[5] Mechanisms to obtain an optimal[6] Lindahl σ have been the subject of an extensive literature in public economics.[7] The main issue is to have consumers reveal truthfully how much they would pay for an additional unit of the nonrival good. It is in each consumer's interest to underreport his willingness to pay (in the limit he does not contribute anything and becomes a free-rider[8]), but this would lead to an inadequate supply. If consumers can anticipate this, they will offer to contribute more. Hence, in the same way as there can be an excess demand for rival goods, eliminated via price adjustments, for nonrival goods, there is an excess price, $\Sigma_i \phi_{ik} - p_k$ eliminated through adjustments in x_{gk}. Thus the intuition of tatonnement applies and has been invoked in theoretical work.[9] However, as we have seen in sections 1.2.4 and 1.2.5, in general equilibrium the convergence of such a procedure is not guaranteed due to the lack of generic properties of excess demand.

Note also that every consumer must know that he has an impact σ_i on the supply of the nonrival commodities, and therefore, however limited his size, his influence must be accounted for. This is not the case for prices of competitive commodities, where neglecting the influence a consumer has on prices does not hamper achievement of Pareto-efficiency.

The consumer must not only be informed on his own contribution; he must also be assured that others will pay their share. If he feels that he has to support more than his Lindahl share, he will restrict his demand. Such a situation is known as the assurance problem. We return to nonefficient allocations in section 9.3.

Finally, once the level of demand for nonrival commodities has been set optimally, and the cost-sharing rules specified, the contribution can be levied as a direct tax, leaving consumer i with the standard decision to choose his consumption according to

5. See section 1.2.5.

6. Note that if utility functions are not differentiable, the equilibrium σ may not be unique.

7. See Laffont (1987) for a survey and references.

8. In the budget constraint of the consumer, it may seem that the quantity x_g of the nonrival commodity is given and that the consumer can decide on the price ϕ_i he is willing to pay. In reality this is not the case. The consumer is never willing to pay anything, for he takes the x_g for granted and free-rides.

9. See, for example, Drèze and de la Vallée-Poussin (1971) and Malinvaud (1971).

$\max u_i(x_i, x_g)$,

$x_i \geqslant 0$,

subject to

$px_i = h_i$,

where $h_i = p\omega_i + \Sigma_j \theta_{ij} \Pi_j(p) - \phi_i x_g$ and p, ϕ_i, and x_g are given.

9.1.2 Interdependent Utilities: Empathy

Up to this point, agents had to agree on a collective choice, which naturally called for an outside institution to eventually set the equilibrium level of the nonrival commodity. Here we consider the more general case in which the choices of agents influence other agents directly, and not only via markets. For example, a consumer may dislike seeing another consumer in poor condition and consider helping him (empathy).

Empathy offers a way to introduce moral issues into economic models, and there is a growing interest in this topic.[10] Consumer i has empathy for other consumers when their utilities affect his: He cares for some of them. Then his utility function (which can be thought of as his "social welfare" function) may be written as $W_i(x_i, v_i)$, where $v_i = (v_{i1}, v_{i2}, \ldots, v_{im})$ is the vector of utilities of all consumers as perceived by him.[11]

Consumer i's utility function has the following properties:

ASSUMPTION CW1 (Utility function with empathy)

1. The utility function $W_i\colon R^r_+ \times R^m_+ \to R_+$, $W_i(x_i, v_i)$ is continuously differentiable, concave, strictly concave in x_i, nondecreasing in v_i, and it satisfies $W_i(0, v_i) = 0$ for all i. For $i = 1$ it is increasing in x_i; for $i = 2, 3, \ldots, m$, it is nonsatiated in x_i.
2. Limited altruism: $\lim_{\|u\| \to \infty} \max_i (u_i - W_i(x_i, u)) > 0$.

In assumption CW1, part 1 adapts assumption CG1 to the case with empathy. Nondecreasingness in v_i is important because it rules out envy and more generally any negative feelings with respect to other consumers. Part 2 ensures that no consumer i will be overaltruistic. Otherwise, for the sake of the others, no one will want to consume, and utilities could go to infinity at fixed consumption.[12]

10. See Hausman and McPherson (1993) for a survey.

11. For notational convenience v_{ii} is an element of v_i, but, by construction, we require $\partial W_i / \partial v_{ii} = 0$ so that v_{ii} does not affect the level of W_i. Alternatively, we could have written $W_i(x_i, v_{-i})$, where v_{-i} is the vector of utilities of all consumers but i.

12. A recursive structure in which consumer 1 cares for 2, etc. (but not vice versa), would also do. See, for example, Hirschleifer (1985).

We now follow the same steps as in section 9.1.1 and consider a welfare problem at given welfare weights, which we subsequently decentralize over consumers, adjusting the welfare weights.

The Welfare Problem

Consider the following welfare problem:

$$\begin{aligned}
&\max \Sigma_i \alpha_i u_i,\\
&u_i \geqslant 0,\ x_i \geqslant 0, \text{ all } i,\ v_{ih} \geqslant 0, \text{ all } i \text{ and } h,\ y_j \text{ all } j,\\
&\text{subject to} \qquad\qquad\qquad\qquad\qquad\qquad\qquad\qquad (9.6)\\
&u_i \leqslant W_i(x_i, (v_{i1}, \ldots, v_{im})) \qquad (\rho_i),\\
&v_{ih} \leqslant u_h \qquad (\mu_{ih}),\\
&\Sigma_i x_i - \Sigma_j y_j \leqslant \Sigma_i \omega_i \qquad (p),\\
&y_j \in Y_j.
\end{aligned}$$

In this program, v_{ih} is the utility of consumer h as perceived by consumer i. The $v_{ih} \leqslant u_h$ constraints indicate that perceived utility cannot exceed realized utility. The constraint set is nonempty, since it is possible to set all variables to zero. Clearly all x_i and y_j are bounded, and part 2 of assumption CW1 ensures that the u_i will be bounded as well. Assumption C2′ ($\Sigma_i \omega_i > 0$) together with nonsatiation guarantees that Slater's constraint qualification holds, since we can allocate consumption to all i so that they have positive W_i while v_i is set at zero. Program (9.6) now has an optimal solution as well as bounded Lagrange multipliers for any $\alpha \in S^r$. Suppose that in the optimum $v_{ih} < u_h$ for some i, h. Then it is possible to increase v_{ih} without creating infeasibilities or changing the objective (by concavity, utility cannot increase once it has become stationary). By CW1, part 2, u_h is bounded; hence we can reach u_h and assume that equality holds in the optimum. Note that a consumer with a zero welfare weight may now have positive expenditure due to the value that other consumers with positive welfare weights derive from his utility.

"Price" Signals under Fixed Welfare Weights

Since $W_i(x_i, v_i)$ is differentiable, the following first-order conditions of (9.6) hold:

$$\mu_{ih} = \frac{\rho_i \partial W_i}{\partial v_{ih}} \qquad \text{if } v_{ih} > 0, \tag{9.7}$$

where μ_{ih} is the marginal value attributed by consumer i to incremental utility of consumer h. First-order conditions with respect to u_i imply that

$$\rho_i = \alpha_i + \Sigma_h \mu_{hi} \qquad \text{if } u_i > 0,$$

and ρ_i can be interpreted as the effective welfare weight of consumer i, which may be positive even if $\alpha_i = 0$ as long as some other consumers with positive ρ care for him. This ρ_i is also the inverse of the marginal utility of income ($\rho_i = 1/\lambda_i$) in the following consumer problem:

$$\max W_i(x_i, (v_{i1}, \ldots, v_{im})),$$

$$x_i \geqslant 0,\ v_{ih} \geqslant 0, \text{ all } h,$$

subject to

$$px_i + \Sigma_h \mu_{ih} v_{ih} = h_i \qquad (\lambda_i),$$

where p, μ_{ih}, and income h_i are given.[13] The value $\mu_{ih} v_h$ is the amount consumer i voluntarily transfers to consumer h.

Price Signals and Financing: The Equilibrium with Empathy

Now consider changing the welfare weights until each consumer is on his budget constraint, where the only transfers are those that are voluntary:[14]

$$px_i + \Sigma_h \mu_{ih} v_{ih} = p\omega_i + \Sigma_i \theta_{ij} \Pi_j(p) + \Sigma_h \mu_{hi} v_{hi}.$$

Prices p are obtained from (9.6), and "prices" μ_{ih} by evaluating (9.7) at the optimum of (9.6). Consumer i freely decides to give $\Sigma_h \mu_{ih} v_{ih}$, while he receives $\Sigma_h \mu_{hi} v_{hi}$ from other consumers. This is an equilibrium with empathy.

DEFINITION 9.2 (Equilibrium with empathy) The allocation y_j^*, all j, x_i^*, u_i^*, all i, v_{ih}^*, all i and h, supported by the "prices" $p^* \geqslant 0$, $\mu_{ih}^* \geqslant 0$, all i and h, is an equilibrium with empathy if:

1. for every producer j, y_j^* solves $\max_{y_j}\{p^* y_j$ subject to $y_j \in Y_j\}$; Π_j^* is his optimal profit;

2. for every consumer i, (x_i^*, v_i^*) solves $\max_{x_i \geqslant 0, v_i \geqslant 0}\{W_i(x_i, v_i)$ subject to $p^* x_i + \Sigma_h \mu_{ih}^* v_{ih} \leqslant h_i^*\}$, where $h_i^* = p^*\omega_i + \Sigma_j \theta_{ij} \Pi_j^* + \Sigma_h \mu_{hi}^* v_{hi}^*$;

3. consumers agree on utilities

$$v_{ih}^* = u_h^*;$$

4. all markets are in equilibrium,

$$\Sigma_i x_i^* - \Sigma_j y_j^* - \Sigma_i \omega_i \leqslant 0 \perp p^* \geqslant 0.$$

13. Note that if $\mu_{ih} = 0$, the problem may look unbounded though this is not the case, since at that point $\partial W_i / \partial v_{ih} = 0$.

14. Out of equilibrium, transfers other than those that are voluntary may be needed.

We state and prove:

PROPOSITION 9.2 (Existence of an equilibrium with empathy) Let assumptions P1′, CW1, and C2′ be satisfied. Then there exists an equilibrium with empathy as stated in definition 9.2.

Proof We proceed as we did for the Negishi format of proposition 3.1, but we first need to prove that as in part 2 of proposition 2.14, optimal consumption is a continuous function of α whenever $\alpha_1 > 0$ or $\alpha_i > 0$. For this we prove that $v_{ih} = u_i$ is uniquely determined for all i. We distinguish two cases. Consider those consumers for whom $x_i \cdot \partial W_i / \partial x_i = 0$ in the optimum. For these $x_i = 0$ (because α_1 or $\alpha_i = 0$), and hence $v_{ih} = 0$ and is unique. For the other consumers, consider the vector function $W(x, u) = (W_1(x_1, u), \ldots, W_m(x_m, u))$, where u is the vector of realized utilities. From concavity and $W(0, 0) = 0$, it follows that

$$u \geqslant x \cdot \frac{\partial W}{\partial x} + u \cdot \frac{\partial W}{\partial u}.$$

Since, for these consumers, $x \cdot \partial W / \partial x > 0$, we have

$$u > u \cdot \frac{\partial W}{\partial u}.$$

Since $\partial W / \partial u \geqslant 0$ and y is nonnegative, this implies that $(I - \partial W / \partial u)$ is nonsingular, and by the implicit function theorem, the system $u = W(x, u)$ has a unique solution. For the remainder, the proof of proposition 3.1 carries through. ∎

Interpreting the Empathy Equilibrium

It is obvious that the sum of transfers received is equal to the sum of transfers given, so the financing of the voluntary scheme is ensured.

As in the competitive and Lindahl equilibria, equity considerations other than those felt by the consumers (and that are reflected in their utility function) are not taken into account. For example, when two consumers have identical initial endowments ω_i and profit shares θ_{ij}, the first may end up receiving transfers without giving (if others like him and he likes no one), while the other may give almost all his income away without receiving anything. This may seem inequitable, but it is in accordance with preferences. In this model altruism is treated purely as a preference, not as a moral duty. Transfers between consumers are decided endogenously and should not be imposed because they could impair on the efficiency properties of the equilibrium.

The assumption of nondecreasingness of W_i with respect to v_i (in assumption CW1) allows one to decentralize the optimum via voluntary transfers.

If decreasingness (envy) is allowed for, imposed transfers may be necessary for decentralization. Conversely, restricting actions to voluntary transfers protects consumers against negative feelings by others, since no one can take commodities away from someone he happens to dislike.

In section 8.4 we considered empathy in the dynastic and the overlapping generations models, but it was of a special, recursive nature: The parent cared about his immediate descendant only, and the descendant did not care about his predecessor's well-being. In the model of definition 9.2 a general structure of relations is allowed for.

Consumers can organize themselves to provide mutual assistance via other schemes than the one of proposition 9.2. For example, individual charity that takes the income of the recipient as given and supplements it, without taking into consideration that others may also give, will lead to overestimated μ_i and therefore to "costly" transfers. If the consumer knows that others will participate, the marginal utility of his gift will be increased.[15] This is the counterpart of the assurance problem for nonrival goods. The free-rider problem exists here too, when the μ_i deviate from their equilibrium values. An agent who feels empathy may still not be willing to give, in the belief that others will take care of the problem, and perhaps that his contribution would be too small to make any difference. Truthful revelation is another problem, since potential recipients will tend to exagerate their poverty as well as the positive effect of contributions. Moreover they may become dependent on aid, and thus anticipate that if they earn more, they will receive less aid.[16]

Finally, note that in the model with empathy, utility comparisons are made by each agent separately, and the approach does not conflict with the idea that utilities cannot be compared across individuals. Yet it may be difficult for each agent to measure other consumers' utilities. Therefore he may resort to simply observing the consumption of other consumers and take this as arguments in his utility function. This leads to the model with interdependent consumptions.

9.1.3 Interdependent Consumptions

Agents do not only interact through altruism. Their consumption choices can directly affect others. A consumer may like to see his neighbor's garden and house well kept, and be willing to share the cost or to decide on maintenance jointly with his neighbor. In this case a "local" institution is created that manages the garden as a local nonrival good. If consumer i's utility depends on other agents' consumption, one speaks of externalities in consumption.

15. See Arrow (1981).

16. Then the marginal adjustment in aid will have the effect of an indirect tax.

Under the empathy approach consumer i was sensitive to consumer k's utility level as measured by k. Here consumer i measures k's utility according to his own criteria and assumes that k does the same. This leads to endowing consumer i with the utility function $W_i(x_i, v_i(x))$, where $x=(x_1, x_2, \ldots, x_m)$.[17] The formulation can be given either of the following interpretations: (1) It is a simplified way to measure other consumers' utilities, through some standard of living index; (2) $v_i(x)$ reflects what i thinks would be good for other consumers, despite knowing that this may not reflect their actual preferences. For example, a parent may want to promote his child's health and education. This is often referred to as paternalism.[18]

A more general formulation considers $W_i(x_i, \xi_i)$, where ξ_i is consumer i's perception of x. This allows two further interpretations: (3) $W_i(x_i, \xi_i)$ is a reduced form of the empathy formulation; (4) it represents the fact that consumer i directly draws utility from the consumption of his fellow consumers. He may like houses in his neighborhood to be well kept.

We make the following assumptions on the utility function:

ASSUMPTION CX1 (Externalities in the utility function) The utility function W_i: $R^r_+ \times R^{rm}_+ \to R_+$, $W_i(x_i, \xi_i)$ is continuously differentiable, concave, strictly concave in x_i, nondecreasing in ξ_i, and satisfies $W_i(0, \xi_i) = 0$ for all i. For $i = 1$, it is increasing in x_i; for $i = 2, 3, \ldots, m$, it is nonsatiated in x_i.

We again consider the following centralized welfare optimum:

$$\max \Sigma_i \alpha_i W_i(x_i, \xi_i),$$

$$x_i \geqslant 0, \text{ all } i, \ \xi_{ih} \geqslant 0, \text{ all } i \text{ and } h, \ y_j \text{ all } j,$$

$$\text{subject to} \tag{9.8}$$

$$\Sigma_i x_i - \Sigma_j y_j \leqslant \Sigma_i \omega_i \qquad (p),$$

$$\xi_{ih} = x_h \qquad (\phi_{ih}),$$

$$y_j \in Y_j,$$

and show how it can be decentralized.

As in program (9.6) we can calculate for every i,

$$\phi_{ih,k} = \frac{\alpha_i \partial W_i}{\partial \xi_{hk}}, \qquad h = 1, 2, \ldots, m, \ k = 1, 2, \ldots, r, \tag{9.9}$$

which, by assumption CX1, is nonnegative; $\phi_{ih,k}$ can be interpreted as the subsidy consumer i is willing to pay to h for each unit of commodity k

17. Consumer i's consumption x_i is included in the vector x for notational simplicity, but it will not contribute to utility.

18. See, for example, Archibald and Donaldson (1976) and Hirschleifer (1985).

consumed by h. This enables us to define for each consumer the following utility maximization problem:

$$\max W_i(x_i, \xi_i),$$

$$x_i \geqslant 0, \xi_i \geqslant 0,$$

subject to

$$\Sigma_k(p_k - \Sigma_h\phi_{hi,k})x_{ik} + \Sigma_k\Sigma_h\phi_{ih,k}\xi_{ih,k} \leqslant h_i,$$

where $p_k, \phi_{ih,k}$ and the income before transfers h_i are given. The value $\Sigma_k\Sigma_h\phi_{ih,k}\xi_{ih,k}$ is the amount consumer i voluntarily transfers to other consumers to promote their consumption, while $\Sigma_h\phi_{hi,k}$ is the subsidy he receives from other consumers per unit of consumption of commodity k. Consumer i now faces "personalized" prices $p_i = p - \Sigma_h\phi_{hi}$.[19] Thus we can write consumer i's budget constraint as

$$p_i x_i + \phi_i\xi_i \leqslant h_i.$$

We define:

DEFINITION 9.3 (Equilibrium with interdependent consumptions) The allocation y_j^*, all j, x_i^*, all i, supported by the "price" vectors ϕ_i^*, p^*, $p_i^* = p^* - \Sigma_h\phi_{hi}^* \geqslant 0$, is an equilibrium with interdependent consumptions if:

1. for every producer j, y_j^* solves $\max_{y_j}\{p^*y_j$ subject to $y_j \in Y_j\}$; Π_j^* is his optimal profit;

2. for every consumer i, (x_i^*, ξ_i^*) solves $\max_{x_i \geqslant 0, \xi_i \geqslant 0}\{W_i(x_i, \xi_i)$ subject to $p_i^*x_i + \phi_i^*\xi_i \leqslant h_i^*\}$, where $h_i^* = p^*\omega_i + \Sigma_j\theta_{ij}\Pi_j^*$;

3. consumers agree on consumptions:

$$\xi_{ih}^* = x_h^*;$$

4. all markets are in equilibrium:

$$\Sigma_i x_i^* - \Sigma_j y_j^* - \Sigma_i\omega_i \leqslant 0 \perp p^* \geqslant 0.$$

Existence follows from proposition 9.3.

PROPOSITION 9.3 (Existence of an equilibrium with interdependent consumptions) Let assumptions P1′, CX1, and C2′ be satisfied. Then there exists an equilibrium with interdependent consumptions as specified in definition 9.3.

Proof The proof is simpler than the one of proposition 9.2, since boundedness is ensured by P1′ and C2′ and uniqueness of ξ_i follows from the

19. It can be verified from the Kuhn-Tucker conditions of (9.8) that $p_k - \Sigma_h\phi_{hi,k}$ is nonnegative.

standard arguments of proposition 2.14. Therefore the proof can proceed as for proposition 3.1 (Negishi theorem). ■

As before, transfers add up, there may be an assurance or a free-rider problem, and more generally, decentralization is problematic.

The decentralized setup leaves it up to every consumer to decide on his own, on the basis of subsidy inclusive prices. However, since all this will usually happen at a local level, only those consumers who are involved need to reach agreement, and individuals can create a local institution (a club) to manage this consumption jointly. As in the case of a local nonrival commodity, the level and the funding is decided upon by a club. The principle that collective choices should be implemented at the lowest possible level is known as the principle of subsidiarity.

Here we have only considered external effects with respect to consumption flows. Debates on the environment have made it clear that the consumer could also value stocks (forests, wildlife, etc.). These may be represented by adding empathy through a valuation of future stocks in the dynamic models of chapter 8.

9.1.4 Interdependence in Production

It is difficult to find examples of commodities that are needed in a nonrival way by all firms. Even a harbor is only needed by some. Yet many firms do share nonrival inputs but only at a local or sectoral level, for example, feeder roads, marketplaces, and fairs.

The research on endogenous growth theory has also stressed the importance of accumulation of knowledge, which is often considered as a capital stock used in a nonrival way. This topic will be further discussed in section 10.3, since most of the literature involves nonconvexities in the production or in the use of the nonrival good.

Nonrivalry is only concerned with a joint use by several producers; jointness in output within a firm can be dealt with in the usual way[20] through the specification of technology, and no special treatment is required as long as there is free disposal of all joint outputs. If the output is produced jointly by different firms, some coordination is required to have the firms agree on the level and on the sharing of revenues and costs. Pareto-efficient decentralization calls for joint profit maximization and joint ownership of firms.

Therefore we focus on nonrivalry of inputs. For convenience we will represent it using transformation functions rather than production sets. Let $F_j(y_j, -g_j)$ be the transformation function of firm j, where y_j is the vector

20. See section 2.1.

of firm specific net outputs while g_j is the vector of nonrival inputs. The transformation function satisfies the following standard properties:

ASSUMPTION **PG1** (Transformation function with nonrival inputs) The transformation function F_j: $R^r \times R^r_- \to R$ is continuous, strictly quasi-convex, nonsatiated in inputs, and it has possibility of inaction ($F_j(0, 0) = 0$); the associated profit function is compact valued.[21]

We assume that the nonrival input is provided by a firm (denoted with subscript g) which produces it without use of nonrival inputs according to a transformation function $F_g(y_g)$. We also distinguish inputs from outputs and write $y_g = y_g^+ - y_g^-$. The optimal level of y_g^+ to be produced can be obtained from the following program, which is the counterpart of (9.2) for externalities in production:

$$
\begin{aligned}
&\max \Sigma_i \alpha_i u_i(x_i), \\
&x_i \geqslant 0, \text{ all } i, \; y_g^-, y_g^+ \geqslant 0, \; g_j \geqslant 0, \; y_j, \text{ all } j, \\
&\text{subject to} \qquad\qquad\qquad\qquad\qquad\qquad\qquad\qquad (9.10) \\
&\Sigma_i x_i - \Sigma_j y_j + y_g^- - y_g^+ \leqslant \Sigma_i \omega_i \qquad (p), \\
&g_j = y_g^+ \qquad\qquad\qquad\qquad\qquad\qquad (\phi_j), \\
&F_j(y_j, -g_j) \leqslant 0, \\
&F_g(y_g^+ - y_g^-) \leqslant 0.
\end{aligned}
$$

As before we can decentralize decisions, this time between a collective of producers and individual consumers. The collective of producers is a single profit-maximizing firm with net output vector $y = \Sigma_j y_j + y_g^+ - y_g^-$. Welfare weights can now be found such that the solution of this program is a competitive equilibrium with a single firm and m consumers, and decisions can be decentralized in this way.[22]

The price signals p and ϕ_j of program (9.10) make it possible to decentralize producer decisions further over the n firms,[23] by having producer j solve

$$\Pi_j(p, \phi_j) = \max_{y_j, g_j \geqslant 0} \{p y_j - \phi_j g_j | F_j(y_j, -g_j) \leqslant 0\}.$$

21. See section 2.1 for a discussion of these assumptions; compact valuedness of the profit function implies that no firm can have an output without providing any input to produce it. The fact that nonrival commodities can only be used as inputs, can be reflected by having the transformation function satisfy $F_j(y_j, -g_j) > 0$ whenever $g_{jk} < 0$ for some k.

22. Note that since inputs can always be disposed of at zero cost, a producer will not object receiving more inputs than he needs as long as he does not have to pay for them. It is thus not restrictive to write $g_j = y_g^+$ instead of $g_j \leqslant y_g^+$.

23. ϕ_j will exist because (9.10) satisfies Slater's constraint qualification; it will be nonnegative because the equalities can be replaced by inequalities.

The cost p of the nonrival input is fully covered ($\Sigma_j \phi_j = p$). This follows from the Kuhn-Tucker conditions of (9.10) with respect to y_g^+. Shares in the cost of producing the nonrival good can be defined as $\sigma_{jk} = \phi_{jk}/p_k$. If these are not unique, additional rules have to be provided to settle the accounts among firms.

Most of the comments made for consumption also apply here. In particular, firms may free-ride or, more generally, deviate from the optimal σ. Unless producers decide collectively, inefficient equilibria will result. Yet it may be easier to achieve agreement here because cooperation leads to higher profits, so firms can be induced to participate. This may lead to collective arrangements, buyouts, mergers, and the like, which can cause imperfect competition.

9.1.5 Endowments Dependent on Consumption: Efficiency Wages

Efficiency wage theory studies the relation between labor supply and consumption of a worker.[24] We will account for this by making the endowment vector ω_i depend on x_i. For this we replace assumption C2′ by CX2:

ASSUMPTION **CX2** (Consumption dependent endowment) The function $\omega_i(x_i)$ is continuous, concave, nondecreasing, and uniformly bounded from above; $\omega_i(x_i) \geqslant 0$, $\omega_i(x_i) \neq 0$; for $i = 1$, $\omega_i(0) > 0$.

This assumption amounts to specifying a production function for endowments, with consumption as an input.[25,26]

The Negishi program is

$$\max \Sigma_i \alpha_i u_i(x_i),$$

$$x_i \geqslant 0, \text{ all } i, \; y_j \text{ all } j,$$

$$\text{subject to} \qquad (9.11)$$

$$\Sigma_i x_i - \Sigma_j y_j - \Sigma_i \omega_i(x_i) \leqslant 0 \qquad (p),$$

$$y_j \in Y_j.$$

24. See Leibenstein (1959), Stiglitz (1976), Mirrlees (1975), and Dasgupta and Ray (1986). Stiglitz does not consider the relation $\omega_i(x_i)$ but makes labor supply depend on real wages. This will not lead to efficient allocations when there are many commodities.

25. The consumer problem can be written max $u_i(x_i)$ subject to $px_i \leqslant p\omega_i(x_i) + \Pi_i$, where $\Pi_i = \Sigma_j \theta_{ij} \Pi_j$. It can be seen as a household model with only some of the x_i affecting utility, the other being inputs in production. However, decentralizing this model into a profit-maximizing firm and a "pure" consumer problem is only possible if commodity k enters either the utility function or the endowment function. Otherwise, the external effects of the jointness would have to be internalized within the firm and the pure consumer.

26. This assumption abstracts from the nonconvexity of the relation; it is usually specified to account for a minimum level of consumption, below which no labor would be supplied (setup cost or subsistence consumption). This concept will be treated in section 10.3.

Under assumptions P1′, C1′, and CX2 this program will have a bounded solution, and Lagrange multipliers p will exist. One might expect that as in the previous sections, some additional signals are needed to decentralize the optimal allocation at equilibrium welfare weights, but we show that this is not necessary. The following equilibrium achieves the decentralization of (9.11):

DEFINITION 9.4 (Equilibrium with consumption dependent endowment) The allocation y_j^*, all j, x_i^*, all i, supported by the price vector $p^* \geqslant 0$ is an equilibrium with consumption dependent endowments if:

1. for every producer j, y_j^* solves $\max_{y_j}\{p^* y_j$ subject to $y_j \in Y_j\}$; Π_j^* is his optimal profit;
2. for every consumer i, x_i^* solves $\max_{x_i \geqslant 0}\{u_i(x_i)$ subject to $p^* x_i \leqslant p^* \omega(x_i) + \Pi_i^*\}$, where $\Pi_i^* = \Sigma_j \theta_{ij} \Pi_j^*$;
3. all markets are in equilibrium,

$\Sigma_i x_i^* - \Sigma_j y_j^* - \Sigma_i \omega_i(x_i^*) \leqslant 0 \perp p^* \geqslant 0.$

Note that this equilibrium is very similar to the competitive one: There are no taxes or subsidies, and all agents take prices as given. Producers maximize profits as usual, and the only distinguishing feature is that consumers do not take their endowments as given. We prove that such an equilibrium exists and, more important, that it is Pareto-efficient.

PROPOSITION 9.4 (Existence of an equilibrium with consumption dependent endowment) Let assumptions P1′, C1′, and CX2 be satisfied. Then, there exists an equilibrium with consumption dependent endowment as specified in definition 9.4.

Proof At positive prices p the consumer problem generates continuous functions $x_i(p, \Pi_i)$ and $\omega_i(x_i(p, \Pi_i))$ by assumptions C1′ and CX2. The remainder of the proof follows immediately from the proof of proposition 3.4 with endowments taken as part of the feedback relation. ∎

PROPOSITION 9.5 (Pareto-efficiency) The equilibrium of proposition 9.4 is Pareto-efficient.

Proof In the proof we show that the competitive equilibrium is a solution of program (9.11) evaluated at equilibrium welfare weights. For this, consider the Lagrangian associated to program (9.11) with welfare weights $\alpha_i^* = 1/\lambda_i^*$, where λ_i^* is consumer i's (positive) marginal utility of consumption in equilibrium. Consider now

$$V_i(\alpha_i, p) = \max_{x_i \geqslant 0}\{\alpha_i u_i(x_i) + p\omega_i(x_i) - px_i\}, \tag{9.12}$$

evaluated at (α_i^*, p^*), which solves the consumer problem and can also be seen as resulting from the Lagrangian inequality corresponding to the welfare problem. This shows that the solution of the consumer problem 2 of definition 9.4 is optimal in program (9.11). For producers the Lagrangian inequality shows, as in proposition 1.9, that profit-maximizing supplies (condition 1) are optimal in (9.11). Condition 3 is obviously satisfied at the optimum of (9.11). ■

Propositions 9.4 and 9.5 show that no transfers are needed and that, though the workers (consumers) work better when they can consume more, there is no need for the firm (or the government) to pay any premium above the market wage rate.[27]

The following comments are in order: First, it may be realistic to assume that endowments also depend on nonrival consumption x_g (education, health conditions), in which case the above treatment applies. In principle, consumers can collectively finance this consumption, and there is no need for those who do not benefit to subsidize the nonrival goods. Second, it may take time for the effects of consumption on endowments to materialize; then loans may be required in the earlier periods (reimbursable fellowships). Finally, the model presented is a generalization of the competitive model. It incorporates the classical notion that labor must be "produced" and makes it possible to account for the specificity of labor as a production factor.

The conclusion that no intervention is needed has to be revised when the assumptions made to prove propositions 9.4 and 9.5 are amended, for example, when $\omega_i(x_i)$ is not concave, but we postpone the discussion of this case until section 10.4. Here we only consider two changes: (1) a proportional income tax is imposed, or (2) there is no market for labor effort but only for labor hours.

Lump-sum Transfers and Proportional Taxes

An equilibrium with lump-sum transfers can be obtained from (9.11). An important deviation from the usual case is that consumer i may have positive expenditure, even when he has zero welfare weight. This can be seen by comparing (9.11) with the case where ω_i is not a function of x_i. This is so because, at zero welfare weight, giving him less consumption, would reduce the value $p\omega_i(x_i)$ of his supply of labor more than it would save on the value of the consumption attributed to him by some central planner. The planner will "exploit" the zero-weight consumer by assigning to him the minimum bundle that is in accordance with the interests of the consumers who do have positive welfare weights.

27. If there is a minimum subsistence level, transfers may be needed.

For the individual consumer this leads to the following problem with lump-sum transfers T_i:

$$\max u_i(x_i),$$

$$x_i \geqslant 0,$$

subject to

$$px_i = p\omega_i(x_i) + \Pi_i + T_i$$

for given $\Pi_i + T_i = -V_i(0, p)$ defined as in (9.12) with zero welfare weight. Then T_i is the highest possible tax, and if $\omega_i(x_i)$ is strictly concave in the optimum of $p\omega_i(x_i) - px_i$, x_i will be uniquely determined by the constraint (together with the requirement $x_i \geqslant 0$). In this case the consumer actually has no choice, but if x_i is set valued in the optimum of $p\omega_i(x_i) - px_i$, he may still have some choice. Clearly any larger value of T_i will make the problem infeasible, so no equilibrium will exist: The consumer-worker cannot produce a larger surplus.

Note that the welfare problem will never generate T_i values that exceed this critical level. Conversely, taxes that are lump sum but do not exceed the critical level will maintain Pareto-efficiency.

Let us now consider a proportional tax τ_i on income as an alternative to a lump-sum tax. The budget constraint is $px_i = (1 - \tau_i)(p\omega_i(x_i) + \Pi_i)$. As long as ω_i does not depend on x_i, this is not distortionary, but in the present model, it will in general be.[28] Such a tax will have to be redistributed to other consumers. It will not make consumer problems infeasible as long as $\tau_i \leqslant 1$, but consumption expenditure may now fall below a critical level, at which it would become beneficial for all to reduce the tax rate. An economy with institutions that overtax labor in this way can be trapped in an underdevelopment equilibrium. The workers have a low productivity because their income is too low, and this keeps everyone's income low.[29] Income transfers will restore efficiency if the efficiency relation is specified as $\omega_i(u_i(x_i))$; if it is specified as $\omega_i(x_i)$, price subsidies for consumption may have to be given as well.

Missing Markets for Labor Effort

Since it is difficult to measure labor effort by objective standards, there will rarely exist a market and a competitive price for it.[30] This will cause

28. The tax operates like a commodity tax on the endowment. Note that if the dependence of the endowment on consumption is specified as $\omega_i(u_i(x_i))$, the income tax will still be nondistortionary (at least as long as τ_i does not exceed the critical level).

29. See Mirrlees (1975).

30. See also the discussion in section 3.6.2.

inefficiencies that can only be eliminated via additional signals, such as premia. Efficiency will be restored when the actual payment per unit of labor effort corresponds to the solution of the competitive model of definition 9.4. Wages per unit of effort will be equalized, but they will be different per labor hour. This leads to the wage dispersion found in models that explain wage differentials among workers on the basis of their personal characteristics.

9.2 Pollution and Negative Externalities

9.2.1 Pollution in the Competitive Model

Pollution may be the first issue that comes to mind in relation to the negative effects that agents may have on each other. Here we argue that for technical reasons pollution should not be viewed as a negative externality but as the use of a natural, possibly unpriced, resource.

Let us consider the case of a producer who generates some industrial waste that is disliked by consumers and possibly hinders other producers. In the competitive model discussed so far, this waste would be a commodity disposed of at no cost, and since there is no demand, its price would be zero. This is achieved through the weak inequality of the commodity balances (free disposal). Even if this inequality were replaced by an equality, producers would be able to buy the surplus at zero price and dispose of it freely.[31] Once this possibility is ruled out, the price of waste can become negative. Then, to prove existence of equilibrium, elaborate modifications of the assumptions are needed.

Instead of changing our assumptions, we will interpret pollution not as a joint output in production but as a use of inputs. Economically what matters is not so much the volume of pollution as the resources it uses up (productive land, fresh water, clean air). These are in fact the commodities that enter the production processes as inputs. With this interpretation the competitive model applies virtually without any change: Instead of focusing on "bads," we deal with the associated "goods."

Polluting processes will be part of the technology sets, where the resources used will appear as inputs. Cleaning activities may also appear as ordinary (and profitable) production processes.

On the consumer side, demand for environmental goods can enter the utility functions like any rival or nonrival commodity. It may also appear as a nonrival demand for stocks (the ozone layer, wildlife, etc.). These resources may be used privately, if outsiders can be excluded (no right to dump into the neighbor's garden), or collectively (the seashore).

31. This means that if $y_j \in Y_j$, then $(y_j - \eta) \in Y_j$ for all $\eta \geqslant 0$. Although this conflicts with the compactness assumption on Y_j, it can be accommodated easily, since it does not violate the compact valuedness of the profit function. See also the discussion in section 2.1.2.

In principle, all these resources can be owned privately. The issue of ownership is important in this context because for previously free resources no one has ever cared to claim property. Government will often be the first to do so to obtain more revenue. However, there is no obvious reason for government to own environmental resources. These could be distributed to consumers as part of their initial endowment (an entitlement). Consumers would have to pay for the possibly nonrival use of resources and be entitled to the income generated by the endowment.

With respect to markets, the main point is that the list of commodities should include all environmental resources. Whenever they are anticipated to become scarce, they should fetch a positive price, so those agents who plan to use them will pay. This is known as the polluter pays principle. It must also be stressed that all users (not only the polluters) should pay. Polluters should pay the capitalized value, while consumers who simply enjoy the presence of the resource should rent it. If these payments are made, it will become profitable for the polluter himself or for others to engage in cleaning activities.

If one leaves aside nonrival commodities, possible nonconvexities, and issues related to the infinite horizon, it would seem that pollution can be treated within the standard competitive framework, so the welfare theorems would apply. Theoretically this is the case, but there is an important difference with respect to excludability between the consumption of, say, caviar and fresh air. Both the producer and the consumer of caviar will meet and come to an exchange. By contrast, the producer of smog will have an incentive to hide this emission, since he has access to the resource "fresh air" without having to wait for an agreement with its owners. This makes it very difficult to protect environmental property rights and organize competitive markets for environmental resources. Unless there are institutions that effectively protect these rights (by measuring emissions and having polluters and users pay to the owner), decentralization will not be Pareto-efficient. Polluters will go free, and victims will not be compensated.

Hence, while decentralization is problematic, theoretically it is not more difficult to model pollution as a demand for natural resources than to model any other commodity. What may, in applied modeling, distinguish this from, say, mining activities is that the technology has only recently come under study, that it has global interdependencies, and that the link between emission and resource depletion is hard to establish (how much fresh air is used up when pollution spreads?). As long as these issues are not elucidated, the best approach may be to define tolerable levels of emission absorption, since the scarce resource for each period by periodically distributing tradable licenses to pollute. This is indeed the direction in which environmental policy has been moving. In terms of the model, this amounts to extending the commodity list with the license, requiring it as an input in the production sets, and distributing the property rights as endowments that can be sold.

9.2.2 The Treatment of Externalities

Since we deal with "goods" instead of "bads," most of the treatment given for positive externalities in sections 9.1.1, 9.1.3, and 9.1.4 applies.

Antipathy, the negative counterpart of the empathy discussed in section 9.1.2, occurs when consumers dislike each other. In contrast to bads which were translated into goods, there is no natural positive expression for negative feelings. At any rate it is questionable whether a social choice problem, which after all is one of cooperation, should account for antipathy and optimize accordingly, maximizing some joint sum of utility functions that include antipathy. Similar issues arise when "negative paternalism" is incorporated in the utility functions $W_i(x_i, \xi_i)$ of section 9.1.3.

9.3 Perspectives for Implementation in Applied Models

In this chapter we have been concerned with decentralization of first-best solutions that lead to models technically not more difficult to implement than the applied models covered in chapters 2 to 7. Nonrival goods, transfers to other agents, consumption-dependent endowments, negative externalities, could be represented in utility functions and technological relations. Given these relations, signals are needed for efficient decentralization, but these are part of the solution that can be computed by standard fixed point methods.

However, in applied modeling, it is difficult to specify the relations in a realistic way: For nonrival goods one could expect that only some aggregates (like "national security") will be of interest to consumers, who will leave it to public services to decide upon details (to reflect this, utility functions will have to possess a nested structure). Empathy is restricted to close relatives or abstract entities like "the poor." This simplifies the functional forms. Nevertheless, there are serious problems in quantifying the willingness to pay for such goods under observed conditions that are far from the first best. There is an expanding literature on attempts to estimate willingness to pay from actual household expenditures (Gertler and van der Gaag 1988) as well as from explicit surveys about the hypothetical situation in which the market would exist (the so-called contingent valuation method; see Carson and Mitchell 1993; Gregory et al. 1993; Mitchell and Carson 1989).

Once such functions have been estimated, the model can be calibrated to reproduce an initial distorted allocation[32] by setting exogenously the values in a decentralized solution. Alternatively one could add distortion terms in the objective of the welfare program (as for the taxes of chapter 5), introduce additional technical and political constraints (e.g., a complementarity between police and defense), or incorporate constraints on transactions (e.g., the impossibility of organizing aid programs for individuals separately). One could subsequently design welfare improving scenarios that reduce these distortions.

External effects are of special relevance in public and development economics. However, so far they have not found their way into applied general equilibrium models. This is unfortunate, for it severely limits the possibilities for welfare analysis of changes in public policies. In developing countries transfers often form a substantial part of household income, but in most applied models they are treated, like income taxes, through functions, and not derived from maximizing behavior. The relevance of efficiency wage theory for developing countries is rather obvious. There is an extensive empirical literature on this topic (measuring the impact of improved nutrition, vaccination, and education on health and on labor productivity; e.g., see Rodgers 1975; Strauss and Thomas 1995), but the results have so far not been incorporated in AGE models.

By contrast, pollution has been studied extensively in a general equilibrium context, usually in dynamic models, sometimes with infinite horizon, and often for carbon dioxide emissions. Pollution is usually represented as a by-product that cannot be freely disposed of. Emission constraints are imposed to limit it, and the impact on allocations is studied under various timings and rules for distributing and trading emission rights. Tax rates are adjusted until constraints defined by emission permits are satisfied and welfare comparisons are made between various timings and rules. These uses of CGE models must be seen as tax reform scenarios. The recognition of emission rights introduces a new commodity used by firms and owned by the government. The tax revenues can thus be considered as proceeds from endowments and are therefore nondistortionary (given the chosen emission level). The scenarios merely compare alternative regimes of taxation.

Whalley and Wigle (1991) formulate a static model, and Burniaux et al. (1991) a recursively dynamic one. Intertemporal models with perfect fore-

32. Note that because of the additional condition that consumers have to agree on the provision of the nonrival good, the demand for nonrival goods involves solving a system of simultaneous equations. This is avoided under particular separability assumptions on the utilities, like $u_i(x_i) + \beta_i u_g(x_g)$ in which case $\beta_i / \Sigma_h \beta_h$ will be the share of consumer i in the cost of the good. An alternative specification is to have $\beta_i \ln\{u_i(x_i)\} + (1 - \beta_i)\ln\{u_g(x_g)\}$. Here the consumer will spend a fixed fraction $1 - \beta_i$ of his income on nonrival goods.

sight are used by Manne and Richels (1990, 1992), Jorgenson and Wilcoxen (1990, 1993), and Goulder (1991). These models were originally developed to obtain efficient investment decisions in the energy sector, which made them well suited to analyze carbon dioxide emissions. They are related to developed countries or for the international economy. Blitzer and Eckhaus (1986) and Blitzer et al. (1994) have constructed intertemporal models for Mexico and Egypt. As stressed by Goulder (1991), one of the problems with these formulations is that pollution does not enter utility, production functions, or welfare criteria. As a result environmental degradation has no effect, and leaving emission rights unused cannot bring a welfare gain. Welfare comparisons between scenarios with different emission rates are meaningless in this case.

10 Nonconvexities

So far the mathematical programs were assumed to be convex in all the models considered. This was ensured via the assumption that commodities were divisible, production sets convex, and utility functions concave. The only exception concerned the competitive model in the first welfare theorem (proposition 1.5). There it was shown that even if convexity is relaxed in assumptions P1′ and C1′, a competitive equilibrium, if it exists, will be Pareto-efficient. However, such an equilibrium may fail to exist due to discontinuities in the excess demand correspondence.[1] Nonconvexity does not always lead to nonexistence of equilibrium. To obtain a characterization of the cases in which problems arise, we distinguish different ways in which nonconvexities can enter the model for production and for consumption; the implications of nonconvexities on producer and consumer behavior were already discussed in chapter 2 (e.g., see figures 2.1c, 2.1d, 2.6, and 2.8).

In this chapter we focus on the issues related to existence and efficiency of equilibrium. Any relaxation of the convexity assumption will lead to nonconvexity. We restrict attention to two special cases from which many others can be obtained: setup requirements (e.g., fixed charges) and increasing returns to scale. We will see that both cases are due to indivisibility.

For production[2] nonconvexity may arise (1) within the firm's production possibilities without causing the net supply to become discontinuous, (2) at firm level within the firm's production possibilities but causing a discontinuity, and (3) for all firms collectively under the form of an external effect.

As an example of the first type of nonconvexity (the within-firm nonconvexity), consider the yield of a single wheat plant that may depend in a nonconvex way on current inputs. Since the farmer has the possibility to decide not only on the input per plant but also on the number of hectares planted, the nonconvexity creates no problems, as we will see. A similar argument holds in a factory, say, a spinning mill where each spinning jenny may have a nonconvexity in technology (e.g., the time it takes for the jenny to reach a constant optimal speed), but where the number of identical jennies is large. If the number of wheat plants, or spinning jennies, is treated as a real variable, there is no indivisibility, and the nonconvexity can be handled within the competitive model. This case is treated in section 10.1.

If the number of machines is small, the switching between on and off will create a jump in net supply, and then discontinuity cannot be avoided. In section 10.2 we discuss the case in which the firm itself is "switched on" or "off," though obviously every machine can be treated as such a firm. Pareto-efficiency will require a central planner to be involved in the setting of these switches, and this is problematic when their number is relatively large (but still too small to treat this number as a real variable).

1. This was already illustrated in figures 1.1c and 1.1d.

2. See also Brown (1991) for a survey.

A third group of nonconvexities occurs at higher than firm level, through "inputs" that firms use in a nonrival way, for example, under economies of scope or when there is learning by doing. Under economies of scope firms can produce more efficiently when the sector (or the economy) as a whole supplies more because this is supposed to generate gains from which all firms can benefit. Learning by doing occurs through accumulation of knowledge. Both topics are discussed in section 10.3.

For consumers nonconvexities can arise because the utility functions are nonconcave or because commodities are indivisible (cars or houses).[3] There may also be a nonconvexity in the efficiency-wage relation $\omega_i(x_i)$ discussed in section 9.1.5, due to setup requirements with respect to food, education, or health below which the consumer cannot survive or work; returns to these consumption items may also be increasing. This is the subject of section 10.4.

As long as the production possibilities of the economy are bounded and the utility functions are continuous, nonconvexities do not pose problems for the existence of a welfare optimum.[4] The issue is, however, that decentralization of this optimum may, apart from lumpsum transfers, also require rationing. In some cases it is possible to decentralize by assuming other rules of conduct than profit (or utility) maximization. Therefore, except for the within-firm nonconvexity which can be treated within the competitive framework, we will proceed, as in chapter 9, by formulating a welfare program which we decentralize as much as possible. For this some variables are set centrally, and there are transfers among consumers and firms. We will also consider equilibria without transfers, but these will be noncompetitive and thus possibly inefficient.

10.1 Within-Firm Nonconvexities

We introduce the within-firm nonconvexity via the producer problem with perfect divisibility in technology:

$$\max\ pF(v, n) - cv,$$

$$0 \leqslant n \leqslant 1,\ v \geqslant 0,$$

where v is the current input (e.g., labor), n is the fixed input (the machine), p and c are output and input prices, and $F(\,.\,)$ is a constant-returns-to-scale

3. We discuss this indivisibility in the context of consumption. It is usually less relevant for producers because they tend to produce the same commodity in large numbers, while the consumer will buy one or two units.

4. Computation may be difficult, since algorithms for solving nonconvex programs are known to be far less efficient than for convex programs: They are not guaranteed to come close to the optimum within reasonable computing time.

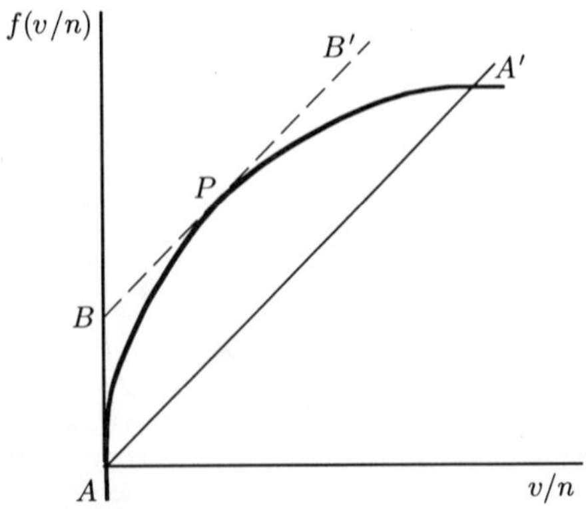

Figure 10.1
The concave production function.

production function. Because of constant returns, $F(v, n) = F(v/n, 1)n$, which we write as $f(v/n)n$. We assume that $f(0) = 0$, $F(v, n)$ is continuous with $F(v, 0) = 0$ and that f is increasing and uniformly bounded above. The program is then

$$\max pf\left(\frac{v}{n}\right)n - cv,$$

$$0 \leqslant n \leqslant 1,\ v \geqslant 0.$$

We can distinguish two cases, the first of which is convex: (1) The function $f(.)$ is strictly concave, as in figure 10.1. The line AA' is the real unit input cost $(c/p)(v/n)$ and the tangent BB' defines the point P where marginal revenue equals marginal cost c/p. If optimal profits are positive, then n will be set equal to 1. If they are zero, v/n will be zero (and $f(v/n) = 0$). Therefore any value of $n \in [0, 1]$ will be optimal but will give the same zero production as when $n = 1$. Without loss of generality, we can thus set $n = 1$ and recover the case of decreasing returns to scale.

(2) Assume now that $f(.)$ is convex-concave, as in figure 10.2, where it is convex at the origin and becomes concave afterward. As long as the line BB' lies above the real unit input cost curve AA' (as in figure 10.2a), profits will be positive; the optimal point will then be P, and n will be set equal to 1. As the ratio c/p increases, AA' will rotate upward until it becomes, as in figure 10.2b, tangent to the production function. Profits will then be zero while production and input demand will be positive. If c/p is increased further, there is a jump to the origin and thus a discontinuity of the optimal choice. However, there is no discontinuity in the optimal choices v and n because, when profits are zero, all n in $[0, 1]$ are optimal and all $v = an$, $n \in [0, 1]$ are optimal. Formally the supply and input demand functions

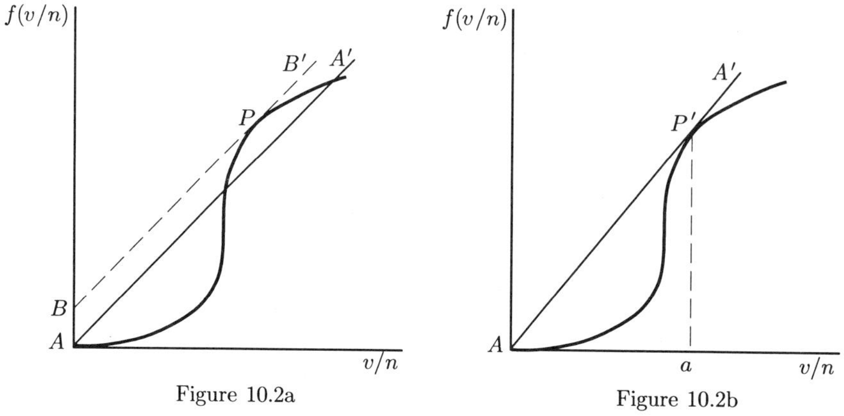

Figure 10.2
The convex-concave production function.

remain uppersemicontinuous in the nonnegative price orthant and convex valued despite the nonconcavity of the production function.

The key assumption here is that we can scale n arbitrarily on the unit interval. This rules out indivisibilities and implies that the use of the fixed factor can be varied at will: The machine can be used at various rates of capacity utilization without changing the "per unit production function." Likewise the response to fertilizer application of a single wheat plant may be nonconcave, but as long as the number of hectares planted to this crop can be adjusted, there will be no discontinuity in production and fertilizer demand.

These two examples illustrate that if indivisibilities only arise within the technology of the firm, they do not necessarily create discontinuities. The resulting net supply functions will remain well-behaved, so existence and efficiency of equilibrium can be proved in the usual way.

In the previous cases there was a single nonconvexity, so only one scaling factor n had to be considered. For the more general case, consider a production set $\tilde{Y}$ that represents the technology of some production process and satisfies the following assumption:

ASSUMPTION PN1 The production set $\tilde{Y}$ is compact and nonempty.

The set $\tilde{Y}$ is more general than the set Y in assumption P1′ because it is not required to contain the origin, and because it may be nonconvex like in Figure 2.1a, 2.1c, and 2.1d. This assumption can be relaxed further: All one needs to ensure is that the maximum profit be compact valued for any nonnegative price vector p.

Though the set $\tilde{Y}$ is very general, perfect divisibility has far-reaching consequences on the set of productions that can be attained. Suppose that we can start an arbitrary[5] number H of processes with net outputs $\tilde{y}_h \in \tilde{Y}$.

The profit maximization problem will be

$\max p\Sigma_h \tilde{y}_h \delta_h,$

$\delta_h \geqslant 0, \tilde{y}_h,$ all h,

subject to

$\tilde{y}_h \in \tilde{Y},$

$\Sigma_h \delta_h \leqslant 1.$

This may be rewritten as

$\max py,$

y

subject to

$y \in \hat{Y},$

where $\hat{Y} = \{y \mid y = \Sigma_h \tilde{y}_h \delta_h, \tilde{y}_h \in \tilde{Y}, \delta_h \geqslant 0, \Sigma_h \delta_h \leqslant 1, h = 1, 2, \ldots, H\}$. In other words, when an arbitrary number of processes can be started, $\hat{Y}$ becomes the convex hull of $\tilde{Y} \cup \{0\}$; $\hat{Y}$ will satisfy assumption P1′ and thus be compact, convex, and have possibility of inaction. This shows in a general way that for production, indivisibility is the issue, not nonconvexity.

10.2 Indivisibilities at Firm Level

Indivisibility will be represented by restricting n, the degree of capacity utilization, to the values 0 and 1, which we denote as $\delta \in \{0, 1\}$. We again consider the compact, nonempty production set $\tilde{Y}$ and the profit maximization program:

$\max p\tilde{y}\delta,$

$\tilde{y}, \delta,$

subject to

$\tilde{y} \in \tilde{Y},$

$\delta \in \{0, 1\}.$

This is a mixed integer program that has a bounded solution, but the supply response is not uppersemicontinuous, convex valued. Hence, if such firms are introduced into the model, an equilibrium may fail to exist.

5. For ease of exposition, we assume H to be finite.

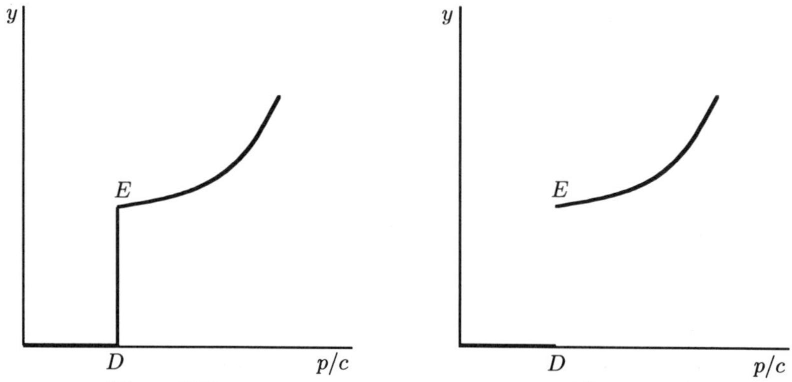

Figure 10.3
The supply response under divisibility and indivisibility.

The complication introduced by indivisibilities is illustrated in figure 10.3, which shows the supply function corresponding to figure 10.2. In figure 10.3a there is no discontinuity because divisibility makes all outputs between D and E attainable. In figure 10.3b output jumps to zero as soon as the firm makes a loss.

The situation in figure 10.3a does not create problems for existence of a decentralized equilibrium, since the net supply schedule drawn will intersect with the demand schedule. In figure 10.3b, the demand schedule may lie between D and E and hence fail to cross the supply curve. The two situations correspond to the fixed point mappings illustrated in figure 1.1b and 1.1c, respectively.

To analyze existence and efficiency of equilibrium, we introduce the common distinction between setup costs and increasing returns:

1. *Setup costs.* The production sets $\tilde{Y}_j$ are nonempty, compact, and convex, but they do not include the origin.

2. *Increasing returns to scale.* Technologies are represented by single output production functions with increasing-returns-to-scale segments but possibly also with constant- or decreasing-returns segments.

These cases are illustrated in figure 10.4. In figure 10.4a there are setup costs requiring an input OA before production q can start but the origin is part of the production set. This set is therefore not connected and hence nonconvex. In figure 10.4b returns to scale are increasing on the interval OB: The production set is connected, but it is not convex.

The second welfare theorem describes a decentralized setting in which transfers among consumers are the only form of intervention. When production sets are nonconvex, more intervention will be required: Some producers may have to be given transfers, while others will be rationed (forced to be

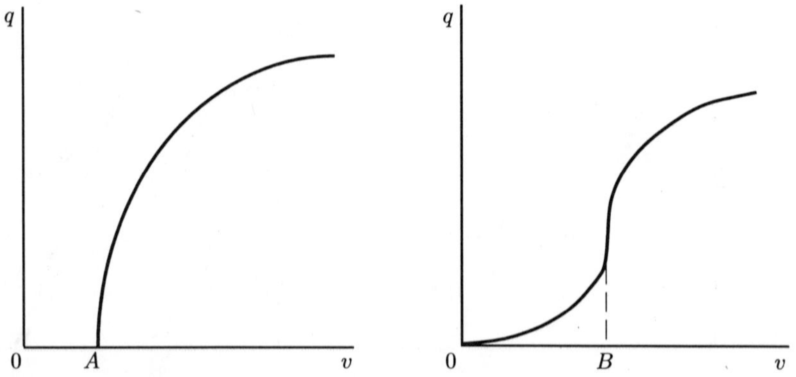

Figure 10.4
Setup costs and increasing returns.

inactive). With convex firms it is possible to use the Negishi approach, adjusting welfare weights, to eliminate transfers among consumers. With nonconvex firms this will not work because budget deficits are discontinuous. However, under additional assumptions it is possible to prove existence of an equilibrium which may, however, be Pareto-inefficient.

10.2.1 Decentralization of a Welfare Optimum: The General Case

In order to analyze decentralization and equilibrium, we start from the following welfare program:[6]

$$\max \Sigma_i \alpha_i u_i(x_i),$$

$$x_i \geqslant 0, \text{ all } i, \tilde{y}_j, \delta_j, \text{ all } j,$$

$$\text{subject to} \tag{10.1}$$

$$\Sigma_i x_i - \Sigma_j \tilde{y}_j \delta_j \leqslant \Sigma_i \omega_i,$$

$$\tilde{y}_j \in \tilde{Y}_j,$$

$$\delta_j \in \{0, 1\}.$$

This program is feasible (since it is feasible for $\delta_j = 0$, all j); it is also bounded. Therefore, by the Weierstrass theorem A.5.1, an optimum exists.[7] Possible outcomes are illustrated in figure 10.5.

6. In (10.1), integer variables are used only to represent setup costs: The firm is either active or not. One can impose further constraints involving integer variables, for example, to indicate that only one of several alternatives processes can be activated ($\Sigma_h \delta_{hj} \leqslant 1$, $\delta_{hj} \in \{0, 1\}$).

7. Note that we do not associate Lagrange multipliers to this program. Though Slater's condition holds (if $\delta_j = 0$ for all j), this is insufficient to guarantee existence of Lagrange multipliers. However, assume that an optimum has been found; then one can keep the optimal δ_j fixed and consider the restricted program with the x_i and the $\tilde{y}_j$ as choice variables. This is an ordinary convex program, and Lagrange multipliers will exist if Slater's condition holds.

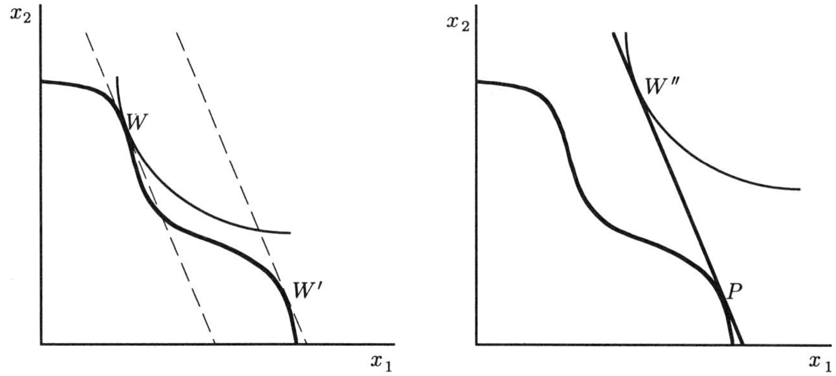

Figure 10.5a

Figure 10.5b

Figure 10.5
Welfare optima for closed and open economies.

Figure 10.5a shows a nonconvex production possibility set for a two-good economy as well as the indifference curves generated by a welfare function with given weights. The dotted line through W is the aggregate budget constraint of the consumers, while the one through W' is the profit line (the line of maximum profits) of the collective of producers.[8] Both have the price ratio as their slope. The point W is the optimal total consumption, obtained as a solution of program (10.1). This solution will not be attained by individual producers separately, since even the collective of profit maximizing producers would choose W', which yields a lower consumer welfare and is not a solution of (10.1). Therefore it may be necessary to ration producers (i.e., to impose δ_j) and to cover their losses.

However, if as in figure 10.5b the economy is open to international trade[9] and faces given world prices (the slope of the line PW'') for the two commodities, then decentralization can be achieved by letting producers maximize collective profits and choose P, allowing for transfers to firms, which make a loss, and distributing the remaining profit to consumers, who will then be able to attain W''. Hence, as long as changes in the δ_j do not affect prices (like in the small open economy), collective[10] profit maximization is an efficient mode of decentralization, even when production sets are

8. If the technology sets are convex for all producers, profit maximization by each producer leads to collective profit maximization. If some production sets are not convex, the two no longer coincide, but the collective of producers can use side payments to induce individual producers to agree with allocations that result from the collective decision.

9. The open economy can be obtained by adding imports m and exports e in the commodity balances of (10.1) as well as a balance of payments constraint $p^w(m - e) \leqslant 0$.

10. We can call this collective profit maximization decentralizable to individual producers because this provides a surplus that is sufficient to induce every individual firm to cooperation.

not convex. In the closed economy case, prices may be affected when producers choose their optimal point, and no equilibrium may exist.

10.2.2 Setup Costs: Decentralization of a Welfare Optimum

To represent setup costs, the production set Y_j can be specified according to the following assumption:

ASSUMPTION PS1 (The production set under setup costs) The production set of producer j is $Y_j = \{y_j \mid y_j = \tilde{y}_j \delta_j,\ \tilde{y}_j \in \tilde{Y}_j,\ \delta_j \in \{0, 1\}\}$, where $\tilde{Y}_j$ is nonempty, compact, and convex.

For the single-output firm, the production set $\tilde{Y}_j$ can be written in terms of a production function:

$$Y_j = \{y_j \mid y_j \leqslant q_j b_j - v_j^1 - v_j^0,\ q_j \leqslant \delta_j f_j^1\left(\frac{v_j^1}{\delta_j}\right),\ \delta_j \leqslant \delta_j f_j^0\left(\frac{v_j^0}{\delta_j}\right),$$

$$q_j \geqslant 0,\ v_j^1,\ v_j^0 \geqslant 0,\ \delta_j \in \{0, 1\}\}. \tag{10.2}$$

Here v_j^1 is the vector of inputs needed to produce the output q_j, v_j^0 is the vector of inputs necessary to activate the firm, $f_j^1(.)$ and $f_j^0(.)$ are standard production functions (concave, nondecreasing, going through the origin and with the property[11] that $\lim_{\delta \downarrow 0} \delta f_j^1(v_j^1/\delta) = 0$ for any given $v_j^1 \geqslant 0$, and similarly for $f_j^0(.)$); b_j is a vector with 1 in the row for the produced commodity and 0 elsewhere. The integer variable δ_j ensures that no production is possible as long as the setup cost v_j^0 has not been incurred.[12]

We now return to program (10.1) under assumptions PS1, C1′, and C2′. For given welfare weights, this program is feasible because $\delta_j = 0$, and hence $y_j = 0$ for all j is feasible. Since the sets $\tilde{Y}_j$ are bounded, the optimum is bounded also. Consider the optimal δ_j, and solve the convex program that takes these optimal δ_j as given parameters rather than as choice variables. Assume now that there exist Lagrange multipliers associated to the balance constraints.[13] Then one can decentralize the solution as follows:

1. Markets clear at prices that are equal to the multipliers.
2. For given δ_j each producer maximizes profits, taking these prices as given.

11. This implies that $F(v, \delta) = \delta f(v/\delta)$ is concave, nondecreasing in (v, δ). Hence v and δ can be treated as ordinary inputs in production. See theorem A.1.5.

12. This specification will lead to an additive cost function: $c_j(p, q_j, \delta_j) = c_j^0(p, \delta_j) + c_j^1(p, q_j)$ where the first term represents the setup cost (or fixed charge).

13. This will be the case, for example, if Slater's constraint qualification holds. So far, assumption C2′ ensures this, but here it may happen that for given optimal δ_j positive consumption is not possible for some commodities because total net supply uses up all the endowments. Note that in the open economy model Slater's constraint qualification will always hold because imports are possible.

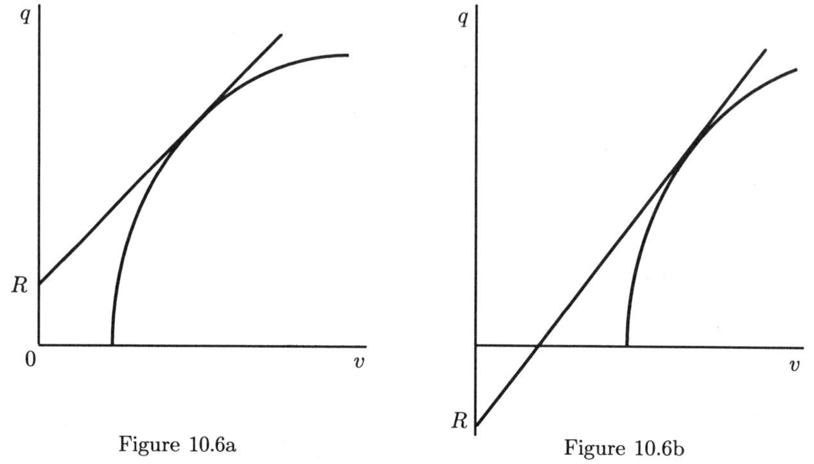

Figure 10.6
Profit maximization of firms with setup costs.

3. Each consumer maximizes utility, subject to his budget constraint at these prices; his income consists of the value of his endowment, his share in profits (or losses), and income transfers T_i.

At this welfare optimum there may be firms that, though they could make positive profits, are not allowed to operate because their δ_j is set to zero in the plan. They face a zero ration (via a bound $\delta_j = 0$) or a prohibitive tax. In figure 10.6 the production function is drawn for two types of firms. In both figures the tangent is the profit line (with slope c/p), which defines optimal production and input levels q and v with, associated to these, the maximum real profit R $(= \Pi/p)$. The firm in figure 10.6a wishes to produce, since it can make positive profits. The firm in figure 10.6b would have to be forced to produce via a bound $\delta_j = 1$, or alternatively, it would have to receive subsidies to cover its setup costs.

When transfers to firms are incorporated into the producer problem, this may be written

$$\Pi_j = \max(T_j + p\tilde{y}_j)\delta_j,$$

$$\tilde{y}_j,\ \delta_j,$$

subject to

$$\tilde{y}_j \in \tilde{Y}_j,$$

$$\delta_j \in \{0, 1\},$$

where $T_j = 0$ for each firm that makes a loss and should not produce in the welfare optimum, as well as for each firm which makes no loss and should produce. Otherwise, T_j should tax away positive profits of firms that should not produce and cover the losses of those that should produce.[14]

According to (10.1) the role of the central planner is to prevent firms that would like to produce from becoming active through a ration. In practice, this is often implemented through licenses used to regulate, for example, the number of pharmacies in a district. The planner may also have to induce nonprofitable firms to producing. He can do so by forcing them, for example, if they are part of a profitable larger firm (a profitable airline company may be forced to operate on a nonprofitable route), but an obvious alternative is to cover the losses through transfers. These transfers can be financed in various ways:

1. They are paid out of taxes collected by a government. This often applies in public transportation.
2. A profitable firm covers the losses of firms which supply its inputs. A firm may, for example, own research units or subsidize them.
3. Users of a commodity pay membership fees in addition to the price of the commodity. For example, users pay a flat connecting fee to the electricity company and a price per unit consumed (nonlinear pricing).
4. The planner allocates endowments ω_j, like labor or land, to firms. These endowments will be used as inputs, or sold as outputs at prevailing market prices, with the additional restriction that they cannot be used productively when the firm does not operate. This amounts to providing a transfer $T_j = p\omega_j$ to firm j, which will produce as long as its profit plus the value of its endowment is nonnegative.

10.2.3 Setup Costs: Equilibrium with Given δ_j and Zero Transfers Among Consumers

Decentralizing the optimal solution of (10.1) requires the planner to organize transfers among consumers and to control the δ_j of many firms. We also should consider setup costs in economies without a central planner, where no transfers are forced upon consumers. Two issues now come up: (1) showing existence of a (rationed) equilibrium with given δ_j but without transfers and (2) describing institutions that make firms willing to produce despite losses.

Equilibrium

In an equilibrium without transfers, consumer i's primary income $p\omega_i + \Sigma_j \theta_{ij} \Pi_j$ may be taxed, though only through a tax function $T_i(.)$ that

14. Consider the welfare optimum in which the integer restrictions are relaxed. We have seen in section 10.1 that this corresponds to the case of perfect divisibility. The program is then convex in (y_j, δ_j), where $y_j = \tilde{y}_j \delta_j$. It is therefore a decentralizable convex program. If in its optimum, it happens that all δ_j are integer valued, the solution can be decentralized without transfers to firms and without need to ration them. Note that the value of welfare obtained by relaxing the integer constraints provides an upper bound of attainable welfare.

depends on prices and satisfies assumption T introduced in chapter 5. Such an equilibrium will require adjustments of the welfare weights. However, discontinuities in the budget deficits can only be avoided if the δ_j are kept constant. The resulting equilibrium will be arbitrary and inefficient, since these δ_j are essentially arbitrary. We only sketch an existence proof for such an equilibrium. It would proceed in two steps:

1. The values of the δ_j should be chosen so that infeasibility is avoided in (10.1). For this it is sufficient to assume that $\Sigma_j v_j^0 < \Sigma_i \omega_i$ holds at the given values of the δ_j. Slater's condition is now satisfied. Then, for any given set of positive welfare weights α_i, one can solve (10.1). In the optimum the Lagrange multipliers p will be positive (increasingness for consumer 1 in assumption C1′), and consumption will be positive for at least one commodity, so $p\Sigma_i x_i > 0$. Hence there is sufficient income to cover losses.
2. Keeping the δ_j constant, one could, as in the Negishi theorem 3.1, adjust the welfare weights until consumer budgets[15] are satisfied.

Step 1 is the critical one, and it is not evident that any choice other than $\delta_j = 0$ for all j will lead to satisfaction of Slater's condition.

Institutions

We have mentioned four ways in which losses can be covered by government, consumers or profitable firms: (1) They are financed out of the government budget, (2) firms pay part of the setup costs of their suppliers of inputs, (3) consumers pay membership fees, and (4) endowments ω_j are allocated to firms.

Case 4 needs further elaboration. If the endowments ω_j of firms cannot be traded, their reward may fall below the market price. This will absorb the losses from setup costs. A firm will then produce as long as the value accruing to this fixed endowment is nonnegative. The producer problem is

$$\Pi_j = \max p(\omega_j + \tilde{y}_j)\delta_j,$$

$$\tilde{y}_j, \delta_j,$$

subject to

$$-\tilde{y}_{jk}\delta_j \leqslant \omega_{jk}\delta_j \qquad \text{for } k \in K_1 \quad [\sigma_{jk}],$$

$$\tilde{y}_j \in \tilde{Y}_j,$$

$$\delta_j \in \{0, 1\},$$

where K_1 is the set of endowments which cannot be resold and σ_{jk} is the

15. The budget constraint of consumer i is $px_i \leqslant p\omega_i + \Sigma_j \theta_{ij}(py_j + T_j\delta_j) + T_i(p, T, \ldots)$, where $T_j = \max\{-py_j, 0\}$ is the loss of firm j, and in equilibrium, $T = -\Sigma_j T_j$.

shadow price (reward) of the unsellable resources, defined only when $\delta_j = 1$. In view of the convexity of $\tilde{Y}_j$ in assumption PS1, the firm will choose to be active ($\delta_j = 1$), whenever $\sigma_{jk}\omega_{jk} > 0$ for some $k \in K_1$. Since the firm freely receives part of the inputs, it may now be able to survive financially, and since it cannot sell these endowments, it will have to use them in production.

In the labor-managed firm, workers own the capital, and the income they receive from the firm is the sum of a wage and a dividend. The system ties workers to firms by limiting their possibility to sell shares. This may, for instance, induce them to accept a wage-dividend contract, in which both the wage rate and the dividend are lower than the market rates but the total income exceeds the market wage rate. Such a restriction on selling the shares clearly introduces an inefficiency. This also happens in the family farm which supplies labor, owns the land it operates, and faces legal or institutional restrictions on its sale and lease. Then the setup cost of the farm will be borne by the rent on land. If, moreover, the family's labor has little alternative employment, the wage will also carry part of the burden. The farm will survive financially but likely under poor conditions.

Finally, markup pricing[16] is equally powerful in avoiding losses, but it also leads to inefficiencies since active firms will restrict supply by choosing their production level in order to keep marginal cost below the price.

10.2.4 Increasing Returns to Scale: Decentralization of a Welfare Optimum

In figure 10.4b returns to scale are increasing on the interval OB. This generates a nonconvexity in the production set, so the treatment through setup costs does not apply directly. However, if one is willing to approximate the production function by linear segments, each segment is concave, and the nonconvexity is shifted to the integer constraint $\Sigma_h \delta_{jh} \leqslant 1$, where δ_{jh} corresponds to segment h of the production function. Such a constraint for firm j can then be added to program (10.1). This is a convenient approximation for solving the welfare program, since for given δ_{jh} this program will be convex. It is not, however, a convenient format for proofs of equilibrium, nor is it used in applications.

To represent increasing returns to scale, the set of firms $J = \{1, 2, \ldots, n\}$ is partitioned into J_1 (containing firms with increasing returns) and J_2 (firms with a convex production set). The production sets satisfy:

ASSUMPTION PI1 (Production sets under increasing returns)

1. Firms $j \in J_1$ (e.g., the first n_1) with increasing returns produce a single output, with a production function $f_j(v_j)$ that is strictly quasi-concave,

16. See section 3.6.2 and chapter 11.

nondecreasing, and such that $f_j(0) = 0$. Their production set is given by $Y_j = \{y_j \mid y_j \leqslant b_j q_j - v_j,\ q_j \leqslant f_j(v_j),\ q_j \geqslant 0,\ v_j \geqslant 0\}$.

2. Firms $j \in J_2$ ($n_1 + 1$ to n) satisfy assumption P1′.

3. The set $X = \{x \geqslant 0 \mid x \geqslant \Sigma_j y_j + \Sigma_i \omega_i,\ y_j \in Y_j\}$ is bounded above.

The welfare program is

$$\max \Sigma_i \alpha_i u_i(x_i),$$

$$x_i \geqslant 0, \text{ all } i,\ q_j \geqslant 0,\ v_j \geqslant 0, \text{ for } j \in J_1,\ y_j, \text{ all } j,$$

subject to (10.3)

$$\Sigma_i x_i - \Sigma_j y_j \leqslant \Sigma_i \omega_i,$$

$$y_j \leqslant b_j q_j - v_j,\ q_j \leqslant f_j(v_j) \qquad \text{for } j \in J_1,$$

$$y_j \in Y_j \qquad \text{for } j \in J_2.$$

Under assumptions PI1, C1′, and C2′, program (10.3) is feasible (since $q_j = 0$ is feasible) and bounded (by assumption PI1, part 3). Hence it has an optimal solution. Consider the optimal q_j for $j \in J_1$, and solve the convex program that takes these optimal q_j as given parameters rather than as choice variables. Assume now that there exist bounded Lagrange multipliers associated to the balance constraints.[17] Then we can decentralize the solution as follows:

1. Markets clear at prices that are equal to the multipliers.

2. Producers $j \in J_1$ minimize costs taking these prices as given, and supply the (possibly zero) output q_j forced upon them; producers $j \in J_2$ maximize profits at given prices.

3. Each consumer maximizes utility, subject to his budget constraint at these prices; his income consists of the value of his endowment, his share in profits (or losses), and income transfers T_i.

Figure 10.7 shows possible tangency points between the profit line (with slope c/p) and the production function of a firm with two increasing returns intervals. At given prices (p, c), point B is profit maximizing for the firm, but it may not be compatible with the optimum of (10.3). If point B happens to be chosen by the planner, this firm's actions can be decentralized, since the tangent at B intersects the q-axis above the origin. If the planner chooses D, decentralization requires imposing a lower bound, so the firm's optimal choice will be D (where profit happens to be positive). However, points A

17. Note that for given optimal q_j, (10.3) is a convex program. If for these optimal q_j the constraint set satisfies Slater's constraint qualification, such multipliers will exist.

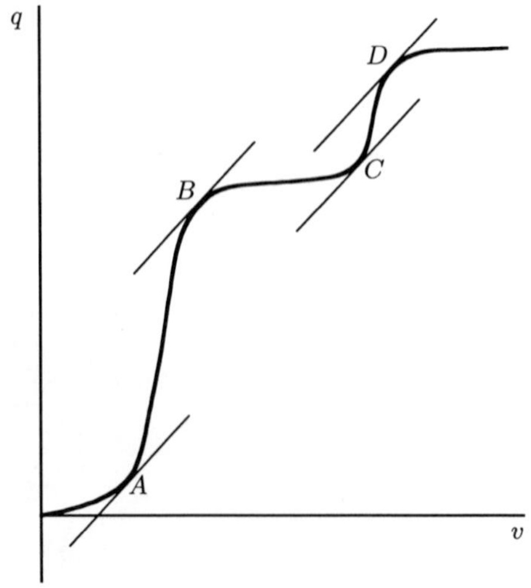

Figure 10.7
Possible optimal points under increasing returns.

or C amount to requiring the firm to minimize profits (at least locally). Hence, if one of these points happens to be chosen by the planner, there is no alternative but to force the firm to produce at these levels. This will create losses for the owners. Yet these owners will not face bankruptcy because they will receive the transfers they need to maintain their welfare weights.

The slope of the profit line will determine whether other points than A give profits or losses. Points in the first convex segment will always lead to losses. If this slope is steeper than in figure 10.7, due to higher costs, all points will eventually give negative profits. If the planner then wants to achieve B, he will face a situation similar to the covering of losses due to setup costs.

10.2.5 Increasing Returns: Equilibrium with Rules to Determine q_j and Zero Transfers among Consumers

It is commonly assumed that increasing returns to scale apply for all positive inputs, as in figure 10.8. Then, if the producer can freely set the scale of production, he will choose it to be infinite; this will yield an infinite profit, which is obviously incompatible with equilibrium. On the other hand, if the producer is forced to choose an output level that makes marginal cost equal to price, like at point A in figure 10.8, he will always incur a loss. For other quantity setting rules (e.g., point E), profits will be positive. This indicates that decentralization requires quantity setting either by the central planner or according to some behavioral rule other than profit maximization.

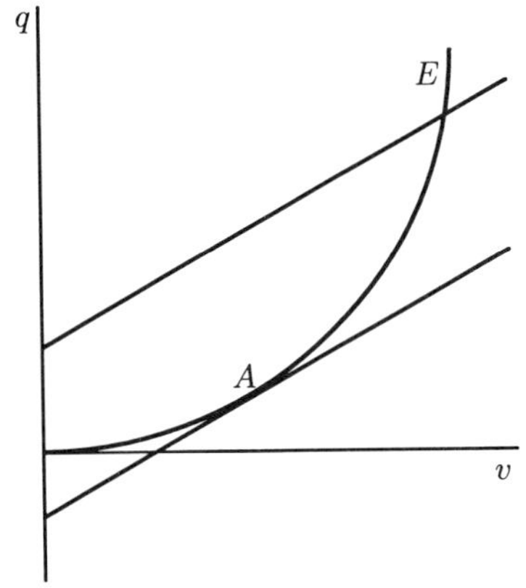

Figure 10.8
Profits under increasing returns.

To focus on the increasing returns aspect, we assume that all firms $j \in J_1$ have increasing returns everywhere; that is, they have cost functions that are strictly concave in q and defined for all $q \geqslant 0$ so that all finite production levels can be attained with finite inputs by each firm. Later on, we return to the more general case of figure 10.7.

In the equilibrium with setup costs and no transfers among consumers, the active and inactive firms (the δ_j values) were selected and kept unchanged, while the welfare weights were being adjusted to achieve zero transfers among consumers. This was needed to avoid the discontinuities that would result from changes in the δ_j. Under increasing returns, the q_j can vary in continuous fashion, and the problem is to set the q_j. These are real variables, so here the number of choices is infinite. This makes it necessary to set q_j endogenously by specifying rule 1 to determine the scale of production of firms and rule 2 to cover the losses.

One could think of several ways to determine the scale of production. We describe three rules: cost minimization by the group of firms that produce the same commodity, setting output so that average cost is equal to (or lower than) price, setting output so that marginal cost is equal to price.[18] We discuss the problems associated with these rules, without giving formal proofs, because there is no standard model as yet.

Cost Minimization by Groups of Firms

Under decentralization, total demand $\bar{q}$ for a commodity will be set by consumers and by the firms that use the commodity as input. If there are

18. See Bonnisseau and Cornet (1991) or Cornet (1988) for surveys.

no further requirements specified on the production by each firm, it is natural to assume that costs will not be minimized by each firm separately (as it is in the welfare problem), but that firms will seek to minimize the sum of their costs.

The associated cost minimization program is

$$\min \Sigma_j C_j(p, q_j),$$

$$q_j \geqslant 0,$$

$$\text{subject to} \qquad (10.4)$$

$$\Sigma_j q_j \geqslant \bar{q},$$

where the index j runs over firms that produce the same commodity. By the strict concavity assumption on the cost function, the program is feasible and has an optimal solution, which is a corner solution, so only one firm will produce the commodity.[19] Hence, among the firms with increasing returns, only one firm will survive. The economic interpretation is that it will be efficient to "specialize" in order to exploit increasing returns.

This choice is not necessarily Pareto-efficient, since it does not result from the planner's program. Since the cost minimization program is not convex, the q_j are not uppersemicontinuous, convex-valued multifunctions of $(p, \bar{q})$, and one cannot prove existence of equilibrium.

However, the specialization result makes it possible to reformulate the cost minimization problem allowing the δ_j to be real valued:

$$\min \Sigma_j \delta_j C_j(p, \bar{q}),$$

$$0 \leqslant \delta_j \leqslant 1, \text{ all } j,$$

$$\text{subject to} \qquad (10.5)$$

$$\Sigma_j \delta_j = 1.$$

This is a convex program, so that there will be no discontinuity, and an equilibrium will exist (assuming that losses can be covered through lump-sum transfers). In the optimum one can recover the q_j, since $q_j = \delta_j \bar{q}$. Whenever there is a single lowest-cost producer at equilibrium values $(p, \bar{q})$, cost minimization will set $\delta_j = 1$ for this producer, and $\delta_j = 0$ for all the others. The solution will then be integer valued and (10.5) will solve the original cost minimization problem (10.4), so the cost-minimizing equilibrium has been found. If there are several lowest-cost producers at given equilibrium values of $(p, \bar{q})$, the solution procedure suggested in (10.5) cannot be used.

19. See theorem A.5.2.

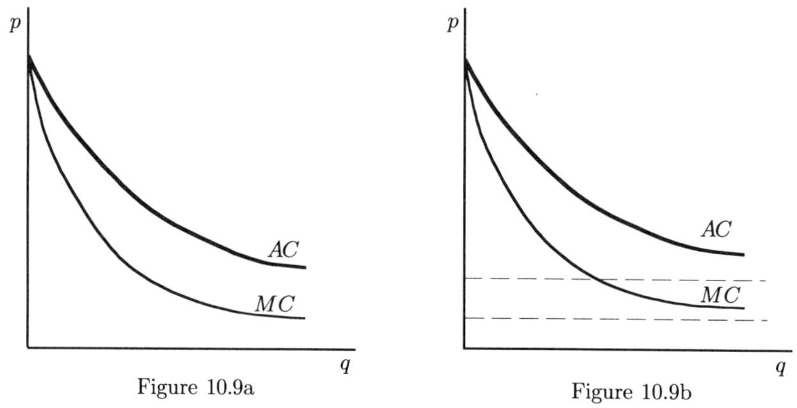

Figure 10.9
Average and marginal cost curves under increasing returns.

Average Cost Pricing

The rule of average cost pricing implies that the producer sets his output to ensure zero losses. He will then expand his output beyond the level at which price equals marginal cost, as can be verified from figure 10.9. Only the producers who can break even at the prevailing price will be active.

We first consider the case in which the average cost curve $C_j(p, q_j)/q_j$ extends to all nonnegative values, as in figure 10.9a. This means that the average cost curve will have an inverse, so there is a unique output choice for each price. The supply function of each producer is then continuous, homogeneous of degree zero in prices. Since there are no losses, consumer incomes remain positive, and their consumption can be calculated. Given the output levels, it is possible to compute input demand that satisfies the same properties as the outputs. Excess demand will now satisfy assumptions Z1 to Z4, proposition 1.4 applies, and an equilibrium exists.[20] Note that since firms do not make losses, there is no difficulty in satisfying positivity of consumer incomes at positive prices and Walras's law at the same time.

In figure 10.9b there is a minimum (the upper dotted line) to the average cost that the firm can reach. When prices fall below this minimum, the firm has to close down. At this point there is a discontinuity, so one cannot work with a (single-valued) supply function. One may instead treat the equality between price and average cost as an equilibrium condition and adjust outputs for firms with increasing returns, simultaneously with prices in order to eliminate excess demands and losses. This amounts to considering the

20. Since some producers $j \in J_2$ may have production sets that are not strictly convex, excess demands are uppersemicontinuous and convex valued rather than continuous. Hence proposition 3.4 applies provided that in the program of definition 3.5, the production set defined is for given q_j levels of the firms with increasing returns. The q_j are now feedback variables.

following (nonlinear complementarity) problem:

$$z(p, q_1, q_2, \dots, q_{n_1}) \leqslant 0 \perp p \geqslant 0, \tag{10.6}$$

$$pb_j \leqslant \frac{C_j(p, q_j)}{q_j} \perp q_j \geqslant 0, \qquad j = 1, 2, \dots, n_1. \tag{10.7}$$

The left-hand side in (10.6) represents aggregate excess demand: Supplies of increasing-returns firms enter as parameters, since they cannot be formulated as functions of prices. Conditions (10.7) are the average cost pricing relations (recall that b_j is a vector with a unit entry for the commodity that is produced, and zero elsewhere): Production is positive only if the price is equal to the average cost. Whenever price is below average cost, there is no production. This is the zero-profit condition that can be interpreted as a rule for entry (and exit) of firms, where firms are treated as indivisible.

Excess demand results by subtracting net supplies by firms in J_1 (those with increasing returns) and in J_2 (the other firms) from net consumer demand that results from standard utility maximization under the budget constraint, where income consists of the value of the endowments and profits of firms in J_2. Hence income is positive at positive prices. Out of equilibrium, firms in J_1 may have negative, zero, or positive profits, but these are not redistributed, so Walras's law is not satisfied. This causes the main difficulty in the existence proof.

Note that when the supply function is continuous (figure 10.9a), one could solve for q_j in (10.7) by setting the price equal to the average cost. This will yield a function $q_j(p)$ that can be substituted in (10.6) and again lead to the standard excess demand formulation.

Finally, if the output is chosen so as to keep price above average cost, say, by a fixed markup rate, there will be profits to be redistributed. Equation (10.6) will have to account for this and (10.7) will incorporate the markup.[21]

Marginal Cost Pricing

Under the rule of marginal cost pricing, the producer adjusts his output in order to equalize his marginal cost to the prevailing price. This is to reproduce the first-order conditions of the welfare program and is obviously a necessary but not a sufficient condition for welfare optimality, since the welfare program is not convex. These conditions will usually be verified at several points which, however, may not even be local optima. Therefore the rationale of the rule is debatable.

Formally, the marginal cost pricing rule is obtained as follows: We have seen that under decentralization of the welfare program, each producer in

21. See also sections 3.6.2, 10.5, and 11.1 to 11.4.

J_1 minimizes costs. Since the production function of these producers is strictly quasi-concave (assumption PI1, part 1), the cost function is differentiable with respect to prices. Shephard's lemma of proposition 2.4 can then be applied, and the continuous input demands $v_j(p, q_j)$ can be recovered. These can be substituted in the optimum of program (10.3) for given Lagrange multipliers p, assuming differentiability of the cost function with respect to q_j.[22] One then obtains the following Kuhn-Tucker conditions:

$$pb_j \leqslant \frac{\partial C_j(p, q_j)}{\partial q_j} \perp q_j \geqslant 0, \qquad j = 1, 2, \ldots, n_1, \tag{10.8}$$

where $C_j(p, q_j) = pv_j(p, q_j)$.

Equation (10.8) is the marginal pricing counterpart of the average pricing rule defined in (10.7): Production is positive only if price is equal to marginal cost. Whenever price is below marginal cost, there is no production.

When the marginal cost curve is shaped as in figure 10.9a, it defines a supply function, but when it appears as in figure 10.9b, there is a discontinuity in supplies (the marginal cost curve never falls below the lower dotted line), and one has to treat (10.8), like (10.7), as an equilibrium condition. The system then consists of (10.6) and (10.8), with (10.6) appropriately redefined to redistribute losses incurred by firms in J_1.

These losses will always be present since $(\partial C_j(\,.\,)/\partial q_j)q_j < C_j(\,.\,)$, by strict concavity of the cost function w.r.t. q_j and because $C_j(p, 0) = 0$. Hence, by (10.6), $(pb_j)q_j < C_j(\,.\,)$. Thus in equilibrium the producer makes a loss, unlike in the case of average cost pricing. These losses can make consumer's income negative and create problems for existence of equilibrium.[23]

10.2.6 Increasing Returns: Equilibrium for Concave-Convex Cost Functions

When the production function is shaped as in figure 10.7, the associated cost function will be concave on some intervals and convex on others. Marginal cost pricing will not lead to a unique choice of output, since for the given prices all the points A, B, C, and D qualify. For average cost pricing the ray through the origin may also have several intersections with the production function, so there are also many possible choices. Yet the formulation of (10.7) and (10.8) applies, since it does not require uniqueness of the solution. This leads to a model that may have multiple equilibria, without any

22. Bonnisseau and Cornet (1991) do not impose this restriction and work with the subgradients of the cost function. (See definition A.1.16 and theorem A.3.4 for the relation between Lagrange multipliers and the subgradients of the cost function.)

23. To avoid negative incomes, one could allocate an endowment ω_j to firm j so that $(pb_j)q_j + p\omega_j \geqslant C_j(p, q_j)$ for all (p, q_j); this may be feasible because the cost curve is concave. Alternatively, one could impose restrictions on the inputs that the firm can use, for example, by assuming that it only uses the endowment of its owners as inputs.

criterion for selecting a particular one. This is also the case of multiple equilibria in the competitive equilibrium, but there all agents maximize the same objective in all equilibria and each equilibrium is Pareto-efficient. Here the marginal pricing firm may find itself maximizing profits in one equilibrium and minimizing these in another (points B and A, respectively). The various equilibria are usually not Pareto-efficient, and it may happen that one of them Pareto-dominates another.

10.2.7 Solving Models with Indivisibilities

We have seen that nonconvexities within the firm can be handled in the competitive framework: When machines are sufficiently small and numerous, their number may be considered a real variable that can be treated as a separate commodity n_j used as input in production. When firms are small relative to the market, this may also be an acceptable representation.

Conversely, large nonconvex firms can have a significant impact on prices, even if they are price takers, by their decision to be active or not (under setup costs) or by their choice to be active on a particular segment of the production function (when there are increasing returns segments). It is this "power" that creates the discontinuities and the possible lack of existence of a fully decentralized equilibrium solution. It signals lack of perfect competition, and it may provide justification for public intervention. Though one should not expect decentralization to work in the presence of setup costs, the central planner can find an efficient allocation by solving a mixed integer program. The computation of an equilibrium for exogenously given values of the integer variables proceeds as in the competitive model provided that all firms can cover their losses. We have discussed institutional arrangements that facilitate this.

If production functions are continuous but have increasing-returns segments, the planner's problem is a nonconvex program for which all the stationary points (local minima, maxima, or saddle points) have to be computed and compared. This involves the use of approximation methods that are currently operational for small-sized models only. The equilibrium version of this problem is somewhat simpler: Due to the continuity of the production functions, it can be formulated without integer variables. We have considered three modes of decentralization: cost minimization by firms producing the same commodity, average cost pricing, and marginal cost pricing. All three were seen to generate a system of nonlinear equalities or a nonlinear complementarity problem, which is computationally tractable. Only the first mode seems to provide a realistic decentralization, for it is in line with profit maximization by producers who face given demand and have to take prices as given. Both the average and the marginal cost pricing rules conflict with profit maximization (but may be seen as forms of markup pricing, and hence of imperfect competition) and go only halfway in the direction of decentralization, since they replace central quantity setting by imposed rules of conduct.[24]

The marginal pricing equilibrium may in some cases be efficient. The system of equations will then correspond to the first-order conditions of the planner's program (supplemented by equations that require zero transfers among consumers). If one can prove uniqueness of a solution to this system, this solution will be a global welfare optimum.

10.3 Nonconvexities at Above-Firm Level

At economy wide or sector level, nonconvexities can occur in two ways. First, there may be large firms with nonconvexities (e.g., a national airport) that supply a rival input to smaller firms. It may be difficult to decentralize the production plan of such national firms in an efficient way, but there is less need for doing so when their number is limited. At any rate, for this type of nonconvexity, the earlier discussion applies.

Second, private firms may need a nonrival input supplied or used in a nonconvex way. This nonrival good may be produced separately (research and development) or as the by-product of activities that take place anyway (e.g., the benefits from experience in learning by doing and economies of scope). Here it is the nonrivalry in relation to the nonconvexity with which we will be concerned. Note that if the nonrivalry is not associated with nonconvexity, the treatment of section 9.1 applies.

We proceed as before, starting from a welfare program that is decentralized through price and quantity signals. As in chapter 9 prices provide sufficient signals if technologies are convex, but we will see that under nonconvexities, existence of an efficient equilibrium without transfers among consumers is not ensured. If the nonrival good has a zero price (because its market is not represented), an equilibrium will exist, though it will be inefficient. Learning by doing and economies of scope may lead to such a representation.

10.3.1 Nonrival Goods Produced Intentionally: Research and Development

We first consider the supply of a nonrival good y_g^+ by a specific firm g. For this we reproduce program (9.10):[25]

$$\max \Sigma_i \alpha_i u_i(x_i),$$

$$x_i \geqslant 0, \text{ all } i, \ y_g^-, \ y_g^+ \geqslant 0, \ g_j \geqslant 0, \ y_j, \text{ all } j,$$

24. Average cost pricing is compatible with profit maximization if there is free entry, and the number of firms can be treated as a real variable. But then there is no indivisibility, and we are back in the case of section 10.1.

25. Here we assume that Lagrange multipliers exist.

subject to (10.9)

$$\Sigma_i x_i - \Sigma_j y_j + y_g^- - y_g^+ \leqslant \Sigma_i \omega_i \qquad (p),$$

$$g_j = y_g^+ \qquad (\phi_j),$$

$$F_j(y_j, -g_j) \leqslant 0,$$

$$F_g(y_g^+ - y_g^-) \leqslant 0.$$

The nonrival production y_g^+ is supplied by firm g according to the transformation function $F_g(y_g^+ - y_g^-)$ and is used as input g_j by any other firm j according to the transformation function $F_j(y_j, -g_j)$. The constraint $g_j = y_g^+$ indicates nonrivalry: Producers have to agree on the level of production of the nonrival good. ϕ_j is the contribution of firm j to the cost of this good.

Various possibilities can now be distinguished:

1. If $F_j(.)$ and $F_g(.)$ are strictly quasi-convex (sqc), we are back in the case discussed in section 9.1.4: The ϕ_j can be used to decentralize the operations of firm g, and at these prices all firms will agree on the provision of the nonrival goods. Also $\Sigma_j \phi_j = p$. Equilibrium without transfers among consumers will exist and be efficient.

2. If $F_j(.)$ is sqc but $F_g(.)$ is not, program (10.9) is nonconvex, and y_g^+ as well as y_g^- have to be set centrally. For this one may consider the convex program with the y_g^+ and y_g^- levels entering as parameters rather than as choice variables so that decentralization can be achieved. It will be optimal for firms to use this output, provided that they are faced with prices ϕ_j. This will not leave them with negative profits, and as before, $\Sigma_j \phi_j = p$. However, this price may be insufficient to cover the cost of firm g which produces the nonrival good. In the absence of transfers among consumers, equilibrium may fail to exist. Rules like marginal cost pricing may be needed to guarantee existence, but as discussed in section 10.2, this equilibrium will often be inefficient.

3. If $F_g(.)$ is sqc but $F_j(.)$ is sqc in y_j only, then program (10.9) is again nonconvex and all the g_j have to be set equal to y_g^+ by the central planner. One can as in case 2 consider (10.9) with the $g_j = y_g^+$ fixed. This is a convex program, so decentralization can follow. For given $g_j = y_g^+$, each firm can maximize profits. Since it has free disposal of inputs, the possibility of inaction ensures nonnegativity of py_j. However, $py_j - \phi_j g_j$ may be negative. Clearly, if the firm receives the nonrival input h freely, ϕ_{jh} will be zero for that input, so it will not cause the negative profits problem, though inefficiency will prevail.

So far, the discussion assumes that firms face known prices ϕ_j and that they do not free-ride. Obviously, if no firm pays for the nonrival good, it will not be produced and can therefore not be used. Since now all nonconvexities are "inactive," equilibrium will exist but be inefficient.

10.3.2 Nonrival Goods Produced as By-products: Economies of Scope and Learning by Doing

There are also cases in which the nonrival good results as a by-product of ordinary production: When many computers (ordinary commodities) are sold, "computer literacy" (the nonrival by-product) spreads also. We represent this by assuming that the output y_g^+ is obtained from the transformation of total output $\Sigma_j y_j^+$ via a transformation function $F_g(.)$ and is available as a nonrival input to all firms j. This leads to the following modification of program (10.9):

$$\begin{aligned}
&\max \Sigma_i \alpha_i u_i(x_i),\\
&x_i \geqslant 0, \text{ all } i,\ y_g^+,\ g_j,\ y_j^+,\ y_j^- \geqslant 0,\ y_j, \text{ all } j,\\
&\text{subject to} \qquad\qquad (10.10)\\
&\Sigma_i x_i - \Sigma_j y_j \leqslant \Sigma_i \omega_i \qquad (p),\\
&g_j = y_g^+ \qquad (\phi_j),\\
&F_j(y_j, -g_j) \leqslant 0,\\
&F_g(y_g^+ - \Sigma_j y_j^+) \leqslant 0,\\
&y_j = y_j^+ - y_j^-,\\
&y_j \in Y_j,
\end{aligned}$$

where y_j^+ and y_j^- represent the outputs and inputs of firm j.

The constraint $g_j = y_g^+$ indicates nonrivalry. Though $\Sigma_j y_j^+$ is only a "by-product," producers will have to agree on its level. The ϕ_j are the associated contributions for y_g^+. Just as for intentionally produced nonrival goods, several possibilities may be distinguished:

1. If $F_j(.)$ and $F_g(.)$ are strictly quasi-convex (sqc), we are back in the case discussed in chapter 9. But decentralization of the optimum is more problematic: Inefficiency from free-riding is even more likely now because the nonrival output will be produced anyway, although not necessarily in the appropriate quantity, and because it may be difficult to exclude nonpaying users (particularly of nonrival goods like knowledge). Note that efficiency requires firm j to pay a contribution ϕ_j for the nonrival goods and to receive a premium for its outputs as a supplement to the market price p.

2. If $F_j(.)$ is sqc but $F_g(.)$ is not, then program (10.10) is nonconvex.[26] y_g^+ has to be set centrally, and pricing has to secure the appropriate supply of input for its production and hence of y_j^+.[27] This is a very strong requirement, but in principle, the comments made for case 2 in section 10.3.1 apply.

3. If $F_g(.)$ is sqc but $F_j(.)$ is sqc in y_j only, then program (10.10) is nonconvex. All the g_j have to be set equal to y_g^+, and this leads to the same difficulties as in case 2.

The discussion again assumes that prices ϕ_j are available and that there is no free-riding. Here, even if there is free-riding by all firms, the nonrival goods will be available. In the extreme case where no firm pays or is rewarded for the nonrival good, one obtains a model in which g_j is taken as given by each firm. This generates production plans (y_j^+, y_j^-) which in turn make $\Sigma_j y_j^+$ available to generate y_g^+. Since the nonconvexities enter through variables given for every firm, they do not create discontinuities. Hence an equilibrium exists[28] where all markets clear and the condition $g_j = y_g^+$ is satisfied for all j. Obviously this equilibrium will, in general, not be efficient.

This is the version that is often used in the context of economies of scope. Learning by doing can also be represented in this way, but it is better thought of as a dynamic process of accumulation of knowledge through a nonconvex technology. The knowledge automatically benefits individual producers as a nonrival input. There is an output, the supply of experience, that results from an increasing-returns production function with total past production levels as inputs. Knowledge then serves as a nonrival input to firms. The central planner does not have to control switches, but his intervention is required in order to provide appropriate price signals to individuals who have to pay for the experience and are rewarded for their contribution to it. Due to the nonconvexity, this will cause losses.

In the specification of learning by doing we make time subscripts explicit, as in chapter 7. We incorporate the use of nonrival goods g_{jt} as inputs into the recursive production function $F_{jt}(.)$ of the producer model (7.1):

$$\max \Sigma_t p_t y_{jt},$$

$$k_{jt}, \tilde{k}_{j,t+1} \geqslant 0, \; y_{jt}, \; t = 1, 2, \ldots, T,$$

26. Program (10.10) can be expressed in terms of y_j^+ and y_j^- (after elimination of y_g^+ and g_j). Therefore the nonconvexity in the nonrival good does not necessarily imply nonconvexity of the constraint set. In the discussion we assume that the nonconvexity cannot be eliminated.

27. This is possible if the production function underlying the transformation function $F_g(.)$ is strictly quasi-concave; in the more general case, even $\Sigma_j y_j^+$ has to be set centrally.

28. For example, if the variable y_g^+ is a continuous function of $\Sigma_j y_j^+$, then $g_j = y_g^+$ can be treated as feedback variables in the fixed point mapping.

subject to

$$F_{jt}(y_{jt}, \tilde{k}_{j,t+1}, k_{jt}, g_{jt}) \leqslant 0 \qquad (\rho_{jt}),$$

$$k_{jt} \leqslant \tilde{k}_{jt} \qquad (\psi_{jt}),$$

where $\tilde{k}_{j1}$ is given.

Under learning by doing, experience g_{jt} is usually specified as a capital stock that accumulates as a function of total production in the economy. This model has gained acceptance in the literature on endogenous growth.[29] The allocation will clearly be inefficient unless firms are rewarded for their contribution to the production of experience.

10.4 Nonconvexities in Consumption

In the consumer problem nonconvexities may appear in the utility function (strict quasi-concavity, or switches between preferences or locations) in the indivisible character of the commodities and in the relation between endowments and consumption.

10.4.1 Nonconvexities in the Utility Functions and Indivisible Commodities

Nonconvexities in consumption can be viewed as if they were occurring in production because the utility function $u_i(x_i)$ may be interpreted as a consumer-specific production function that yields a nonmarketed commodity called "utility." This is illustrated in the following welfare program, where, as usual, subscripts i, j, and k refer to consumers, producers, and commodities, respectively, while $h = 1, 2, \ldots, H$, refers to consumers' lifestyles:

$$\max \Sigma_i \alpha_i \tilde{u}_i,$$

$$x_i \geqslant 0, \text{ all } i, \ \kappa_{ik}, \text{ all } i \text{ and } k, \ \varepsilon_{ih}, \text{ all } i \text{ and } h, \ y_j, \text{ all } j,$$

subject to (10.11)

$$\tilde{u}_i \leqslant \Sigma_h \varepsilon_{ih} u_{ih}(x_i),$$

$$\Sigma_h \varepsilon_{ih} = 1, \qquad \text{(a)}$$

$$\varepsilon_{ih} \in \{0, 1\},$$

$$C x_i = \kappa_i, \qquad \text{(b)}$$

$$\kappa_{ik} \in \{0, 1, 2, \ldots\},$$

$$\Sigma_i x_i - \Sigma_j y_j \leqslant \Sigma_i \omega_i,$$

$$y_j \in Y_j.$$

29. See Romer (1991) for a survey.

In constraints (a), $u_i(.)$ is not necessarily concave in x_i. The integer variables ε_{ih} also cause a nonconvexity. They operate like parameters that perform switches between utility functions (given the restriction $\Sigma_h \varepsilon_{ih} = 1$, ε_{ih} will be equal to unity for only one h). Depending on the ε_{ih} chosen, a particular utility function $u_{ih}(.)$, associated with one of the H lifestyles or locations will apply. The ε_{ih} may also relate to the activities of the consumer: having employment or not, migrating to town or living in the countryside, and so on. In constraints (b) the possible indivisibility of purchases is represented: C is a matrix with entries 0 or 1, and any indivisible commodity k will have a nonzero entry in its column; it has a row for each commodity with indivisibilities. The two sets of constraints may be viewed as a consumer-specific nonconvex production system that supplies $\tilde{u}_i$ from inputs x_i and ε_i. Consumer i owns a firm with setup costs (the integer-valued purchases) and with a nonconcave "production" function $\Sigma_h \varepsilon_{ih} u_{ih}(x_i)$.

For positive welfare weights α_i, the solution of (10.11) defines a Pareto-efficient allocation,[30] so decentralization can, in principle, proceed as in production. The central planner sets the integer variables and the real variables that appear in nonconcave functions, while the other variables can be left to individual agents. Agents' decisions can be coordinated on competitive markets. This may leave the impression of an almost fully decentralized economy, but actually consumers have to be rationed whenever their utility function is not concave.

Fortunately (10.11) describes an extreme case. Nonconvexity appears everywhere in the consumer problem: Commodities are indivisible, the utility function generates nonconvex indifference curves, and it contains switches. We will argue that it may be acceptable to make nonconvexities in consumption less pervasive.

Nonconcave Utility Functions

In program (10.11) the utility function $\Sigma_h \varepsilon_{ih} u_{ih}(x_i)$ is not concave because the ε_{ih} are integers and because $u_{ih}(x_i)$ may lack concavity. Note, however, that since only one ε_{ih} will be nonzero, actual consumer behavior will be determined by the associated $u_{ih}(.)$. The nonconcavity of $u_{ih}(.)$ can be ignored if one accepts the generalized axiom of revealed preference.[31,32]

30. The program is feasible and bounded and thus defines an optimal allocation. Suppose that this allocation were not Pareto-efficient. Then it would be possible to increase the utility of some consumer, without reducing it for some other. But since all α_i are positive, this contradicts optimality of the allocation. Note that this argument requires positivity of all α_i.

31. See proposition 2.9 in section 2.2.4.

32. However, this does not extend to nonrival commodities, since revealed preference cannot be tested there. If rival consumption involves no indivisibilities, the consumer problem is convex in x_i, once nonrival consumption x_g has been set by the planner and the discussion on nonrivalry in section 10.3 applies. If x_g has to be produced intentionally, an efficient equilibrium will exist but needs transfers; if x_g is a by-product, an equilibrium without transfers exists but is inefficient.

We therefore assume that $u_{ih}(x_i)$ is a strictly concave subutility function. Let $e_{ih}(p, u_i)$ be the associated subexpenditure function. The resulting expenditure minimization problem is

$$e_i(p, u_i) = \min \Sigma_h \varepsilon_{ih} e_{ih}(p, u_i),$$

$$\varepsilon_{ih} \in \{0, 1\},$$

subject to

$$\Sigma_h \varepsilon_{ih} = 1.$$

As in the cost minimization problem (10.5) the integer restriction on the ε_{ih} can be relaxed, since the consumer will then choose the cheapest way of achieving his utility, and this will be integer valued if the choice is unique. This linear minimization problem can be incorporated within the equilibrium model. If it is the only nonconvexity, an equilibrium without transfers among consumers will exist, and if the equilibrium ε_{ih} are integer valued, a single lifestyle is chosen by each consumer. This equilibrium thus solves the mixed integer-valued model and is Pareto-efficient.

Indivisibility

Virtually all commodities are purchased in integer numbers, even flour and milk, but such indivisibilities may be neglected for commodities purchased on a regular basis. This is the counterpart for consumption of within-firm indivisibilities in production. Obviously, for purchases of consumer durables, this representation is less appropriate because the indivisibility matters.

Indivisibilities generate discontinuities in the set of feasible allocations. They can be disregarded as long as they are too small to alter significantly the Lagrange multipliers of the commodity balances in (10.11).[33]

This argument has been formalized by Aumann (1964, 1966) (and applied in Hildenbrand and Hildenbrand 1986, Härdle et al. 1991) who consider a continuum of consumers. Assume that income is distributed according to a given continuous density function $f(\theta)$, where θ is a parameter (e.g., the consumer's endowment or his age) that characterizes a consumer with income $h(p, \theta)$ and with utility function $u(x, \theta)$.[34] Then aggregate demand $X(p)$ is given by

$$X(p) = \int x(p, h(p, \theta), \theta) \, f(\theta) d\theta. \tag{10.12}$$

33. See also the discussion of figure 10.4.

34. Note that the continuity assumption on $f(\theta)$ implies that all consumers are different. Although this argument is not restrictive in theory, it is difficult to apply, since this requires very detailed data on individual consumers.

Discontinuities in $x(.)$ will not create discontinuity of the aggregate demand function if they only appear for a finite number of values of θ.[35] The economic intuition behind this property is that consumers are sufficiently different to ensure that those who have a discontinuity in demand at the given prices are so few that they do not affect aggregate demand. Obviously it is only possible to design specifications of consumer models that satisfy this assumption if the discontinuities occur at very few points. This means that the number of indivisible commodities should be limited, the vector θ should be of small dimension (usually a scalar), and the functions $h(p, \theta)$ and $f(\theta)$ should be tractable analytically.[36]

10.4.2 Nonconvexity and Efficiency Wages

Nonconcavity of the efficiency-wage relation can take two forms: Either the relation $\omega_i(x_i)$ itself is nonconcave, or there is an additional parameter δ_i, the job, that is integer valued so that one has $\omega_i(x_i, \delta_i)$ with $\delta_i \in \{0, 1\}$ and where ω_i is concave in x_i. We discuss both cases but will not be able to prove existence of a Pareto-efficient equilibrium without transfers among consumers. We subsequently make the assumption that agents are infinitesimally small, and this will make it possible to prove such properties.

Increasing Returns of Endowments with Respect to Consumption

As in section 9.1.5 we make the labor endowment of consumer i dependent on his consumption as $\omega_i(x_i)$, but here the relation is not required to be concave because there is a level of subsistence below which no labor can be supplied. Hence we replace assumption CX2 by CX2′:

ASSUMPTION CX2′ (Nonconcave efficiency-wage relation) The function $\omega_i(x_i)$ is continuous, nondecreasing, and uniformly bounded from above; $\omega_i(x_i) \geqslant 0$, $\omega_i(x_i) \neq 0$. Moreover $\omega_1(0) > 0$.

It may happen that the lack of concavity of $\omega_i(.)$ is sufficiently mild to maintain convexity of both the planner and the consumer problems. This will be the case as long as at given welfare weights and for the associated Lagrange multipliers p in the welfare program, $\alpha_i u_i(x_i) + p\omega_i(x_i)$ is concave in x_i for all consumers. Then equilibrium exists and is Pareto-efficient, as in section 9.1.5. We now discuss cases in which the nonconvexity matters.

We first consider the planner's problem. To facilitate exposition, we disregard profits and restrict ourselves to the special case in which ω_i

35. This is a sufficient condition; the necessary condition is that the integral, taken over values of θ at points where a discontinuity occurs should be zero.

36. Note that in estimating the density function $f(\theta)$, nonparametric methods are used (e.g., kernel estimators), but currently such methods are only available for scalar θ. See Härdle et al. (1991).

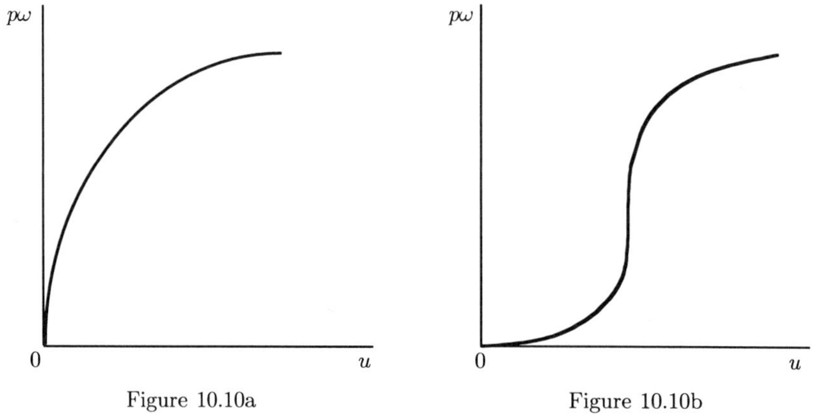

Figure 10.10
Concave and nonconcave efficiency-wage relations.

depends on $u_i(x_i)$ rather than on x_i directly.[37] The relation is represented in figure 10.10 for the concave and nonconcave situations.

If, as in figure 10.10a, $\omega_i(x_i)$ is concave, propositions 9.4 and 9.5 on existence and Pareto-efficiency hold.[38] In the nonconcave case a complication arises. Indeed, the planner's problem is not a convex program, but it has an optimal solution. As before this solution cannot always be decentralized without the planner's intervention. Since the nonconcavity is in $\omega_i(u_i(x_i))$, he may decide to set the x_i centrally.

Decentralization is also possible when the planner sets incomes, but we show that in this case lump-sum taxation will not achieve this objective. If the consumer is faced with given income h_i, he will solve $\max\{u_i(x_i) \mid px_i \leqslant h_i, x_i \geqslant 0\}$, and the optimal u_i will determine $\omega_i(u_i)$. Figure 10.11 illustrates how decentralization can be implemented in the concave and the nonconcave cases.

In the concave case of figure 10.11a, utility maximization will lead the consumer to select the intersection of $p\omega(u)$ and $e(p, u)$ as his optimal point, if he does not have to pay taxes. The planner can force him to utility level C by levying a tax AB. In the nonconcave case shown in figure 10.11b, the untaxed consumer will also choose the intersection of the endowment and the expenditure function.[39] Suppose now that the planner wants him to be at utility level D; then levying a tax EF, which happens to be equal to AB,

37. Here the welfare and the consumer programs will remain convex if $p\omega_i(u_i) - e_i(p, u_i)$ is concave in u_i.

38. These propositions are specified for the more general function $\omega_i(x_i)$.

39. Note that in the nonconcave case there may be several such intersections. The consumer will always choose the one at the utmost right because it gives him the highest utility.

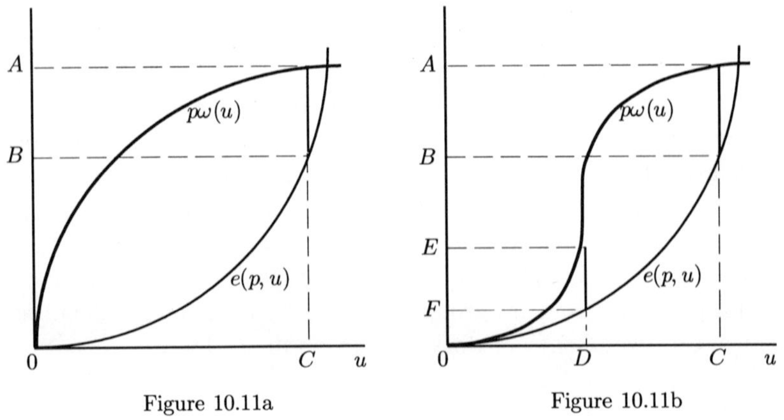

Figure 10.11
Decentralization under an efficiency-wage relation.

will not force the consumer to choose D because C is better for him. This shows that lump-sum taxation is not sufficient to achieve decentralization. The planner has to control total income $h_i = p\omega_i(.) + T_i$, reached at F.

In the general case with endowments depending on consumption rather than on utility, the planner has to set not only income but also consumer subsidies at $p\partial\omega_i/\partial x_i$. Again the consumer will solve a standard problem, and his optimal consumption will lead to socially optimal endowments $\omega_i(x_i)$.

An equilibrium without transfers exists if an institution is created that makes the consumer's income independent of his supply of labor power. The consumer will then solve a standard utility maximization problem and the supply of labor power will follow from optimal consumption, but again this equilibrium may not be Pareto-efficient.

Integer-Valued Jobs

The case of integer-valued jobs can be represented by integer switches that describe either setup costs or a convex segment of increasing returns to consumption. The case with setup costs can be treated along the same lines as for technology;[40] the efficiency-wage relation is $\omega_i(x_i, \delta_i)$, which the central planner has to set (δ_i is an integer variable equal to 1 if consumer i finds employment and 0 otherwise). He can communicate the optimal values through rations or premia: Some workers will receive an employment opportunity, but they may have to be induced to use it. Others will be left unemployed. For given δ_i an equilibrium with zero transfers among consumers can be obtained as in section 10.2.3.

40. See section 10.2.

Infinitesimally Small Agents

It is also possible to specify a model with a continuum of consumers, and this will lead to an efficient competitive equilibrium (without transfers) provided that consumers are sufficiently different, as in (10.12), or are identical within groups.[41] The assumption that all agents are different (in terms of their supply and demand behavior) may not be very realistic in the efficiency-wage model because unemployed workers who live under the same poor conditions will have similar $\omega_i(x_i, 0)$ functions.[42] Here we specify a model that assumes that all consumers within one group i ($i = 1, 2, \ldots, m$) are infinitesimally small but identical. Their behavior does not depend on θ. Hence

$$\int x_i(p, \delta_i(p, \theta), \theta) f_i(\theta) d\theta = \mathrm{x}_i(p, \delta_i(p)) n_i,$$

where $n_i = \int f_i(\theta) d\theta$ is a measure of the "number" of consumers. The model may then be formulated in the standard Negishi format. Each consumer group i is subdivided into employed consumers (e) and unemployed consumers (f, for free). The variables e_i and f_i measure the number of employed and unemployed consumers in group i, ($e_i + f_i = n_i$). For given $\delta_i = \{0, 1\}$, the endowment function satisfies concavity (assumption CX2), but we allow for setup costs (negative $\omega_{ik}(x_i^e, 1)$ for some k). Profits from firms are shared equally among group members. The resulting welfare program may be written

$$\begin{aligned}
&\max \Sigma_i(\alpha_i^e e_i u_i(x_i^e) + \alpha_i^f f_i u_i(x_i^f)),\\
&e_i, f_i \geqslant 0,\ x_i^e, x_i^f \geqslant 0, \text{ all } i,\ y_j, \text{ all } j,\\
&\text{subject to} \qquad\qquad\qquad\qquad\qquad\qquad\qquad\qquad (10.13)\\
&\Sigma_i(e_i x_i^e + f_i x_i^f) \leqslant \Sigma_i(e_i \omega_i(x_i^e, 1) + f_i \omega_i(x_i^f, 0)) + \Sigma_j y_j,\\
&e_i + f_i = n_i,\\
&y_j \in Y_j.
\end{aligned}$$

In equilibrium the welfare weights α_i^e and α_i^f are such that in each group i the budget constraints for employed (e) and unemployed (f) consumers in group i are satisfied:

$$px_i^e = p\omega_i(x_i^e, 1) + \frac{\Sigma_j \theta_{ij} \Pi_j(p)}{n_i}, \tag{10.14a}$$

41. Dasgupta and Ray (1986) and Dasgupta (1993) assume that agents are different.

42. Moreover the representation through a continuum of agents leads to a model that lacks transparency (because measure theoretic concepts have to be employed) and to inconclusive propositions on existence and efficiency (because only generic results can be proved). It is also difficult to develop applied models along this line because no standardized algorithms are available that solve such models.

$$px_i^f = p\omega_i(x_i^f, 0) + \frac{\Sigma_j \theta_{ij} \Pi_j(p)}{n_i}. \tag{10.14b}$$

Denote by $\tilde{x}_i^e = e_i x_i^e$ and by $\tilde{x}_i^f = f_i x_i^f$ total consumption of the employed and the unemployed in group i, respectively. Under the conditions of theorem A.1.5, the extended functions $\tilde{u}_i(\tilde{x}_i^e, e_i) = e_i u_i(\tilde{x}_i^e/e_i)$ and similarly for $\tilde{\omega}_i$ and for the unemployed will be concave, nonstationary, nondecreasing (and homogeneous of degree one) in total consumption and the number of consumers. Therefore program (10.13) can be rewritten as a convex program with y_j, e_i, f_i, and total consumptions $\tilde{x}_i^e$ and $\tilde{x}_i^f$ as choice variables. Hence the Negishi proof of proposition 3.1 can be used, and a solution exists that is Pareto-efficient if employed and unemployed are treated as different agents.

This solution can be decentralized with job rationing. First, every consumer is given a status as employed or unemployed worker. This allocation is not achieved voluntarily through supply and demand on a market but it is forced upon individuals by the planner. While the number of people who find employment results from the equilibrium conditions, the actual selection of the employed is arbitrary, so identical individuals are treated differently and reach different utility levels. Second, each consumer maximizes his utility subject to a budget constraint (10.14a) or (10.14b). To make him "freely" choose the employment $\delta_i = 0$ or 1, the unemployed worker is faced with a zero employment ration and the employed worker with a unit ration, possibly supplemented by a self-financed wage subsidy. This shows that the process of job allocation actually generates a social stratification.

An important property of the resulting allocation of jobs follows from Pareto-efficiency itself. No unemployed worker can gain by paying the employed for giving him his job, and vice versa. A further property of the equilibrium is that $\alpha_i^e u_i(x_i^e) = \alpha_i^f u_i(x_i^f)$, whenever e_i and f_i are positive: Hence identical individuals will reach equal "money metric utility"[43] but not necessarily equal utility. The employed may be allowed to consume more than the inactive because it makes them more productive. This can lead to cases in which it is efficient to let the unemployed live in extreme poverty because scarce consumer goods should be reserved for those who are employed. Note that if endowments are not a function of consumption, all individuals which belong to the same group i will reach the same utility level,[44] but then, as a rule, either e_i or f_i will be zero.

In an alternative specification, each group may be left to decide for itself as to whom within the group will work. It then faces a single budget

43. This follows from the first-order conditions of (10.13) evaluated in equilibrium.

44. More generally, this will also be the case if $\partial\omega_i(x_i, 0)/\partial x_i = \partial\omega_i(x_i, 1)/\partial x_i$.

constraint, and there will be only one welfare weight for the group ($\alpha_i^e = \alpha_i^f$). The resulting equilibrium will involve within-group transfers (e.g., from the employed to the unemployed), but money metric utilities may differ. This may be interpreted as follows: The individual faces uncertainty as to whether he will become employed. His likelihood to become employed is e_i/n_i. Hence individuals within the same group i could organize a social security scheme with transfers, and this would lead to Pareto-superior outcomes as compared with the model with different welfare weights for employed and unemployed.

The discussion shows that this model may be well suited to represent the process of social stratification. Though it involves several nonconvexities like setup costs and job indivisibility, and a continuum of consumers ending up in different social strata, the model has been cast in a standard setting so that it can be analyzed and solved by standard methods.

10.5 CGE Applications

In CGE models nonconvexities enter only in production, through single output production functions. Each good is produced by a single firm, which is always active, since there is demand for that good at all prices. Setup costs and losses from increasing returns are usually covered through average cost or even markup pricing in the context of imperfect competition. Two approaches can be distinguished. One with fixed number of identical firms, the other with variable (but real as opposed to integer-valued) number of identical firms.

For short-run analysis the number of firms is fixed (with n_k firms producing good k). It is assumed that each firm produces a fixed share in output and that prices are set either at average or at marginal cost. Alternatively, markup pricing is used. As in the standard CGE model of section 3.4.1, supply of goods is determined by demand. The balance equations are

$$\Sigma_i x_i^g(p^g, p^f, h_i) + A^g(p^g, p^f)q^g = q^g \qquad \text{for goods,}$$

and

$$\Sigma_i x_i^f(p^g, p^f, h_i) + A^f(p^g, p^f)q^g = \Sigma_i \omega_i^f \qquad \text{for factors,}$$

where, for producing good k, the vector of input demand coefficients for goods and factors are

$$A_k^g = \frac{\partial C_k(p^g, p^f, q_k^g/n_k)}{\partial p_k^g} \cdot \frac{n_k}{q_k^g},$$

$$A_k^f = \frac{\partial C_k(p^g, p^f, q_k^g/n_k)}{\partial p_k^f} \cdot \frac{n_k}{q_k^g}.$$

The price and income equations now depend on the price setting rule that is adopted:

1. Under average cost pricing, the price equation for goods reads as

$$p^g = p^g A^g(p^g, p^f) + p^f A^f(p^g, p^f),$$

and the income equation is

$$h_i = p^f \omega_i^f.$$

2. Under markup pricing, the relations become

$$p^g = p^g A^g(p^g, p^f) + p^f A^f(p^g, p^f) + \mu^g(p^g, p^f, q^g),$$

$$h_i = p^f \omega_i^f + \Sigma_k \theta_{ik} \mu_k^g q_k^g,$$

where θ_{ik} is the share consumer i owns in firms of sector k and $\mu_k^g q_k^g$ are the proceeds from markups in that sector.

3. Under marginal cost pricing, for good k, the price is set according to

$$p_k^g = \frac{\partial C_k(p^g, p^f, q_k^g/n_k)}{\partial q_k^g/n_k}.$$

This generates a (usually negative) markup:

$$\mu_k^g = \frac{\partial C_k(p^g, p^f, q_k^g/n_k)}{\partial q_k^g/n_k} - \frac{C_k(p^g, p^f, q_k^g/n_k)}{q_k^g/n_k},$$

as difference between marginal and average cost and leads to the same income equation as for markup pricing. Obviously this may cause losses and negative incomes. It may thus happen that no solution exists.

In CGE models the long run is often represented by allowing n_k, the number of firms, to vary in a continuous fashion. Every firm is then assumed to set its scale of production in order to minimize its average cost, and average cost pricing holds. Obviously this is possible only if, eventually, decreasing returns take over like in figure 10.2b, where average cost is minimized at P'. This will lead to a continuous supply function $\tilde{q}_k^g(p^g, p^f)$ if the average cost function is strictly (quasi)convex in output. The input coefficients for this firm are

$$A_k^g = \frac{\partial C_k(p^g, p^f, \tilde{q}_k^g)}{\partial p_k^g} \cdot \frac{1}{\tilde{q}_k^g},$$

$$A_k^f = \frac{\partial C_k(p^g, p^f, \tilde{q}_k^g)}{\partial p_k^f} \cdot \frac{1}{\tilde{q}_k^g}.$$

The now variable number of firms n_k will follows from $n_k = q_k^q/\tilde{q}_k^q$, and the markup will be zero.

The convexity assumption on the average cost function required for long-run analysis, may be problematic for the short run because it implies that there are two output levels with the same average cost, so one of these has to be selected. If, on the other hand, minimal average cost supply is large relative to q (this happens when the increasing returns segment in figure 10.2b is relatively wide), the outcome is not meaningful since the number of firms is small (maybe even less than one) and, except by chance, noninteger.

Theoretically the variable number of firms approach is only valid if one assumes perfect divisibility of firms, which means that firms are infinitesimally small and numerous, like the spinning jenny of section 10.1. If the number is large but not infinite, one may view the outcome of the model as a reasonable approximation of integers, but then again, the increasing returns at firm level are of less importance.

The CGE model with nonconvexities in production (usually setup costs) has been applied to study tariff reforms and economic integration when nontariff barriers are removed. These applications tend to find welfare gains that are larger than with constant or decreasing returns to scale. The explanation given is that larger markets allow for lower costs, since the scale of production can increase, so the gains from specialization are more important than they would be under convex technologies.

The CGE literature tends to analyze nonconvexities only in conjunction with imperfect competition, as a means to cover losses arising from nonconvexities. The approach was pioneered by Harris (1984) and pursued by Burniaux and Waelbroeck (1992, 1994), de Melo and Roland-Holst (1994), Devarajan and Rodrick (1991), Dixon (1990), Mercenier (1994) and Gasiorek, Smith, and Venables (1992a, 1992b) among others. This is described in more details in section 11.5.

11 Imperfect Competition

There is imperfect competition as soon as any agent in the model no longer takes market prices as given and anticipates the effect of his own actions on these prices. Such an agent is said to be noncompetitive. Leaving the setting of perfect competition raises three basic issues:[1]

1. What is a rational rule of conduct for a noncompetitive agent? For example, is profit maximization rational from the perspective of the owner of the firm? Recall that under perfect competition, profit maximization follows from the second welfare theorem and leads to Pareto-efficient allocations. Therefore it is rational for the owner of the firm to impose this as a criterion. But when prices are no longer taken as given, the criterion loses its validity, as will be seen.
2. How does an agent perceive his environment? Under perfect competition, prices (and in some cases, rations or tax rates) are the only signals that reach the agent. He does not need any further signals nor knowledge of his environment other than his budget constraint or his technology set to make a rational decision. Under imperfect competition, he will also need to know the demand schedules of his customers for the commodities he supplies, and possibly more. For example, does he take into account the impact of his own actions on all markets including those on which he does not buy or sell, or does he consider only a subset, while taking prices on the remaining markets as given?
3. What are the coordination mechanisms that lead to consistent anticipations and allocations? When agents have an explicit representation of their environment, they necessarily appear in each others' models. This leads to serious conceptual problems unless it is assumed that every agent takes actions of the others as given or uses a simplified representation of their reactions to his own behavior.

Imperfect competition opens a Pandora box of possible specifications, since every combination of assumptions with respect to the above issues leads to a different model. The main purpose of this chapter will be to present some of these combinations in a systematic way. Existence proofs can only be given under very specific assumptions on the behavior of consumers and producers, which are too restrictive for applied modeling (such as Cobb-Douglas utility functions, which ensure uniqueness of Walrasian equilibrium, when there is no production). Therefore we will not give such proofs but focus on the difficulties that arise in the general setting. Nonetheless, an existence result is obtained, which introduces simplifying assumptions on the agents' representation of their environment, as opposed to simplifying assumptions on preferences and technology.

1. See also d'Aspremont et al. (1992).

In section 11.1 we introduce the main issues using a traditional partial equilibrium model and identify mathematical problems related to the lack of convexity of decision models as well as economic problems related to rationality of agents. We also argue that it is essential to distinguish between a stage in which noncompetitive agents anticipate the consequences of their actions (the anticipation stage) and a stage in which they use this knowledge for making their optimal decisions and in which markets clear (the market clearing stage).[2] In section 11.2 we turn to a general equilibrium representation by means of a Cournot-Walras model (Cournot, because in the market-clearing stage each noncompetitive agent takes the quantity supplied by other noncompetitive producers as given; Walras, because for given decisions by all noncompetitive agents the economy is assumed to operate competitively, i.e., in the Walrasian way). Alternatives are considered in section 11.3. In section 11.4 simplified representations of the anticipation stage are introduced (e.g., Negishi's subjective demand formulation) via markup functions. By forcing the agents' programs to be convex, these alternatives make it possible to ensure existence of equilibrium at the expense of microeconomic rigor. Yet we indicate how these markup functions can be made compatible with true excess demand functions. Finally, section 11.5 discusses applications and possible extensions.

11.1 The Main Issues in Partial Equilibrium

We start with the case of a Cournot duopoly under partial equilibrium. Though simple and well-known, this model has a structure that is sufficiently rich to highlight the issues that are problematic in partial equilibrium and become even more so in general equilibrium.

11.1.1 An Example: Cournot Duopoly

Suppose that there is a single industry with two producers $j = 1, 2$. Each of them produces a single commodity in quantity q_j sold at price p^q and uses other commodities as inputs with a cost function $c_j(p^v, q_j)$ that is convex in q_j; p^v is the vector of input prices; here, inputs are distinguished from the commodity supplied by the duopolists.[3] There is a market demand (by consumers) represented by the function $D(p^q)$ that is known to both producers. The market equilibrium condition is

$$D(p^q) = q_1 + q_2. \tag{11.1}$$

2. This distinction can be found in Marschak and Selten (1974) and in d'Aspremont et al. (1992), where the stages are called "planning" and "implementation" stages, respectively. We use an alternative terminology to avoid confusion with the terms planning and implementation in the planning literature.

3. Thus the price vector p^v has no entry for the commodity produced by the duopolists.

Under Cournot oligopoly, producer j (whose supply is denoted by q_j) takes the supply of the other producer (denoted by q_{-j}) as given and solves:[4]

$$\Pi_j(p^v, q_{-j}) = \max p^q q_j - c_j(p^v, q_j),$$

$$q_j \geqslant 0, \ p^q \geqslant 0,$$

subject to (11.2)

$$D(p^q) = q_j + q_{-j}.$$

Under appropriate concavity conditions, this problem defines a function $Q_j(q_{-j})$ for producer j, and a similar function $Q_{-j}(q_j)$ for the other one. A Cournot duopoly equilibrium is a solution such that

$$q_j^* = Q_j(q_{-j}^*), \tag{11.3a}$$

$$q_{-j}^* = Q_{-j}(q_j^*). \tag{11.3b}$$

This may also be written in the form of a Nash equilibrium,[5] namely as a solution that satisfies

$$p^q(q_j^*, q_{-j}^*)q_j^* - c_j(p^v, q_j^*) \geqslant p^q(q_j, q_{-j}^*)q_j - c_j(p^v, q_j), \tag{11.4a}$$

$$p^q(q_{-j}^*, q_j^*)q_{-j}^* - c_{-j}(p^v, q_{-j}^*) \geqslant p^q(q_{-j}, q_j^*)q_{-j} - c_{-j}(p^v, q_{-j}). \tag{11.4b}$$

Equation (11.4) specifies that in equilibrium the actions of each producer are optimal provided that he takes the actions of his competitor as given.

If the demand function is monotonic in q_j and q_{-j} (i.e., if the inverse demand function exists), program (11.2) may be written

$$\max p^q(q_j + q_{-j})q_j - c_j(p^v, q_j), \tag{11.5}$$

$$q_j \geqslant 0.$$

If moreover the cost and demand functions are differentiable, the associated first-order conditions are

$$p^q\left(1 + \frac{\partial p^q(D)}{\partial D} \cdot \frac{\partial D}{\partial q_j} \cdot \frac{q_j}{p^q}\right) = \frac{\partial c_j(p^v, q_j)}{\partial q_j}.$$

Since the supply of the other producer is taken as given, $\partial D/\partial q_j = 1$, and if we denote $(-\partial p^q(D)/\partial D)(D/p^q)$ by η, and the market share q_j/D by σ_j, this

4. This notation is often used and is especially convenient when there are more than two producers; in that case q_{-j} denotes the vector of supplies by all producers except j.

5. See section 11.3.2 for the more general definition.

reduces to the well-known Amoroso-Robinson formula:

$$p^q(1 - \eta\sigma_j) = \frac{\partial c_j(p^v, q_j)}{\partial q_j}. \tag{11.6}$$

Note that if $D(\cdot)$ is decreasing in p^q, the parameter η will be positive and the price p^q will exceed marginal cost by a value $\eta\sigma_j p^q$, known as the markup.[6] If σ_j is small, the firm has little impact, and p^q will be close to marginal cost. Monopoly prevails if there is only one producer, that is, if $\sigma_j = 1$.

Finally, condition (11.6) also shows that (11.2) can be decentralized between the owner and the manager of each firm. The owner solves a "master problem" that computes the markup. The manager is only required to solve a subproblem in which he sets q_j so as to maximize profits $p^q(1 - \eta\sigma_j)q_j - c_j(p^v, q_j)$ at given markup-ridden prices $p^q(1 - \eta\sigma_j)$ for the output and input prices p^v. He does not need to know the demand function and behaves like a competitive producer.

11.1.2 Main Issues

Due to its partial nature this model neglects the redistribution of the profits of the duopolists to consumers, and hence the effect of such profits on consumer demand $D(\cdot)$. Otherwise, it contains in a basic form, the elements of the more sophisticated general equilibrium versions:

1. Each agent has full knowledge of his environment which here is reflected in the demand constraint (11.1); this constraint is a market-clearing condition that presupposes the existence of a coordination mechanism by an auctioneer who adjusts the output.
2. Market equilibrium exists for all nonnegative (q_j, q_{-j}).
3. Under Cournot competition agent j takes his competitor's output as given and competitors do not communicate on their setting of outputs.

The knowledge of the structure in conditions 1 and 2 is acquired in the so-called anticipation stage; it is used to compute optimal plans q_j, under the structure of 3, in the so-called market-clearing stage. The above conditions pertain to the economic logic. For the correspondence generated by program (11.2) to be well-behaved, a technical assumption has to be satisfied:

4. Either the constraint set of program (11.2) has to be convex or the profit function in (11.5) must be continuous and quasi-concave. This requires

6. See also sections 3.4.2 and 10.5.

uniqueness of equilibrium prices (i.e., the possibility to invert the demand function D).

Within a general equilibrium setting, it will prove extremely difficult to guarantee satisfaction of the requirements spelled out in condition 4. Moreover the assumption that producers know the demand function (condition 1) is far-reaching, especially in a general equilibrium context, where the demand schedule for each commodity depends on the whole model, including the redistribution of profits. Condition 2 may require special assumptions with respect to the inputs that are needed by noncompetitive producers. Finally, condition 3 can be relaxed in three ways.

First, communication between agents could be introduced via threats, promises, or collusion. Though these are the central topics in current applications of game theory to economics, they have so far hardly been considered in general equilibrium models. The discussion of the next section will show that even in the simple Walrasian setting, several issues are as yet unresolved, and these are even less tractable when communication structures are made more complex. Therefore in this chapter we largely disregard the possibility of direct communication between agents.[7] Second, each producer could assume that the others will react to his setting of q_j in accordance with subjective reaction functions, which are constants in the Cournot case. Third, as is well-known, there are also alternatives to Cournot quantity setting, for example, Bertrand-Nash behavior with price-setting producers; we will discuss alternatives and see that the classification of market structures is most conveniently done in terms of the variables that each producer takes as given, rather than in terms of the instruments (quantities, prices) he uses.

It is important to note that the anticipation stage of this partial equilibrium model only contains demand functions for the commodity with price p^q; it takes the prices p^v of all inputs as given, hence disregarding the possible effects the duopolists can have on the input markets and, by the same token, the possibility that the demand for their inputs exceeds supply. For simplicity, we will call a market "competitive" (though "nonstrategic" would be a more appropriate term) if its clearing does not appear in the anticipation stage model, and its price is taken as given by all agents both in the anticipation and in the market-clearing stage. Otherwise, it is "noncompetitive." Thus, in the model of section 11.1.1, input markets are "competitive," while the output market is "noncompetitive."

11.1.3 The Stages of Anticipation and Market Clearing

In a programming model that describes the optimal decision of an agent, it is natural to incorporate, among the constraints, relations that represent the

7. We consider this topic briefly in section 11.5.3.

agent's perception of his environment. Under perfect competition these perceived relations are the technology set for the producer and the budget set for the consumer. They are assumed to coincide with the "true" relation, and the issue as to how the agent gets to know his constraint set is disregarded. Apart from his private technology, income, and utility function, the price vector is the only information that he needs to make his decisions. Under imperfect competition an agent will have to know other aspects of his environment, such as the demand function of his customers.

To learn about this demand function is to learn about the behavior of other agents if they take his actions as given (Nash) or possibly anticipate his reaction to their own actions. In the real world it is obviously impossible for a producer to anticipate all the general equilibrium effects correctly, and he will only be able to assess an approximate demand schedule, which ignores some relations that must, however, be satisfied in equilibrium. For instance, the demand schedule which he faces may be characterized out of equilibrium, that is, without requiring all consumers to meet their budget constraints and all markets to be in equilibrium. It may also be characterized without taking into account the fact that consumers who own the firm will receive the profit which is generated and which should appear as an argument in the demand schedule (the so-called Ford effects). Finally, the model of the anticipation stage will often only represent the market clearing in the producer's own "industry" and not the full excess demand function. These limitations will have to be reflected in the model.

It is therefore practical to draw a clear distinction between the anticipation stage and the market-clearing stage. Whereas in the market-clearing stage, all markets must be in equilibrium and all agents must have made their optimal decision under constraints (budgets, technologies, etc.), the anticipation stage may abstract from some of these "realities." It is only in the market-clearing stage that all general equilibrium conditions have to be met (in particular, budgets and commodity balances). In the anticipation stage some simplifications may be introduced, some feedbacks neglected, and some variables not be at their equilibrium level. Moreover, agents may take as fixed some variables that may indirectly be influenced by their decisions.

It must be emphasized that the anticipation stage does not precede the market-clearing stage in a temporal sense. The anticipation stage has a feedback relation with the market-clearing stage. It sends functions and parameter values to it and may use equilibrium values from it. The framework that is considered is a static one in which all agents simultaneously make a single move.[8] Obviously the criticism against the atemporal

8. In section 11.5.3 we discuss how to reinterpret this framework in a dynamic sense.

nature of the process by which prices are obtained in the competitive equilibrium (tatonnement) now applies for both stages.[9]

11.2 The Main Issues in General Equilibrium

We are now ready to incorporate Cournot competition in a general equilibrium model, and we will do so by means of an example[10] for an economy with two commodities and m consumers whose utility functions and endowments satisfy assumptions C1 and C2.[11] Two of these m consumers are noncompetitive, and there are no producers. Production is only introduced in section 11.2.4 because, under imperfect competition, the usual criterion of profit maximization is questionable and introduces an additional difficulty that may obscure the main aspects.

11.2.1 An Example with Two Strategic Consumers

In the competitive economy, each consumer comes to the market with his total endowment $\omega_i = (\omega_{i1}, \omega_{i2})$ and chooses a consumption $x_i(p, \omega_i)$ such that

$$x_i(p, \omega_i) = \underset{x_i \geqslant 0}{\operatorname{argmax}} \{u_i(x_i) | px_i \leqslant p\omega_i\}. \tag{11.7a}$$

Individual decisions are coordinated by an auctioneer who announces prices; an equilibrium is obtained if normalized prices p are such that on each market k,

$$\Sigma_i x_{ik}(p, \omega_i) - \Sigma_i \omega_{ik} = 0, \qquad k = 1, 2. \tag{11.7b}$$

The auctioneer is supposed to perform a price normalization according to

$$N(p) = 1, \tag{11.7c}$$

where $N(\cdot)$ is any continuous, concave function that is homogeneous of degree 1, for example, $N(p_1, \ldots, p_r) = \Sigma_k p_k$, the usual normalization on the simplex (here $r = 2$). An equilibrium consists of allocations $x_i(p, \omega_i)$ for every i and a price vector p such that (11.7) holds.

As a transition to imperfect competition, assume now that, instead of bringing to the market their entire endowment of commodity 1, consumers $i = 1$ and 2 decide to supply the quantity q_{i1} only, keeping $r_{i1} = \omega_{i1} - q_{i1}$

9. We will see that there is no need for coordination in the market-clearing stage if the anticipation stage coordinates all actions (involves all markets).

10. The first example is inspired by the Gabszewicz and Michel (1992) model. See also Codognato and Gabszewicz (1991).

11. This is only for simplicity of exposition.

in reserve; consumers $i = 3, \ldots, m$, set $q_{i1} = \omega_{i1}$. For commodity 2 all consumers are price takers and do not keep any quantity in reserve. Hence $0 \leqslant q_{ik} = \omega_{ik} - r_{ik}$ with $r_{ik} \geqslant 0$ for $i = 1, 2$ and $k = 1$, and $r_{ik} = 0$ otherwise.

Then, if consumers $i = 1$ and 2 choose a strategy $q_i = (q_{i1}, \omega_{i2})$, the utility level they can achieve if they eventually consume $(x_i + r_i)$ is $u_i(x_i + r_i)$. If both were now behaving in the competitive way, that is, as ordinary Walrasian consumers, taking p and q_i as given, they would solve

$$x_i(p, q_i) = \underset{r_i, x_i \geqslant 0}{\operatorname{argmax}}\{u_i(x_i + r_i) | p(x_i + r_i) \leqslant p\omega_i, r_i = \omega_i - q_i\}. \tag{11.8a}$$

The auctioneer will, as in (11.7b) and (11.1), compute prices p so that commodity balance holds on each market k:

$$\Sigma_i x_{ik}(p, q_i) - \Sigma_i q_{ik} = 0, \qquad k = 1, 2. \tag{11.8b}$$

Even if both noncompetitive consumers decide to sell nothing of the first commodity, such a Walrasian equilibrium will exist because, by assumption C2, the competitive consumers will still supply that commodity in positive amounts. Hence, for any feasible choice of q_i, equilibrium prices p which depend on the choice of q_i will be well defined. For further use, we define the vector q as $q = (q_i, q_{-i})$, where $q_{-i} = (q_1, q_2, \ldots, q_{i-1}, q_{i+1}, \ldots, q_m)$ and denote the equilibrium price correspondence for consumer i by $p^i(q_i, q_{-i})$,[12] which, by construction, satisfies the normalization $N(p) = 1$.

The specification of $p^i(q_i, q_{-i})$, that is, the way in which prices will react upon variations of the strategic variable q_i given q_{-i}, provides consumers 1 and 2 with a full description of their environment. As in the partial equilibrium model (11.2), the market equilibrium conditions (11.8b) describe the environment, and the function $p^i(q_i, q_{-i})$ is a representation of the "anticipation stage" for the noncompetitive agents. Competitive consumers, on their part, only need prices p as information to make their consumption decision.

Since consumers $i = 3, \ldots, m$ behave competitively, for the market-clearing stage we only have to define the behavior of noncompetitive consumers. They act as follows: given q_{-i}, the vector of supplies chosen by all other consumers, and p^i, the resulting equilibrium price vector computed via the Walrasian coordination mechanism as depending on the q_i, consumers

12. This does not mean that price expectations are different, but we have to use the superscript to distinguish the correspondence by agent because the ordering of the arguments is not the same for $i = 1$ and for $i = 2$.

$i = 1, 2$ choose their optimal supplies Q_i so that

$$Q_i = \underset{q_i,\, r_i,\, \xi_i \geqslant 0}{\operatorname{argmax}}\{u_i(\xi_i + r_i) \mid \xi_i = x_i(p^i(q_i, q_{-i}); q_i), r_i = \omega_i - q_i\}. \tag{11.9}$$

The important point here is that $\xi_i = x_i(p^i(q_i, q_{-i}); q_i)$ is the demand function of consumer i,[13,14] derived from the Walrasian behavior of the consumers as obtained from the anticipation stage by solving (11.8a–11.8b). In this function, prices $p^i(q_i, q_{-i})$ are such that markets clear, and given these prices, consumers act in a Walrasian way and solve (11.8a). Note that they do not deviate from this rule of conduct, though it might be in their interest to do so.

As in (11.3), this defines functions $Q_1(q_2)$ and $Q_2(q_1)$ and hence the vector correspondence $Q(q)$. The Nash equilibrium is a fixed point of this correspondence. To meet the requirements of the Kakutani theorem A.4.2, this correspondence has to be uppersemicontinuous, convex valued; this imposes continuity and concavity requirements on the objective functions $u_i(x_i(p^i(q_i, q_{-i}); q_i) + r_i)$, which do not follow from microeconomic assumptions C1 and C2. Indeed, (1) $p^i(q_i, q_{-i})$ may not be a function, since there may be several equilibria for given q; (2) even if it is a continuous function, it will lack further general properties;[15] and (3) the function $x_i(p^i; q_i)$ involves the Marshallian demand function, which, for the same reason, has no concavity property either. Hence the composition $u_i(x_i(p^i(q_i, q_{-i}); q_i) + r_i)$ can hardly fulfill the requirements of the composition theorem A.1.4, and it will in general lack (quasi-)concavity.

In terms of a Nash equilibrium, the solution is, as in (11.4), defined as a pair of optimal plans (q_1^*, q_2^*) with associated (r_1^*, r_2^*), such that for every $i = 1, 2$:

$$u_i(x_i(p^i(q_i^*, q_{-i}^*); q_i^*) + r_i^*) \geqslant u_i(x_i(p^i(q_i, q_{-i}^*); q_i) + r_i), \tag{11.10}$$

for all nonnegative r_i satisfying $q_i = \omega_i - r_i \geqslant 0$. Expression (11.10) merely states that, given the optimal choices q_{-i}^* made by the other noncompetitive consumer, consumer i's best choice is q_i^*. No deviation from this value can make him better off by increasing his utility. The q_i are often called strategies, and the utilities as functions of the strategy pairs (q_i, q_{-i}) are payoff functions.

13. The Marshallian demand function would read as $x_i(p^i(q_i, q_{-i}), p^i(q_i, q_{-i})q_i)$, where $p^i(q_i, q_{-i})q_i$ is consumer i's income. For brevity of notation, we write it as $x_i(p^i(q_i, q_{-i}); q_i)$.

14. The consumer demand may actually be a correspondence; indeed, $p^i(q_i, q_{-i})$, may be set valued, since for given q's there may be several equilibria.

15. See sections 1.2.4 and 2.2.3 on the lack of general properties of the excess demand function.

To summarize, in this example the anticipation stage was characterized by Walrasian coordination, through which all noncompetitive agents acquire the capacity to anticipate each others reactions. This notion goes at the expense of individual rationality in the market-clearing stage because noncompetitive consumers are constrained to meet their Walrasian demands, which may not be in their best interest. This discussion of the economic issues obviously presupposes existence of equilibrium. Such existence is problematic for the market-clearing stage due to the possible lack of quasi-concavity of the objective in program (11.9). Because we represented all balances in the anticipation stage and maintained anticipation stage consumer demands in the market-clearing stage, there was no need to add commodity balances to clear the markets in the market-clearing stage; the $x_i(p^i(q_i, q_{-i}); q_i)$ were market-clearing demands already.

11.2.2 A Representation That Avoids Inverse Demand Functions

For later reference we write program (11.9) of the noncompetitive consumer in an alternative form by including among the constraints he faces the market excess demand function as well as a price normalization constraint,[16] forcing him as before to have his demand x_i chosen in a Walrasian way. We denote by $z^i(p^i, q_i, q_{-i})$ the function faced by the noncompetitive consumer i,[17] and define his problem as

$$\begin{aligned}
&\max\ u_i(\xi_i + r_i),\\
&p^i, q_i, r_i, \xi_i \geqslant 0,\\
&\text{subject to} \qquad\qquad (11.11)\\
&p^i\xi_i \leqslant p^i q_i,\\
&q_i + r_i = \omega_i,\\
&\xi_i = x_i(p^i; q_i),\\
&z^i(p^i, q_i, q_{-i}) = 0,\\
&N(p^i) = 1.
\end{aligned}$$

An equilibrium consists of a price vector p^* (since all excess demand functions "agree" among noncompetitive agents, $p^{i*} = p^*$ will hold), supplies q_i^*, reservations r_i^*, and additional consumptions ξ_i^* for all i satisfying

16. This constraint $N(p^i) = 1$ is needed here. Since all prices are endogenous, their boundedness should be ensured explicitly.

17. Although this market excess demand function is the same for all noncompetitive consumers, it has a superscript i. Each i has a different ordering of arguments, as in the price function $p^i(\cdot)$.

(11.11), and such that markets are in equilibrium: $\Sigma_i x_i(p^*; q_i^*) = \Sigma_i q_i^*$. The requirements are the same as for (11.9), except that because p^i is now a variable, the use of the price correspondence is avoided, and only continuous functions appear in the program. Nonetheless, the program is nonconvex due to both the Marshallian demand functions $x_i(\cdot)$ and the excess demand function $z^i(\cdot)$. Hence the uppersemicontinuity of the correspondence $Q_i(q_{-i})$ is problematic, as before. Clearly the allocation will be unaffected by the price level because prices do not enter the objective and because the constraints where they appear are homogeneous in prices.

Since program (11.11) is not convex, and each noncompetitive consumer chooses an optimal q_i as well as an optimal p^i it may happen that even if the q_i agree and a Nash equilibrium in q is obtained, the solution is not a general equilibrium. This is because due to the possible multiplicity of equilibria, the individual prices p^i may not coincide. To avoid this problem, one would have to modify the model of the anticipation stage. This can be achieved either by making its equilibrium unique (through specific assumptions on utility functions or by reducing the number of noncompetitive markets) or by explicitly letting the coordination mechanism choose a unique price.[18]

11.2.3 A First Sketch of Alternative Specifications for the Anticipation Stage

The behavior of the noncompetitive consumers in section 11.2.1 is only one possible specification, imported from partial equilibrium analysis. There are three alternative possibilities: First, different variables could be taken as given in the market-clearing stage, for example, prices (Bertrand behavior) instead of quantities (Cournot behavior, here q_{-i}).

Second, a different description of the economy might be considered in the anticipation stage; for example, an economy with rationing (or any model described in the other chapters) could be used instead of the Walrasian (competitive) equilibrium.

Third, more partial representations of the anticipation stage could be envisaged. One could for example, ignore the clearing on market 2. Then there would be no commitment for commodity 2 and clearing would have to be incorporated within the market-clearing stage, where even noncompetitive agents would act as price takers for commodity 2.[19] This alternative is in sharp contrast with what we have assumed so far, that noncompetitive agents set quantities only on the market for commodity 1 but anticipate the effects of this on market 2, whose price is indirectly influenced by the

18. See d'Aspremont et al. (1992).

19. Shifting market clearing on these markets to the market-clearing stage avoids infeasibilities in the anticipation stage in the event that producers need to purchase inputs from these markets.

noncompetitive quantity setting on market 1, and that they also commit themselves in the market-clearing stage to consume for all commodities, the anticipated quantities.

Every combination of these modifications will lead to a different model and obviously to a different equilibrium.

11.2.4 Introducing Producers

In competitive equilibrium it is assumed that every producer maximizes his profits. This is more than an assumption on producers' behavior, since the first welfare theorem provides a social welfare justification for this rule of conduct. The second welfare theorem shows that it is also in the interest of the shareholders of a firm to ask the managers of this firm to maximize profits (and, by the Negishi theorem, this also holds in a competitive equilibrium, i.e., when transfers are zero). Thus, in the competitive model, profit-maximizing producers act in the interest of consumers. However, this is no longer so under imperfect competition. As pointed out by Gabszewicz and Vial (1972), the owner of a firm may prefer lower profits distributed to him and a much lower price in purchasing some desired commodity. This can be illustrated by extending program (11.11) to include producers.

In the example we assume the following: (1) only noncompetitive consumers $i = 1, 2$ own firms $j \in J^i$; (2) they do not share ownership of firms; (3) firms obtain all their inputs from the endowment of the consumer who owns them or from other firms belonging to this consumer;[20] (4) production sets Y_j satisfy assumption P1.[21] Consumer i's program (11.11) can be reformulated as

$$\max u_i(x_i(p^i; q_i) + r_i),$$

$$p^i, q_i, r_i, \xi_i \geqslant 0, y_j, \text{ all } j \in J^i,$$

$$\text{subject to} \tag{11.12}$$

$$p^i \xi_i \leqslant p^i q_i,$$

$$q_i + r_i = \omega_i + \Sigma_{j \in J^i} y_j,$$

$$y_j \in Y_j,$$

$$z^i(p^i, q_i, q_{-i}) = 0,$$

$$N(p^i) = 1,$$

20. We make this highly restrictive assumption to keep this model as close as possible to the pure exchange model. Later we will allow firms to buy inputs only of competitively priced commodities. The Walrasian anticipation stage could become infeasible if noncompetitive firms bought from other noncompetitive firms.

21. Here, to ensure that supply plans y_j will be unique, we make the assumption that firms produce under decreasing returns to scale.

where now, the excess demand function $z^i(\cdot)$ includes production. Program (11.12) is feasible because (1) $0 \in Y_j$, (2) $q_i, r_i \geqslant 0$, and (3) by assumptions C1 and C2 there exists a Walrasian solution to the system $p^i \geqslant 0$, $z^i(p^i, q_i, q_{-i}) = 0$, $N(p^i) = 1$, but as before, the program is not convex so that the Nash equilibrium in the q may fail to exist.

Since consumer i's utility function is nonsatiated, he will ask the managers of his firms not to waste and to produce efficiently but not necessarily instruct them to maximize profits. To verify this, let us consider the reduced version of (11.12), in which the y_j are given and satisfy $y_j \in Y_j$ and $0 \leqslant \omega_i + \Sigma_{j \in J^i} y_j$,[22] while, as before, consumer i takes q_{-i} as given. This defines in (11.12) a value function $U_i((y_j)_{j \in J^i}, q_{-i})$. One may now think of consumer i as maximizing this value function with the variables $y_j \in Y_j$ for all $j \in J^i$ as his choice variables. To make shadow prices explicit, we introduce the additional constraint $\tilde{y}_j \leqslant y_j$ and write this problem as

$$\max U_i((\tilde{y}_j)_{j \in J^i}, q_{-i}),$$

$$y_j, \tilde{y}_j, \text{ all } j \in J^i,$$

$$\text{subject to} \tag{11.13}$$

$$\tilde{y}_j \leqslant y_j,$$

$$0 \leqslant \omega_i + \Sigma_{j \in J^i} y_j,$$

$$y_j \in Y_j.$$

If the function $U(\cdot)$ happens to be continuous and (quasi-)concave in its optimum, there will be prices p^i_j associated to the constraints $\tilde{y}_j \leqslant y_j$. The associated Lagrange inequality (theorem A.2.1, assertion 7) implies that the optimum of (11.13) also solves $\max p^i_j y_j$ subject to $y_j \in Y_j$. However, the prices p^i_j at which these profits are maximized are not necessarily equal to the Walrasian prices p^i. Thus consumer i transacts at prices p^i but instructs the managers of his firms to maximize profits at prices p^i_j which may be linked to p^i via variable markups as in (11.6):

$$p^i_j = p^i - \mu^i_j. \tag{11.14}$$

If several consumers own a firm j, this scheme will not work unless all the consumers who own the firm have the same prices p^i_j, which is unlikely.[23] Joint optimization by the subset of consumers who own firm j is possible and suggests a Negishi-type formulation whereby all noncompetitive consumers who jointly own firms decide to collude and act like a single

22. This means that firms obtain their inputs from the endowment of the consumer who owns them or from firms that belong to this consumer.

23. A similar problem will arise when asset markets are incomplete. See section 12.2.6.

noncompetitive consumer except that, within the group, welfare weights are set so as to maintain individual budgets. Obviously this requires intensive communication.

11.2.5 Producers as Noncompetitive Agents

Many theoretical models of imperfect competition[24] assume profit maximization despite the limitations of this criterion; they treat producers as the only noncompetitive agents while consumers are taken to be competitive. This adapts to the common perception of what imperfect competition is about and is in accordance with the partial equilibrium approach to industrial organization. It also avoids the rather artificial requirement that noncompetitive consumers should, in the market-clearing stage, maintain their behavior of the anticipation stage, for example, their Walrasian demand $x_i(p^i; q_i)$. We have seen earlier that such a commitment is problematic with respect to individual rationality.

For producers the issue of commitment is only relevant if they buy inputs on markets that are explicitly made to clear in the anticipation stage and their input requirements are not their strategic variables. If the input demands are explicit functions of the output level (the strategic variable), as would be the case under a Leontief technology, they have to commit for technological reasons and not because the model structure requires them to adhere to a particular rule of conduct, such as cost minimization at prevailing prices.

We now return to the general setting with $r > 2$ commodities, m competitive consumers, and n producers. We partition the set $K = \{1, 2, \ldots, r\}$ of commodities into subsets K_s and K_c for "noncompetitive" and "competitive" commodities, respectively. Recall that the markets that clear in the anticipation stage of the model are the noncompetitive ones. Prices will be denoted accordingly as p_s and p_c. For the "competitive" commodities, markets clear only in the market-clearing stage, and prices p_c are taken as given in both stages by all agents. Similarly the set $J = \{1, 2, \ldots, n\}$ of producers is partitioned into subsets J_s for noncompetitive producers and J_c for competitive ones. We assume that the noncompetitive producer j uses quantities as strategic variables (Cournot behavior) and that these quantities relate to outputs only. His vector of outputs will be denoted by q_j so that $q_{jk} = y_{jk}$ for $k \in K_s$.

For every noncompetitive producer $j \in J_s$, the model of the anticipation stage is

$$z_s^j(p_s^j, p_c, q_j, q_{-j}) = 0, \tag{11.15}$$

24. Gabszewicz and Vial (1972) were the first to consider such a full general equilibrium version of this model. See also Mas-Collel (1982), Hart (1985), Gary-Bobo (1989), and Bonanno (1990).

where p_c, q_j, and q_{-j} are given.[25] The excess demand function for producer j, $z_s^j(p_s^j, p_c, q_j, q_{-j})$ is obtained as follows: For given q_{jk}, $k \in K_s$, and given prices $p^j = (p_s^j, p_c)$, every noncompetitive producer solves

$$\Pi_j(p^j, q_j) = \max_{y_j} p^j y_j,$$

subject to (11.16)

$$y_j \in Y_j,$$

$$y_{jk} = q_{jk} \quad \text{for } k \in K_s.$$

Hence, by proposition 2.1, assertion 4, and assumption P1,[26] we can, for noncompetitive producers, recover net supplies as $y_j(p^j, q_j) = \partial \Pi_j(p^j, q_j)/\partial p^j$.

Competitive producers and consumers behave in the standard way, taking profits of competitive and noncompetitive firms as given, and the q_j will enter the excess demand function in two ways: first, via profits[27] of the noncompetitive firms that are redistributed to consumers and, second, via net supply $y_j(p^j, q_j)$ as a (negative) entry in the excess demand function.

In the market-clearing stage the noncompetitive producer j solves[28]

$$\max \Sigma_{k \in K_s}(p_{sk}^j y_{jk}) + \Sigma_{k \in K_c}(p_{ck} y_{jk}),$$

$$p_{sk}^j \geqslant 0, \; q_{jk} \geqslant 0, \; k \in K_s, \text{ and } y_{jk}, \text{ all } k,$$

25. As in the earlier sections we write the excess demand function in an agent-specific form not because the information they contain is different, but to indicate that the arguments q_j and q_{-j} have different orderings.

26. We maintain the assumption of strict convexity of the production set in order to have unique production plans and the option of ensuring that each oligopolist will produce at a positive scale in equilibrium. Recall from the discussion in section 2.1.3 that decreasing returns may be viewed as constant returns with at least one fixed input. If, for each firm, the marginal productivity of this fixed input is infinite when output is zero, the firm will produce, but if there are no such fixed inputs and all firms use the same inputs under constant returns, then a monopoly could result in which only the least-cost firm would be active.

27. Note that for arbitrarily given q_j, even if these are restricted to keep the associated y_j feasible ($y_j \in Y_j$), profits can be negative. However, what matters is whether this problem arises when the excess demand function is introduced in the producer maximization problem of the market-clearing stage. In this stage, this does not happen, since all producers have possibility of inaction.

28. For convenience, the column vector y_j is partitioned as $y_j = (y_s^j, y_c^j)$, which separates noncompetitive from competitive commodities. If the production set of producer j does not allow him to produce the noncompetitive commodity k, the optimum will set $q_{jk} = 0$. Under monopoly, q_{jk} is nonzero for a single producer j; a single-output producer will have it nonzero for at most one k. This notation avoids having to distinguish commodity sets by producer.

subject to (11.17)

$y_j \in Y_j$,

$y_{jk} = q_{jk}$ for $k \in K_s$,

$z_s^j(p_s^j, p_c, q_j, q_{-j}) = 0$.

Other producers and consumers behave in a competitive way, and market conditions are introduced for $k \in K_c$:

$$\Sigma_i x_{ik} - \Sigma_j y_{jk} - \Sigma_i \omega_{ik} = 0 \qquad \text{for } k \in K_c. \tag{11.18}$$

If, to the Nash equilibrium on the choices variables q_j, there corresponds a unique Walrasian equilibrium, prices p^j will coincide:

$$p^j = p, \qquad \text{all } j \in J_s. \tag{11.19}$$

Finally, prices have to be normalized:

$$N(p) = 1. \tag{11.20}$$

Note that, since the anticipation stage takes prices p_c as given, there is an implicit normalization. Hence the normalization constraint should not be part of the anticipation stage nor of the producer program (11.17).

11.2.6 The Normalization Issue

The issue of normalization under imperfect competition was raised by Gabszewicz and Vial (1972) and has received much attention since. In section 11.2.1 we saw that if all markets are noncompetitive (i.e., are represented in the model of the anticipation stage), it is necessary to introduce a price normalization within the program of every noncompetitive agent in order to avoid unboundedness of the price set. One could choose any commodity as numéraire provided that its price is nonzero in equilibrium.[29] This is clearly seen in programs (11.11) and (11.12) where the allocations are homogeneous of degree zero in prices.

In the model where markets are competitive for some commodities (i.e., absent from the anticipation stage) as in (11.17)–(11.20), all agents (producers or consumers) take the prices on these competitive markets as given, and hence there is a natural normalization (e.g., $\Sigma_{k \in K_c} p_k = 1$) provided, of course, that some of these prices are nonzero.[30]

This leaves open a third possibility in which there is profit maximization by producers and all markets are noncompetitive. This is the case

29. Obviously, if one imposes that both relative prices, and absolute prices should coincide in equilibrium, all agents should use the same normalization.

30. See Cripps and Myles (1988).

considered by Dierker and Grodal (1986, 1994), Böhm (1990), Grodal (1992, 1993), and Ginsburgh (1994), among others. We have already seen in section 11.2.4 that profit maximization at market prices is a questionable criterion under imperfect competition. If there are no competitive markets, the excess demand constraint of problem (11.17) is $z_s^j(p_s^j, q_j, q_{-j}) = 0$ and is homogeneous of degree zero in p_s^j. If any positive profit is attainable, it will be unbounded unless a normalization restriction is introduced as an additional constraint within the program. This restricts choices and will obviously affect the allocation. Therefore this model is at odds with the usual approach in general equilibrium, where agents are not subject to money illusion.

11.2.7 Free Entry

The endogenous representation of the number of firms that was already introduced in section 7.4.2 is closely linked to imperfect competition. If new producers can enter at zero cost, and with the same technology as the producers that are already active, there can be no profits in equilibrium. Then a profit-maximizing firm has no possibility of improving or anticipating its market share, since all producers are indifferent with respect to scale. The only objective that will then remain is avoiding losses, namely cost minimization, and since other producers have the same technology, it will be cost minimization at given prices. Clearly this reasoning does not apply for nonproduced commodities for which suppliers try to maintain a rent. There can be no perfect entry for such commodities. More important, program (11.12) shows that monopolistic consumers may even accept losses from the firms they own if this helps them in acquiring other commodities at a lower price. Then, while the argument holds that under free entry there can be no profits in production, it does not imply that there must be perfect competition. Firms that produce at a loss may keep competitors out. Entry by new firms is often represented in applied models, and the issue will be considered again in section 11.5.1.

11.2.8 Pareto-Efficiency

It is important to note that whatever the degree of sophistication of anticipations and the complexity of the market-clearing mechanism, the welfare achieved in the economy cannot be larger than the level reached under a competitive equilibrium with transfers, and it will often be lower.[31] A state of imperfect competition is for that reason, very vulnerable, at least theoretically, since consumers will always have an incentive to pay transfers to imperfectly competitive agents, inducing them to give up their distortionary

31. The only case where no loss of efficiency can arise is the zero-sum game. For example, when the sum of utilities is constrained to remain constant (pie-sharing), any feasible allocation will be Pareto-efficient. However, this game structure is not used in a general equilibrium context and will not be discussed further.

actions. By the same token, it is rational for individual consumers to engage in imperfect competition only if the alternative, competitive behavior with compensatory payments is not an available option. However, if there are nonconvexities in technology, imperfect competition may lead to a Pareto-improvement with respect to perfect competition, since the profits from imperfect competition can be used to cover the losses of socially desirable activities.

11.3 Alternative Specifications for the Anticipation and the Market-Clearing Stages

In this section we restrict attention to imperfect competition among producers and characterize it accordingly:

1. The number of noncompetitive producers of a particular commodity: monopoly if there is one; oligopoly if there are several. This is the traditional meaning of the term, which, in the industrial organization literature, also describes a group of producers who supply close but imperfect substitutes. We avoid this use because in general equilibrium each such producer would be considered as a monopolist supplying a different commodity on his own market.[32]
2. The actions of the other producers that are taken as given when producer j makes his decision: quantities in the Cournot case of the previous section, prices in Bertrand oligopolistic, and Chamberlin monopolistic competition.
3. The economy of the anticipation stage is often taken to be Walrasian, but other representations are possible, especially in connection with price-setting behavior, where rationing of agents can be introduced.
4. The market clearing stage is not necessarily of the Nash- type;[33] a special case arises if there is only one noncompetitive agent (leader-follower equilibrium). Producers may also choose to maximize joint profits (collusion) or form conjectures about the reactions of their competitors on their own strategy (conjectural variations).

11.3.1 The Anticipation Stage

Cournot Quantity Setting

The Cournot case was covered in section 11.2. For every noncompetitive producer j, it is represented by the excess demand constraint

$$z_s^j(p_s^j, p_c, q_j, q_{-j}) = 0. \tag{11.21}$$

32. In section 11.5.1 we will discuss the product differentiation approach introduced by Dixit and Stiglitz (1977), which is used in many applied models.

33. Here we use the Nash concept in the restricted sense that it is the actual choice variables (quantities or prices) of the competitors that are taken as given and not, for instance, their strategy spaces.

Bertrand Price Setting

Here the noncompetitive producer j takes the prices of his competitors as given. The issue is now how his market share will change if he announces prices that are higher. If this means that he sells nothing, he has no choice and becomes a price taker, and the competitive equilibrium is the outcome. This is known as the Bertrand paradox. Alternatively, he may be able to sell at a higher price, but this can be interpreted as his commodity being differentiated. In general equilibrium these are then two separate commodities with a separate market for each (Chamberlin monopolistic price setting), a situation that should not be described as a Bertrand oligopoly. Finally, there is the option to combine Cournot with price-setting behavior by allowing differences in (11.21) among producers, so announced prices may become different. This requires incorporating an explicit price coordination scheme (a function of announced prices and quantities) to enforce the law of one price on each market.[34]

Chamberlin Price Setting

Producer j is the only one to produce a given commodity. This general equilibrium model was explored by Marschak and Selten (1974), Nikaido (1975), and more recently by Benassy (1988, 1991). In this setting every producer is a monopolist on his market and faces the excess demand constraint

$$z_k^j(p_k^j, p_{-k}^{-j}, p_c, q_{jk}, q_{-j,-k}) = 0 \qquad \text{for all } k \in K_j, \tag{11.22}$$

where K_j is a subset of K_s, to be specified below. In this constraint p_{-k}^{-j} represents the vector of prices of noncompetitive commodities that are produced by firms other than j, while $q_{-j,-k}$ is their output level. For the remainder, the notation is as before. The specification of the variables that are taken as given and the set K_j complete the description of the agent's environment.

In the simplest case the producer takes p_{-k}^{-j}, p_c, and $q_{-j,-k}$ as given, and K_j contains a single element. He adjusts his supply q_{jk} only to his strategic variable p_k^j. The income effects on market k of his profits and purchases of inputs on competitive markets are taken into account. A further simplification would abstract from these effects but would require income to appear as a separate variable in (11.22) with an exogenously given value.

A more sophisticated approach is to assume that the producer anticipates the effects of his price setting not only on his own output but also on the outputs of the other noncompetitive firms. Then the set K_j extends to all

34. See d'Aspremont et al. (1992).

commodities for which he is supposed to have anticipations. The model is now in a regime of output quantity adjustment with all prices fixed (the own price because it is a decision variable, and all others because they are taken as given). Therefore it may happen that no equilibrium exists unless quantity rationing is performed, and since all prices are fixed, Benassy rationing is a natural concept to use.[35]

11.3.2 The Market-Clearing Stage

To keep the presentation general and independent of the specification of the anticipation stage, we describe the market-clearing stage in terms of strategic actions a_j taken by agent j, instead of using quantities or prices.

Nash Equilibrium

In a Nash equilibrium every agent j faces given values of variables $a_{-j} = (a_1, a_2, \ldots, a_{j-1}, a_{j+1}, \ldots, a_n)$ set by other agents and makes his optimal choice a_j, choosing from elements belonging to a set $A_j(a_{-j})$. Denoting the vector of actions during the market-clearing stage by $a = (a_1, a_2, \ldots, a_j, \ldots, a_n)$ while a^* denotes optimal choices, the model may be written

$$a_j^* = \operatorname*{argmax}_{a_j \in A_j(a_{-j})} f_j(a_j, a_{-j}), \tag{11.23a}$$

for every noncompetitive agent j. There is a coordination mechanism such that

$$a^* = a. \tag{11.23b}$$

As in section 11.2 for Cournot competition in a model that is written as a Nash equilibrium, the issue of existence of equilibrium essentially hinges on requirements that allow to apply the maximum theorem A.3.1. Each agent's objective $f_j(a_j, a_{-j})$ should be continuous and (quasi)concave in a_j. The constraint set should be nonempty, compact, and convex for given a_{-j}, and the correspondence $A_j(a_{-j})$ should be uppersemicontinuous. These requirements are problematic when the anticipation stage follows a general equilibrium representation.

In a Nash equilibrium, agent j conjectures that decisions made by other agents are unaffected by his own decision. If the payoff function of j is strictly (quasi)concave, then the optimal decision is unique and is a function of the other agents' decisions, denoted by $a_j = G_j(a_{-j})$. The function $G_j(a_{-j})$ which associates the best response of agent j to decisions of others is called the reaction function of j. This makes it possible to give an alternative representation of a Nash equilibrium as a fixed point $a = G(a)$.

35. See section 6.2.4 and Benassy (1988, 1991).

Leader-Follower Equilibrium

If there is only one noncompetitive agent (or if all noncompetitive agents collude), the model can be of the leader-follower kind. Here one of the agents, say, $j = 1$ is the so-called leader, while others are the followers. The leader takes into account that the followers will base their decision on his move. Agent $j \neq 1$ takes thus a_1 as given and reacts according to $q_j = h_j(a_1, q_{-j})$, where q_j is his action. This function is assumed to be known to the leader. The optimal decision a_1^* of the leader is then

$$a_1^* = \operatorname*{argmax}_{a_1 \in A_1(q_2, \ldots, q_n)} f_1(a_1, h_2(a_1, q_{-2}), \ldots, h_n(a_1, q_{-n})). \tag{11.24}$$

As long as the functions $f_1(\cdot)$ and $h_j(a_1, q_{-j})$ are continuous and the correspondence $A_1(q_2, \ldots, q_n)$ is nonempty, compact valued, this program has an optimal solution. Hence existence of a leader-follower equilibrium is guaranteed under rather weak conditions.

While this setting may be appropriate to describe the relations between producers within one industry (and indeed it originates from partial equilibrium), it can hardly be used to represent the whole economy. It is possible to use the leader-follower model for separate sectors, disregarding general equilibrium effects. This would amount to applying (11.24) in the market-clearing stage, possibly extending it with given variables coming from other parts of the model, and this yields a program similar to (11.12). However, since (11.24) is in general not a convex program, it is not suited for the anticipation stage.

Cooperation and Collusion

If several agents make optimal decisions by maximizing a joint objective function, say, joint profits, they are said to collude and form a cartel. Since they operate as a single noncompetitive agent, the earlier discussion applies. Whether participation in this cartel is rewarding will depend on how much every member would earn in an alternative coalition or on his own. The welfare difference between the cartel and the noncartel alternatives is the cooperative surplus. If it is positive, a sharing rule has to be agreed among members that redistributes it to the consumers who own the firms, but then the given θ_{ij} shares are not necessarily appropriate.[36] If the surplus is negative, no cartel should form.

Conjectural Variations

Each of the previous forms of competition makes an implicit assumption on agent j's conjectures about the behavior of other producers. In a Nash

36. See, for example, Moulin (1985). We return to the issue of cooperation in section 11.5.3.

equilibrium, for example, producer j assumes that other producers do not react to his decision a_j, and he takes a_{-j} as given. Under the conjectural variations approach, one can specify any other rule according to which agents anticipate reactions by their competitors. This can be incorporated via a subjective reaction function $\hat{a}_{-j}(a_j)$ that is continuous.[37] Instead of (11.23a), every noncompetitive agent j then solves

$$a_j^* = \underset{a_j \in A_j(\hat{a}_{-j}(a_j))}{\operatorname{argmax}} f_j(a_j, \hat{a}_{-j}(a_j)).$$

11.4 Simplified Representations of the Anticipation Stage

In sections 11.2 and 11.3.1 we saw that (1) full excess demand representations for the anticipation stage require every noncompetitive agent to have access to and process extremely detailed information and (2) due to the lack of generic properties of the excess demand function, the programs in the market-clearing stage may not be convex. Therefore, for existence proofs, it is necessary to impose additional assumptions that ensure this. One approach is to replace the excess demand function by a well-behaved approximation. Negishi's (1961) subjective demand formulation (section 11.4.1) was an early attempt in this direction, and we will show that it can be generalized to a more flexible markup formulation (sections 11.4.2 and 11.4.3) and even fully represent the anticipation stage (section 11.4.4), but we do not give any existence proof.

11.4.1 Negishi's Subjective Demand Approach

Negishi considers a Cournot model in which each producer j makes a linear approximation of the excess demand constraint around Walrasian equilibrium values, with respect to the own price only. For the noncompetitive commodity k that he produces,[38] the excess demand function in the anticipation stage is then

$$a_k^j(p^\circ, q^\circ)(p_k^j - p_k^\circ) = q_{jk} - q_{jk}^\circ, \tag{11.25}$$

where the $^\circ$ indicates values in the Walrasian model[39] and the negative slope $a_k^j(p^\circ, q^\circ)$ is continuous and homogeneous of degree minus one.[40] From

37. Such functions may also be used to make the program strictly convex, thus ensuring existence of equilibrium.

38. To simplify presentation, we assume here that firm j produces a single commodity k.

39. That is, in the model solved for prices at given values of the strategic variables.

40. This constraint ensures that price normalization has no effect on allocations.

(11.25) it follows that

$$a_k^j(p^\circ, q^\circ)p_k^j + b_k^j(p^\circ, q^\circ) = q_{jk}, \tag{11.26}$$

where $b_k^j(p^\circ, q^\circ) = \overset{\circ}{q}_{jk} - a_k^j(p^\circ, q^\circ)\overset{\circ}{p}_k$. Equation (11.26) is a linear approximation (with coefficients depending on given values (p°, q°)) of the demand function (11.1) at the point (p°, q°). For a single output the technology can, as in (11.2), be represented by a cost function. This yields a producer model similar to (11.2):

$$\max\ p_k^j q_{jk} - c_j(p^\circ, q_{jk}),$$

$$q_{jk} \geqslant 0,\ p_k^j \geqslant 0,$$

subject to

$$a_k^j(p^\circ, q^\circ)p_k^j + b_k^j(p^\circ, q^\circ) = q_{jk}.$$

As in (11.6) the first-order conditions lead to a markup formulation:

$$p_k^j(1 - \eta_k^j(p^\circ, q^\circ)\sigma_k^j(p^\circ, q^\circ)) = \frac{\partial c_j(p^\circ, q_{jk})}{\partial q_{jk}}. \tag{11.27}$$

The manager of firm j can now be viewed as maximizing profits at a given markup-ridden output price $p_k^j(1 - \eta_k^j(\cdot)\sigma_k^j(\cdot))$, where the ad valorem markup rate is $\eta_k^j(\cdot)\sigma_k^j(\cdot)$. Since the manager takes the price as given, we can equivalently write this in terms of nominal markups:

$$p_k^j = \mu_k^j(p^\circ, q^\circ) + \frac{c_j(p^\circ, q_{jk})}{q_{jk}}, \tag{11.28}$$

where the unit value of output is equal to markup plus average cost and $\mu_k^j(p^\circ, q^\circ)$ is continuous in (p°, q°) and homogeneous of degree one in p°. The manager will now maximize the profit:

$$\Pi_k^j = (p_k^j - \mu_k^j(p^\circ, q^\circ))q_{jk} - c_j(p^\circ, q_{jk}),$$

and this function is clearly continuous, concave in q_{jk}, so equilibrium can be shown to exist.

11.4.2 Generalizing Negishi's Approximation

The Negishi subjective demand approach is probably the most far-reaching simplification of the excess demand constraint that makes the producer program convex. Less restrictive formulations could be envisaged, for example, including cross effects among commodities provided that they maintain convexity of the producer program. Moreover, it is possible to generalize further by focusing on the (price-taking) profit-maximizing problem of the manager of the firm rather than on the full producer problem. This makes it possible to use more flexible forms of markup functions.

It should by now be clear that many variants can be considered. Here we keep with the assumption that the Cournot oligopolist j produces commodity k only, and we suppose that he takes all other prices as given (i.e., he disregards the clearing on other markets). We start with the following Walrasian excess demand constraint:[41]

$$g_k^j(p_k, p_{-k}, q_j, q_{-j}) = D_k^j(q_{jk}, q_{-jk}) \qquad \text{for } k \in K_s, \tag{11.29}$$

where $D_k^j(q_{jk}, q_{-jk}) = q_{jk} + \Sigma_{-j} q_{-jk}$ and $\Sigma_{-j} q_{-jk}$ is the supply by all firms other than j that produce commodity k; $g_k^j(\cdot)$ is the net demand by consumers and other firms for commodity k; the terms q_j and q_{-j} enter as arguments as a consequence of the Ford effects, which were ignored in the partial equilibrium approach of section 11.1.1. If the inverse demand function exists, solving (11.29) for p_k leads to a price function

$$p_k = P_k^j(p_{-k}, D_k^j(q_{jk}, q_{-jk}), q_j, q_{-j}).$$

Maximizing firm j's profit as in (11.5) leads to

$$\max P_k^j(p_{-k}, D_k^j(q_{jk}, q_{-jk}), q_j, q_{-j}) q_{jk} - c_{jk}(p_{-k}, q_{jk}), \tag{11.30}$$

$$q_{jk} \geqslant 0.$$

The first-order conditions imply that

$$p_k(1 - \eta_{jk}\sigma_{jk} + \zeta_{jk}) = \frac{\partial c_{jk}(p_{-k}, q_{jk})}{\partial q_{jk}}, \tag{11.31}$$

where, as before, $\eta_{jk} = -(\partial P_k^j / \partial D_k^j)(D_k^j / P_k^j)$ is the demand elasticity, $\sigma_{jk} = q_{jk}/D_k^j$ is the market share of firm j, and $\zeta_{jk} = (\partial P_k^j / \partial q_{jk})(q_{jk}/P_k^j)$ is the elasticity due to the Ford effects. If these effects are ignored (as is often the case), equation (11.31) reduces to the partial formulation of the Amoroso-Robinson condition.

Note that if all producers in the oligopoly are identical, $\sigma_{jk} = 1/n$ (and if the owners are identical so that the ζ_{jk} are the same for all j, or if the Ford effects are ignored), the number of producers appears explicitly in (11.31). Furthermore, if q_{-jk} is taken to depend on q_{jk}, we have conjectural variations and the term $\eta_{jk}\sigma_{jk}$ has to be multiplied by an additional term $\partial D_k^j / \partial q_{jk}$. This term is equal to unity under Cournot competition.

Of course these are only formulas for elasticities and derivatives that yield parameters taken as given by the managers of the firms in the market-clearing stage (thus ensuring convexity of the manager's program). In the

41. The notation q_{-jk} denotes the output of commodity k by producers other than j.

full model of the market-clearing stage, as may be seen from (11.30), these parameters depend on virtually all other variables of the model (and should be written like in the Negishi formulation as $\eta_k^j(p^\circ, q^\circ)$, $\sigma_k^j(p^\circ, q^\circ)$, etc.). These functions should appear explicitly among the equations of the model. In applications, their continuity and homogeneity properties should be checked. It is not sufficient, as is often done in applications,[42] to compute elasticities by perturbations around (p°, q°) because such perturbations are an inadequate description of the anticipation stage model.

11.4.3 A Markup Representation

While the subjective demand approach has the appeal of being a relatively close approximation to an excess demand function derived from the usual set of microeconomic assumptions, the markup specification may lack this transparency. One could therefore argue that postulating such functions amounts to giving up the microeconomic foundations of general equilibrium theory. Thus the alternatives are clear: One could remain within micro-based theory but be unable to prove existence of equilibrium (though a solution might be computed), or one could ensure existence by accepting markup formulations at the expense of simplifying the model in the anticipation stage. Markups have the additional advantage that they can be measured statistically and their functions estimated. There is also a wide body of literature that advocates markups as a way of representing social structures within economic models.[43]

In this section we further motivate the markup method by showing that any efficient allocation can be represented through an equilibrium with markup-ridden prices. Subsequently we prove that if the markup functions are well behaved, a general equilibrium with markups will exist. We shall operate under our standard assumptions P1′, C1′, C2′, and this implies that the production technology can now accommodate constant returns.

Implementing an Allocation through an Equilibrium with Markups or Rationing

Building on definition 5.1 and proposition 5.1, we state and prove:

PROPOSITION 11.1 (Markup market clearing of arbitrary allocations) Let assumption P1′ hold for producers with a convex transformation function representation $F_j(y_j) \leqslant 0$. Let assumptions C1′ and C2′ hold for consumers, and assume also that utility functions are differentiable. Consider an

42. See, for example, Mercenier (1994).

43. See, for example, Robinson and Eatwell (1973) and Taylor (1990).

arbitrary allocation x_i^* for all i, y_j^* for all j, such that

1. x_i^* is nonnegative for all i,
2. y_j^* lies on the transformation curve, $F_j(y_j^*) = 0$ for all j,
3. the allocation is feasible, $\Sigma_i x_i^* = \Sigma_j y_j^* + \Sigma_i \omega_i$.

Then there exist (possibly negative) equilibrium markup rates τ_i^c, τ_i^+, τ_i^- on consumption prices $p_{ik}^c = (1 + \tau_{ik}^c)p_k^*$, production prices $p_{ik}^+ = (1 - \tau_{ik}^+)p_k^*$, and input demand prices $p_{ik}^i = (1 + \tau_{ik}^-)p_k^*$, respectively, which are zero for, say, commodity r and transfers T_i^* among consumers such that the allocation is an equilibrium. Consumers maximize utility subject to their budget constraint and producers maximize profits on their technology, while taking markup-ridden prices as given.

If moreover there exists a positive p^* such that $p^* x_i^* = p^* \omega_i + \Sigma_j \theta_{ij} p^* y_j^*$ for all i, and markups are imposed on commodity r as well, then there exists an equilibrium with virtual markups, that is, with $T_i^* = 0$ for all i.

Proof The first-order conditions of consumer i's problem, evaluated at given x_i^*, determine $\lambda_i p_i^c$, where λ_i is the marginal utility of income and p_i^c the markup-ridden price for consumer i. Without loss of generality, we may assume that $p_{ir}^c = p_r^*$ for all i (there is no markup on commodity r). This determines λ_i also, and since utility functions are increasing in all commodities, p_i^c and λ_i will be positive. For production, the markup-ridden output and input prices p_j^+ and p_j^- can likewise be determined from the derivatives of the transformation function. We can now arbitrarily choose a clearing price p_k^* for $k \neq r$ and compute markup rates from the ratio between markup-ridden and clearing prices.[44] For p_j^+ and p_j^- computed in this way, profits will be nonnegative for all j, due to convexity of $F_j(y_j)$ and because $F_j(y_j^*) = 0$, by property 2 above. Finally, transfers T_i^* among consumers can be computed as balancing items of every consumer's budget. By property 3 they will add up to total proceeds from markups. An equilibrium with virtual markups is obtained by allowing for markups on commodity r and treating p^* as market-clearing prices. ■

This proposition shows, for example, that the case of a consumer who decides to withhold part of his endowment could be represented through a subsidy on consumption financed out of own income. The proposition does not claim that an arbitrary allocation can be achieved without "forced trade," that is, without subsidizing positive net sales or purchases. A result on market clearing through virtual markups that excludes forced trade was given in proposition 6.2, assertion 4a and in section 6.2.5 for "no forced trade" and virtual taxation.

44. See section 5.1.

Equilibrium with Markup Pricing

Since a markup may be viewed as a commodity tax whose proceeds accrue to the firm, and through the profit shares, eventually to consumers, the treatment can proceed exactly as for taxes in section 5.2.3. The only difference is that there the proceeds were distributed to the consumers according to a tax distribution function implemented by the government, so nonnegativity of income was an issue. Here the manager will never choose negative profits because he has the possibility of inaction and the issue of negative income does not arise. This leads to the following representation of the behavior for the manager of firm j:

$$\max(p - \mu_j^+)y_j^+ - (p + \mu_j^-)y_j^-,$$

$$y_j^+, y_j^- \geqslant 0, y_j,$$

subject to

$$y_j^+ - y_j^- = y_j,$$

$$y_j \in Y_j.$$

The μ are markups on outputs and inputs y^+ and y^-; they will often be proportional to the associated price but may also, as we have been in section 11.4.1, be functions of any other variable in the model. More generally, they have to satisfy the following assumption:

ASSUMPTION **PM (Markup functions)** Markup functions $\mu_j^+(\cdot)$, $\mu_j^-(\cdot)$ are continuous, nonnegative, and homogeneous of degree one in prices.

As before, the homogeneity assumption ensures that price normalization has no real effects.[45] We now define and prove the existence of an equilibrium with markups.

DEFINITION **11.1** (Equilibrium with producer markups) The allocation y_j^*, all j, x_i^*, all i, supported by the price vector $p^* \geqslant 0$ is an equilibrium with producer markups if the following conditions are satisfied:

1. For every producer j, $y_j^* \in \text{argmax}\ \{(p^* - \mu_j^+)y_j^+ - (p^* + \mu_j^-)y_j^-$ subject to $y_j = y_j^+ - y_j^- \in Y_j$, $y_j^+, y_j^- \geqslant 0\}$, where the μ may be functions of other variables.
2. For every consumer i, $x_i^* = \text{argmax}\ \{u_i(x_i)$ subject to $p^*x_i \leqslant h_i^*$, $x_i \geqslant 0\}$, where $h_i^* = p^*\omega_i + \Sigma_j\theta_{ij}p^*y_j^*$.
3. All markets are in equilibrium, $\Sigma_i x_i^* - \Sigma_j y_j^* - \Sigma_i\omega_i \leqslant 0 \perp p^* \geqslant 0$.

45. In applied models monetary effects and money illusion are often introduced by relaxing this homogeneity assumption. See also section 12.3.5.

PROPOSITION 11.2 (Existence of an equilibrium with producer markups) Let assumptions P1′, PM, Cl′, and C2′ be satisfied. Then an equilibrium with producer markups exists.

Proof The proof closely follows the arguments of proposition 3.4 but for one difference. In proposition 3.4 consumers receive profits $\Pi_j(p)$ which, being the optimal value of producer j's objective, are continuous functions. Here consumers still receive py_j, but this is no longer producer j's value function and is therefore not necessarily continuous. This difficulty can be tackled by defining the following program in open economy format:

$$\max p^e e - p^m m - \Sigma_j(\hat{\mu}_j^+ y_j^+ + \hat{\mu}_j^- y_j^-),$$

$$e, m, y_j^+, y_j^- \geqslant 0, \text{ all } j,$$

subject to

$$\Sigma_j y_j^+ - \Sigma_j y_j^- + m - e \geqslant \Sigma_i \hat{x}_i - \Sigma_i \omega_i \qquad (p),$$

$$(y_j^+ - y_j^-) \in Y_j.$$

The optimal values of y_j^+ and y_j^-, the multipliers p and the markups $\hat{\mu}_j^+$, $\hat{\mu}_j^-$ are then used in the feedback equations to set consumption as

$$\hat{x}_i = \text{argmax}\{u_i(x_i) \text{ subject to } px_i \leqslant h_i\},$$

where $h_i = p\omega_i + \Sigma_j \theta_{ij}(\Pi_j + T_j)$, with profit and markup revenue defined as

$$\Pi_j = \max(p - \hat{\mu}_j^+)y_j^+ - (p + \hat{\mu}_j^-)y_j^- \text{ subject to } (y_j^+ - y_j^-) \in Y_j,\ y_j^+, y_j^- \geqslant 0,$$

$$T_j = \hat{\mu}_j^+ y_j^+ + \hat{\mu}_j^- y_j^-.$$

The updating is completed with the markup adjustments

$$\hat{\mu}_j^+ = \mu_j^+(\cdot) \quad \text{and} \quad \hat{\mu}_j^- = \mu_j^-(\cdot).$$

Note that income h_i is positive since (1) p cannot fall below p^e which is positive, (2) endowments are nonzero for all i, (3) Π_j is nonnegative (possibility of inaction), and (4) T_j is nonnegative because markups are nonnegative (assumption PM). This defines the fixed point mapping, and the rest of the proof is similar to the one given for proposition 3.4. The closed economy model is obtained, as in proposition 3.4, by bringing p^e to zero. ■

11.4.4 Objective Markups

In proposition 11.2 noncompetitive producers were faced with given markup functions. We now show how to generate markups that are consistent with the objective demand formulation of section 11.2.5, in particular, with program (11.17).[46] The limitation will be that we only characterize the fixed point but cannot show that it exists.

We assume that consumers behave competitively, while noncompetitive producer $j \in J_s$ faces the following problem:

$$\max_{p^j \geqslant 0,\, y_j} p^j y_j,$$

subject to

$$y_j \in Y_j,$$

$$z^j(p^j, y_j, y_{-j}) \leqslant 0,$$

$$N(p^j) = 1,$$

where $z^j(p^j, y_j, y_{-j}) = \Sigma_i x_i(p) - \Sigma_{h \in J_c} y_h - \Sigma_{h \in J_s} y_h - \Sigma_i \omega_i$.

Rather than solving this highly nonconvex program as part of the market-clearing stage of the imperfect competition model, we decompose the equilibrium model in two parts: (1) a markup-ridden equilibrium in which producers take markups as given and (2) an artificial construct which updates the markups on the basis of the outcome of (1). An equilibrium with imperfect competition obtains if the markups in (1) agree with those generated by (2).

For every producer j, the program in (1) reads as

$$\max_{y_j} (p - \mu_j) y_j,$$

subject to (11.32)

$$y_j \in Y_j,$$

while the associated artificial construct in (2) is

$$\max_{p \geqslant 0,\, y_j} p \tilde{y}_j,$$

subject to (11.33)

$$y_j \in Y_j,$$

$$z^j(p, y_j, y_{-j}) \leqslant 0,$$

$$y_j = \hat{y}_j \qquad (v_j),$$

$$N(p) = 1.$$

46. See Keyzer (1995) who also treats the case of noncompetitive consumers.

In (11.33), v_j is a Lagrange multiplier, and $\hat{y}_j = \tilde{y}_j$ is the value obtained from (11.32) in the markup equilibrium. Though in (11.33), $\hat{y}_j$ and $\tilde{y}_j$ are given the same value, we distinguish the two to emphasize that v_j is the marginal contribution associated with the constraint, while keeping constant the parameter $\tilde{y}_j$ in the objective function. Program (11.33) is also highly nonconvex, but it has a single feasible y_j, so we can initialize it with this value and with the (usually unique) markup equilibrium price p. Thus an optimization algorithm is needed not to solve the primal of (11.33) but only to compute the marginal value v_j. It can be checked that in the fixed point, $\mu_j = -v_j$, and this suggests updating the markups through the value generated by (11.33).[47]

Though program (11.33) does not represent the behavior of any agent, it has as natural interpretation. It evaluates the marginal contribution to profit of a change in net supply by the noncompetitive producer j, taking into account his technology and the effect this change will have on prices through the excess demand function.

This approach holds promise from a computational point of view, though it will in general not lead to an existence proof. The algorithm iterates between markup equilibria and the construct (11.33). Both are known to have a solution, but the process may fail to converge. Yet, since even out of equilibrium, all allocations are feasible, budget constraints are met, and profits are maximized in (11.32), the process can still be given a real time interpretation in which producers learn about their environment.

11.5 Practices in Applied Modeling

The discussion has made it clear that it is very difficult to represent imperfect competition without resorting to markups. It is also difficult to compute a solution for the more general formulations of section 11.2 because of nonconvexities in the decision problems of noncompetitive agents. Hence it is no surprise that most general equilibrium applications reduce to markup formulations.

Recall from the discussion in chapters 3 and 10 that markups may also be due to the fact that (1) a commodity has been left out of the model or is represented in the model but there is no market for it (section 3.6.2), and (2) firms supply quantities so as to achieve equality between their marginal cost and the price of the output (chapter 10). The CGE model with markups was already presented in sections 3.6.2 and 10.5 and does not need further treatment here. For specifications of producer technology other than the

47. It is also possible to test the local optimality of the resulting producer decisions and to model collusion without side payments by defining (11.33) for the coalition of producers. See Keyzer (1995).

single-output firm in the CGE model, the treatment of indirect taxes of chapter 5 applies, except that the proceeds of the markup (tax) are part of the firm's profits and are simply redistributed to consumers according to the profit shares θ_{ij}.

Therefore in section 11.5.1 we merely discuss issues that are specific to imperfect competition: nonconvexities in technology, product differentiation, and number of firms. We postpone our survey of applications to section 11.5.2, since most of these deal with all issues within a single model. In section 11.5.3 we explore possible extensions.

11.5.1 Issues

Imperfect Competition and Increasing Returns

Although constant returns and decreasing returns to scale are compatible with imperfect competition (e.g., when there is licensing), in applied models imperfect competition is often studied in conjunction with increasing returns or setup costs because these are thought to limit the number of competing firms. Under increasing returns this has the practical advantage of making the constraint set of the firm convex more easily, so the producer problem can be formulated in terms of profit maximization. Indeed, if in the profit $p^q(q)q - c(p^v, q)$, the cost function is concave in q (because of increasing returns), the concavity of $p^q(q)q$ in q may be sufficient to ensure (quasi) concavity of the profit function. Moreover, if there are setup costs, $c(p^v, 0) > 0$, then $p^q(q)q$ may make the price sufficiently high when q is small to produce at a positive scale without losses.

Product Differentiation

In general equilibrium there is a one-to-one correspondence between commodities, markets, and balance constraints. Two differentiated commodities, like hard and soft wheat or cars that differ only by their color, are treated as different commodities: Each is fully represented in the consumer problem, the producer problem, and the balance constraints. Every commodity (e.g., blue cars) is, by definition, homogeneous, and there is no difficulty in computing its aggregate supply and its aggregate demand by summation of individual supplies and demands.

Two commodities are said to be vertically differentiated if all consumers objectively agree on their ordering. Typically, if the differentiation is based on quality, consumers will agree that for bakery, hard wheat is superior to soft wheat. Two commodities are horizontally differentiated if there is no such ordering based on quality, but consumers have different tastes, and some prefer blue cars, while others prefer red ones. Both types of differentiation can be represented in general equilibrium within the setting of perfect competition: The vertical one via the Lancaster characteristics approach of section 3.6.2 and the horizontal one via nested production and utility functions and the Armington specification of sections 2.1.4, 2.2.5, and 4.4.2, respectively.

Horizontal differentiation may have special implications in the context of imperfect competition. Consider the case of an "industry" that sells to the consumer a composite commodity X, supplied by n firms where every firm j supplies an amount q_j of a different quality. As in the Armington specification, assume that the composite commodity is "produced" according to the decreasing returns to scale function $X(q_1, \ldots, q_n)$. Let P be the price of this composite commodity paid by the consumer (or by any other buyer, e.g., the exporter or the firm that uses it as input) and assume that (1) each firm faces a constant or decreasing returns to scale technology and (2) the manager of firm j faces a cost function $c_j(p^v, q_j)$. Then the industry solves the following profit maximization problem:

$$\Pi(P, p_1, \ldots, p_n) = \max PX(q_1, \ldots, q_n) - \Sigma_j c_j(p^v, q_j),$$

$$q_j \geqslant 0, \text{ all } j,$$

while each firm j minimizes the cost to produce q_j. The first-order conditions are

$$P\frac{\partial X}{\partial q_j} = \frac{\partial c_j}{\partial q_j}.$$

This is as if the manager of firm j faced a "markup-ridden" price $p_j = P(\partial X/\partial q_j)$ at which he is supposed to maximize profits. The owner of the firm will receive $p_j q_j - c_j(p^v, q_j)$ plus some (here, unspecified) share of the profit of the industry, $PX(q_1, \ldots, q_n) - \Sigma_j p_j q_j$.

This profit results from the appropriation of consumer surplus (producer rent) and is thus an expression of imperfect competition, caused by the fact that the consumer's preference for variety reduces competition among producers. This preference is represented by the imperfect substitution between q_j in the function $X(q_1, \ldots, q_n)$. Indeed, under perfect competition, there would be no industry profit, and the consumer would solve

$$\max u(x_0, X(x_1, \ldots, x_n)),$$

$$x_j \geqslant 0, \text{ all } j,$$

$$\text{subject to} \tag{11.34}$$

$$p_0 x_0 + \Sigma_j p_j x_j \leqslant h,$$

where x_0 represents an aggregate of all other commodities and p_0 is their price. The market balances would read

$$x_j - q_j \leqslant 0 \perp p_j \geqslant 0, j = 1, 2, \ldots, n.$$

The case of imperfect competition treated by Dixit and Stiglitz (1977) can be understood as following either from the consumer model max $u(x_0, X)$ subject to $p_0 x_0 + PX \leqslant h$, where the surplus is appropriated by the industry or, equivalently, from (11.34) where the p_j are noncompetitive prices equal to $(1 + \tau)P(\partial X/\partial q_j)$, and $(1 + \tau)$ is, by definition, equal to $X/\Sigma_j(\partial X/\partial q_j)q_j$.[48]

Welfare maximization would leave all the surplus to the consumer, since the activity of producing the composite commodity does not require inputs, other than the commodities q_j themselves. In section 4.4.2 for the Armington case, the composite commodity consisted of imports from different origins. There it was "produced" under constant returns, so the issue of appropriation of profits did not arise. Here we can also use a constant returns technology $X(q_0, q_1, \ldots, q_n)$ with a fixed factor q_0 to which the profits from the mixing accrue. If this factor is allowed to vary without cost, profits will disappear, and the formulation is equivalent to the one used by Armington.

Variable Number of Firms (or of Differentiated Commodities)

Assume now that all firms are identical in terms of their cost structure and produce a horizontally differentiated commodity. Each of these commodities enters into the X function in the same way. In the CES form this can be represented as $X(q_0, q_1, \ldots, q_n) = [q_0^\rho + (\Sigma_j q_j^\theta)^{\rho/\theta}]^{1/\rho}$ for ρ and $\theta < 1$. Then in equilibrium, if the consumer is only interested in purchasing the aggregate commodity X, all the firms that contribute to it will face the same price p and produce the same output q. Hence the composite function becomes $X = [q_0^\rho + (nq^\theta)^{\rho/\theta}]^{1/\rho}$. If the fixed factor q_0 is taken to represent the number of firms, this reduces to $X = [n^\rho + n^{\rho/\theta} q^\rho]^{1/\rho}$ which, in terms of "aggregate supply" $Q = nq$ and number of firms n, may be rewritten

$$\Xi(Q, n) = [n^\rho + n^{\rho(1/\theta - 1)} Q^\rho]^{1/\rho}. \tag{11.35}$$

$\Xi(Q, n)$ can be interpreted as a production function with Q and n as inputs. Two cases have now to be distinguished. First, under perfect competition the consumer with a preference for variety will like to increase n. If $\Xi(\cdot)$ is increasing in n and n can be produced at zero cost, it will go to infinity. On the other hand, if the consumer is satiated with respect to variety or if there is, say, a setup cost for every firm (i.e., inputs are needed to "produce" n), n will be finite in the optimum, and the consumer surplus from the composition process will vanish. Second, under

48. Dixit and Stiglitz (1977) show that if $X(\cdot) = (\Sigma_j x_j^\rho)^{1/\rho}$, and if the number of firms is large, the demand function for commodity j has an approximately fixed elasticity and can be written $x_j/x_0 = \gamma_j(p_j/p_0)^\varepsilon$ and $\varepsilon = (\partial x_j/x_j)(p_j/\partial p_j)$; the demand elasticity for product j is equal to $1/(1 - \rho)$ and is identical for all producers j. This elasticity formulation can, as discussed in section 11.4, be represented through a markup.

imperfect competition where producers appropriate the consumer surplus, it could be assumed that new entrants will be treated in the same way as incumbent managers and charged the standard markup. Then the situation is the same as if there were only one manager producing the aggregate commodity in quantity X. Therefore, if n is variable, the profit of the manager will vanish, and the consumer will be charged a price p equal to the unit cost incurred by the manager, incremented by the markup from the composition process. Here the barrier to entry is caused by the markup from composition, in contrast to the case discussed in section 11.2.7 where it was a negative profit for the manager that prevented entry.

Note also that under Cournot oligopoly (where products are not differentiated), if all n firms are identical, the market share σ_{jk} is equal to $1/n$ in (11.31). Then under free entry the number of firms will increase until profits (and therefore markups) are zero so that perfect competition is the outcome.

11.5.2 Applications

Applied models usually assume that every firm supplies a different commodity, though Cournot oligopoly can be accommodated naturally (unlike Bertrand oligopoly). Modelers often compute both Cournot-Nash and Bertrand-Nash solutions and find that the results differ; some authors introduce conjectural variations. Experiments are also performed with variable number of firms and free entry.

Most studies deal with the effects of tariff reductions and economic integration between countries. Besides the standard gains from trade liberalization (tariff reductions and tax harmonization), welfare effects basically originate from three sources:

1. *Scale effects.* On the supply side the main effect originates from savings on setup costs in producer technologies. As markets become larger, average costs can fall.

2. *Reduced monopoly power.* In each country there is a monopolist for every noncompetitive commodity who sells on his own market. Hence there are country-specific prices and markups based on market shares of unity. After integration there will be, apart from transportation costs, a single price for each commodity on the integrated market, and every monopolist will have turned into an oligopolist with a market share smaller than unity, and this will by itself reduce the markup.

3. *Preference for variety.* After integration more varieties become available, and this is assumed to increase utility.

Harris (1984) and Cox and Harris (1985) study trade liberalization between Canada and the United States in a model that uses a fixed markup rate (derived from an isoelastic demand) and a cost function that is linear in scale with setup costs as intercept. Since a tariff reduction lowers the input cost of a sector with increasing returns, it allows the sector to reduce its markup and therefore its price, which leads to an increase in demand. The efficiency gain then comes from the resulting exploitation of scale economies, and trade liberalization has much larger specialization effects than under constant or decreasing returns with perfect competition.

Cox and Harris were the precursors of many studies applied to other international agreements, like the completion of the European Internal Market (Burniaux and Waelbroeck 1992, 1994; Gasiorek et al. 1992a, b; Mercenier 1994; the influential and early study by Smith and Venables 1988 was of a partial equilibrium nature), the U.S.–Mexico Free Trade Agreement (de Melo and Tarr 1992), and South American integration (Flores 1993). Most of these studies have setup costs but also try to endogenize the markup through the Dixit and Stiglitz (1977) formulation. This makes it possible to show not only the specialization gains attributable to returns to scale but also the consumer welfare gains due to an increase in the variety of commodities and a reduction of market shares of monopolists. Finally, other studies focus on country-specific applications that consider the consequences of unilateral tariff reductions on a particular economy (Delorme and van der Mensbrugghe 1990 for Canadian agriculture; Devarajan and Rodrick 1991 for Cameroon; de Melo and Roland-Holst 1994 for Korea).

In all these studies the managers of firms are assumed to maximize profits, given some specification of the markup functions. Mercenier (1994) goes a step further by approximating the elasticity functions discussed in section 11.4.2. Berthelemy and Bourguignon (1994) are closer to the spirit of the model described in section 11.2.4 and consider a dynamic model in which there is a representative consumer in each country, who maximizes utility.

From a theoretical point of view the general weakness of applied models is that they do not conform to the distinction between the anticipation and the market-clearing stage. For those that use fixed elasticities or markup functions, this leads to a misrepresentation of the environment in the agents' decision models. In the case of Mercenier (1994) and Berthelemy and Bourguignon (1994), there is an attempt to go further with respect to the representation of the environment, but the elasticity functions are not explicitly derived from an anticipation stage; as a result issues of commitment, convexity of the decision models, and continuity properties of the elasticity functions are obscured. The approach followed in section 11.4.4 makes it possible to represent the environment fully and in a transparent way. It avoids specifying elasticities because dual variables are used to generate the markups.

11.5.3 Possible Extensions and Directions for Future Research

In recent years the literature on industrial organization and applications of game theory has been expanding at an impressive rate. Most of these new results have not as yet found their way into general equilibrium models, and quite naturally so. They deal with very partial bargaining problems, say, between the manager and the owners and have little to gain from broadening the perspective beyond the direct concerns of the players in the game. However, game-theoretical concepts are being increasingly used in the study of international trade, transfrontier pollution, labor negotiations, sector models, and tax competition among countries,[49] where economywide effects can obviously not be ignored. Though originating from the need to describe how agreements are reached,[50] the theory tends to concentrate on normative implications and hence on Pareto-efficiency, which also calls for analysis at the economy level. The difficulties that have to be addressed in the relatively simple framework of the Cournot-Walras model probably explain why more advanced game theory shuns the domain of general equilibrium. Yet it may be fruitful to perform applied general equilibrium experiments with some of its new concepts. We mention possibilities for extending the treatment in sections 11.1 to 11.4.

Extending the Strategies to Other Variables Than Quantities and Prices

Quantities (and prices) were the only strategies included in this chapter. This is a fairly general formulation already, since many strategic actions such as advertising, research and development, or acquisition of other firms can be represented as purchases of (usually competitive) inputs. However, other instruments have been used, for example, tax rates, tariffs, and labor regulations.

There are also variables less common in general equilibrium models that could be introduced, such as promises of payments or punishments to deter entry. These may be zero in equilibrium but affect it nonetheless, for they force or rule out certain actions. Although incorporation of such variables may create problems (e.g., bankruptcy) with respect to the feasibility of the anticipation stage, it can proceed in the same way as for other strategic variables.

Multistage Games

The model discussed so far is static, but as we have seen in section 7.2 for the T-period equilibrium model, the static model can be reinterpreted as a

49. See, for example, Helpman and Krugman (1989, 1990) or Salvatore (1993) for international trade, Kaitala, Mäler, and Tulkens (1992) and Chander and Tulkens (1994) for transfrontier pollution, Svejnar (1986) for labor negotiations, Smith and Venables (1988) for sector models, and Mintz and Tulkens (1986, 1990) for tax competition.

50. See, for example, the well-known book by Tirole (1988).

multiperiod one by mere labeling of variables and equations. This also applies to models with imperfect competition. The industrial organization literature[51] studies a wide variety of multiperiod models under the heading of multistage games.[52]

As an illustration, consider a producer model with two periods. In period 1 a noncompetitive producer j buys inputs (of research, advertising, etc.) at competitive prices, in quantities q_{1j}, his strategic variables in this period. This produces capacities according to the transformation function of the first period. The capacities can be used as inputs in the second-period transformation function when producer j decides on q_{2j} his level of output, the strategic variable in the second period. Under Cournot behavior the payoff function of producer j is $\Pi_j(p_1; q_{1,j}; q_{1,-j}; q_{2,j}; q_{2,-j})$.[53] This is now a Nash equilibrium problem with $(q_{1,j}, q_{2,j})$ as strategic variables for producer j. Here, the recursive structure allowed the simplification that the anticipation stage (the Walrasian clearing) had to be specified for period 2 only.

While such models can be solved as multiperiod models, in the static way, their time recursivity generates a bloc-diagonal structure that can be exploited to obtain a more tractable representation of the anticipation stage, yielding, for instance, analytical forms for the inverse demand functions. Alternatively, it is possible to incorporate uncertainty in period 2. This can be done by distinguishing, say, S states of nature that are uncertain in period 1 when the first-period decisions have to be made, and revealed in period 2. This also leads to a payoff function that can be written as $\Pi_j(p_1; q_{1,j}; q_{1,-j}; q_{2,j}; q_{2,-j})$ in terms of first- and second-period strategies. The difference is that the anticipation stage will have to account for market clearing in each of the S states and that q_{2j} will be a vector with an entry for every state.[54]

Anticipation Stages Departing from the Walrasian Setting

Models of imperfect competition commonly use the Walrasian concept for the anticipation stage. We have seen that in price competition, Benassy rationing is a natural alternative. The earlier discussion in this section has indicated that international trade, taxation, dynamics, and so on, can also be incorporated within the anticipation stage. Moreover nonrivalry may

51. Among the early contributions on multistage games, one should cite Hotelling (1929). For other cases, see Tirole (1988).

52. Here of course, the term "multistage" should not be viewed as an extension of the two-stage procedure of anticipation and market clearing.

53. Note that p_1, the competitive price for inputs in period 1 is given, while p_2 is absent, since it is the Walrasian clearing price for the output in period 2.

54. The q_{2j} are the strategies that can be adjusted after the state of nature has been revealed. Therefore they are usually called coping strategies.

have to be accounted for, since many of nonconventional inputs, such as advertising and punishments, are of this nature. They change a stock variable that represents reputation. This stock will affect in a nonrival way the utility of the commodity with that reputation for the consumer. This effect may even spill over to close substitutes supplied by competing firms.

Communication among Noncompetitive Agents

In chapter 3 (and in almost all other chapters) we have shown that a competitive equilibrium can be represented in several perfectly equivalent formats. The excess demand format gives the impression of a fully decentralized economy in which each agent acts on his own, taking prices as given, without any further communication. The Negishi format conveys the opposite impression of a central plan in which all agents collude and act jointly, but they are nevertheless competitive in the sense that they take the welfare weights as given and accept that these are set so as to satisfy all budgets. In equilibrium the welfare program can thus be interpreted as the program of the grand coalition of consumers. This is what makes the fully cooperative model equivalent to the noncooperative one.

In the welfare sections of previous chapters, other variables, such as indirect taxes, were also represented within the Negishi format. Therefore, since markups are mathematically equivalent to price-dependent taxes, equilibria of a model with imperfect competition can also be specified in this way, yielding welfare programs similar to those in section 5.3. In the case of imperfect competition models, the equivalence still holds if in the anticipation stage the excess demand constraints are replaced by the welfare program supplemented by the appropriate budget constraints (i.e., the full format). Indeed, in the Cournot competition model of section 11.2, consumers set quantities in order to influence prices to their benefit. One can similarly envisage that they set quantities to affect welfare weights in order to improve their utility. However, since they have also to satisfy budget constraints, the outcome will be the same.

This equivalence breaks down if agents no longer use quantities as their strategies or take other variables than their competitors' outputs as given. They may, for example, treat the welfare weights of other agents as given or incorporate lower bounds on other agents' utilities (based on the possible threats by their competitors). If these bounds are obtained from indirect utility functions, the budgets will automatically be satisfied; otherwise, transfers may be needed.

The previous discussion shows that in general equilibrium there is no obvious borderline between cooperation and noncooperation. Therefore the distinction drawn in the game-theoretic literature between cooperative games (where there is an exploration of the feasible space of joint actions

(a_j, a_{-j})) and noncooperative games (where every agent acts on his own) is not particularly helpful here. However, using in the anticipation stage the Negishi format, possibly supplemented by some additional bounds or more general constraints on utilities, makes it possible to represent a rich variety of cooperative situations, while the construction of the associated models can proceed along the same lines as in sections 11.2 to 11.4. This approach also makes normative analysis easier.[55]

Avoiding the Distinction between the Anticipation and the Market-Clearing Stage

Finally, the distinction between the anticipation and the market-clearing stage is a strenuous one. Ideally one would like to avoid committed behavior altogether and, within the program of the noncompetitive agent in the market-clearing stage, use the constraint that accurately reflects his environment in the market-clearing stage itself, and not one that represents a different economy such as the Walrasian one. The difficulty of achieving this goal is that such a constraint would no longer be an exogenously given function. It would have to be "learned" by the agent and adjusted until it is in full accordance with the reactions of the environment.[56] This involves finding a fixed point in a functional space (i.e., choosing a function instead of a point)[57] and seems to offer a challenging area for further research.

55. See, however, Keyzer (1995) for a computational approach that allows for collusion.

56. The solution is comparable to the one encountered in section 8.1.2 (equation 8.6) where the value function of the program enters the program itself.

57. See Stokey and Lucas (1989, 50–52).

12 Money and Incomplete Asset Markets

Transportation, storage, and marketing facilities are needed to permit trade flows between agents. In the competitive model these activities are hidden in the production sets and the locations (and dates) attached to commodities. Trades are assumed to be frictionless: There are enough costless "trucks" and enough costless "roads" that link all locations.

In chapter 4 the competitive model was extended to account for transportation costs of commodities, but transactions may likewise be subject to further constraints and require "financial" inputs and channels of transaction. If the inputs are provided in adequate quantity (the number of trucks) and recognized by all agents (channels of the road network), the exchange of commodities is not constrained by them, and hence not affected by the financial sphere, which therefore does not have to be represented explicitly in this case. In this chapter we will introduce such financial constraints.

In section 12.1 we formulate a model with cash in advance in which the availability of the financial instrument (medium of exchange) may constrain transactions. This is only one of the many ways in which money demand can be represented. It is meant to illustrate the effect of a binding constraint on money availability and does not address the issue of risk. In section 12.2 we introduce an explicit network of possible transactions (the roads). When this network limits transactions (e.g., because of missing links), asset markets are said to be incomplete. Most of this section presents a model that has not been applied so far but offers useful insights on the functioning of financial markets. CGE applications, discussed in section 12.3, largely pertain to static models with demand functions for money and financial assets.

12.1 Cash in Advance

In section 3.2 we introduced the competitive model in a format that incorporates financial constraints (the full-format). These constraints were seen to be nonbinding in equilibrium. If, as in chapter 7, we reinterpret this as a two-period model, the solution will involve some agents saving and others dissaving in period 1. The planner may issue "paper money" which enables the agents who dissave to cover their first-period deficit. At the start of period 2, those who have saved will hold the money. Hence, money serves as a store of value but does not carry any rent. Since it has no other function, it is a financial asset, as opposed to the capital goods considered in chapters 7 and 8 which also were stores of value. Obviously in this setup the financial system does not constrain welfare.

The store of value aspect of money is obscured when one aggregates over time. Then it is as if agents need money within the period and money demands will become complementary to demand for commodities. This is the transactions demand for money. In this sense the distinction between transactions and asset demand is an artifact that results from time aggregation. In this section we specify such a single-period model. If the money supply adjusts in a flexible way, money will not affect allocations. If it becomes constraining, a positive interest rate will result, and money will have real effects. Finally, if money supply is set in nominal terms (fiat money), the price level can adjust to support a zero rate of interest, and there will be no real effects. The three cases are dealt with in sections 12.1.1 and 12.1.2. Section 12.1.3 deals with welfare aspects. Applications are discussed in section 12.3.

12.1.1 Flexible Supply of Money

Consider an economy in which producers and consumers need coupons (cash[1]) for buying. The producer needs this cash to buy inputs y_j^- in order to produce outputs y_j^+. He solves

$$\max \Pi_j,$$

$$m_j \geqslant 0, y_j^+, y_j^- \geqslant 0, \Pi_j,$$

$$\text{subject to} \tag{12.1}$$

$$\Pi_j \leqslant p(y_j^+ - y_j^-),$$

$$(y_j^+ - y_j^-) \in Y_j,$$

$$m_j = py_j^-.$$

The last constraint can obviously be left out of the problem, and m_j can be computed once the optimal decision has been made.

Likewise the consumer has an entitlement for cash that cannot exceed his income $h_i = p\omega_i + \Sigma_j \theta_{ij} \Pi_j$, and he needs this cash for all purchases. He solves

$$\max u_i(x_i),$$

$$m_i \geqslant 0, x_i \geqslant 0,$$

$$\text{subject to} \tag{12.2}$$

$$m_i \leqslant h_i,$$

$$px_i = m_i.$$

Since the provision of cash is freely adjusting and costless, the two constraints can be combined to obtain the standard budget constraint. The amount of cash m_i can be computed outside the program.

1. We will denote cash by m (for money), but this variable should not be confused with the m for imports used in earlier chapters.

For the economy as a whole, (12.1) and (12.2) generate a demand for cash $m = \Sigma_j m_j + \Sigma_i m_i$, whose supply is supposed to adjust flexibly.[2] Note that there are no cash holdings appearing in the income of the consumer. Cash is only a medium of exchange, not an asset. Moreover cash is required for total consumption (including consumption out of the own endowment) and for total inputs (even if they are produced by the firm itself). Relaxing this constraint involves only minor changes in the treatment that follows.[3]

In the full-format framework of section 3.2, we had not represented producers and consumers in the same detail. While consumer budgets are incorporated, producer budgets are not. This limitation can easily be overcome by introducing in the full-format intermediate variables Π_j for profits and defining the constraints for them. This only makes transactions of all agents explicit, and it will now serve to articulate the role of money as medium of exchange in program (12.3). We also incorporate an equation defining aggregate demand for cash for which the shadow price will be zero:

$$\max \Sigma_i \alpha_i u_i(x_i),$$

$$m_i \geqslant 0, x_i \geqslant 0 \text{ all } i, m_j \geqslant 0, y_j^+, y_j^- \geqslant 0, \Pi_j, \text{ all } j, m \geqslant 0,$$

$$\text{subject to} \tag{12.3}$$

$$\Sigma_i x_i - \Sigma_j(y_j^+ - y_j^-) \leqslant \Sigma_i \omega_i \quad (p),$$

$$(y_j^+ - y_j^-) \in Y_j,$$

$$\Pi_j \leqslant \bar{p}(y_j^+ - y_j^-),$$

$$\bar{p}x_i \leqslant \bar{p}\omega_i + \Sigma_j \theta_{ij} \Pi_j \qquad (\rho_i),$$

$$m_j = \bar{p}y_j^- \quad \text{and} \quad m_i = \bar{p}x_i,$$

$$m = \Sigma_j m_j + \Sigma_i m_i.$$

As in the full-format of proposition 3.3, if the α and $\bar{p}$ are chosen so that $\rho_i = 0$ and p is proportional to $\bar{p}$, the solution of program (12.3) is a competitive equilibrium.

By dropping the budget constraints and by substituting out m_i and m_j in the demand for money equation, (12.3) can also be written in a more

2. The cash will have to be expressed in some unit, the numéraire, for which one takes any weighted sum of prices. For example, if prices are normalized to lie on the unit simplex, the unit of cash is $\Sigma_k p_k$. In this case cash is called commodity money, since it is expressed in terms of a commodity bundle (here with unit weights). If the bundle consists of a single commodity (e.g., $k = 1$), this commodity serves as the standard, like gold. When there is no such normalization and prices are expressed in an arbitrary unit like dollars, one speaks of fiat money.

3. In this model all transactions are assumed to occur simultaneously. This is unrealistic and the total amount of cash will be only a fraction of $\Sigma_i px_i + \Sigma_j py_j^-$. The inverse of this fraction is called the velocity of money.

compact form as

$$\max \Sigma_i \alpha_i u_i(x_i),$$

$$m_i \geqslant 0,\ x_i \geqslant 0,\ \text{all } i,\ m_j \geqslant 0,\ y_j^+,\ y_j^- \geqslant 0,\ \text{all } j,\ m \geqslant 0,$$

$$\text{subject to} \qquad (12.4)$$

$$\Sigma_i x_i - \Sigma_j(y_j^+ - y_j^-) \leqslant \Sigma_i \omega_i \qquad (p),$$

$$(y_j^+ - y_j^-) \in Y_j,$$

$$m_j = \bar{p} y_j^- \quad \text{and} \quad m_i = \bar{p} x_i,$$

$$m = \Sigma_j m_j + \Sigma_i m_i.$$

This is the original Negishi format, expanded to include an equation that balances demand and supply for money, and it can be considered as an additional balance equation for a "commodity" produced in quantity m and used by consumers and producers. Prices $\bar{p}$ play the role of "input coefficients" for money.

12.1.2 Scarcity of the Medium of Exchange: General Equilibrium with Cash in Advance

We now constrain the supply of money through an upper bound $\bar{m}$, a function of $\bar{p}$ to be specified below. Money will become scarce; under competitive conditions its price will become positive. One should specify to whom money belongs and who is able to change its supply. For simplicity we assume that consumer 1 (the government) owns the stock of money $\bar{m}$ as initial endowment and that supply is given and fixed. This implies that an upper bound has to be introduced in program (12.4):

$$m \leqslant \bar{m} \qquad (v), \qquad (12.5)$$

where the Lagrange multiplier v may become positive. This multiplier measures the rent that will accrue to consumer 1 who owns $\bar{m}$.

The model describes a Hicksian week, at the beginning of which producers and consumers borrow money in amounts m_j and m_i from the bank. They immediately spend the money to pay inputs, dividends, and consumption. The bank's payment operations take one week, at the end of which producers receive payment for their outputs and consumers receive their income from endowment and dividends. On these cash loans, agents pay interest to the owner of the cash. This is the so-called cash-in-advance approach.[4]

4. See, for example, Lucas and Stokey (1987) where, however, the demand for cash originates from decisions under uncertainty.

In this model producer j solves the following problem:

$$\Pi_j = \max \bar{\zeta}\bar{p}_j^+ - \bar{p}y_j^-,$$

$$m_j \geqslant 0,\ y_j^+,\ y_j^- \geqslant 0,$$

$$\text{subject to} \tag{12.6}$$

$$(y_j^+ - y_j^-) \in Y_j,$$

$$m_j = \bar{p}y_j^-,$$

where $\bar{\zeta} = 1/(1 + \bar{v})$ is the discount factor charged on the value of outputs, $(\bar{p}, \bar{v})$ is the value of (p, v), normalized on the simplex S^{r+1}.

Consumer i's problem is

$$\max u_i(x_i),$$

$$m_i \geqslant 0,\ x_i \geqslant 0,$$

$$\text{subject to} \tag{12.7}$$

$$\bar{p}x_i \leqslant h_i,$$

$$m_i = \bar{p}x_i,$$

where

$$h_i = \bar{\zeta}\bar{p}\omega_i + \Sigma_j\theta_{ij}\Pi_j + \sigma_i\bar{\zeta}\bar{v}\bar{m}, \tag{12.8}$$

and $\sigma_i = 1$ for consumer 1 and zero otherwise. Note that the income of consumer i is also discounted with factor $\bar{\zeta}$ (i.e., at the rate $\bar{v}$). Hence cash in advance discounts revenue to the beginning of the period when the expenditure takes place. The discount rate $\bar{v}$ operates like an indirect tax (on outputs and on endowments) whose proceeds accrue to consumer 1.

It can be verified from the profit equations, the budget equations, and the money demand equations in (12.6)–(12.8) that allocations are homogeneous of degree zero in $(\bar{p}, \bar{m})$. Also $\bar{p}$ and m are measured in the same units and so must be $\bar{m}$. Therefore, if we want to keep the allocations independent from the unit in which money is measured, we must also adjust $\bar{m}$ whenever we normalize $\bar{p}$, say, to the simplex. This implies that $\bar{m}$ cannot be kept constant and should be set according to a money supply function that depends on $\bar{p}$ and has the following properties:

ASSUMPTION M (Money supply) The money supply function $M(\bar{p})$ is positive, continuous, and homogeneous of degree one in $\bar{p}$ for all nonzero, nonnegative $\bar{p}$.

Money supply is thus set according to

$$\bar{m} = M(\bar{p}). \tag{12.9}$$

We substitute (12.9) within (12.5)–(12.8) and define an equilibrium with cash in advance as follows:[5]

DEFINITION 12.1 (Equilibrium with cash in advance) The allocation y_j^*, m_j^*, all j, x_i^*, m_i^*, all i, supported by the price vector $\bar{p}^* \geqslant 0$ and the discount rate $\bar{v}^*$ is an equilibrium with cash in advance if the following conditions are satisfied:

1. For every producer j, y_j^*, m_j^* solve (12.6).
2. For every consumer i, x_i^*, m_i^* solve (12.7) with income given by (12.8).
3. All commodity markets are in equilibrium, $\Sigma_i x_i^* - \Sigma_j y_j^* - \Sigma_i \omega_i \leqslant 0 \perp \bar{p}^* \geqslant 0$.
4. The market for cash in advance is in equilibrium, $\Sigma_i m_i^* + \Sigma_j m_j^* \leqslant M(\bar{p}^*) \perp \bar{v}^* \geqslant 0$.

We are now ready to define a full-format program for a monetary economy:

$$\max \Sigma_i \alpha_i u_i(x_i),$$

$$h_i, m_i \geqslant 0,\ x_i \geqslant 0, \text{ all } i,\ m_j \geqslant 0,\ y_j^+,\ y_j^- \geqslant 0, \text{ all } j,\ m \geqslant 0,$$

$$\text{subject to} \qquad (12.10)$$

$$\Sigma_i x_i - \Sigma_j (y_j^+ - y_j^-) \leqslant \Sigma_i \omega_i \qquad (p),$$

$$(y_j^+ - y_j^-) \in Y_j,$$

$$\Pi_j = \bar{\zeta} \bar{p} y_j^+ - \bar{p} y_j^-,$$

$$h_i = \bar{\zeta} \bar{p} \omega_i + \Sigma_j \theta_{ij} \Pi_j + \sigma_i \bar{\zeta} \bar{v} M(\bar{p}),$$

$$\bar{p} x_i \leqslant h_i \qquad (\rho_i),$$

$$m_j = \bar{p} y_j^- \quad \text{and} \quad m_i = \bar{p} x_i,$$

$$m = \Sigma_j m_j + \Sigma_i m_i,$$

$$m \leqslant M(\bar{p}) \qquad (v),$$

where $\bar{\zeta} = 1/(1 + \bar{v})$ and $(\bar{p}, \bar{v})$ lies on the simplex S^{r+1}.

As in proposition 3.3, if the α_i, $(\bar{p}, \bar{v})$ are chosen so that $\rho_i = 0$ and $(\bar{p}, \bar{v})$ is proportional to (p, v), then the solution of the program is a cash-in-advance equilibrium. We state and prove:

PROPOSITION 12.1 (Existence of an equilibrium with cash in advance) Under assumptions P1′, C1′, C2′, and M there exist positive welfare weights

5. We give the definition for $(\bar{p}, \bar{v})$ on the simplex S^{r+1}; this normalization will be changed below.

$\alpha^* \in S^m$, positive prices $\bar{p}^*$, and a nonnegative discount rate $\bar{v}^*$ in program (12.10) such that the resulting allocation is an equilibrium with cash in advance as specified in definition 12.1.

Proof The proof proceeds as for proposition 3.3, but the fixed point mapping is defined on the parameter set $C = A \times \bar{B}$, where $A = S^m \times S^{r+1}$ is the set containing $(\alpha, \bar{p}, \bar{v})$ and the $(r + m + 1)$-dimensional set $\bar{B}$ is the convex hull of the set of Lagrange multipliers of (12.10).[6] We perturb program (12.10), adding a constant $\varepsilon > 0$ to the budget constraints, so that we have $\bar{p}x_i \leqslant h_i + \varepsilon$ for all i; this ensures constraint qualification.

Step 1 Construction of the Fixed Point Correspondence

The perturbed program defines a Lagrange multiplier correspondence that maps into $\bar{B}$, a compact, convex set that does not contain the origin (by nonsatiation of the utility functions). The updating of the α_i is as in step 1 of the proof of proposition 3.3. To update $\bar{p}$ and $\bar{v}$, we store their current values $\hat{p} = \bar{p}$ and $\hat{v} = \bar{v}$ and define

$$\bar{p}_k = \lambda(p_k + \rho\hat{p}_k), \qquad \text{all } k, \tag{12.11}$$

$$\bar{v} = \lambda v,$$

where $\rho = \max_i \rho_i$ and the positive scalar λ is set in such a way that $\Sigma_h \bar{p}_h + \bar{v} = 1$. Since $0 \notin \bar{B}$, it cannot happen that $(p + \rho\hat{p}) = 0$ and $v = 0$ at the same time. Hence the normalization is feasible and (12.11) maps into S^{r+1}.

We can now define the fixed point mapping. The ε perturbed program (12.10) defines, together with the adjustment rules for welfare weights and (12.11), a correspondence $G: C \Rightarrow C$. The parameter set C is compact, convex. By the maximum theorem A.3.1, G is uppersemicontinuous in C and convex valued. Therefore, by Kakutani's theorem A.4.2, a fixed point exists.

Step 2 The Fixed Point is an Equilibrium

By the same arguments as in proposition 3.3, all the budget constraints are nonbinding due to the introduction of ε so that $\rho_i = 0$ for all i.[7] We still have to prove that $\bar{p} \neq 0$ in the fixed point.

To verify this, consider (12.10) with $\bar{p} = 0$. Then all m_i, m_j are zero, so there is no cash in advance needed, and hence v must be zero. Since, by monotonicity of $u_i(.)$ and $\alpha_i > 0$ for some i, some multiplier must be positive, and p cannot be zero. But $\bar{p}$ is obtained by normalizing p, and therefore $\bar{p}$ cannot be zero, a contradiction.

6. We write $\bar{B}$ instead of the usual $\bar{M}$ for this set to avoid confusion with the money supply function.

7. For notational convenience, we will, in the remainder of the proof, drop the * superscript for values in the fixed point.

Then, as in step 2 of the proof of proposition 3.3, both α and $\bar{p}$ will be positive, so income and utility are positive for all consumers. We can, in equilibrium, bring ε to zero without affecting allocations. Also, since $\rho = \max_i \rho_i = 0$,

$$\bar{p}_k = \lambda p_k, \qquad \text{for all } k,$$

and

$$\bar{v} = \lambda v,$$

which shows that $(\bar{p}, \bar{v})$ is proportional to (p, v).

Since $\rho_i = 0$, the budget constraints no longer determine the x_i and the y_j. From the Lagrange inequality of program (12.10), it may be verified that every producer solves (12.6) and that every consumer solves (12.7) with given income (12.8). Equilibrium on the commodity and the cash-in-advance markets follow from the Kuhn-Tucker conditions of the program. ∎

Several Monies and Exchange Rates

The model is easily extended to incorporate several media of exchange. If they are perfect substitutes, they will all fetch the same discount rate $\bar{v}$, and they can be aggregated into a single one. However, in an international context, for example, not all agents will be able to use the same money and monies will be country specific. The country-specific restrictions on money supply will generate country-specific $\bar{v}$. Even under pure exchange these restrictions will create distortions.

Exchange rates will measure the ratio of prices of (the same bundle of) commodities between countries. It is now straightforward to consider endogenous exchange rate adjustments. For this we return to the small-country model of sections 4.2.2 and 4.3.2, including a balance of payments constraint, and incorporate a domestic transaction demand for foreign money whose supply may be limited. If this limit is binding, it will carry a rent in foreign currency, an interest rate on foreign exchange, which accrues to its owner. Foreign exchange will now be binding through the balance of payments as well as through transactions demand, and import prices will adjust upward to cover the interest charge. This adjustment may be interpreted as a devaluation of the exchange rate.[8]

Fiat Money

The two price normalizations used in proposition 12.1 do not represent measurements in terms of "commodity money": The normalization of

8. There is a special case in which the initial stock of foreign exchange is equal to the value of exports plus the trade deficit. Then the balance of payments and the transactions constraint will coincide (assuming a velocity of demand for foreign exchange equal to unity).

shadow prices p depends on $\Sigma_i \alpha_i = 1$, while the normalization of $\bar{p}$ depends on $\Sigma_h \bar{p}_h + \bar{v} = 1$. To recover prices expressed in commodity money, with, say, commodity 1 used as monetary standard, it is sufficient to divide all prices as well as $\bar{m}$ by $\bar{p}_1$. This is possible since all prices are positive at equilibrium.

If $M(\bar{p})$ is only a function of $\bar{p}_1$, the supply of money expressed in this standard will be fixed, but absolute prices are expressed in terms of commodity 1 and can therefore not be scaled freely to make the availability of money unconstraining. When prices are expressed in a unit that is not tied to a commodity, this unit is called fiat money. Prices $\hat{p}$ measured in fiat money will now adjust to the availability of money $\hat{m}$ and be obtained as $\hat{p} = (\hat{m}/\bar{m})\bar{p}$. This is equivalent to solving for $\hat{m} = M(\hat{p})$ under the restriction that $\hat{p}$ be proportional to $\bar{p}$. A reduction in the absolute price level will not succeed in eliminating the prevailing scarcity of money ($\bar{v}$ will remain constant). This may be considered as a policy equation of the monetary authority.

Under the alternative specification in which the money supply is not a function of the price level, prices can adjust to make money supply unbinding. Therefore one obtains an equilibrium with $\bar{v} = 0$; this is the so-called Cambridge model.

12.1.3 Welfare Properties of the Equilibrium

Proposition 12.1 shows that there exist α_i and $\bar{p}$ such that the solution of (12.3) is a competitive equilibrium (and is therefore Pareto-efficient) for $\bar{m}$ sufficiently large. At such an equilibrium, $v = 0$, and all the ρ can be taken to be zero, so money does not affect the allocation. This refects the classical dichotomy between the real and the monetary spheres. However, when the supply of money $\bar{m}$ is binding, the constraint operates like a technological constraint (there are not enough trucks available). This will affect allocations and restrict welfare. Nonetheless, since $p > 0$, it is clear that the supply of money, however tight, cannot by itself be the only effective determinant of allocations, and thus of welfare (no pure monetarism).

Note that in this model the specification is equivalent to one in which producers are taxed on outputs, while consumers are only taxed on endowments. Therefore, if there are no firms, the cash-in-advance economy only redistributes incomes through lump-sum transfers so that its equilibria are Pareto-efficient.

Finally, in this economy, money does not require physical inputs in its production or its use, nor does it provide utility. One could easily introduce a medium, like a silver coin, that needs inputs (by adding the input requirements into the commodity balances) and provides utility. Then, when the government introduces besides silver coins a new medium, say, paper bills, which are cheaper to produce and do not provide utility, welfare will increase: Some resources will be freed, and consumers will no longer have to use silver coins for transactions. They can keep them for their private collections (a version of Gresham's law).

12.2 Incomplete Asset Markets

In the T-period competitive equilibrium of chapter 7, the financial assets (net worths) V_{it}, V_{jt} merely serve as an accounting device. They make it possible to obtain the entries of each period's budgets but do not affect allocations, since they do not restrict transactions. One can therefore say that there is no economic need for financial assets. Equilibrium prices on commodity markets that are available for all periods provide all the information necessary to achieve a competitive equilibrium that is Pareto-efficient.

In the temporary equilibrium model of section 7.2.2, financial assets appeared as bonds. It was necessary to introduce an additional price, which was given two alternative interpretations: one as the change in the absolute level of prices, the other as the interest rate on single-period bonds, which we called the financial asset. There was no need for more financial assets, since all agents had access to this market and only two periods were considered. Had we introduced more than one financial asset, all would have carried the same rate of interest in equilibrium and would have been perfect substitutes. Then there would have been no way, and no need, to specify the holdings for each asset separately (the portfolio). We have seen that the temporary equilibrium is not Pareto-efficient and that it does not lend itself to a study of conditions under which efficiency would be restored, since agents' expectations only depend on past prices.

In the model that will be presented now, the requirements for efficiency can be made explicit and the properties of the financial assets that ensure efficiency can be studied. There are two periods. States of nature are uncertain in the second period, but there is perfect information on the prices to be realized in each state of the second period. Whereas inefficiency may occur, one can investigate the requirements on the structure of the financial assets (the road network) to ensure efficiency. To introduce the subject, we start in section 12.2.1 with a single-period example in which access to some commodities is restricted. Sections 12.2.2 to 12.2.4 deal with issues related to existence of (an approximate) equilibrium in a model with incomplete asset markets, without firms. Sections 12.2.5 and 12.2.6 treat efficiency and other properties of the equilibrium as well as the introduction of firms. Finally, section 12.2.7 presents a case in which an equilibrium exists and there is no need for approximation.

12.2.1 An Example of Restricted Access: The Case of Commodity Markets

Though money could so far be scarce, it gave access to all markets. We now discuss a situation in which there is no scarcity in supply (the number of trucks is sufficient), but two types of money coexist (dollars and rubles). There are also two groups of commodities. Dollars allow us to buy all commodities in both groups, but purchases in group 2 can only be paid for in dollars. Dollar prices are the same in all shops. There is also a rule that specifies that the amount of dollars may not exceed a share β of income. This will restrict the ratio of dollar and ruble purchases but does not impose an absolute bound on money supply (which, here, is adjusting to demand). Then the consumer is faced with the following utility maximization problem, involving two budget constraints expressed in dollars:

$$\max u(x_1, x_2),$$

$$x_1 \geqslant 0,\ x_2 \geqslant 0,$$

subject to

$$p_1 x_1 + p_2 x_2 \leqslant h,$$

$$p_2 x_2 \leqslant \beta h.$$

To make more explicit how such an arrangement can be viewed as a technological constraint (a restricted network of "one-way roads"), we rewrite the problem faced by the consumer as

$$\max u(x_1, x_2),$$

$$v_1 \geqslant 0, x_1 \geqslant 0, x_2 \geqslant 0,$$

subject to

$$v_1 \leqslant h,$$

$$p_1 x_1 + p_2 x_2 \leqslant v_1,$$

$$p_2 x_2 \leqslant \beta v_1.$$

The scalar variable v_1 can be given the interpretation of a financial instrument that makes it possible to allocate income over the two groups of commodities. The consumer cannot buy as much as he wants of group 2. Constraints of this type can be incorporated into program (12.3). They will clearly cause welfare losses, but it can be shown that an equilibrium exists.

To reduce this welfare loss, one could allow the consumer to buy v_2 coupons at a unit price of ϕ, which gives him additional access to dollar commodities of group 2. This amounts to introducing an additional financial instrument, and the consumer problem is

$$\max u(x_1, x_2),$$

$$v_1, v_2 \geqslant 0, x_1 \geqslant 0, x_2 \geqslant 0,$$

subject to

$$v_1 \leqslant h,$$

$$p_1 x_1 + p_2 x_2 \leqslant v_1 - \phi v_2,$$

$$p_2 x_2 \leqslant \beta v_1 + v_2.$$

The two last constraints can be written in matrix form as

$$\begin{bmatrix} p_1 & p_2 \\ 0 & p_2 \end{bmatrix} \begin{bmatrix} x_1 \\ x_2 \end{bmatrix} \leqslant \begin{bmatrix} 1 & -\phi \\ \beta & 1 \end{bmatrix} \begin{bmatrix} v_1 \\ v_2 \end{bmatrix}.$$

The matrix $\begin{bmatrix} 1 & -\phi \\ \beta & 1 \end{bmatrix}$ has rank two and enables the consumer to reach both "states" in any desired proportion as long as he does not spend more than his income h. This extends his possibilities with respect to the earlier situation, when only a single financial instrument was available (and the associated matrix had rank one only).

In this example the access to commodity markets was constrained by a regulation (the coefficient β). In an intertemporal context, constrained access may, for example, be due to a structure of financial assets that provides revenues in fixed proportions over the various future periods, thus restricting the consumer in his intertemporal choice. Similarly in a two-period context with, say, three uncertain states of nature in the second period (high, medium, and low rainfall), it may happen that the contract specifies returns in fixed proportions across states. These proportions may not be in accordance with what the consumer would like to receive in each state. In such a situation asset markets are said to be incomplete. This is the model to which we now turn.

12.2.2 Concepts and Assumptions

We start with the two-period model of chapter 7, adding uncertainty, by distinguishing in period 2, S states of nature, indexed by $s = 1, 2, \ldots, S$. For notational convenience we drop the subscript for the period and denote first-period outcomes by the subscript $s = 0$. We consider a pure exchange economy with m consumers, r commodities, and we also add C financial assets (indexed by $c = 1, 2, \ldots, C$).[9]

9. We follow the presentation given in Ginsburgh and Keyzer (1994).

As in the competitive equilibrium, consumers take all prices as given. The full price vector for commodities is $\bar{p} = (\bar{p}_0, \bar{p}_1, \ldots, \bar{p}_S)$, where $\bar{p}_s$ is the price vector in state s. Financial asset c is purchased in the initial period at price $\bar{\phi}_c$ and provides a return $\bar{p}_s D_{sc}(\bar{p}, \bar{\phi})$ in state $s \neq 0$; $\bar{\phi}$ is the vector of $\bar{\phi}_c$, and $D_{sc}(.)$ is a vector function whose elements are the quantities of each commodity that serve as indexing rule to compute the return $\bar{p}_s D_{sc}(.)$. The vectors D_{sc} form the columns of the matrix D_s. The following assumption is made:

ASSUMPTION R The return functions $D_{sc}(\bar{p}, \bar{\phi})$ are nonnegative, continuous, and homogeneous of degree zero in $(\bar{p}, \bar{\phi})$. For given c, $D_{sc}(\bar{p}, \bar{\phi})$ is different from zero in some state s.

Assumption R allows for several specifications of the return functions. We give a few examples, but obviously many other formulations are possible:

1. $D_{sck}(\bar{p}, \bar{\phi}) = B_{sck}\bar{\phi}_c/\bar{p}_{sk}$ (k is a commodity) will yield a fixed nominal return equal to $\bar{\phi}_c \Sigma_k B_{sck}$ for a purchase of $\bar{\phi}_c$ in state 0.[10] When $\Sigma_k B_{sck} = 1$, the financial asset is a (perfectly safe) cash holding.
2. $D_{sck}(\bar{p}, \bar{\phi}) = B_{sck}\bar{\phi}_1/\bar{p}_{sk}$. Here we assume that asset 1 is cash; the financial asset is a bond purchased at price $\bar{\phi}_c$ in state 0, and the return is $\bar{\phi}_1 \Sigma_k B_{sck}$.
3. D_{sc} is constant. This means that returns are units of commodities. In state s this can be interpreted as letting the consumer receive (or give) these commodities, or as a device to index the amount $\bar{p}_s D_{sc}$ that he receives (or pays).
4. More generally, $\bar{p}_s D_{sc}(.)$ may depend on prices of state s only.

Consumer i decides on his consumption $x_i = (x_{i0}, x_{i1}, \ldots, x_{iS})$, where x_{is} is a vector of dimension r in state s. He also decides on his net demand v_i for financial assets in the initial period (v_i is his portfolio; the entries v_{ic} of the vector v_i are not constrained in sign, since assets can be purchased or sold). For one unit of asset c he pays a price $\bar{\phi}_c$ and expects to receive unit returns $\bar{p}_s D_{sc}(\bar{p}, \bar{\phi})$ in state s. He has access to all assets, knows commodity prices $\bar{p}$, asset prices $\bar{\phi}$, and expected returns $D(\bar{p}, \bar{\phi})$.

Consumer i will maximize his utility function $u_i(x_{i0}, x_{i1}, \ldots, x_{iS})$ under budget constraints that are specified for each state (this utility function can be interpreted as incorporating a subjective assessment of the probabilities

10. This function is not continuous when $\bar{p}_{sk}$ goes to zero. In the model one could use $B_{sck}\bar{\phi}_c/\max\{\bar{p}_{sk}, \gamma(\bar{p}, \bar{\phi})\}$, for a scalar function $\gamma(.)$ that is continuous, homogeneous of degree one, and has a small value for $(\bar{p}, \bar{\phi})$ on the simplex. However, increasingness for consumer 1 and $\omega_1 > 0$ (assumptions CA1 and CA2 that follow) will ensure that $\bar{p}_{sk} > 0$ for all s and k so that for γ small enough, $\max\{\bar{p}_{sk}, \gamma(\bar{p}, \bar{\phi})\} = \bar{p}_{sk}$.

of states $s \neq 0$). Consumer i's problem is

$$\max u_i(x_{i0}\, x_{i1}, \ldots, x_{iS}),$$

$$x_{i0}, x_{is} \geqslant 0, \text{ all } s \neq 0,\ v_i,$$

$$\text{subject to} \tag{12.12}$$

$$\bar{p}_0 x_{i0} + \bar{\phi} v_i \leqslant \bar{p}_0 \omega_{i0},$$

$$\bar{p}_s x_{is} \leqslant \bar{p}_s \omega_{is} + \bar{p}_s D_s(\bar{p}, \bar{\phi}) v_i, \qquad \text{all } s \neq 0.$$

In (12.12) consumer i faces a budget constraint in the initial period. He owns an endowment ω_{i0} of commodities but holds no assets (from earlier periods). He can buy an asset ($v_{ic} > 0$) on which he will get a return in period 2. He can also borrow ($v_{ic} < 0$) and repay in period 2: $\bar{\phi} v_i$ will be his net saving. In each state s of the second period, he also faces a budget constraint. His income will originate from two sources, the endowment ω_{is} and the returns $\bar{p}_s D_s(\bar{p}, \bar{\phi}) v_i$ on his portfolio v_i.

A two-period exchange equilibrium with financial assets can now be defined as follows:[11]

DEFINITION 12.2 (Equilibrium with incomplete markets) The allocation x_i^*, v_i^*, all i, supported by commodity prices $\bar{p}^* \geqslant 0$ and asset prices $\bar{\phi}^*$ is an exchange equilibrium with financial assets if the following conditions are satisfied:

1. For every consumer i, x_i^*, v_i^* solve (12.12).
2. All commodity markets are in equilibrium in every state, $\Sigma_i x_i^* - \Sigma_i \omega_i \leqslant 0$ $\bar{p}^* \geqslant 0$, and $\bar{p}^*(\Sigma_i x_i^* - \Sigma_i \omega_i) = 0$.
3. Markets for financial assets are in equilibrium, $\Sigma_i v_i^* \leqslant 0$, $\bar{\phi}^* \geqslant 0$ and $\bar{\phi}^* \Sigma_i v_i^* = 0$.

In this definition there is a list of states s common to all consumers, though some consumers may find some states irrelevant, as will be seen in assumption CA1 below. Markets exist and clear in the initial period as well as in each state of period 2. All prices are known to all consumers. What distinguishes this equilibrium from a two-period competitive equilibrium is the (possibly) limited option consumers have to save for each state separately. An asset bought in the initial period will give some return in each state, and there may be fewer assets than states. The consumer is facing incomplete financial markets because he is constrained in his control over the portfolio returns in state s.[12]

11. We give the definition for $(\bar{p}, \bar{\phi})$ on the simplex; this normalization will be relaxed below.

12. In the model to be discussed, we allow all consumers to have access to all assets. Restrictions on participation can easily be incorporated by putting bounds on the v_{ic} for some i and some c. For instance, in a multicountry model, consumers of a certain country may not have access to other countries' capital markets.

For symmetry with the equilibrium conditions for commodity markets, we allow for excess supply of financial assets. However, as will be seen, all assets will have zero excess supply in equilibrium.

12.2.3 Issues and Approaches Related to the Existence Proof

In a seminal paper Hart (1975) has shown that when asset markets are incomplete, equilibrium may fail to exist, and this has stimulated an extensive strand of research on the genericity of equilibrium. This literature has established that for "most" initial endowment values, there will exist an equilibrium (technically, an equilibrium exists on an open and dense set of endowments); see, for example, Geanakoplos and Polemarchakis (1986), Geanakoplos (1990), Polemarchakis (1990a, 1990b). Duffie (1991), and Magill and Shafer (1991).

Here we do not consider genericity and only ask whether equilibrium exists in an approximate sense. Though all markets will clear, budget constraints may only be satisfied approximately, namely involve arbitrarily small, endogenous income transfers.[13] The approach is general in the sense that it uses a nonlinear specification for returns on assets.[14] We show that the possibility of nonexistence of equilibrium is linked to the violation of a constraint qualification condition in the welfare program. To ensure constraint qualification, we allow for arbitrarily small income transfers among consumers. These transfers may be interpreted as a perturbation of initial endowments.[15]

Three difficulties have to be addressed: (1) Out of equilibrium the consumer problem is not compact-valued, (2) one wants to avoid equilibria where in some state s all prices are zero, and (3) asset returns could become negative.

The Problem of the Consumer Not Being Compact Valued

Since in (12.12), v_i is unconstrained in sign and can be produced at zero cost, the consumer problem may not be compact valued. There are two reasons for this: (1) Income, and thus the consumption of some commodity can become infinite at positive prices $(\bar{p}, \bar{\phi})$ because asset transactions are profitable and not bounded; (2) even if consumption and utility remain

13. In the literature on rational expectations equilibria, a distinction is also made between generic and approximate existence. There, in the approximation, either expectations are approximately correct and markets clear or expectations are correct and markets only clear approximately. See Radner (1982, 1001–1002).

14. Differentiability of utility functions is not required in the proofs, but the welfare programming approach uses concavity rather than quasi-concavity of the utility functions.

15. In section 12.2.6 we show that under some additional assumptions, and in particular, if returns in a certain state of nature depend on prices prevailing in that state only, no such transfers are needed, and an equilibrium always exists (see also Chae 1988). This is a special case of the more general existence result (with transfers) given in proposition 12.3.

bounded, the level of asset transactions may not remain within a compact set.[16]

We avoid the first difficulty by using a welfare formulation that embodies the constraints $\Sigma_i x_{is} \leqslant \Sigma_i \omega_{is}$ for all s. However, this does not eliminate the difficulties due to the absence of a given bound on the v_i. This is illustrated in the following example.

Consider a two-period single-consumer economy, with the possibility of transferring the commodity between periods. The utility function is $u(x_1, x_2) = \min\{x_1, x_2\}$. The consumer solves

$\max u,$

$x_1, x_2 \geqslant 0, u, v,$

subject to

$u \leqslant x_1,$

$u \leqslant x_2,$

$x_1 \leqslant \delta v + \omega_1,$

$x_2 \leqslant -\delta v + \omega_2,$

where ω_1 and ω_2 are given endowments, v is the intensity of the "transferring" activity and δ is a technical coefficient (the number of "trucks" necessary to transfer one unit of commodity). Clearly as long as δ is positive, the number of trucks is bounded by the endowments of the economy. However, as δ goes to zero, one needs ever more trucks, and v increases (this does not matter, because these use no inputs, and their number is not constrained). But in the limit no connection is possible anymore (v is not uniformly bounded). Technically the transactions matrix (here the column vector $(\delta, -\delta)$) which was of rank one is now of rank zero. This difficulty is avoided by Radner (1972) who suggests to impose bounds on the v_i in each consumer problem.[17] Indeed, if we were to impose such bounds in the above program, its constraints set would become compact, so that standard tools could be used to construct fixed point correspondences. However, such bounds are artificial and obscure many of the results in the financial literature.

In this chapter, we address this second difficulty by introducing a normalization, and a scaling factor n and by expressing the x_{is}'s and the v_i's

16. The creation of a financial asset in period 1 does not bring any revenue in that period. The period 1 value purely depends on the revenue in period 2. This is different from the OLG model in chapter 8, where increased expenditure could be financed by money creation, and from the model in section 12.1, where the interest rate on the increased money supply accrued to the holder of the cash balance.

17. One could also allow for transactions that need commodities as inputs (transactions costs).

as ratios $x_{is} = y_{is}/n$ and $v_i = w_i/n$, where y_{is}, w_i and n are normalized and hence bounded. We also show that there exists a positive value for n in the fixed point, so that finite values for x_{is} and v_i can be recovered. This approach has its equivalent in the standard competitive model, where one replaces $(r - 1)$ relative prices by r absolute prices normalized on the r-dimensional simplex and uses a nonsatiation property to show that p_r is positive in the fixed point; it is then possible to recover relative equilibrium prices, which are bounded, though out of equilibrium they may not be since p_r may be zero.

Avoiding All Prices to Be Zero in Any Relevant State

This is obviously a minimal requirement for an equilibrium to be meaningful. Moreover problems in normalizing returns arise when the prices of commodities used as numéraire become zero. To avoid this, we will, in assumption CA1, impose the requirement that consumer 1's utility is increasing in all commodities in all (relevant) states.

Avoiding these two problems is nevertheless insufficient to prove existence of equilibrium, and robust examples have been constructed of models that have no equilibrium.[18] Here we will show that there exists an "approximate" equilibrium with state-specific transfers $\bar{T}_{is}$ that do not exceed an arbitrarily small positive value ε and satisfy $\Sigma_i \bar{T}_{is} = 0$. The resulting transfers can be interpreted as an (endogenous) arbitrarily small state-specific redistribution of initial endowments. Moreover there are cases where this approximation is not required, and a true equilibrium can be shown to exist. We return to this issue at the end of section 12.2.4 and in section 12.2.7.

Negative Returns

The third difficulty, negative returns, was excluded by assumption R. When negative returns are possible for some asset c in some state s, consumers may all wish to have negative demands (i.e., "sell" the asset in $s = 0$). This may lead to an excess supply of assets, meaning that $D_{sc}\Sigma_i v_{ic} \neq 0$, so the receipts and payments from financial assets do not balance. This imbalance will spill over to other asset markets and possibly also to commodity markets.[19] Though somewhat restrictive, nonnegative returns have the advantage that a consumer cannot be forced to pay in the second period for an asset that he has already purchased in the first.

18. Hart (1975) was the first to construct such an example, but his assumptions are specific: Consumers only possess endowments in period 2, they do not derive utility from consumption in period 1 and only in some states in period 2, and the utility functions are of the CES type.

19. This imbalance can be avoided by specifying budget constraints and balances for assets as equalities, but this could lead to negative shadow prices and complications in the proofs.

In section 12.2.4 we prove existence of an approximate equilibrium, in which consumer i solves (12.12′) instead of (12.12), where (12.12′) is defined, in equilibrium, as follows:

$$\max u_i(x_{i0}, x_{i1}, \ldots, x_{iS}),$$

$$x_{i0}, x_{is} \geqslant 0, \text{ all } s \neq 0, v_i,$$

$$\text{subject to} \qquad (12.12')$$

$$\bar{p}_0^* x_{i0} + \bar{\phi}^* v_i \leqslant \bar{p}_0^* \omega_{i0} + \bar{T}_{i0}^*,$$

$$\bar{p}_s^* x_{is} \leqslant \bar{p}_s^* \omega_{is} + \bar{p}_s^* D_s(\bar{p}^*, \bar{\phi}^*) v_i + \bar{T}_{is}^*, \qquad \text{all } s \neq 0,$$

where $\bar{T}_{is}^*$ is an income transfer. This leads to the following definition of an approximate equilibrium:

DEFINITION 12.2 (Approximate equilibrium with incomplete markets) The allocation x_i^*, v_i^*, all i, supported by commodity prices $\bar{p}^* \geqslant 0$ and asset prices $\bar{\phi}^*$ is an exchange equilibrium with financial assets if the following conditions are satisfied:

1. For every consumer i, (x_i^*, v_i^*) solves (12.12′).
2. All commodity markets are in equilibrium in every state, $\Sigma_i x_i^* - \Sigma_i \omega_i \leqslant 0$, $\bar{p}^* \geqslant 0$, and $\bar{p}^*(\Sigma_i x_i^* - \Sigma_i \omega_i) = 0$.
3. Markets for financial assets are in equilibrium, $\Sigma_i v_i^* \leqslant 0, \bar{\phi}^* \geqslant 0$, and $\bar{\phi}^* \Sigma_i v_i^* = 0$.
4. Transfers satisfy $\Sigma_i \bar{T}_{is}^* = 0$ for all s and $\bar{T}_{is}^* \leqslant \varepsilon$, for an arbitrarily small positive ε.

12.2.4 Approximate Equilibrium with Financial Assets: An Existence Proof

Existence of an approximate equilibrium will be proved under the following assumptions:

ASSUMPTION CA1 (Utility function) The utility function $u_i : R_+^{r(1+S)} \to R_+$, $u_i(x_{i0}, x_{i1}, \ldots, x_{iS})$ is continuously differentiable and strictly concave for all i. For $i = 1$, it is increasing in all commodities in all states, including $s = 0$ and $\partial u_1/\partial x_{1ks} \geqslant \delta_0$ for some positive constant δ_0 and $x_{1ks} \leqslant \Sigma_h \omega_{hks}$. For $i = 2, 3, \ldots, m$, it is nonsatiated in state $s = 0$. Moreover $u_i(0) = 0$, and $\lim_{x \to x_0, n \downarrow 0} n u_i(x_i/n) = 0$ for every $x_0 \geqslant 0$.

Consumer 1 attributes positive probability to each state (increasingness in assumption CA1), but other consumers may find some states irrelevant and not expect any utility from consumption in such states. The assumption will be used to have all prices positive in equilibrium. In the limit condition the variable n is the same as the scaling factor mentioned earlier. It relates the utility function $u_i(x_i)$ to the "extended utility function" $v_i(x_i, n) = nu_i(x_i/n)$ which is homogeneous of degree one and is, by the limit condition, continuous at $n = 0$. This function will play an essential role in the proof.[20] Since in equilibrium n will be positive, the limit condition is purely technical.

We also assume the following:

ASSUMPTION **CA2** The endowment ω_{is} of consumer i in state s is nonnegative, nonzero for all i and all s; $\Sigma_i \omega_{is} > 0$ for all s.

We first define the basic welfare program that is used in the existence proof. Next we describe the properties of optimal solutions of this program, and finally we turn to the existence proof itself.

The Basic Welfare Program

Our basic tool is the following mathematical program:

$$J(\bar{p}, \bar{\phi}) = \max \Sigma_i \alpha_i u_i(x_{i0}, x_{i1}, \ldots, x_{iS}),$$

$$x_{i0} \geqslant 0, x_{is} \geqslant 0, s \neq 0, v_i, \text{ all } i,$$

$$\text{subject to} \qquad (12.13)$$

$$\Sigma_i x_{i0} \leqslant \Sigma_i \omega_{i0} \qquad (p_0),$$

$$\Sigma_i x_{is} \leqslant \Sigma_i \omega_{is} + D_s(\bar{p}, \bar{\phi}) \Sigma_i v_i, \quad \text{all } s \neq 0 \qquad (p_s),$$

$$\bar{p}_0 x_{i0} + \bar{\phi} v_i \leqslant \bar{p}_0 \omega_{i0} + \varepsilon, \quad \text{all } i \qquad (\rho_{i0}),$$

$$\bar{p}_s x_{is} \leqslant \bar{p}_s \omega_{is} + \bar{p}_s D_s(\bar{p}, \bar{\phi}) v_i + \varepsilon, \quad \text{all } s \neq 0 \text{ and all } i \qquad (\rho_{is}),$$

$$\Sigma_i v_i \leqslant 0 \qquad (\phi),$$

for given $\alpha_i = 1$ all $i, \bar{p} \geqslant 0$, and $\bar{\phi} \geqslant 0$.

As can be seen, this format incorporates among the constraint set all the budgets and market equilibrium conditions. The objective function of the program is a weighted sum of utilities. Lagrange multipliers appear at the right-hand side of the various constraints. In equilibrium prices $(\bar{p}, \bar{\phi})$ will be proportional to the associated Lagrange multipliers (p, ϕ).

Note that we have artificially adjusted the commodity balance of definition 12.2 to include the term $D_s(\bar{p}, \bar{\phi}) \Sigma_i v_i$ for every state $s \neq 0$. We will see that in equilibrium $\Sigma_i v_i$ will be equal to zero, so the term will cancel, yielding the usual commodity balance constraints in state s. The role of this term is to ensure that the asset prices $\bar{\phi}$ appearing in the consumers budget equations for $s = 0$ will be proportional to the Lagrange multiplier ϕ.[21]

20. See theorem A.1.5 for a discussion of the properties of the "extended" function.

21. Introducing this term in the commodity balance can be avoided, but this requires mappings more complicated than proportionality to determine $\bar{\phi}$.

The constants ε in the budget constraints ensure that the program will satisfy Slater's constraint qualification. If in equilibrium, any of the constraints with ε is binding, it only is an approximate equilibrium in the sense that it requires transfers or a small redistribution of endowments. Proposition 12.3 will establish existence of such an approximate equilibrium for a general asset structure.

The above program does not prevent some of the v_i from becoming unbounded. To bound the program, we proceed as follows:[22]

1. We define new variables v_i^+ and v_i^- such that

$$v_i = v_i^+ - v_i^-, \text{ all } i,$$

with $v_i^+ \geqslant 0, v_i^- \geqslant 0$.

2. We introduce an additional normalization variable $n \geqslant 0$ as well as variables y_{is}, w_i^+, and w_i^-. We then define $x_{is} = y_{is}/n$ for all i and s as well as $v_i^+ = w_i^+/n$ and $v_i^- = w_i^-/n$ for all i. We will show that in equilibrium n is positive.

3. We bound the program by adding a "normalization constraint"

$$\Sigma_i e^C(w_i^+ + w_i^-) + \Sigma_i e^r(y_{i0} + \Sigma_s y_{is}) + n \leqslant 1,$$

where e^C and e^r are vectors with unit elements of dimension C and r, respectively. This constraint normalizes all the nonnegative variables of the program to lie on a simplex.

4. To make this constraint ineffective, and ensure that the normalization constraint does not affect allocations in equilibrium, we penalize the normalizing variable at a rate θ to be adjusted appropriately.

We are now ready to extend program (12.13), introducing the additional variable n and keeping all variables bounded via the normalization

$$V(\theta; \bar{p}, \bar{\phi}) = \max \Sigma_i \alpha_i n u_i\left(\frac{y_{i0}}{n}, \frac{y_{i1}}{n}, \ldots, \frac{y_{iS}}{n}\right) - \theta n,$$

$$y_{i0} \geqslant 0,\ y_{is} \geqslant 0,\ s \neq 0,\ w_i^+ \geqslant 0,\ w_i^- \geqslant 0, w_i, \text{ all } i, n \geqslant 0,$$

subject to (P1)

$$\Sigma_i y_{i0} - n\Sigma_i \omega_{i0} \leqslant 0 \qquad (p_0),$$

$$\Sigma_i y_{is} - n\Sigma_i \omega_{is} - D_s(\bar{p}, \bar{\phi})\Sigma_i w_i \leqslant 0, \qquad \text{all } s \neq 0 \qquad (p_s),$$

$$\bar{p}_0 y_{i0} + \bar{\phi} w_i - n(\bar{p}_0 \omega_{i0} + \varepsilon) \leqslant 0, \qquad \text{all } i \qquad (\rho_{i0}),$$

22. Such a framework is explicitly used by Morgenstern and Thompson (1976) to prove existence of an equilibrium in the von Neumann model. There the model assumes constant returns to scale, and the production levels are scaled on the simplex, which does not affect allocations since profits are zero.

$$\bar{p}_s y_{is} - n(\bar{p}_s \omega_{is} + \varepsilon) - \bar{p}_s D_s(\bar{p}, \bar{\phi}) w_i \leqslant 0, \qquad \text{all } s \neq 0 \text{ and all } i \quad (\rho_{is}),$$

$$\Sigma_i w_i \leqslant 0 \qquad (\phi),$$

$$\Sigma_i e^C(w_i^+ + w_i^-) + \Sigma_i e^r(y_{i0} + \Sigma_s y_{is}) + n \leqslant 1 \qquad (\mu),$$

$$w_i = w_i^+ - w_i^-,$$

for given $\alpha > 0$, $\bar{p} \geqslant 0$, and $\bar{\phi} \geqslant 0$.

We focus attention on a value θ^0 of θ obtained as

$$\theta^0(\bar{p}, \bar{\phi}) = \min\{\theta \mid V(\theta; \bar{p}, \bar{\phi}) = 0\}. \qquad \text{(P2)}$$

Characterizing the Solution(s) of Program P1–P2

In proposition 12.2 we essentially show that for any given choice of the parameters $(\bar{p}, \bar{\phi}) \geqslant 0$, an optimum exists for program P1–P2. Such an optimum has two properties:

1. It contains a positive value for n, so we can recover bounded values for the original variables in (12.13).
2. The normalization constraint $\Sigma_i e^C(w_i^+ + w_i^-) + \Sigma_i e^r(y_{i0} + \Sigma_s y_{is}) + n \leqslant 1$ is nonbinding, so it does not affect the allocation.

PROPOSITION 12.2 (Properties of program P1–P2 for given $(\bar{p}, \bar{\phi})$) Let assumptions R, CA1, and CA2 be satisfied. Then

1. The constraints set in program P1 is nonempty, compact, and satisfies Slater's constraint qualification, so an optimal solution as well as nonnegative and bounded Lagrange multipliers will exist;
2. the penalty parameter $\theta(\bar{p}, \bar{\phi})$ is positive, continuous, and homogeneous of degree zero;
3. the shadow price μ of the normalization constraint is zero, so the constraint does not affect the resulting allocation;
4. in the optimum, $n > 0$ belongs to the optimal solution set of P1–P2; at this value all variables $x_{is} = y_{is}/n$ and $v_i^+ = w_i^+/n$ and $v_i^- = w_i^-/n$ are bounded;
5. there is a unique value y_{is}°, n° such that all optimal solutions satisfy $y_{is} = \lambda y_{is}^\circ$ and $n = \lambda n^\circ$ for $\lambda \geqslant 0$;
6. Lagrange multipliers are not all zero.

Proof

Part 1: $nu_i(y_{i0}/n, y_{i1}/n, \ldots, y_{is}/n)$ is continuous, concave, and has strictly convex isoquants (theorem A.1.5, assertions 1 and 2). Hence the objective function of P1 is continuous and concave. Since $D_s(.)$ is nonnegative, and $\Sigma_i \omega_{i0}$ and $\Sigma_i \omega_{is}$ are strictly positive, the constraint set is nonempty. Due to the normalization constraint, it is also compact. Now pick values $x_{i0} = 0$ and $x_{is} = 0$ for all i and s, a value $n > 0$ and some $v_i^- > 0$ sufficiently small. Since $\varepsilon > 0$, and $\Sigma_i \omega_{i0}$ and $\Sigma_i \omega_{is}$ are also positive, Slater's constraint qualification holds. Program P1 will therefore always have an optimal solution as well as bounded and nonnegative Lagrange multipliers p_0, p_s, ρ_{i0}, ρ_{is}, ϕ, and μ.

Part 2: The value function $V(\theta; \bar{p}, \bar{\phi})$ is continuous, convex, and nonincreasing in θ (theorem A.3.2, assertion 2). Since the constraints in P1 are satisfied if all variables are set to zero, $V(\theta; \bar{p}, \bar{\phi}) \geqslant 0$. Moreover it may be verified that $V(0; \bar{p}, \bar{\phi}) > 0$, due to constraint qualification. Also, comparing the Lagrange inequalities for programs P1 and (12.13) shows that $V(\theta; \bar{p}, \bar{\phi}) = 0$ for any $\theta \geqslant J(\bar{p}, \bar{\phi})$, where $J(\bar{p}, \bar{\phi})$ is the value function of (12.13). Hence $V(\theta^0; \bar{p}, \bar{\phi}) = 0$ for some value θ that is uniformly bounded from above for all $(\bar{p}, \bar{\phi})$. Therefore $\theta^0(\bar{p}, \bar{\phi})$ in P2 is well defined and positive and, by the maximum theorem A.3.1, assertion 3, is a continuous function of the parameters $(\bar{p}, \bar{\phi})$. Moreover it is homogeneous of degree zero. Hence problem P1–P2 has a solution.

Part 3: Since the objective function is homogeneous of degree one in (x, n) (theorem A.1.5, assertion 4) and the constraints are linear, all value accrues to the right-hand side of the constraints, so $V(\theta^0; \bar{p}, \bar{\phi}) = \mu = 0$.[23] Therefore the Lagrange multiplier μ is equal to zero. The normalization constraint is thus nonbinding and does not affect the allocations (the primal optimal solution).

Part 4: We now show that there is a positive value n^0 that is optimal in P1–P2. Introduce the constraint $n \leqslant n^0$ into program P1, and drop θn from the objective function. Note that θ can then be interpreted as the Lagrange multiplier associated to the constraint $n \leqslant n^0$. Assume that $n^0 = 0$. By part 2, the multiplier θ is positive. Therefore increasing n^0 increases the value of the objective of P1 as well as $\Sigma_i \alpha_i n u_i(y_{i0}/n, y_{i1}/n, \ldots, y_{iS}/n)$ (see theorem A.1.5, assertion 3). Hence, by homogeneity of degree one of the objective (see theorem A.1.5, assertion 2), such an increase is also optimal in P1–P2. Thus $n^0 > 0$ is also optimal in P1–P2.

Part 5: Since the value of the objective is zero in any optimum and since it is homogeneous of degree one and the constraints are linear, all optimal solutions lie on a ray. Uniqueness follows from strict convexity of isoquants of $\Sigma_i \alpha_i n u_i(y_{i0}/n, y_{i1}/n, \ldots, y_{iS}/n)$, which implies that for any given value of n, the optimal y_{is} are unique (though the w_{is} may not be).

23. This result can be verified as follows: θ^0 is not only homogeneous of degree zero in $(\bar{p}, \bar{\phi})$ but also in the 1 appearing on the right-hand side of the normalization constraint. Hence increasing this 1 keeps the value function $V(.)$ at zero, and the associated Lagrange multiplier will be zero as well.

Part 6: For $n^0 > 0$, $nu_i(y_{i0}/n, y_{i1}/n, \ldots, y_{iS}/n)$ is monotonic (theorem A.1.5, assertion 2), so some commodity balance must have a positive multiplier.■

The Existence Result

We now prove that there exists a fixed point in which the parameters $(\bar{p}, \bar{\phi})$ appearing in the budget constraints of P1–P2 are proportional to (p, ϕ), the Lagrange multipliers of P1–P2. In the fixed point an equilibrium with ε-transfers (i.e., transfers $T_{is} \leqslant \varepsilon$, with $\Sigma_i T_{is} = 0$ for each state s) is obtained. The existence proof uses the mathematical program P1–P2 which is compact; therefore the maximum theorem can be invoked.

PROPOSITION **12.3** (Existence of equilibrium) Let assumptions R, CA1, and CA2 be satisfied. Then there exist positive prices $\bar{p}^*$, nonnegative asset prices $\bar{\phi}^*$, such that (1) $(\bar{p}^*, \bar{\phi}^*)$ is proportional to the Lagrange multipliers (p^*, ϕ^*) and (2) the solution is an approximate exchange equilibrium with financial assets, with (possibly zero) transfers $\bar{T}_{is}^* \leqslant \varepsilon$ for every i and every s for any given positive ε.

Proof We distinguish two steps:

1. *Existence of a fixed point.* Since all the allocations and the penalty parameter $\theta^0(\bar{p}, \bar{\phi})$ are homogeneous of degree zero in $(\bar{p}, \bar{\phi})$, we may constrain $(\bar{p}, \bar{\phi})$ to lie on the simplex S^a, where $a = r(1 + S) + C$. Then, by proposition 12.2, assertion 2, $\theta^0(\bar{p}, \bar{\phi})$ is continuous, and we can, for any $(\bar{p}, \bar{\phi}) \in S^a$ compute it and enter its value into program P. By proposition 12.2, assertion 1, P1–P2 then define a Lagrange multiplier correspondence from $(\bar{p}, \bar{\phi})$ to (p, ϕ, ρ), which by the maximum theorem A.3.1 is uppersemicontinuous, compact, and convex valued and defines a compact, convex hull B^d, where $d = (r + m)(S + 1) + C$.

Now we update $\bar{p}_s$ for $s = 0, 1, \ldots, S$ and $\bar{\phi}$ by defining a function $(\bar{p}, \bar{\phi}) : S^a \times B^d \to S^a$. Let $\bar{p}_s, \bar{\phi}$ be the current value of prices. We set $\hat{p}_s = \bar{p}_s$ and $\hat{\phi}_c = \bar{\phi}_c$ and define the functions

$$\bar{p}_{ks} = \lambda \max(p_{ks} + \psi \hat{p}_{ks}, \delta_1), \tag{12.14}$$

$$\bar{\phi}_c = \lambda(\phi_c + \psi \hat{\phi}_c),$$

where $\psi = \max_{is} \rho_{is}$, the constant δ_1 satisfies $0 < \delta_1 < \delta_0$, $(p, \phi, \rho) \in B^d$, and the positive scalar λ is such that $\Sigma_k \Sigma_s \bar{p}_{ks} + \Sigma_c \bar{\phi}_c = 1$. By construction, (12.14) maps into the simplex S^a.

The Lagrange multiplier correspondence defined by P1–P2 with θ adjusted according to $\theta^\circ(\bar{p}, \bar{\phi})$, and the functions (12.14) define a correspondence $G : S^a \times B^d \Rightarrow S^a \times B^d$. The parameter set $A = S^a \times B^d$ is compact, convex, and G is uppersemicontinuous in A and convex valued. Therefore, by Kakutani's theorem, there exists a fixed point.

Since, by proposition 12.2, assertions 3, 4, and 5, we can take n to be positive without affecting the allocation nor the multipliers in the fixed point, we can conclude that bounded values exist for the original variables $x = y/n$ and $v = w/n$, and we return to the basic welfare program (12.13).

2. *Properties of the fixed point.*

2a. *Proportionality of prices.* In the fixed point $\bar{p}^*_{ks} = \hat{p}^*_{ks}$ and $\bar{\phi}^*_c = \hat{\phi}^*_c$. By assumption CA1, $\delta_0 \leqslant \partial u_1/\partial x_{1ks} \leqslant p^*_{ks} + \rho^*_{1s}\bar{p}^*_{ks}$. Hence $p^*_{ks} + \psi^*\bar{p}^*_{ks} > \delta_1$, and (12.14) reduces to

$$\bar{p}^*_{ks} = \lambda^*(p^*_{ks} + \psi^*\bar{p}^*_{ks}) \qquad \text{for all } k, s, \tag{12.15}$$

$$\bar{\phi}^*_c = \lambda^*(\phi^*_c + \psi^*\bar{\phi}^*_c) \qquad \text{for all } c. \tag{12.16}$$

Since p^* and $\bar{p}^*$ are nonnegative, the product $\lambda^*\psi^*$ cannot exceed unity. Suppose that it is equal to unity. This implies that $p^* = 0$, so in program (12.13) the first-order condition in $s = 0$ reads

$$\frac{\partial u_i}{\partial x_{i0}} \leqslant \rho_{i0}\bar{p}_0 \qquad \text{for every } i.$$

Moreover ρ_{i0} must be zero for at least one i, since when all commodity and asset balances hold, transfers add up to zero and one of the budget equations is therefore nonbinding and has $\rho_{i0} = 0$. However, this implies that for this consumer i, $\partial u_i/\partial x_{i0} = 0$, which contradicts nonsatiation in assumption CA1 for $s = 0$. It follows that $\lambda^*\psi^*$ is smaller than unity, and then (12.15)–(12.16) show that $(\bar{p}^*, \bar{\phi}^*)$ is proportional to (p^*, ϕ^*). Since $\alpha_1 = 1$ and utility of consumer 1 is increasing in all states, equilibrium prices $\bar{p}^*$ are positive.

2b. *Approximate equilibrium.* We must now show that in the fixed point, commodity and asset balances hold and that every consumer solves (12.12′), where we can use equalities in the budget constraints, since the slack can be included in the transfers.

Commodity balances obviously hold in every optimum of (12.13) for state 0. For state $s \neq 0$, they also hold, but they include the term $D_s(\bar{p}^*, \bar{\phi}^*)\Sigma_i v_i$ which can be shown to be zero. However, we give this proof in a separate proposition (Proposition 12.4), because it gives an important and well known result on asset pricing.

To show that allocations solve (12.12′), we pick one equilibrium value for $(p^*, \phi^*, \rho^*) \in B^d$ from the set of equilibrium Lagrange multipliers and determine the associated $(\bar{p}^*, \bar{\phi}^*)$. We now renormalize the welfare weights α to $\bar{\alpha}^*$, multiplying them by the $\lambda^*/(1 - \lambda^*\psi^*)$ value obtained from (12.15)–(12.16) so that the equilibrium allocation and the chosen

$(\bar{p}^*, \bar{\phi}^*)$ will be unchanged and be equilibrium Lagrange multipliers. The multipliers $\bar{\rho}_{is}^*$ associated to the budget constraints will be subject to the same scaling.

For individual consumers we know from (12.13) that

$$\bar{p}_0^* x_{i0} + \bar{\phi}^* v_i \leqslant \bar{p}_0^* \omega_{i0} + \varepsilon, \tag{12.17}$$

$$\bar{p}_s^* x_{is} \leqslant \bar{p}_s^* \omega_{is} + \bar{p}_s^* D_s(\bar{p}^*, \bar{\phi}^*) v_i + \varepsilon. \tag{12.18}$$

Since commodity balances hold with equality, we can premultiply each of them by $\bar{p}_s^*$ and obtain:

$$\bar{p}_0^* \Sigma_i x_{i0} = \bar{p}_0^* \Sigma_i \omega_{i0}, \tag{12.19}$$

$$\bar{p}_s^* \Sigma_i x_{is} = \bar{p}_s^* \Sigma_i \omega_{is} + \bar{p}_s^* D_s(\bar{p}^*, \bar{\phi}^*) \Sigma_i v_i. \tag{12.20}$$

We can, in the fixed point, calculate the $\bar{T}_{is}^*$ as

$$\bar{T}_{i0}^* = \bar{p}_0^* x_{i0}^* + \bar{\phi}^* v_i^* - \bar{p}_0^* \omega_{i0}, \qquad \text{all } i,$$

$$\bar{T}_{is}^* = \bar{p}_s^* x_{is}^* - (\bar{p}_s^* \omega_{is} + \bar{p}_s^* D_s(\bar{p}^*, \bar{\phi}^*) v_i^*), \qquad \text{all } s \neq 0, \text{ all } i,$$

where by (12.17) and (12.18), $\bar{T}_{is}^* \leqslant \varepsilon$; by (12.19) and (12.20), $\Sigma_i \bar{T}_{i0}^* = 0$ (recall that $\bar{\phi}^* \Sigma_i v_i^* = 0$) and $\Sigma_i \bar{T}_{is}^* = 0$.

Because of the new normalization, $(p^*, \phi^*) = (\bar{p}^*, \bar{\phi}^*)$. In the fixed point we may now replace in (12.13) the constraints (12.17) by $\bar{p}_0^* x_{i0} + \bar{\phi}^* v_i = \bar{p}_0^* \omega_{i0} + \bar{T}_{i0}^*$ and (12.18) by $\bar{p}_s^* x_{is} = \bar{p}_s^* \omega_{is} + \bar{p}_s^* D_s(\bar{p}^*, \bar{\phi}^*) v_i + \bar{T}_{is}^*$. Since the Lagrange multipliers are now known, the Lagrangian of (12.13) may be written

$$\begin{aligned} L(x, v, p, \phi, \rho) = {} & \Sigma_i \bar{\alpha}_i^* u_i(x_{i0}, x_{i1}, \ldots, x_{iS}) - \bar{p}_0^*(\Sigma_i x_{i0} - \Sigma_i \omega_{i0}) \\ & - \Sigma_{s \neq 0} \bar{p}_s^* (\Sigma_i x_{is} - \Sigma_i \omega_{is} - D_s(\bar{p}^*, \bar{\phi}^*) \Sigma_i v_i) \\ & - \Sigma_i \bar{\rho}_{i0}^* (\bar{p}_0^* x_{i0} + \bar{\phi}^* v_i - \bar{p}_0^* \omega_{i0} - \bar{T}_{i0}^*) \\ & - \Sigma_i \Sigma_{s \neq 0} \bar{\rho}_{is}^* (\bar{p}_s^* x_{is} - \bar{p}_s^* \omega_{is} - \bar{p}_s^* D_s(\bar{p}^*, \bar{\phi}^*) v_i - \bar{T}_{is}^*) - \bar{\phi}^* \Sigma_i v_i \end{aligned}$$

for $x_{is} \geqslant 0$, all i and s.

Regrouping of terms leads to

$$\begin{aligned} L(x, v, p, \phi, \rho) = {} & \Sigma_i [\bar{\alpha}_i^* u_i(x_{i0}, x_{i1}, \ldots, x_{iS}) \\ & - (1 + \bar{\rho}_{i0}^*)(\bar{p}_0^* x_{i0} + \bar{\phi}^* v_i - \bar{p}_0^* \omega_{i0} - \bar{T}_{i0}^*) \\ & - \Sigma_{s \neq 0} (1 + \bar{\rho}_{is}^*)(\bar{p}_s^* x_{is} - \bar{p}_s^* \omega_{is} - \bar{p}_s^* D_s(\bar{p}^*, \bar{\phi}^*) v_i - \bar{T}_{is}^*)] \end{aligned} \tag{12.21}$$

for $x_{is} \geqslant 0$, all i and s. This is the sum of Lagrangians of the individual consumer problems (12.12′) (to see this, set $\bar{\lambda}_i^*$, the marginal utility of income of consumer i, equal to $1/\bar{\alpha}_i^*$).[24] This sum can be decomposed so that in the fixed point the allocation satisfies (12.12′) for all i. ■

The ε-approximation has served to ensure boundedness of Lagrange multipliers. There are two cases in which an equilibrium with $\varepsilon = 0$ can be shown to exist so that no approximation is necessary: (1) if returns are expressed in terms of state-specific commodities (this well-known case is treated in section 12.2.7), and (2) if special assumptions are made on the slope of the utility function; because all the constraints are linear, Lagrange multipliers will exist, even if Slater's condition is not satisfied, but may be unbounded. Uniform boundedness can be ensured through assumptions on the slope of the utility functions at points where some consumption is zero (note that this condition is not fulfilled by the CES utility functions in Hart's (1975) counterexample). By concavity of the utility functions, boundedness of the slope at such points will guarantee boundedness of multipliers in any optimum.

Moreover in the proof we have used the nonnegativity of returns in assumption R, which ensured that asset prices $\bar{\phi}$ are nonnegative. If we relax this assumption, we must replace the condition $\Sigma_i v_i \leqslant 0$ by an equality; otherwise, the objective of program (12.13) could become unbounded. Obviously the associated multipliers ϕ will then no longer have to be nonnegative, but they will exist and be bounded because the constraint is linear and has full row rank. Prices may now be normalized according to $\Sigma_k \Sigma_s \bar{p}_{ks} + \Sigma_c |\bar{\phi}_c| = 1$. This normalization is possible because, by monotonicity of utility functions, not all $\bar{p}$ are zero, and the previous proof can be adjusted to show that an equilibrium exists.

In the sequel we consider the general case of proposition 12.3, assuming that, if necessary, appropriate redistribution of endowments has already been performed so there is no need for further transfers, and ε can be set to zero in program (12.13). We assume that the ρ_{is} and α are renormalized as in part 2 of the proof of proposition 12.3. We also redefine the welfare weights as $\beta_i = \bar{\alpha}_i^*/(1 + \bar{\rho}_{i0}^*)$. At these renormalized welfare weights, state 0's budget constraints can be dropped without affecting allocations. Hence the values $\rho_{i0} = 0$, all i, are among the values associated with the budget constraint for $s = 0$. Finally, to lighten notation, we drop the * notation for equilibrium values.

12.2.5 Properties of the Equilibrium: Pricing of Assets, Nonarbitrage, and Efficiency

PROPOSITION 12.4 (Pricing of assets) In any exchange equilibrium with financial assets,

$$\bar{\phi} = \Sigma_{s \neq 0}(1 + \bar{\rho}_{is})\bar{p}_s D_s(\bar{p}, \bar{\phi}) \qquad \text{for all } i, \bar{\phi} > 0, \tag{12.22}$$

so that $\Sigma_i v_i = 0$.

24. Note that here we make use of the term $D_s(\bar{p}, \bar{\phi})\Sigma_i v_i$ introduced in the commodity balances of (12.13).

Proof In equilibrium, first-order conditions of the Lagrangian (12.21) with respect to v_i yield (12.22) (recall that $\bar{\rho}_{i0} = 0$ for all i and that v_i is unrestricted in sign).

Since $\bar{p}_s > 0$ in equilibrium, it follows from assumption R that $\Sigma_{s \neq 0}(1 + \bar{\rho}_{is})\bar{p}_s D_s(\bar{p}, \bar{\phi}) > 0$. Therefore $\bar{\phi} > 0$, and by complementarity slackness for the last constraint in program (12.13), $\Sigma_i v_i = 0$. ■

Equation (12.22) is well-known in financial analysis, where it appears as part of the so-called martingale property.[25] The condition holds as an equality because it is derived as a first-order condition for an unconstrained variable v_i. The optimality condition is particularly simple because v_i does not enter utility and because there are no transaction costs. The equilibrium possesses a further property that is also well-known, the nonarbitrage condition.

PROPOSITION 12.5 (Nonarbitrage) In equilibrium there is no asset portfolio a such that $\bar{\phi}a < 0$ and $\bar{p}_s D_s(\bar{p}, \bar{\phi})a \geqslant 0$ for any $s \neq 0$.

Proof Recall that $\bar{p}_s > 0$. Then by assumption R ($D_{sc} \geqslant 0, D_{sc} \neq 0$), and by (12.22), $\bar{\phi}_c = \Sigma_{s \neq 0}(1 + \bar{\rho}_{is})\bar{p}_s D_{sc}(\bar{p}, \bar{\phi}) > 0$. Define $\bar{\psi}_{is} = 1 + \bar{\rho}_{is}$ and $A_s = \bar{p}_s D_s$ for all $s \neq 0$. Define also the vector $\bar{\psi}_i$ and the matrix A with elements $\bar{\psi}_{is}$ and rows A_s, respectively. By proposition 12.4, we have $\bar{\phi} = \bar{\psi}_i A, \bar{\psi}_i > 0$. It follows from the Farkas lemma, that there can be no solution to the system $\bar{\phi}a < 0$ and $Aa \geqslant 0$. ■

Proposition 12.5 shows that it is not possible to earn income from assets in the initial period without having to incur an expenditure in at least some states of period 2. This simply is a no-free-lunch property of equilibrium allocations. It is obtained as a consequence of equilibrium and is not an assumption imposed on the model.

The nonarbitrage condition uniquely determines asset prices in terms of commodity prices and marginal utilities. At given prices consumer allocations are also determined uniquely, but asset portfolios may not be. The nonarbitrage condition may make many portfolios equivalent, so consumers will be indifferent between them. Moreover the real allocations $x_{is} = y_{is}/n$ are bounded by the aggregate endowment $\Sigma_i \omega_{is}$ but we have not derived uniform bounds on the portfolios $v_i = w_i/n$, and while finite portfolios support the equilibrium, one cannot rule out that some v_{ic} could also be arbitrarily large in equilibrium.[26]

The equilibrium with financial assets is supported by the welfare program; by a suitable renormalization of welfare weights, the budget constraint of

25. See, for example, Duffie (1991).

26. Both problems can be avoided by making returns depend on the size of the portfolio, with decreasing returns to scale. The second difficulty does not arise when there are transaction costs.

state $s = 0$ can be made nonbinding in equilibrium. The budget constraints in states $s \neq 0$ and the balances for assets can be viewed as technological constraints. They are characterized by a number C of assets, a number S of destinations, and a "technology" matrix D.

Therefore, at given equilibrium prices, the resulting allocation is locally Pareto-efficient: At these prices and with the associated technology, there is no Pareto-superior allocation. However, even in a close neighborhood other equilibrium prices may exist that support Pareto-superior allocations. The equilibrium is globally Pareto-efficient if markets are effectively complete, as shown in proposition 12.6.

PROPOSITION 12.6 (Effectively complete markets) If for each state s the $\bar{\rho}_{is}$ are equal across consumers, then the resulting allocation is a competitive equilibrium.

Proof We use the same renormalization as in (12.21) as well as the renormalized welfare weights β_i (ensuring that $\rho_{i0} = 0$). The assumption that the $\bar{\rho}_{is}$ are equal across consumers ($\bar{\rho}_{is} = \bar{r}_s$) makes it possible to write the Lagrangian of (12.13), after dropping the ε's as

$$L(.) = \Sigma_i \beta_i u_i(x_{i0}, x_{i1}, \ldots, x_{iS}) - \bar{p}_0(\Sigma_i x_{i0} - \Sigma_i \omega_{i0})$$
$$- \Sigma_{s \neq 0} \bar{p}_s(1 + \bar{r}_s)(\Sigma_i x_{is} - \Sigma_i \omega_{is}) - \Sigma_i [\bar{\phi} - \Sigma_{s \neq 0}(1 + \bar{r}_s)\bar{p}_s D_s(\bar{p}, \bar{\phi})] v_i.$$

By proposition 12.4, the term in square brackets is zero. Therefore we are left with the Lagrangian of a Negishi welfare program with prices $\bar{p}_0$ and $\bar{p}_s(1 + \bar{r}_s)$ and with welfare weights such that the following "across states" budget constraints are satisfied:

$$\bar{p}_0 x_{i0} + \Sigma_{s \neq 0} \bar{p}_s(1 + \bar{r}_s) x_{is} = \bar{p}_0 \omega_{i0} + \Sigma_{s \neq 0} \bar{p}_s(1 + \bar{r}_s)\omega_{is} \qquad \text{for every } i.$$

The resulting allocation is a competitive equilibrium and is thus Pareto-efficient. ■

Using the definitions introduced in the proof of proposition 12.5, it is possible (and will prove convenient) to express (12.22) in matrix form, as a linear equation for each i:

$$\bar{\phi} = \bar{\psi}_i A, \qquad i = 1, 2, \ldots, m, \tag{12.23}$$

where the matrix A is evaluated at equilibrium prices $(\bar{p}, \bar{\phi})$.

PROPOSITION 12.7 (Complete markets) Consider the set S^a of states s for which some consumer i has positive utility at equilibrium; denote by A^a the associated rows of the matrix A and by $\bar{\psi}_i^a$ the vector of associated elements of $\bar{\psi}_i$. Then, if A^a has full row-rank, the $\bar{\psi}_i^a$ are equal across consumers; the resulting allocation is a competitive equilibrium and markets will be said to be complete.

Proof Since the matrix has full row-rank, we can select S^a columns from A^a such that the resulting $S^a \times S^a$ matrix Δ is nonsingular. Then by (12.23), we have

$$\bar{\psi}_i^a = \bar{\phi}\Delta^{-1},$$

which shows that the $\bar{\psi}_i^a$ are equalized across consumers. Since the other elements of $\bar{\psi}_i$ as well as the $\bar{p}_{i0}$ are zero, all $\bar{p}_i$ are equalized, and proposition 12.6 applies. ■

Let $\bar{\psi}_s$ denote the equalized values of $\bar{\psi}_{is}$; let also $q_{is} = A_s v_i$ denote the equilibrium revenue from financial assets accruing to consumer i in state s. By (12.23), $\bar{\phi} = \Sigma_{s \neq 0}\bar{\psi}_s A_s$, so in the initial period budget, $\bar{\phi}v_i = \Sigma_{s \neq 0}\bar{\psi}_s(A_s v_i) = \Sigma_{s \neq 0}\bar{\psi}_s q_{is}$. This makes it possible to write the budget constraints of consumer i as

$$\bar{p}_0 x_{i0} + \Sigma_{s \neq 0}\bar{\psi}_s q_{is} \leqslant \bar{p}_0 \omega_{i0},$$

$$\bar{p}_s x_{is} \leqslant \bar{p}_s \omega_{is} + q_{is} \qquad \text{for all } s \neq 0,$$

which shows that i can freely access all states: He can purchase in the initial period q_{is} at *state-prices* $\bar{\psi}_s$ and "transport" this to obtain revenues q_{is} in state s. The portfolio computed as $v_i = \Delta^{-1}q_i$ makes it possible to achieve this.

This proposition makes it clear that every state $s \neq 0$ can be accessed and transactions are unconstrained. Hence markets are complete, and the allocation is globally Pareto-efficient.

12.2.6 Introducing Firms

Carrying over Commodities to the Second Period

As a simple case of a model with real as well as financial assets, we introduce the possibility for each consumer to carry over commodities to the second period. The full-format version can be written

$$\max \Sigma_i \alpha_i u_i(x_{i0}, x_{i1}, \ldots, x_{iS}),$$

$$x_{i0}, x_{is} \geqslant 0, s \neq 0, q_i, v_i, \text{all } i,$$

subject to

$$\Sigma_i x_{i0} + \Sigma_i q_i \leqslant \Sigma_i \omega_{i0} \qquad (p_0),$$

$$\Sigma_i x_{is} \leqslant \Sigma_i \omega_{is} + \Sigma_i q_i + D_s(\bar{p}, \bar{\phi})\Sigma_i v_i, \quad \text{all } s \neq 0 \qquad (p_s),$$

$$\Sigma_i q_i \geqslant 0 \qquad (\chi),$$

$$\bar{p}_0 x_{i0} + \bar{p}_0 q_i + \bar{\phi} v_i \leqslant \bar{p}_0 \omega_{i0}, \quad \text{all } i \qquad (\rho_{i0}),$$

$$\bar{p}_s x_{is} \leqslant \bar{p}_s \omega_{is} + \bar{p}_s q_i + \bar{p}_s D_s(\bar{p}, \bar{\phi}) v_i, \quad \text{all } s \neq 0 \text{ and all } i \qquad (\rho_{is}),$$

$$\Sigma_i v_i \leqslant 0 \qquad (\phi).$$

In this program q_i is the (possibly negative) carryover of commodities (real assets) by consumer i. These commodities are purchased at prices $\bar{p}_0$ in the initial period and sold at $\bar{p}_s$ in state s. The total volume of carryovers cannot become negative.

Much of the earlier discussion applies. In equilibrium the pricing rule for real assets satisfies

$$p_0 - \Sigma_{s \neq 0} p_s - \chi + \rho_{i0} \bar{p}_0 - \Sigma_{s \neq 0} \rho_{is} \bar{p}_s = 0.$$

Since in equilibrium $\rho_{i0} = 0$ after normalization, this reduces to

$$\bar{p}_0 = \Sigma_{s \neq 0}(1 + \bar{\rho}_{is})\bar{p}_s + \bar{\chi}. \qquad (12.24)$$

The equalization of the $\bar{\rho}_{is}$ across consumers will now depend on the rank properties of (12.22) and (12.24) combined. Note that a sufficient condition for complete markets is that the rank condition is satisfied in either system.

This model can be interpreted as one in which each consumer with positive q_{ik} owns a constant returns-to-scale firm that keeps an inventory of commodity k. The profit of this firm is

$$\Pi_i = (\Sigma_{s \neq 0}(1 + \bar{\rho}_{is})\bar{p}_{sk} + \bar{\chi}_k) q_{ik} - \bar{p}_{0k} q_{ik}. \qquad (12.25)$$

Equation (12.25) implies that, in the first period, the firm maximizes expected returns minus cost. Because of (12.24), profit is zero.

This treatment can be generalized by having the consumer-specific firm buying several inputs to produce outputs (and not only to carry over inventories) in the second period.

Incomplete Markets with Firms

In the specification of the more general model with production, four issues deserve attention:

1. Who owns the shares of each firm?
2. What does the firm maximize?
3. When does the firm pay dividends?
4. Should bankruptcy be allowed for in the model?

To conclude, we briefly discuss two cases for which we assume that the production sets satisfy the usual assumptions of the competitive model, with the additional restriction that no outputs are produced in the initial period: $(y_{j0}, y_{js}) \in Y_{js}, s \neq 0$. The technology does not allow y_{j0} to have positive entries.

The first case is a straightforward extension of the earlier model of this section, with consumer-specific firms buying inputs in the initial period and selling outputs in each state $s \neq 0$ (the input is nonnegative for all i). This gives a return in period 2 equal to the value of the outputs. Here (1) only one consumer owns the shares issued by a firm; (2) the firm maximizes profits which can be expressed as in (12.25), where only consumer i's $\bar{\rho}_{is}$ appear; (3) because all returns are in period 2, it is natural to assume that this is when dividends are paid, and (4) the consumer who owns the firm carries the losses, so there is no bankruptcy. In the second case several consumers own a firm; then profit maximization is no longer the natural objective: The $\bar{\rho}_{is}$ of the different consumer-owners will matter. This can be analyzed in the full-format formulation.[27]

12.2.7 Returns in State s Depend on Prices in State s Only

We now show that under some additional assumptions, a "true" equilibrium exists, so there is no need for approximation. For this we specify a model in which returns in state s depend on prices realized in state s only instead of depending on the full price vector $\bar{p}$ as well as the purchasing prices $\bar{\phi}$.[28] We strengthen assumption R, CA1, and CA2 as follows:

ASSUMPTION R′ Returns in state s depend on $\bar{p}_s$ only. $D_{sc}(\bar{p}_s)$ is nonnegative, continuous, and homogeneous of degree zero in $\bar{p}_s$. For given c, $D_{sc}(\bar{p}_s)$ should not be zero in all states.

ASSUMPTION CA1′ The utility function $u_i: R_+^{r(1+S)} \rightarrow R_+, u_i(x_{i0}, x_{i1}, \ldots, x_{iS})$ satisfies CA1 with the additional requirement that for consumers $i = 2, 3, \ldots, m$, it is nonsatiated in all states.

ASSUMPTION CA2′ $\omega_{is} > 0$ for all i and s.

Assumption CA1′ will imply that in each state $s \neq 0$, $\bar{p}_s$ will be nonzero, so prices can be normalized for each state separately. The ε-perturbations appearing in the right-hand side of the budget constraints of program P1–P2 can then be brought to zero so that an equilibrium exists for the original economy without need to redistribute income.

PROPOSITION 12.8 (Existence of equilibrium when returns in state s depend only on prices in state s) Let assumptions R′, CA1′, and CA2′ be satisfied. Then there exists an exchange equilibrium without transfers.

27. This issue was raised by Drèze (1974); see also Drèze (1982).

28. This specification is more general than is usual, except that we exclude negative returns.

Proof In the proof of propositions 12.2 and 12.3, the parameter ε is used to ensure Slater's condition for the budget constraints. Since $D_s(.)$ only depends on $\bar{p}_s$, we can consider prices $\bar{p}_s$ normalized for every state $s \neq 0$ separately. We also normalize $\bar{p}_0$ on the simplex. To ensure Slater's condition, we do not need to use ε, since assumption C2′ and the normalization of prices in each state already ensure that Slater's condition holds. Program P1–P2 with $\varepsilon = 0$ will therefore always have an optimal solution as well as bounded Lagrange multipliers p_0, p_s, ρ_{i0}, ρ_{is}, ϕ, and μ.

1. *Existence of a fixed point.* We again define the Lagrange multiplier correspondence and adjustment functions for prices $(\bar{p}, \bar{\phi})$. For the Lagrange multipliers this proceeds as in proposition 12.3 except that the convex hull B^d will, by nonsatiation in assumption CA1′, be such that prices $p_s \neq 0$ in all states including 0.

The adjustment function for prices is specified so as to keep prices $\bar{p}_s$ positive. We update $\bar{p}_s$ and $\bar{\phi}$, storing the current values as $\hat{p}_s = \bar{p}_s$ and $\hat{\phi} = \bar{\phi}$ according to

$$\bar{p}_{ks} = \lambda_s(p_{ks} + \rho_s \hat{p}_{ks}) \qquad \text{for } s \neq 0,$$

$$\bar{p}_{k0} = \lambda_0(p_{k0} + \rho_0 \hat{p}_{k0}), \tag{12.26}$$

$$\phi_c = \lambda_0(\bar{\phi}_c + \rho_0 \hat{\phi}_c),$$

where $\rho_s = \max_i \rho_{is}$ and the positive scalars λ_s are such that $\Sigma_k \bar{p}_{ks} = 1$ for $s \neq 0$, and λ_0 is such that $\Sigma_k \bar{p}_{k0} = 1$. By definition of B^d, it cannot happen that all $p_{ks} + \rho_s \hat{p}_{ks}$ are zero at the same time in any s (including $s = 0$); hence the scalings are feasible. We define $E^0 = \{\bar{p}_0, \bar{\phi} \mid \Sigma_k \bar{p}_{k0} = 1 \text{ and } \bar{p}_{k0} \geqslant 0;\ \bar{\phi} \geqslant 0, \bar{\phi} \in \Phi\}$, where Φ is the convex hull of all possible scaled values for $\bar{\phi}$. This set is compact because B^d and λ_0 are bounded. For $s \neq 0$ we define $E^s = \{\bar{p}_s \mid \Sigma_k \bar{p}_{ks} = 1 \text{ and } \bar{p}_{ks} \geqslant 0\}$. Since all the scalings are feasible, (12.26) maps into the set $E = E^0 \times E^1 \times \cdots \times E^S$.

The Lagrange multiplier correspondence together with the adjustment functions for prices define a correspondence $G: E \times B^d \Rightarrow E \times B^d$. The parameter set $A = E \times B^d$ is compact, convex so a fixed point exists.

2. *Properties of the fixed point.* This part of the proof proceeds as in proposition 12.3.

Since now $\varepsilon = 0$, the fixed point is an equilibrium without transfers. ∎

12.3 Practices in Applied Modeling

The main effect of introducing money and financial assets is that monetary neutrality may be lost. The absolute price level has effects on allocations, and commodity trades may be constrained by other variables than commodity prices, taxes, rations, and quotas. However, nonneutrality may also be lost as a consequence of lags and money illusion in the adjustment of policy variables such as nominal taxes. In applied models lags and nonhomogeneities are usually the main channels, through which nonneutrality is incorporated. This is discussed in section 12.3.1 and has important implications for the representation of exchange rates (section 12.3.2). The subsequent sections are more closely related to the main subject of this chapter and discuss transactions demand for money and financial assets.

12.3.1 Determination of the Absolute Price Level: Nonhomogeneities

The closed economy model only determines relative prices. In the small open economy absolute prices can be determined from given world prices. Alternatively, one could choose a single domestic price and keep it fixed. In both cases the fixed price acts as a scaling factor that does not affect equilibrium allocations. Values are measured in terms of domestic or of foreign currency, but "money" serves as a unit of account only. There is neither transaction nor asset demand for money.

In some applied models the homogeneity requirement on the specification of the tax, tariff, and markup functions is abandoned, while a homogeneous specification is maintained for the consumer and producer problems (no money illusion). Existence of equilibrium will obviously depend on the properties of these functions.[29] In such models all prices are determined in absolute terms, so there is no scope for any scaling factor.

Nonhomogeneities may arise in various ways. First, it may be an empirical fact that contracts cannot be adjusted instantaneously, so some wages may, for example, be nominally rigid in the short run. Second, the income of consumers may contain nominally fixed components, positive or negative, say, as a result of debt servicing (but this is more adequately represented in a model with financial assets). Third, sometimes the government intervenes in a market by setting a nominally fixed price through a variable indirect tax; more generally, prices may be set within some nominally fixed bands. Fourth, direct tax schedules are rarely fully indexed to prices and therefore contain nonhomogeneities. Finally, markup behavior may involve money illusion.[30]

If such a specification is incorporated within a dynamic model, the rules for dynamic adjustment of the price rigidities (indexation, lags) will cause variations in the absolute price level, and hence deflation or inflation.

29. The equilibrium format of section 3.4 still applies, with some modifications: The indirect taxes and markups should enter the objective function as in the Negishi program. Hence a solution still exists, but for export prices approaching zero, exports could remain positive. Then the economy would export a commodity that should not be traded in equilibrium.

30. See section 11.4.3.

Many of these specifications are not unrealistic, but they are only reduced forms and current research aims at a more explicit representation of the process that generates them. Since nonhomogeneities destroy the classical dichotomy between the real and financial spheres, it is difficult to pursue an analysis of the consequences of nonhomogeneities without an explicit representation of financial assets and transaction demand for money.

12.3.2 Exchange Rates

When the exchange rate is the only monetary variable (as in some of the models in Dervis, de Melo, and Robinson 1982), we can distinguish three cases. In the simplest one, the exchange rate is merely a conversion factor that allows to express foreign and domestic values in their respective currencies. In the second case, some basket of domestic goods is chosen as numéraire, and the exchange rate is expressed as the value of the basket measured in international currency. It then is only an indicator that changes endogenously but does not affect other endogenous variables. In both situations this amounts to no more than a renormalization, and in the absence of a money market, one should not take this indicator as the market exchange rate. No policy implications can be derived from changes in its value.

In the third case, the adjustment of the exchange rate is part of a closure rule to satisfy, say a balance of payments equation.[31] The numéraire is again specified as the price of some basket of commodities. But the exchange rate will affect allocations because only the domestic price of tradables will be influenced directly by changes in the exchange rate; the prices of non-tradables will remain constrained by the normalization. Sometimes the numéraire is a rigid price that is obtained from an estimated equation.[32] This approach has two shortcomings. First, there is the general criticism of using closure rules: The exchange rate is not the equilibrating variable on the money market (in fact this market is usually not represented in the model). Second, if the choice of numéraire is only meant as a normalization, then changes in the exchange rate should have no real effects. If it is more than a normalization, it must represent a nominal price rigidity. It is better to model explicitly the mechanism that generates this rigidity.

12.3.3 Transactions Demand for Money

In practice, one cannot avoid time aggregation, since the Hicksian week is usually at least a year long. Therefore the representation of money has to

31. See Dewatripont and Michel (1987).

32. See Condon, Corbo, and de Melo (1990) for an example of a wage equation.

include a transactions demand component. It is an empirical question whether money should be modeled as commodity money. Typically, under a golden standard, this will be appropriate, but even under fiat money, the representation may be correct if the money supply adjusts to commodity prices.

Transactions demand is sometimes presented like in section 12.1.1, through the Cambridge equation (Lewis 1994). For a given money supply, and if all other equations are homogeneous, this will only determine the absolute price level without affecting the allocations. Usually more general money demand functions are introduced to arrive at a specification inspired by macroeconomics. This leads to money demand functions that depend on interest rates and are not homogeneous in prices. Hence money is non-neutral even if there are no other nonhomogeneities in the model. To go beyond this one must change the problem of the consumer and firm to incorporate money demand as well as demand for financial assets.

12.3.4 Financial Assets

The model with incomplete asset markets can accommodate a large variety of financial assets in a theoretically satisfactory way, and thus it seems attractive for applications, especially when it is written in the full-format. The model can also be extended to incorporate other constraints on transactions than asset market incompleteness. The flexible specification of the utility functions allows to account for differences in probability judgments between consumers (asymmetric information), so it becomes possible to compute the welfare gains from improved information, namely when the probability judgments are correct.

However, the main problem in applying the model is to characterize the states of nature adequately. This might involve specifying joint distributions of all the stochastic events as well as the consumer-specific valuation of these events. As yet there have been no applications in general equilibrium. One explanation is that the number of states can rapidly become very large. We might also mention the fact that on the basis of observed behavior it is difficult to determine what is the consequence of risk aversion and what is simply due to preference and habit.

In practice, modelers try to capture uncertainty in a different way that can be interpreted as a simplification of the model with incomplete markets. Uncertainty does not appear explicitly, and portfolio diversification is obtained by introducing additional constraints on the substitutability between assets in the two-period model without uncertainty. Nonhomogeneities and transactions demand for money are then introduced as well.

12.3.5 Models with Nonhomogeneities, Transactions Demand for Money, and Financial Assets

In applied models with money and financial assets, demand functions that depend on income (or wealth) and prices link the financial and the real spheres. These models are often static or recursively dynamic though Feltenstein (1986, 1988) and Bovenberg and Goulder (1989) have built dynamic models.

In static models, initial transactions holdings and demand sometimes appear in the budget constraint, so money serves as a store of value. Rosenzweig and Taylor (1990) and Bourguignon et al. (1992) have developed this notion further. They specify a function that distributes wealth over four different assets: money, domestic bonds, foreign bonds, and real capital. This makes it possible to increase the number of policy variables that the model can accommodate. For example, in the context of structural adjustment, it permits one to distinguish fiscal policies, exchange rate adjustment, and external borrowing. Wealth is measured as the discounted value of bonds at an expected interest rate that differs between assets. The expectation is purely static over an infinite horizon, which is of course in contradiction with the simulation path generated (time inconsistency). The allocation depends on relative prices, income, and wealth levels according to behavioral equations not derived from utility maximization let alone intertemporal optimization of the model with incomplete asset markets.

The introduction of financial assets makes it possible to derive financing costs of firms related to their debt, working capital requirements, dividend and tax obligations, and the like. Similar models have been implemented by de Janvry et al. (1991), Morrison (1991), and Thorbecke (1992).[33] The work by Easterly (1990) is closely related. Financial assets can introduce nonhomogeneities, namely when debt is fixed in nominal terms. Obviously in such cases a change in the price level, for example, resulting from an adjustment in the money supply or from a change in the exchange rate will necessarily have real effects. However, these effects are due to the nonhomogeneity of initial stocks of money (see Collier and Gunning 1992), or other unindexed claims such as bonds, rather than to the introduction of financial portfolio decisions. For example, when working capital requirements enter the markup equation, then, if the model is homogeneous, a change in the money supply will lead to a proportional increase in the nominal cost of capital and of all prices. The real interest rate remains unaffected. Similarly, when initial money holdings are given, an increase in

33. Keyzer et al. (1992) perform counterfactual simulations for structural adjustment policies in Indonesia using a model without a financial sector or nonhomogeneities. Their findings do not differ very much from the ones in Thorbecke (1992), which does contain demand functions for financial assets.

the price level will have a real balance effect. McMahon (1990) has reviewed the results of models with money and financial assets and reports that such effects dominate. Bourguignon and Morrison (1992) see it as the main advantage of the portfolio approach that it allows to represent constraints on transactions explicitly. This improves the model's capacity to address issues of phasing of reform (see also Goldin and Winters 1992). They generally find that gradual reforms are superior to shock therapy.

In the category of dynamic models, asset portfolios are represented through constraints on portfolio shares and not through explicit demand functions (e.g., Feltenstein 1986, 1988; Bovenberg and Goulder 1989).

A Mathematical Background

A.1 Sets and Functions

This section contains basic definitions for a set $X \subset R^n$ (definitions 1.1–1.9) and for continuous concave (or convex) function $f(x)$ defined on a nonempty convex set X in R^n, to be denoted as $f: X \subset R^n \to R$ (definitions 1.10–1.16).[1] It also contains five basic theorems (theorems 1.1–1.5).

DEFINITION 1.1 (Open set) Consider a vector $x \in R^n$ and a positive real number λ, and define an open ball $B_\lambda(x)$ of radius λ, centered around $x: B_\lambda(x) = \{y \in R^n \mid |y - x| < \lambda\}$. A set $X \subset R^n$ is open iff for every $x \in X$, there is some ball $B_\lambda(x)$ that is contained in X.

DEFINITION 1.2 (Closed set) A set $X \subset R^n$ is closed iff $R^n - X$ is open.

DEFINITION 1.3 (Bounded set) A set X is bounded iff it can be included in an open ball; that is, there exists a point $y \in R^n$ and a positive real number λ such that $X \subset B_\lambda(y)$.

DEFINITION 1.4 (Compact set) A set $X \subset R^n$ is compact iff it is closed and bounded.

DEFINITION 1.5 (Convex set) A set $X \subset R^n$ is convex iff for any two points x and y contained in X, every point z given by $z = \lambda x + (1 - \lambda)y$, $0 \leqslant \lambda \leqslant 1$ lies in X.

DEFINITION 1.6 (Convex hull)[2] The convex hull of a set $X \subset R^n$ is the intersection of all convex sets containing X.

DEFINITION 1.7 (Boundary point) *A* point $z \in R^n$ is a boundary point of the set X iff any neighborhood of z contains a point in X and a point not in X. The points in X which are not boundary points are interior points.

DEFINITION 1.8 (Extreme point) A point z is an extreme point of the convex set X iff it is a boundary point of X and there are no two distinct boundary points x and y such that $z = \lambda x + (1 - \lambda)y$ for some $0 < \lambda < 1$.

DEFINITION 1.9 (Strictly convex set)[3] A convex set $X \subset R^n$ is said to be strictly convex iff all its boundary points are extreme points.

DEFINITION 1.10 (Increasing) A function $f: X \subset R^n \to R$ is increasing at a point $x \in X$ iff there is (1) no direction of variation $d_1 \geqslant 0$, $d_1 \neq 0$, such that

1. We explicitly require continuity because the set X may not be open. If the set is open, then concavity ensures that $f(x)$ will be continuous on its relative interior.

2. The convex hull of a set $X \subset R^n$ may also be defined as

$\text{co}(X) = \{x \mid x = \Sigma_k \beta_k x^k, x^k \in X, \Sigma_{k=1}^p \beta_k = 1, \beta_k \geqslant 0, k = 1, \ldots, p$, where p is any positive integer$\}$.

3. There are other definitions of the strictly convex set. Here we use Avriel et al. (1988, 79).

$(x + d_1) \in X$ and $f(x + \lambda d_1) \leqslant f(x)$ for all $\lambda \in [0, 1]$, and (2) no $d_2 \geqslant 0$, $d_2 \neq 0$ such that $(x - d_2) \in X$ and $f(x - \lambda d_2) \geqslant f(x)$ for all $\lambda \in [0, 1]$.

DEFINITION 1.11 (Nondecreasing) A function $f: X \subset R^n \to R$ is nondecreasing at a point $x \in X$ iff there is (1) no direction of variation $d_1 \geqslant 0$, $d_1 \neq 0$, such that $(x + d_1) \in X$ and $f(x + \lambda d_1) < f(x)$ for all $\lambda \in [0, 1]$, and (2) no $d_2 \geqslant 0$, $d_2 \neq 0$ such that $(x - d_2) \in X$ and $f(x - \lambda d_2) > f(x)$ for all $\lambda \in [0, 1]$.

The definitions of "decreasing" and of "nonincreasing" follow from "increasing" and "nondecreasing," respectively.

DEFINITION 1.12 (Nonsatiated) A function $f: X \subset R^n \to R$ is nonsatiated at a point $x \in X$ iff it is increasing in one variable at least.

Nonsatiated at a point is sometimes also called locally nonsatiated.

DEFINITION 1.13 (Stationary) A function $f: X \subset R^n \to R$ is stationary at a point $x \in X$ iff it is differentiable and $\partial f(x)/\partial x = 0$.

DEFINITION 1.14 (Monotonic) A function $f: X \subset R^n \to R$ is monotonic at a point $x \in X$ iff it is nonstationary and nondecreasing.

Monotonicity means that the function is nondecreasing and that at every point there exists a direction of change Δx along which the function increases, though it may not change in other directions. A function has the property of, say, stationarity if the property holds everywhere in the domain X. A function that is not satiated anywhere is said to be nonsatiated, and similarly for the other properties.

DEFINITION 1.15 ((Strict/strong) (quasi)concavity) Let $X \subset R^n$ be a nonempty convex set. The function $f: X \subset R^n \to R$ is:

1. concave iff $f(\lambda x + (1 - \lambda)y) \geqslant \lambda f(x) + (1 - \lambda) f(y)$ for all $x, y \in X$, $0 \leqslant \lambda \leqslant 1$;

2. strictly concave iff $f(\lambda x + (1 - \lambda)y) > \lambda f(x) + (1 - \lambda) f(y)$ for all $x, y \in X$, $x \neq y$, $0 < \lambda < 1$;[4]

3. quasi-concave iff $f(x) \geqslant f(y)$ implies that $f(\lambda x + (1 - \lambda)y) \geqslant f(y)$ for all $x, y \in X$, $0 \leqslant \lambda \leqslant 1$;

4. strictly quasi-concave iff $f(x) \geqslant f(y)$ implies that $f(\lambda x + (1 - \lambda)y) > f(y)$, for all $x, y \in X, x \neq y$, and $0 < \lambda < 1$;[5]

4. See Takayama (1985, 240, 250–52).

5. Here we follow Takayama (1985, 113). Sometimes (e.g., in Avriel 1976) the weaker definition, is used: $f(x) > f(y)$ implies that $f(\lambda x + (1 - \lambda)y) > f(y)$ for all $x, y \in X$ and $0 < \lambda < 1$. We note that a function with strictly convex isoquants that is homogeneous of degree one (e.g., the Cobb-Douglas production function with constant returns to scale) is concave and strictly quasi-concave but not strictly concave.

5. strongly concave iff it is twice differentiable with a negative definite Hessian H ($z^T Hz < 0$ for all $z \neq 0$);[6]

6. strongly quasi-concave iff it is twice differentiable and $z^T Hz < 0$ for all $z \neq 0$ such that $\partial f/\partial x \cdot z = 0$.[7]

If $f(x)$ is (strictly/strongly) (quasi)concave, then $-f(x)$ is (strictly/strongly) (quasi)convex.

A real-valued concave function with an open domain is continuous, but if the domain is closed, it may not be. For example, the function $f: R_+ \rightarrow R$, $f(x) = 0$ at $x = 0$ and $x + 1$ at $x > 0$ is concave but not continuous.

DEFINITION 1.16 (Subdifferential) Consider a function $f: R^n \rightarrow R$. If $f(x)$ is convex, its subdifferential at a point x, denoted by $\partial f(x)$, is the set

$$\partial f(x) = \{\xi \in R^n \mid f(x) \leqslant f(y) + \xi \cdot (y - x) \text{ for all } y \in R^n\}.$$

If $f(x)$ is concave, its subdifferential at a point x is the set

$$\partial f(x) = \{\xi \in R^n \mid f(x) \geqslant f(y) - \xi \cdot (y - x) \text{ for all } y \in R^n\}.$$

Each element of a subdifferential is referred to as a subgradient.

The subdifferential is a convex set. When this set contains one subgradient only, then the function is continuously differentiable, and this subgradient is the gradient of $f(x)$. Strict convexity (or strict concavity) does not ensure differentiability, since the function may have kinks. For a concave (or convex) function $f(x)$, definitions 1.10–1.14 reduce to the following definition:

DEFINITION 1.17 (Sign of subgradient of concave function) Consider a concave (or convex) function $f: R^n \rightarrow R$, with nonempty, compact subgradient $\partial f(x)$. At point $x \in X$ this function is:

1. increasing iff $\xi > 0$,
2. nondecreasing iff $\xi \geqslant 0$,
3. nonsatiated iff $\xi_k > 0$ for some k,
4. nonstationary iff $\xi \neq 0$,
5. monotonic iff $\xi \geqslant 0$ and $\xi \neq 0$ for all $\xi \in \partial f(x)$.

THEOREM 1.1 (Maximum on a strictly convex set) Let $X \subset R^n$ be a nonempty, compact, strictly convex set. Then, for any nonzero $c \in R^n, c \cdot x$ attains its maximum on X at one point, which is an extreme point.

6. Clearly a strongly concave function is also strictly concave, but the converse does not hold. For example, the scalar function $y = -x^4$ is twice differentiable and strictly concave, but its second derivative is zero at $x = 0$. Thus its Hessian H is singular at that point, and it cannot be negative definite.

7. A variation z such that $\partial f/\partial x \cdot z = 0$ is a variation along the isoquant.

THEOREM 1.2 (Convexity of upper-level sets) The upper level set $\{x \mid x \in X, f(x) \geqslant \alpha\}$ of a quasi-concave function is convex if it is not empty. For a strictly quasi-concave function it is strictly convex.[8]

THEOREM 1.3 (Uniqueness of a maximum) A strictly (quasi)concave function f defined on the convex set $X \subset R^n$ attains its maximum on X at no more than one point.

In setting up a convex program, one often makes use of composite mappings. This may lead to the difficulty that a composition of concave functions is not necessarily concave. The following theorem applies:[9]

THEOREM 1.4 (Composition and concavity) Let $X \subset R^n$ be a convex set, and let $f: X \to R^m$, $g: X \to R^k$, $h: R^m \times R^k \to R$. Define the composite mapping $H(x) = h(f(x), g(x))$, and let any of the following assumptions hold:

i. f is concave, g is convex, h is weakly increasing-decreasing.[10]

ii. f is linear, g is linear.

iii. f is concave, g is linear, h is weakly f-increasing.[11]

iv. f is convex, g is linear, h is weakly f-decreasing.

Now,

1. if h is concave, then H is concave;
2. if h is strictly quasi-concave, then H is strictly quasi-concave;
3. if h is quasi-concave, then H is quasi-concave.

Moreover,

4. if f and g are strictly concave, then $H(x_1, x_2) = f(x_1) + g(x_2)$ is also strictly concave;
5. if f is strictly concave and g is concave, then $H(x) = f(x) + g(x)$ is strictly concave.

It must be emphasized that the sum of strictly quasi-concave functions is not necessarily strictly quasi-concave.

8. See Avriel et al. (1988, 80).

9. See Avriel (1976, 154) for properties 1–3. Property 4 is obvious; for property 5, use definition 1.15, part 2).

10. A function $h(f, g)$ is weakly increasing-decreasing on $T \subset R^m \times R^k$ if and only if, for every $(f_1, g_1) \in T$ and $(f_2, g_2) \in T$, the inequalities $f_1 \leqslant f_2$ and $g_1 \geqslant g_2$ imply that $h(f_1, g_1) \leqslant h(f_2, g_2)$.

11. A function $h(f, g)$ is weakly f-increasing on $T \subset R^m \times R^k$ if and only if, for every $(f_1, g) \in T$ and $(f_2, g) \in T$, the inequality $f_1 \leqslant f_2$ implies that $h(f_1, g) \leqslant h(f_2, g)$. It is weakly f-decreasing if and only if, for every $(f_1, g) \in T$ and $(f_2, g) \in T$, the inequality $f_1 \leqslant f_2$ implies that $h(f_1, g) \geqslant h(f_2, g)$.

Concavity of $f(x)$ has implications for the concavity of the extended function $\tilde{f}(x, n) = nf(x/n)$, where n is a positive number usually representing the number of consumers or of firms. It is used in chapter 7 in models with entry/exit of firms and migration of consumers. It also serves to compute consumer surplus (chapters 2 and 8).

THEOREM 1.5 (Extended concave function) Define the extended function $\tilde{f}: R^n \times [b, c] \to R^r_+$, $\tilde{f}(x, n) = nf(x/n)$ for given $c \geqslant 1$ and $b \in (0, c]$. Assume that $f(x)$ is continuous, concave, monotonic on R^r_+ and that $f(0) = 0$.

Then

1. $\tilde{f}(x, n)$ is concave, monotonic, and homogeneous of degree one;
2. if there exists some value d such that $\lim_{x \to a, n \downarrow 0} \tilde{f}(x, n) = d$ for all $a \geqslant 0$, then $\tilde{f}(x, n)$ is concave, monotonic, and homogeneous of degree one on $R^r_+ \times [0, c]$;
3. if concavity of $f(x)$ is strict, then $\tilde{f}(x, n)$ is concave and strictly quasi-concave in (x, n) and, for $x \neq 0$, it is increasing in n.

Moreover, if $f(x)$ is differentiable, then

4. $\tilde{f}(x, n) = (\partial \tilde{f}/\partial x) \cdot x + (\partial \tilde{f}/\partial n)n$.

Proof The properties can be verified as follows:

Part 1. Concavity. We choose two points (x_1, n_1) and (x_2, n_2) (since $n_1 > 0$ and $n_2 > 0$). Defining $y_1 = x_1/n_1$ and $y_2 = x_2/n_2$, we consider a convex combination (x, n) of the two points and evaluate the extended function at (x, n) as $\tilde{f}(x, n) = (\lambda n_1 + (1 - \lambda)n_2)$ $f((\lambda x_1 + (1 - \lambda)x_2)/(\lambda n_1 + (1 - \lambda)n_2))$. Now we may also write this as $\tilde{f}(x, n) = (\lambda n_1 + (1 - \lambda)n_2)f(\phi y_1 + (1-\phi)y_2)$, where $\phi = \lambda n_1/(\lambda n_1 + (1 - \lambda)n_2)$. Then, by concavity of $f(x)$, we have $\tilde{f}(x, n) \geqslant (\lambda n_1 + (1-\lambda)n_2)(\phi f(y_1) + (1-\phi) f(y_2))$, so that $f(x, n) \geqslant \lambda n_1 f(x_1/n_1) + (1 - \lambda)n_2 f(x_2/n_2)$, which proves concavity of $\tilde{f}(x, n)$. *Monotonicity*: For $n > 0$, monotonicity in x is inherited from monotonicity of $f(x)$. Then, to prove monotonicity in (x, n), it suffices to verify nondecreasingness in n; for this, consider a convex combination of (x_1, n_1) and $(0, n_2)$, $n_1, n_2 > 0$. Concavity of $f(x)$ implies that $f(\lambda y)/\lambda \geqslant f(y)$ for $0 < \lambda < 1$. Hence $n'f(x/n') \geqslant nf(x/n)$ for all $n' = n/\lambda$, that is, for any $n' > n > 0$. Therefore $f(x, n)$ is nondecreasing in n. *Homogeneity*: This holds by definition of $\tilde{f}(x, n)$.

Part 2. Extension to zero. The limit condition implies that $\tilde{f}(x, n)$ is continuous at $n = 0$, and the properties specified in part 1 also apply at $n = 0$.

Part 3. Strict quasi-concavity. If $f(x)$ is strictly concave, inequalities are strict provided that $y_1 \neq y_2$, namely that (x_1, n_1) and (x_2, n_2) are not proportional (do not lie on a common ray or have $n_1 = 0$ or $n_2 = 0$). Hence the set $\{(x, n) \mid \tilde{f}(x, n) \geqslant \alpha\}$ is strictly convex and $\tilde{f}(x, n)$ strictly quasi-concave. If part 2 holds then, by continuity, this strict quasi-concavity extends to $n = 0$. Increasingness in n is implied by strict concavity of f along rays: $f(\lambda y)/\lambda > f(y)$ for all $0 < \lambda < 1$ for $n > 0$. If part 2 holds, $\tilde{f}(x, n) = 0$, so the function is increasing in n at all nonnegative (x, n) for which $x \neq 0$.

Part 4. Adding up. This follows from homogeneity of $\tilde{f}(x, n)$ by Euler's theorem. ∎

A.2 The Convex Program

A convex program can be written as $\phi = \max\{f(x) \mid g(x) \leqslant 0, x \geqslant 0\}$, where $f(x)$ is concave and $g(x)$ is convex. Obviously the constraint set must be nonempty, but we also want the value function ϕ to be bounded. For this, additional assumptions are needed. One possibility is to assume that the set $X = \{x \geqslant 0 \mid g(x) \leqslant 0\}$ is compact, but this is very restrictive, since producers (and countries) should be given the possibility of buying (importing) infinite quantities if they wish to do so. Another possibility is to distinguish between choice variables x and y, where only x enters the objective, and to assume that $f(x)$ goes to minus infinity if any element of x goes to infinity. Therefore we will consider the constraint set:

$$Q = \{x \geqslant 0, y \in Y \mid g(x, y) \leqslant 0, f(x) \geqslant 0, \ x \geqslant 0\}$$

where the set Y is closed and convex but not necessarily bounded. The nonnegativity restriction on $f(x)$ is arbitrary, and zero could be replaced by any finite, possibly negative, value. We only require boundedness of the set Q.

THEOREM 2.1 (Convex program)[12] Define the nonempty, closed convex set $Y \subset R^m$, the continuous functions $f: R^n_+ \rightarrow R, g: R^n_+ \times Y \rightarrow R^k$ and the program:

$$\phi = \max_{x \geqslant 0, y}\{f(x) \mid g(x, y) \leqslant 0, y \in Y\}. \tag{A.1}$$

Assume that

i. $f(x)$ is concave;

ii. $g(x, y)$ is convex;

iii. the set $Q = \{x \geqslant 0, y \in Y \mid g(x, y) \leqslant 0, f(x) \geqslant 0\}$ is nonempty and compact.

Then,

1. the optimum value ϕ exists and is bounded;

12. See Avriel (1976, 95–100).

2. the set of optimal solutions, $Q^\circ = \{x \geqslant 0, y \in Y \mid g(x, y) \leqslant 0, f(x) \geqslant \phi\}$ is nonempty, compact, and convex;

3. if $f(x)$ is strictly concave, then the optimal x° is unique;

4. if $g(x, y)$ is strictly convex (or if f is nonsatiated, $g(x, y)$ is decreasing in y and Y is compact and strictly convex), then the optimal y° is unique as well.

Assume also that

iv. the set $Q^s = \{x > 0, y \in Y \mid g(x, y) < 0\}$ is nonempty.

Then,

5. the set M° of Lagrange multipliers is nonempty, nonnegative valued, compact, and convex;

6. if $f(x)$ is nonsatiated, then $0 \notin M^\circ$;

7. for every optimal $x^\circ, y^\circ, \mu^\circ$ the following Lagrange inequality holds:

$L(x, y, \mu^\circ) \leqslant L(x^\circ, y^\circ, \mu^\circ) \leqslant L(x^\circ, y^\circ, \mu)$,

for all feasible (x, y) and nonnegative μ, where $L(x, y, \mu) = f(x) - \mu \cdot g(x, y)$ is the Lagrangian (so that $\mu^\circ \cdot g(x^\circ, y^\circ) = 0$, by the right-hand side inequality);

8. if $f(x)$ and $g(x, y)$ are also differentiable and if $Y = R^m$, then any triple (x, y, μ) that satisfies the following Kuhn-Tucker conditions

$$\frac{\partial L(x, y, \mu)}{\partial x} \leqslant 0 \perp x \geqslant 0,$$

$$\frac{\partial L(x, y, \mu)}{\partial y} = 0, \tag{A.2}$$

$$\frac{\partial L(x, y, \mu)}{\partial \mu} \geqslant 0 \perp \mu \geqslant 0,$$

is optimal;

9. finally, there exists a finite constant ρ° such that for all $\rho \geqslant \rho^\circ$, the convex program

$\max_{x \geqslant 0, y \in Y}\{f(x) - \rho \Sigma_h \max(g_h(x, y), 0)\}$,

has the same solution as (A.1).

Condition iv is known as Slater's constraint qualification. It says that the feasible set should have a strict interior and that it is a sufficient condition for the Lagrange multipliers to exist. We have only considered the case with inequality constraints because it is the form we mainly use. If there are also linear[13] equality constraints, say, $Cx + By = d$, Slater's condition reads as:

13. Note that a nonlinear equality constraint can make the Lagrangian nonconvex if the associated multiplier becomes negative.

iv′. The set $Q^s = \{x > 0, y \in Y \mid g(x, y) < 0, Cx + By = d\}$ is nonempty, and the matrix (C, B) has full row rank.[14]

Then Lagrange multipliers v associated with this linear constraint exist, but they are unconstrained in sign.

The program in assertion 9 is known as the exact penalty function.[15] It basically says that in any program that satisfies Slater's constraint qualification, we can bring the inequality constraints into the objective without having to know the value of the Lagrange multipliers. We will use this result when proving theorem 3.1.

We note that the program generates a set of primal optimal solutions Q° and a set of Lagrange multipliers M°. There is no pairwise association between the elements q°, μ° from these sets: Assertion 7 implies that for every Lagrange multiplier $\mu^\circ \in M^\circ$, the problem $\max_{x \geqslant 0, y \in Y}\{f(x) - \mu^\circ \cdot g(x, y)\}$ has the same optimal values x° and y° as program (A.1).

Here and in theorem 3.1 below, the functions $f(x)$ and $g(x, y)$ may also be taken as strictly quasi-concave and strictly quasi-convex, respectively, with the additional requirement that they are differentiable and that the submatrix of $\partial g(x, y)/\partial x$ which belongs to active constraints in (A.2) has full rank and that $Y = R^n$.[16] If $g(x, y)$ depends on y only, it is also possible to treat the constraint $g(x, y) \leqslant 0$ within the definition of the set Y, which will be convex. The Lagrange multipliers associated with the other constraints will exist provided that, for them, Slater's constraint qualification is satisfied.

A.3 The Parametric Convex Program

DEFINITION 3.1 (Uppersemicontinuity in X)[17] The real-valued correspondence $F(x)$ defined over a compact set X in R^n (also denoted as $F: X \subset R^n \Rightarrow Y \subset R^m$ and also called multifunction or set-valued mapping) is uppersemicontinuous in X iff its graph $\{(x, y) \mid x \in X, y \in F(x)\}$ is compact.

The maximum theorem for convex programs describes how optimal allocations in the convex program adapt to parametric variations. In the literature various formulations of this theorem can be found. We will prove a version that is tailored to our needs, which are (1) to allow for parameters in constraints (rather than only in the objective or in the right-hand side of constraints if there is one) and (2) to show how Lagrange multipliers depend

14. See Avriel (1976, 97).

15. See Avriel (1976, 389).

16. See Avriel (1976, 150).

17. Several definitions of uppersemicontinuity coexist. Here we use uppersemicontinuity in X as defined in Takayama (1985, 240, 252).

on these parameters. The proof will use the following version of the maximum theorem.[18]

LEMMA (Maximum theorem) Let $A \subset R^r$ and $Q \subset R^s$ be nonempty compact sets. Consider the continuous function $h: Q \times A \to R, h(q, a)$ and, for given parameter value $a \in A$, the program

$$\phi(a) = \max_{q \in Q} h(q, a). \tag{A.3}$$

Let this program define the correspondence $Q^\circ(a) = \{q \in Q \mid h(q, a) \geqslant \phi(a)\}$ of optimal choices. Then

1. $\phi(a)$ is continuous;
2. $Q^\circ(a)$ is uppersemicontinuous in A;
3. if $Q^\circ(a)$ is single-valued, it is continuous.

THEOREM 3.1 (Maximum theorem for a convex program) Let $A \subset R^r$ be a nonempty compact set and $Y \subset R^m$ a nonempty closed convex set. Consider the continuous functions $f: R^n_+ \times A \to R$, $g: R^n_+ \times Y \times A \to R^k$ and, for given parameter value $a \in A$, the program

$$\phi(a) = \max_{x \geqslant 0, y} \{f(x, a) \mid g(x, y, a) \leqslant 0, y \in Y\}. \tag{A.4}$$

Assume that, for all $a \in A$,

i. $f(x, a)$ is concave in x;

ii. $g(x, y, a)$ is convex in (x, y);

iii. the set $Q(a) = \{x \geqslant 0, y \in Y \mid g(x, y, a) \leqslant 0, f(x, a) \geqslant 0\}$ is nonempty and compact;

iv. the set $Q^s(a) = \{x \geqslant \delta, y \in Y \mid g(x, y, a) + \varepsilon \leqslant 0\}$ is nonempty, for some positive constants δ and ε valid for all $a \in A$.

Then,

1. $\phi(a)$ is continuous;

2a. the set of optimal solutions, $Q^\circ(a) = \{x \geqslant 0, y \in Y \mid g(x, y, a) \leqslant 0, f(x, a) \geqslant \phi(a)\}$ is nonempty, compact, and convex;

2b. the correspondence $Q^\circ(a)$ is uppersemicontinuous in A;

3. if $f(x, a)$ is strictly concave in x, then the optimal $x^\circ(a)$ is single valued and continuous;

18. See Varian (1992, 506). There and in most other references the set Q is allowed to depend on a: $Q(a)$ is a continuous correspondence. Obviously the constant set Q is a special case. We avoid the formulation $Q(a)$ because it is, in general, difficult to verify continuity of a correspondence.

4. if $g(x, y, a)$ is strictly convex in (x, y) (or if $f(x, a)$ is nonsatiated in x, $g(x, y, a)$ is decreasing in y, and the set Y is compact and strictly convex), then the optimal $y^\circ(a)$ is single-valued and continuous as well;

5a. the set $M^\circ(a)$ of Lagrange multipliers associated with the constraint $g(x, y, a) \leqslant 0$ is nonempty, nonnegative valued, convex, and compact;

5b. the correspondence $M^\circ(a)$ is uppersemicontinuous in A, and it is continuous if it is single-valued;

6. if $f(x, a)$ is nonsatiated in x, then $0 \notin M^\circ(a)$.

Proof

Part 1. Primal (properties of $Q^\circ(a)$): Property 2a follows from theorem 2.1, assertion 2. To prove the other properties, we will bring the program (A.4) into the form (A.3) in the lemma above. Since, in assumption iii, $Q(a)$ is compact and assumption iv holds, we may invoke assertion 9 of theorem 2.1 and write the program (A.4) as

$$\phi(a) = \max_{x \geqslant 0, y \in Y}\{f(x, a) - \rho^{\circ\circ}\Sigma_h \max(g_h(x, y, a), 0)\}, \tag{A.5}$$

where the constant $\rho^{\circ\circ}$ is such that it exceeds the minimal value $\rho^\circ(a)$ for all $a \in A$. Since A is compact such a value exists and will be finite. Note that program (A.5) is a convex program without parameters in the constraint set. We denote (x, y) as q, define $h(q, a) = f(x, a) - \rho^{\circ\circ}\Sigma_h \max(g_h(x, y, a), 0)$, and define the constraint set $Q = \{(x, y) \in \bar{Q} \mid x \geqslant 0, y \in Y\}$ where $\bar{Q}$ is the convex hull of $Q(A)$. Since the sets $Q(a)$ and A are nonempty and compact, the set $\bar{Q}$ is nonempty and compact. Therefore Q is nonempty and compact as well. Hence we may invoke the lemma, and properties 1, 2b, 3, and 4 follow from properties 3 and 4 of theorem 2.1 and the lemma.

Part 2. Dual (properties of $M^\circ(a)$): Assertion 5 of theorem 2.1 implies that $M^\circ(a)$ is nonempty, convex, and compact, proving property 5a. For property 5b, let $\bar{M}$ be the convex hull of $M^\circ(A)$. Since the sets $M^\circ(a)$ and A are nonempty and compact, so is $\bar{M}$. By the Lagrange inequality (assertion 7 of theorem 2.1), we may write the Lagrangian of the program (A.4) as

$$\min_{\mu \in \bar{M}} \max_{(x,y) \in Q}\{f(x, a) - \mu \cdot g(x, y, a)\}. \tag{A.6}$$

Let $f^\circ(\mu, a)$ be the value function of the max problem in (A.6). By the lemma, the function is continuous (since the objective is continuous and the set Q is nonempty, compact). Now (A.6) reduces to

$$\min_{\mu \in \bar{M}}\{f^\circ(\mu, a)\}. \tag{A.7}$$

Since $\bar{M}$ is nonempty, compact, and $f^\circ(\mu, a)$ is continuous, we can again apply the lemma, this time to conclude that $M^\circ(a)$ is uppersemicontinuous at a (and continuous if it is single-valued), proving property 5b. Finally, property 6 follows directly from property 6 of theorem 2.1. ■

Note that strict concavity ensures continuity of choice variables but not of Lagrange multipliers. Theorem 3.1 shows that primal variables q are determined according to the correspondence $A \Rightarrow Q, Q^{\circ}(a)$, and Lagrange multipliers according to $A \Rightarrow \bar{M}, M^{\circ}(a)$.

The perturbation theorem, the envelope theorem, and the subdifferential of the value function give a further characterization of the value function.

THEOREM 3.2 (Perturbation theorem)[19] Define the nonempty, compact parameter sets $A \subset R^r$ and $B \subset R^n$ and the nonempty, closed, convex set $Y \subset R^m$. Consider the continuous functions $f: R^n_+ \times A \to R$ and $g: R^n_+ \times Y \times A \to R^k$, and for given parameter values $a \in A$ and $b \in B$, define the program

$\phi(a, b) = \max_{x \geqslant 0, y}\{f(x, a) + b \cdot x \mid g(x, y, a) \leqslant 0, y \in Y\}$.

Assume that for all $a \in A$ and $b \in B$,

i. $f(x, a)$ is concave in (x, a),

ii. $g(x, y, a)$ is convex in (x, y, a),

iii. the set $Q(a, b) = \{x \geqslant 0, y \in Y \mid g(x, y, a) \leqslant 0, f(x, a) + b \cdot x \geqslant 0\}$ is nonempty and compact.

Then $\phi(a, b)$ is

1. concave in a,
2. convex, nondecreasing in b.

THEOREM 3.3 (Envelope theorem) Let $A \subset R^r$ be a nonempty set, and let $X \subset R^n$ be a nonempty compact convex set. Consider the continuous function $f: X \times A \to R$ and, for a given parameter value $a \in A$, the program

$\phi(a) = \max_x\{f(x, a) \mid x \in X\}$.

Assume that

i. $f(x, a)$ is continuously differentiable,

ii. the optimal $x(a)$ is single-valued and continuous on A.

Then $\phi(a)$ is continuously differentiable, and $\partial\phi(a)/\partial a = \partial f(x^{\circ}, a)/\partial a$, for given optimal $x^{\circ} = x(a)$.

Proof By continuous differentiability of $f(x, a)$, in the optimum the total differential of ϕ is

$$d\phi = \frac{\partial f(x^{\circ}, a)}{\partial x} \cdot dx + \frac{\partial f(x^{\circ}, a)}{\partial a} \cdot da$$

19. See Avriel (1976), for convexity w.r.t. b, p. 107; for concavity w.r.t. a, p. 115.

for given $x^\circ = x(a)$ and feasible variations $(dx + x^\circ) \in X, (da + a) \in A$. Optimality of x° implies that $d\phi = 0$ whenever $da = 0$. Hence $\partial f(x^\circ, a)/\partial x \cdot dx = 0$ for all feasible variations dx. This implies that $\partial\phi/\partial a = \partial f(x^\circ, a)/\partial a$. Finally, this derivative is continuous because of continuous differentiability of $f(x, a)$ and because $x(a)$ is continuous. ■

Sometimes the function $f(x, a)$ is taken to be the Lagrangian of the program rather than its objective function. The theorem applies provided that the associated Lagrange multipliers are unique. In the more general case Lagrange multipliers are subgradients of the value function.

THEOREM 3.4 (Subdifferential of the value function)[20] Consider the parameter value $a \subset R^r$, and define the continuous functions $f: R^n_+ \to R$ and $g: R^n_+ \to R^k$ and the program:

$$\phi(a) = \max f(x),$$

$$x \geqslant 0,$$

subject to

$$g(x) \leqslant a \qquad (\mu).$$

Assume that

i. $f(x)$ is concave,

ii. $g(x)$ is convex,

iii. the set $X(a) = \{x \geqslant 0 \mid g(x) \leqslant a, f(x) \geqslant 0\}$ is nonempty and compact,

iv. the set $X^s(a) = \{x > 0 \mid g(x) < a\}$ is nonempty.

Then every Lagrange multiplier μ is a subgradient of $\phi(a): \mu \in \partial\phi(a)$.

For the Lagrange multipliers to be unique, $\phi(a)$ must be differentiable, and this can be ensured if $f(x)$ and $g(x)$ satisfy additional assumptions, which can only be verified in specific applications.

A.4 The Fixed Point Mapping

Consider again the correspondences $A \Rightarrow \bar{Q}, Q^\circ(a)$, and $A \Rightarrow \bar{M}, M^\circ(a)$ of theorem 3.1, where A is now also taken to be convex, and assume that the parameters a are determined according to continuous (feedback) function $A \times \bar{Q} \times \bar{M} \to A$, $h(a, x^\circ, y^\circ, \mu^\circ)$. Then the convex program and its feedback function define, for $C = A \times \bar{Q} \times \bar{M}$, a correspondence $C \Rightarrow C$, where C is a nonempty, compact, convex set. To prove that this correspon-

20. See Avriel (1976, 119).

dence has a fixed point, we can use the Kakutani fixed point theorem. In the special case that x°, y°, and μ° are unique, then $x^\circ(a)$, $y^\circ(a)$, and $\mu^\circ(a)$ will be continuous functions and Brouwer's fixed point theorem can be used.

THEOREM 4.1 (Brouwer's fixed point theorem) Let $C \subset R^s$ be a nonempty compact convex set and G a continuous function $C \to C$. Then there exists a value c^* such that $c^* = G(c^*)$.

The set C is often a simplex or a product of unit simplices.

DEFINITION 4.1 (Simplex) The simplex is a set $S^s = \{c \in R^s_+ \mid \Sigma_h c_h = 1\}$.

Usually, if G is not single valued, the Kakutani fixed point theorem can be used:

THEOREM 4.2 (Kakutani's fixed point theorem) Let $C \subset R^s$ be a nonempty compact convex set and G a correspondence $C \Rightarrow C$ which is uppersemicontinuous in C and which is such that for all $a \in A$, $G(a)$ is nonempty and convex. Then there exists a value c^* such that $c^* \in G(c^*)$.

In proofs of existence of equilibrium the dimensions of the set C do not matter. However, fixed point mappings are also used in computation (see also appendix 8 and 9). Then one tries to make the dimension of C as small as possible by eliminating some variables and the properties of the underlying mathematical program(s) determine the extent to which this reduction is possible. Two cases are of special relevance. First, if the program is single valued both in primal variables q and in Lagrange multipliers μ, then we may construct the composite function $A \to A, h(a, q^\circ(a), \mu^\circ(a))$, so in this case $C = A$. Second, note that in general the feedback function $h(a, q, \mu)$ depends on a. It is a so-called adjustment rule that updates the parameter a and in the fixed point the value generated should be equal to the value entered as argument. This is ensured by defining $C = A \times \ldots$. However, if the feedback function does not depend on the parameter a, we can eliminate this parameter and consider the fixed point mapping $Q \times \bar{M} \Rightarrow Q \times \bar{M}, G(q, \mu)$, where G is defined in partitioned form as $G_1(q, \mu) = Q^\circ(h(q, \mu))$ and $G_2(q, \mu) = M^\circ(h(q, \mu))$. The following theorem ensures that $G(q, \mu)$ will not lose its essential properties.

THEOREM 4.3 (Composition and Cartesian product)[21] Let $A \subset R^r, B \subset R^n$ be nonempty compact sets, and F and G be correspondences that are uppersemicontinuous in A, B, respectively.

1. The composition $F \circ G$ is uppersemicontinuous in A.

21. See Takayama (1985, 252).

2. The Cartesian product of F and G is uppersemicontinuous in $A \times B$.

This theorem shows that uppersemicontinuity is maintained under composition and Cartesian product operations. However, convex valuedness is in general not maintained under composition, but it is maintained under Cartesian product operations (since the Cartesian product of convex sets is convex). Therefore, while we can apply Kakutani's theorem to the correspondences $G(q, \mu)$ defined above, we cannot apply it to, say, the correspondence $G(a) = h(a, Q^\circ(a), M^\circ(a))$.

A.5 The Nonconvex Program

The following theorem applies to nonlinear programs in general and covers the nonconvex case.

THEOREM 5.1 (Nonlinear program) Define $Q \subset R^n$ and $f: Q \to R$, and consider the program

$$\phi = \max_x \{f(x) \mid x \in Q\}.$$

Assume that

i. $f(x)$ is continuous,

ii. the set $X = \{x \mid f(x) \geqslant 0, x \in Q\}$ is nonempty and compact.

Then

1. ϕ is nonnegative and bounded,
2. the set of optimal solutions, $X^\circ = \{x \mid f(x) \geqslant \phi, x \in Q\}$ is nonempty and compact.

Moreover,

3. for $Q = \{x \mid g(x, y) \leqslant 0, x \geqslant 0\}$, if $f(x)$ and $g(x, y)$ are differentiable and the row-rank of the submatrix of $\partial g(x, y)/\partial x$ that belongs to active constraints is full, then locally unique Lagrange multipliers exist and the optimum satisfies the Kuhn-Tucker conditions as defined in theorem 2.1. ■

Properties 1 and 2 of this theorem are the Weierstrass theorem. When the set Q over which one optimizes is a bounded polyhedron, a further characterization is obtained:

THEOREM 5.2 (Maximizing a convex function over a bounded polyhedron)[22] Let $Q \subset R^n$, and let $f: Q \to R$. Define the polyhedron

22. See Pardalos and Rosen (1987, thm. 1.4.4).

$P = \{x \mid x = By, y \geqslant 0, y \cdot e = 1\}$, with B a real-valued matrix and e a vector of ones. Consider the program

$$\phi = \max_x \{f(x) \mid x \in P\}.$$

If $f(x)$ is convex, then every local maximum (and therefore also the global maximum ϕ) will be attained at some extreme point of P.

This theorem implies that when the polyhedron P is a simplex, any optimum will have at most one $x_k = 1$. Such a case arises if, as in chapter 10 (on nonconvexity), there are n firms with cost functions that are concave in output, produce the same commodity, and minimize the total cost of producing a given output. The theorem also illustrates how nonconvexities can force integer solutions and, conversely, how integer programs can be used to solve nonlinear programs with nonconvexities.

A.6 Infinite Dimensions

When the time horizon is infinite, there is an infinite number of commodities and possibly also of agents. Then the allocations $\{x_t\}_{t=1}^{\infty}$ are infinite sequences, so they lie in an infinite dimensional (real) vector space. Functions of these variables will have such a space as domain, so theorems specified in finite real dimensional space no longer apply. Models with infinite horizons are discussed in chapter 8. The following two theorems are used in relation to the model with overlapping generations:[23]

THEOREM 6.1 (Tychonoff theorem) The product of infinitely many compact spaces is compact in the product topology.

THEOREM 6.2 (Continuity) Consider the function

$$f : A \to \Pi_{j \in J} Y_j,\ f(x) = (f_1(x), f_2(x), \ldots, f_j(x), \ldots),$$

and let $\Pi_{j \in J} Y_j$ have the product topology. Then $f(x)$ is continuous if and only if $f_j(x)$ is continuous for all j.

A.7 Multiplicity of Solutions of $F(x) = 0$

We consider the number of solutions to the system of n nonlinear equations in n unknowns, $F(x) = 0$. Here we call x^* an equilibrium if $F(x^*) = 0$ and x^* is bounded. Note that if x^* is a solution of $F(x) = 0$, it is a fixed point of the mapping $G(x) = x + F(x)$ (and of $G(x) = x - F(x)$), and vice versa.

23. See Munkres (1975).

We first discuss uniqueness results which hold if $F(x)$ is continuous and then introduce differentiability.[24]

DEFINITION 7.1 (Monotone mapping) A mapping $F: D \subset R^n \to R^n$ is monotone on $D_0 \subset D$ iff $(F(x) - F(y))(x - y) \geqslant 0$ for all $x, y \in D_0$. It is strictly monotone iff strict inequality holds whenever $x \neq y$ and uniformly monotone iff there is a γ such that $(F(x) - F(y))(x - y) \geqslant \gamma(x - y)(x - y)$ for all $x, y \in D_0$.

Intuitively, for a monotone function, $F_k(x) - F_k(y)$ must have the same sign as $x_k - y_k$ when $x_h = y_h$ for $h \neq k$. Note that if $x \cdot F(x) = 0$ for all $x \in D$ and $y \cdot F(x) \geqslant 0$ for all $x, y \in D$, this condition holds.[25]

DEFINITION 7.2 (Homeomorphism) A mapping $F: D \subset R^n \to R^n$ is a homeomorphism of D onto $F(D)$ iff it is one to one[26] on D and F and F and its inverse F^{-1} is continuous on D and $F(D)$, respectively.

Clearly, if $F(x)$ is a homeomorphism of D onto $F(D)$ and $0 \in F(D)$, then $F(x) = 0$ has a unique solution in D.

THEOREM 7.1 (Uniqueness for the monotone mapping) Let $F: D \subset R^n \to R^n$ be continuous and monotone on the open, convex set $D_0 \subset D$. Then, for any $b \in R^n$, the solution set $\Gamma = \{x \in D_0 \mid F(x) = b\}$ is convex if it is not empty. If $F: R^n \to R^n$ is continuous and uniformly monotone, then $F(x) = b$ has a unique solution and is a homeomorphism of R^n onto itself.

Propositions on uniqueness usually assume continuous differentiability of $F(x)$ in the solution point. The Weierstrass approximation theorem suggests that if $F(x)$ is known to be continuous in such a point, it may not be restrictive also to assume continuous differentiability.

THEOREM 7.2 (Weierstrass's approximation) Let $F: D \subset R^n \to R^n$ be continuous on the compact set $C \subset D$. Then there exists for any $\varepsilon > 0$ a continuously differentiable function $G: R^n \to R^n$ such that $\|F - G\| < \varepsilon$ for all $x \in C$.

Here $\|\cdot\|$ notes an arbitrary vector-norm on R^n. For example, the l_∞-norm is equal to the element of the vector that is largest in absolute value: $\|x\|_\infty = \max_{1 \leqslant i \leqslant n} |x_i|$. One may now consider the following theorem on uniqueness:

24. Theorems 7.1, 7.2, 7.3, 7.5, 7.6, 7.7 are taken from Ortega and Rheinboldt (1970, 143–45, 171, 137, 130, 150, 128, respectively).

25. The first equation is similar to Walras's law and the second to the weak axiom of revealed preference.

26. One to one means that $F(x) \neq F(y)$ whenever $x, y \in D$, $x \neq y$.

THEOREM 7.3 (Uniqueness under nonsingularity and norm coerciveness) Let $F: R^n \to R^n$ be continuously differentiable on R^n, and assume that $F'(x)$ is nonsingular for all $x \in R^n$ and that it is norm coercive on R^n, namely that $\lim_{\|x\| \to \infty} \|F(x)\| = \infty$. Then $F(x) = b$ has a unique solution for any $b \in R^n$.

This theorem needs two comments. First, norm coerciveness does not have to hold for the system of equations under study, say, for $G(x) = 0$. If we know that (i) the solution lies in a compact set X, (ii) $\det(G'(x)) \neq 0$ for all $x \in X$, and (iii) $G(x)$ is bounded from below, then we can define a barrier function around this set as follows: Let $F(x) \equiv G(x) + K(x)$, where $K(x)$ is the barrier function. It is assumed to possess with five properties: (1) $K(x) = 0$ for all $x \in X$, (2) it is norm coercive, (3) it is continuously differentiable on R^n, (4) its Jacobian is diagonal for all $x \notin X$, and (5) $\det(K'(x))\det(F'(x)) > 0$ for all $x \neq X$. The function $F(x)$ will be norm coercive on R^n, and its Jacobian inherits the nonsingularity of the Jacobian of $G(x)$. Second, consider a system $G(y) = 0$ that is only defined on the nonnegative orthant and for which $\|G(y)\|$ goes to infinity whenever any element of y goes to zero. We can now define $F(x) = 0$ by introducing the mapping $y(x)$ with elements $y_k(x_k) = \exp(x_k)$. The function $F(x) = G(y(x))$ will then be defined on R^n and is norm coercive in x. A transformation of this kind can be performed when y represents prices (or welfare weights) and G is excess demand (or budget deficit).

Norm coerciveness is thus relatively easy to obtain, and it is the nonsingularity of $F'(x)$ that appears to impose the main restriction in theorem 7.3. In the economic literature specific nonsingular matrices are often mentioned.

DEFINITION 7.3 (P matrix) The real-valued $n \times n$ matrix A is a P matrix iff it has positive principal minors.

DEFINITION 7.4 (Strict diagonal dominance) The real-valued $n \times n$ matrix A has strict diagonal dominance iff there exist n positive numbers d_j such that $d_j|A_{jj}| > \Sigma_{i \neq j} d_i |A_{ij}|$, $j = 1, \ldots, n$.

DEFINITION 7.5 (Gross substitutability) The real-valued $n \times n$ matrix A has gross substitutability iff it has nonpositive off-diagonal elements and there exists a nonnegative vector x such that $Ax > 0$.

THEOREM 7.4 (Properties of P matrices) Consider the $n \times n$ real-valued matrix A.

1. If A has gross substitution, then it has a positive diagonal, is strictly diagonally dominant, and its inverse is nonnegative, with positive diagonal.

2. If A is strictly diagonally dominant with a positive diagonal, then it is a P matrix, and all its eigenvalues have positive real parts.

3. If A is a P matrix, then it is nonsingular, and the diagonal of its inverse is positive.

Strict diagonal dominance is usually satisfied in the neighborhood of an equilibrium solution, since the direct effect of a price change on the own market (or of a welfare weight on the own budget deficit) will be more important than indirect effects. However, due to the Debreu-Mantel-Sonnenschein theorem (lack of general properties of excess demand other than continuity, homogeneity, and Walras's law, discussed in chapter 1, section 1.2.3), this property cannot be proved to hold globally (out of equilibrium), and models can be constructed where it never holds. Since singularity of the Jacobian can occur, theorem 7.3 cannot be used in a constructive way to prove uniqueness.

In accepting multiplicity of solutions, one might want to know whether the occurrence of an infinite number of solutions is "rare." One says that an occurrence is "rare" if it has measure zero, and Sard's theorem shows that singularity is rare indeed.

DEFINITION 7.6 (Measure zero) A set $S \subset R^n$ has measure zero iff for any $\varepsilon > 0$, there is a countable number of hyperrectangles Q_j with volume q_j such that $S \subset \cup_{j=1}^{\infty} Q_j$ and $\Sigma_{j=1}^{\infty} q_j \leqslant \varepsilon$.

For example, m points have measure zero in R^1 (the real line), m lines have measure zero in R^2 (the plane), m planes have measure zero in R^3, and so on.

THEOREM 7.5 (Sard's theorem) Let $F: D \subset R^n \to R^n$ be continuously differentiable on the open set D. Let S be a compact subset of D, and let $C = \{x \in S \mid \det(F'(x)) = 0\}$. Then $F(C)$ has measure zero.

Since the set of values of $F(x)$ for which $F'(x)$ is singular has measure zero, it would be a "coincidence" if $F(x)$ happened to be equal to zero and to have a singular Jacobian at the same point. Therefore one often disregards solutions where the Jacobian is singular.

THEOREM 7.6 (Finite number of solutions) Let $F: D \subset R^n \to R^n$ be continuously differentiable on the open set D, and let C be an open bounded set with closure $\bar{C} \subset D$. Consider the set $\Gamma = \{x \in C \mid F(x) = y\}$ for given $y \notin F(\bar{x})$ where $\bar{x}$ is any point on the boundary of C. Then there is at most a finite number of points in Γ at which $F'(x)$ is nonsingular.

Note that in theorems 7.5 and 7.6 the function $F(x)$ does not have to be continuously differentiable everywhere, as long as it is so on an open set D (which may consist of disjoint open subsets $D_k, k = 1, \ldots, m$). The theorems establish properties of points in D. Before applying the theorems, one has to prove that Γ is nonempty (existence of a solution) and that $F(x)$ is continuously differentiable on a set D that contains Γ. Theorem 7.6 says that the set R of solutions for which $F'(x)$ is regular, contains a finite number of points, if it is not empty. In economic applications, y is usually equal to zero and the requirement $y \notin F(\bar{x})$ for any $\bar{x}$ on the boundary means that excess demand (or excess budgets) should not be zero when relative prices (or relative welfare weights) are zero, which amounts to a weak form of nonsatiation. Under additional restrictions it is possible to prove that the number of solutions is odd. This is discussed in section A.8.

Since the set R contains a finite number of solutions, these solutions, if they exist, are necessarily isolated (locally unique). One can then analyze the consequences for x of a variation in some parameters, say, the vector ω, by applying the implicit function theorem.

THEOREM 7.7 (Implicit function theorem) Suppose that $F: D \subset R^n \times R^r \to R^n$ is continuous on an open neighborhood $D_0 \subset D$ of a point (x_0, ω_0) for which $F(x_0, \omega_0) = 0$. Assume that $F(x, \omega)$ is continuously differentiable with respect to x at this point and that $\partial F(x_0, \omega_0)/\partial x$ is nonsingular. Then there exist open neighborhoods $S_1 \subset R^n$ and $S_2 \subset R^r$ of x_0 and ω_0, respectively, such that, for any $\omega \in \bar{C}$, the equation $F(x, \omega)$ has a unique solution $x = H(\omega) \in \bar{S}_1$ and the mapping $H: S_2 \to R^n$ is continuous. Moreover, if F is differentiable with respect to ω at (x_0, ω_0), then H is differentiable and

$$H'(\omega_0) = -\left[\frac{\partial F(x_0, \omega_0)}{\partial x}\right]^{-1} \frac{\partial F(x_0, \omega_0)}{\partial \omega}.$$

A.8 Solving a System of Equations $F(x) = 0$

The design of a computational scheme that solves the system $F(x) = 0$ of n nonlinear equations in n unknowns basically proceeds in two stages. In the first stage one defines a differential equation $x' = G(x(t))$, $x(0) = x_0$ given (where $x' \equiv dx(t)/dt$) which converges to a solution $G(x^*) = 0$ such that $F(x^*) = 0$. In the second stage one approximates this differential equation through discretization, for example, by some difference equation in order to obtain an algorithm which approximately solves $F(x) = 0$ within a finite number of iterations. We first discuss the specification of the differential equation.

The Differential Equation

One strategy to solve $F(x) = 0$ is to adjust x in accordance with $-F(x)$, the deviation from "equilibrium." Hence $G(x) \equiv -F(x)$ or a monotonic transformation thereof. This is referred to as the tatonnement approach, represented by the differential equation $x' = -F(x(t))$, $x(0) = x_0$ given. The solution curve $x(t)$ of this equation will converge to a unique solution if all the eigenvalues of $-F'(x)$ have negative real parts, such as will happen if $F'(x)$ is diagonally dominant or positive definite along the path. However, this is a special case, and convergence is in general not ensured for tatonnement rules: $x(t)$ may cycle or leave the feasible set.

For this reason a more reliable strategy is to generate a path along which $x(t)$ remains within the feasible set (through norm coerciveness), and which ensures that $F(x) = 0$ is satisfied by construction when t reaches, say, unity. This is called the homotopy or continuation method.[27] In theorems 8.1 to 8.3 we give some examples, starting with the case in which the Jacobian matrix is nonsingular everywhere, as specified in theorem 7.3.

THEOREM 8.1 (Continuation path for nonsingular Jacobian) Let $F: R^n \to R^n$ be continuously differentiable on R^n, and assume that $F'(x)$ is nonsingular for all $x \in R^n$ and norm coercive. Then for any $x_0 \in R^n$ there exists a unique mapping $x: [0, 1] \to R^n$ such that $H(x(t), t) = 0$ for all $t \in [0, 1]$, where $H(x, t)$ is the homotopy mapping defined as $H(x, t) = F(x) + (t - 1)F(x_0)$. Moreover $x(t)$ is continuously differentiable and solves the differential equation $x' = -F'(x(t))^{-1}F(x_0)$ given $x(0) = x_0$.

Note that when $t = 1$, $H(x(t), t) = F(x(t))$ and the system is solved. In this theorem, along the path defined by the differential equation, every $F_k(x)$ is reduced by the same factor. This differential equation provides the basis for Newton's method of solving $F(x) = 0$.

The theorem of course assumes $F'(x)$ to be nonsingular along the path. We now discuss how singularity can be coped with by adding a diagonally dominant function $K(x, t) = (1 - t)(D(x) - D(x_0))$ that vanishes for $t = 1$ but, for t sufficiently far from 1, ensures nonsingularity of the Jacobian as well as norm coerciveness along the path.

THEOREM 8.2 (Diagonally dominant start of continuation path) Let $F: R^n \to R^n$ be continuously differentiable on R^n and norm coercive. Then for any $x_0 \in R^n$ there exists a value $t_1 \in (0, 1]$ and a unique mapping $x: [0, t_1] \to R^n$ such that $H(x(t), t) = 0$ for all $t \in [0, t_1]$, where $H(x, t)$ is the homotopy mapping defined as $H(x, t) = F(x) + (t - 1)(F(x_0) + D(x_0) - D(x))$ and $D(x)$ is a diagonal, continuously differentiable, and uniformly bounded function. Moreover $x(t)$ is continuously differentiable and solves the differential equation $x' = -[F'(x(t)) + (1 - t)D'(x(t))]^{-1}(F(x_0) + D(x_0) - D(x(t)))$ for given $x(0) = x_0$, $t \in [0, t_1]$.

Proof Since $D(x)$ is uniformly bounded, norm coerciveness of $F(x)$ implies norm coerciveness of $H(x, t)$. Hence $x(t)$ will be bounded for all $x(t)$ that

27. See Ortega and Rheinboldt (1970, 230–39).

maintain $H(x, t) = 0$. Total differentiation of H yields

$$0 = (F'(x) + (1 - t)D'(x))dx - (F(x_0) + D(x_0) - D(x))dt.$$

At $t = 0$ we can choose the derivative of D arbitrarily large so as to make $(F'(x) + (1 - t)D'(x))$ diagonally dominant and hence nonsingular. This defines the differential equation at $t = 0$. Adjusting $x(t)$ according to the differential equation, we can compute the absolute value of the determinant of $(F'(x(t)) + (1 - t)D'(x(t)))$ and choose as t_1 the value of t where it reaches some critical value ε. ■

The continuation path of theorem 8.2 "skips over" singularities that occur far from the solution. The approach is easily implemented and appears to work well in many applications.[28] We know from Sard's theorem that singular solutions are rare. However, we do not know whether solutions with an almost singular Jacobian are rare, so we could be forced to interrupt the path at $t = t_1 < 1$. Then the norm of $F(x)$ may not have been reduced, since the differential equation does not reduce the norm everywhere along the path. Hence it may happen that $\|F(x(t_1))\| > \|F(x(0))\|$, particularly if t_1 lies close to zero that is, at a point where the function $K(x, t)$ matters. However, since singularities are rare in practice, it will usually be possible to proceed sufficiently far with an increase in t so as to achieve norm reduction at t_1. If t_1 is not equal to unity already, we can, if a singularity occurs, set $x_0 = x(t_1)$ and start a new round, hopefully reaching $t_1 = 1$, eventually.

An alternative route[29] for treating singularities is to define the homotopy $H(x, \theta) = x - (1 - \theta)x_0 - \theta(F(x) + x)$ with $\theta \in [0, 1]$ and, instead of assuming that $\theta(t) = t$, to allow for specifications in which t is defined on $[0, \infty)$. For this one defines $y(t) = (x(t), \theta(t))$. Total differentiation of $H(y)$ then yields $\partial H(y(t))/\partial y y' = 0$. This is a system of n linear equations in $n + 1$ unknowns that has an infinite number of solutions. Assume that $H(x, \theta)$ is norm coercive with respect to x for any $\theta \in (0, 1]$ and that 0 is a regular value of $H(x, \theta)$, namely that $\partial H(y)/\partial y$ has rank n. Now let $H'_{-i}(y)$ denote the $n \times n$ matrix that results after dropping the ith column of $\partial H(y)/\partial y$. Regularity implies that we always can find an index value i for which this matrix has rank n. We solve the system for $k \neq i$, setting (arbitrarily) $y_i' = (-1)^{n-i+1} \det H'_{-i}(y)$. Cramer's rule then yields $y_k' = y_i' \det H'_{-k}(y)/\det H'_{-i}(y)$ for $k \neq i$. Thus

$$y_k' = (-1)^{n-i(t)+1} \det H'_{-k}(y(t)) \qquad \text{for all } k \ (k = 1, \ldots, n + 1).$$

28. The approach is described in Keyzer (1989) where it is applied to the nonlinear complementarity problem discussed in section A.9.

29. A discussion along these lines can be found in Kehoe (1991, 2062–2063), but the resulting differential equation is not the same. See also Smale (1976).

This defines a system of (piecewise) continuous differential equations $y' = G_i(y)$. It starts at $\theta = 0$ and $i(0) = n + 1$ with $\theta' = \det I = 1 > 0$. By norm coerciveness, this path does not leave the compact set; it cannot return to $\theta = 0$ (this would contradict the single-valuedness of the differential equation at $\theta = 0$). Hence it must eventually reach $\theta = 1$, a solution.[30]

The argument is easily pursued to show that the number of solutions is odd. The solution attained starting from $\theta = 0$ must have $\theta' > 0$ at the beginning and at the end. Other solutions come in pairs,[31] with each one the end point of a path that starts and ends on the boundary where $\theta = 1$. At one end point $\theta' > 0$, and at the other, $\theta' < 0$ (a path started on the boundary $\theta = 1$ does not return to $\theta = 0$ and does not leave the compact set, so it must return to $\theta = 1$).

Note that at $\theta = 1, \theta' = \det(-F'(x))$. Hence, if one denotes positive determinants with the index $+1$ and negative determinants with -1, then the sum over all solutions is $+1$. This result is known as the index theorem. The above properties are summarized in the following theorem.

THEOREM 8.3 (Homotopy path with singularity) Let $F: R^n \to R^n$ be continuously differentiable on R^n, and for given $x_0 \in R^n$, let the associated homotopy mapping be defined as $H(x, \theta) = x - (1-\theta)x_0 - \theta(F(x)+x)$ with $\theta \in [0, 1]$. Denoting (x, θ) by y, define the system of differential equations

$$y'_k = (-1)^{n-i(t)+1} \det H'_k(y(t)), \qquad k = 1, \ldots, n+1, t = [0, \infty), y(0) = (x_0, 0),$$

with $i(0) = n + 1$, where $H'_{-k}(y(t))$ is the matrix of derivatives of $H(y(t))$ excluding column k and $i(t)$ is constant unless the current $H'_{-i(t)}(y(t))$ becomes singular. In that case, for $i = 1, \ldots, n + 1$, choose the smallest i for which the matrix is regular.

Assume that

i. $H(x, \theta)$ is norm coercive with respect to x for any $\theta \in [0, 1]$,

ii. 0 is a regular value of $H(x, \theta)$.

Then, the following properties hold:

1. The differential equations are piecewise continuous and define a continuous mapping $y: [0, \infty) \to R^{n+1}$ such that $H(y(t)) = 0$ for all $t \in [0, \infty)$.

2. The equations converge to a solution $y^* = (x^*, 1)$, with $\det(-F'(x^*)) > 0$.

3. The number of solutions is odd, and all solutions, except the one reached from $\theta = 0$, appear in pairs.

4. For each pair, $\det(-F'(x))$ is positive in one solution and negative in the other.

30. See the Poincaré-Bendixon theorem in Hirsch and Smale (1973, ch. 11).

31. This matching depends on the specification of the homotopy $H(x, \theta)$.

5. When initialized at the solution with $\det(-F'(x)) < 0$, the differential equation will lead to the other solution.

Discretizing the Differential Equation

Many techniques exist to discretize differential equations. Their numerical precision and speed of convergence has been analyzed in detail by mathematicians. The most simple method of discretization is to use the discrete analogue of $x' = G(x)$ and write the difference equation

$$x^{k+1} = x^k + \omega G(x^k) \qquad \text{for } k = 0, 1, \ldots, x^0 \text{ given},$$

where ω is a positive constant (the relaxation factor) that should be chosen sufficiently large to "move fast" but not so large that the approximation of the differential equation becomes so inaccurate that the algorithm does not converge. The effective adjustment of the relaxation factor ω during iterations is often the decisive factor in achieving a fast converging algorithm. Many algorithms try a large value for ω, reducing it only if norm reduction is insufficient, with a restart at the last point where norm reduction had been achieved.[32] For a further discussion, see Ortega and Rheinboldt (1970, chs. 7–14).

The methods that discretize a differential equation generate a solution curve that can, in principle, lie anywhere in a compact space S. Alternatively, one could discretize S itself, approximating it by a grid of points. This is the approach introduced by Scarf (1967, 1973) to solve fixed point problems. The direction of change x' of the differential equation is replaced by a pivoting rule (as in the simplex algorithm of linear programming) which selects the (adjacent) point that should enter the basis and drive some other variable out of the basis. If the choice of the incoming point is always unambiguous and always leads to a point inside S, then, under some regularity assumption, the choice of the variable that leaves the basis will be unique as well. The grid is usually represented by the vertices of a triangulation. Van der Laan and Talman (1980) have improved on work by Eaves, Merrill, Todd, and many others, incorporating homotopy-based ideas.

Like a continuous differential equation, a pivoting algorithm that does not leave S and does not break down must either converge or cycle. Its path is like a chain with segments of fixed length (the mesh of the triangulation). This chain must either form a loop or end somewhere in S; otherwise, it would not fit within this compact set. In particular, a path that starts on a boundary and does not return to it has to end in S, and so does a path

32. Note that at such points it is also possible to shift from simple, tatonnement rules to more sophisticated algorithms.

started inside S that does not cycle and that does not leave S. Since all the segments of the chain have the same length, a path that does not cycle must lead to the end of the chain within a finite number of iterations. This is known as the Lemke-Howson argument. The approach can be applied to nondifferentiable continuous mappings as well as to set-valued (convex-valued, uppersemicontinuous) mappings. See Allgower and Georg (1980) for a survey of early algorithms and Kehoe (1991) for later developments.

A.9 Solving a Nonlinear Complementarity Problem

Most general equilibrium models consist of (1) convex programs that describe the behavior of individual agents and (2) equality constraints and complementary slackness conditions which define the equilibrium. Recall that Kuhn-Tucker conditions of theorem 2.1 as they appear, for example, in (A.2) can be written as a nonlinear complementarity problem (NCP) of the form: $F(x) \geqslant 0 \perp x \geqslant 0$. Therefore, if the convex programs 1 satisfy differentiability properties so that first-order conditions can be represented by Kuhn-Tucker conditions, then the full model 1 and 2 can also be written as an NCP.[33]

Transforming the NCP into an Equality-Constrained Problem

One way to solve an NCP is to transform it into an equality constrained system.[34] This is done as follows: Define a slack variable s and an adjustment variable y, where s and $y \in R^n$, and solve the equality constrained problem:

$$F_i(x_1(y_1), \ldots, x_n(y_n)) + s_i(y_i) = 0 \qquad \text{for } i = 1, \ldots, n, \tag{A.8}$$

where the functions $x_i(y_i)$ and $s_i(y_i)$ are defined as $x_i(y_i) = \max(y_i, 0)$ and $s_i(y_i) = \max(-y_i, 0)$. As in sections A.7 and A.8 a further characterization of the solution will often require F to be continuously differentiable; obviously the functions $x_i(y_i)$ and $s_i(y_i)$ are not differentiable at $y_i = 0$. To maintain differentiability, one could define a function g_i such that $g_i(y_i) = 0$ for $y_i \leqslant 0$, and derivatives $g_i'(0) = 0$ and $g_i'(y_i) > 0$ for $y_i > 0$. The functions $x_i(y_i)$ and $s_i(y_i)$ can then be defined as $x_i(y_i) = g_i(y_i)$ and $s_i(y_i) = g_i(-y_i)$. One could also distinguish between the g_i functions of x_i and of s_i when the two variables are expressed in different units. Note that $\partial F(x(y))/\partial y_k$ will necessarily be zero at $y_k = 0$, so there will be a singularity at this point.[35]

33. See Cottle and Dantzig (1968) for a discussion of the relation among (linear) complementarity problems, mathematical programming, and equilibrium models, and see Rutherford (1994) for a description of MPSGE, a GAMS subsystem used to solve applied general equilibrium models as nonlinear complementarity problems.

34. This is illustrated in the model of Drèze-rationing; see figure 6.2 and proposition 6.2.

The system of equations (A.8) is of dimension n. Assume, without loss of generality, that $s_i = 0$ for $i = 1, \dots, n_1$ and $s_i > 0$ for $i = n_1 + 1, \dots, n$, and let x and $F(x)$ be partitioned accordingly so that x_1 denotes the variables to be solved for, namely the x_i in (A.8) for which $s_i = 0$; clearly $x_2 = 0$. Then the solution satisfies

$$F_1(x_1, 0) = 0, \tag{A.9a}$$

$$F_2(x_1, 0) \geqslant 0, \; x_1 \geqslant 0. \tag{A.9b}$$

One way to compute a solution is to solve (A.9a) for all possible partitionings, and choose the solution which satisfies (A.9b). This may be an efficient procedure when the dimension of x_2 is known to be very small. Alternatively, one could apply a continuation method as described in theorem 8.1 (or theorem 8.2)[36] and define for a given kth partitioning the homotopy

$$F_1^k(x_1^k, 0) + (t-1)F_1(x_{1,0}^k, 0) = 0, \qquad t \in [0, 1], \; x_{1,0}^k \text{ given.}$$

As in theorem 8.1 this defines a differential equation that can be followed until a (group of) element(s) of $x_1^k(t)$ or of $F_2^k(x_1^k(t), 0)$ becomes zero. Then one performs a pivot. That is, one keeps fixed the element(s) that reached zero, and this defines a new partitioning. Now, if one can prove that the choice of the partitioning is unique and that with the new partitioning it is feasible to increase t further (i.e., that this will lead away from the active bound), then under the assumptions of theorem 8.1 this procedure will eventually lead to a solution.[37] This approach is practical when the number of variables in x_2 is relatively large (e.g., when many prices are given).

Solution by Nonlinear Programming in GAMS

Since powerful and user-friendly nonlinear programming packages such as GAMS (Brooke et al. 1992) have become available, many AGE modelers have chosen to apply these tools, even though these are not guaranteed to

35. Specification of the NCP depends on which differentiability properties one wants to maintain. Consider, for example, the competitive model of pure exchange in which consumers have a twice continuously differentiable utility function. The equilibrium may be defined in excess demand format as $-z(p) \geqslant 0 \perp p \geqslant 0$. This is an NCP, but excess demand $z(p) = \Sigma_i x_i(p) - \Sigma_i \omega_i$ may not be continuously differentiable everywhere because $x_i(p)$ will usually not be differentiable at a point where $x_{ik} = 0$ for some commodity k. One can recover differentiability by treating the Kuhn-Tucker conditions $(\lambda_i p - \partial u_i(x_i)/\partial x_i) \geqslant 0 \perp x_i \geqslant 0$ of the consumer problem as constraints of the NCP rather than the consumer demand functions.

36. See Keyzer (1989) and Elbers (1992).

37. This of course does not establish uniqueness. However, if it can be proved that the solution curve $x(t, x_0)$ is continuous for all x_0, then there exists only one regular solution. Suppose that there are two, one reached from $x_0 = a_0$ and another from b_0. Since both solutions are regular, they must be isolated (theorem A.7.6). Since $x(t, x_0)$ is single valued, it will, for given x_0, lead to only one of the two equilibria. Finally, since $x(t, x_0)$ is continuous, it cannot lead to isolated solutions. Hence $x(t, x_0)$ can only lead to one regular solution, though we know that it could lead to every solution point for x_0 equal to that point. Therefore there is only one such point.

yield an equilibrium solution. This means that heuristic interventions can seldom be avoided, and to use the words of the GAMS authors, "as in all cases of difficulties with nonlinear programming, care and patience are needed."

The most direct approach (which is implemented in the CGE model for Cameroon that is supplied as a standard example) would be to treat the NCP as a system of nonlinear constraints in a program with an arbitrary "dummy" objective, say, minimizing $\|x\|$. Obviously this is not a convex program, and it has been our experience that if the initial values are not very close to the equilibrium solution, the GAMS nonlinear (convex) program algorithm (NLP) is often unable to find a solution, even in the Cameroon example, and terminates with the diagnostic that the program is infeasible, unbounded, or that the basis is structurally singular. In such cases the following scheme has proved more effective:

$$\min x \cdot y + \|s^l\| + \|s^u\|,$$

$$x, y, s^l, s^u \geqslant 0,$$

subject to

$$y = F(x) + s^u - s^l,$$

where s^l and s^u are artificial slacks. Using the norms $\|\cdot\|_1$ (sum of absolute values) and $\|\cdot\|_\infty$ (largest absolute value) is suggested to avoid creation of local optima and loss of effectivity in the neighborhood of the optimum. These norms should also include a scaling factor, to ensure that all contributions to the objective are of a similar order of magnitude. For linear constraints it is usually safe to drop the slack variables (slacks should, in general, be dropped whenever this does not lead to infeasibility). Leaving too many slacks makes it difficult to avoid scaling problems and will often cause the algorithm to end with the diagnostic that the problem is "unbounded (or badly scaled)."

Note that whenever the original NCP is known to possess a solution, the objective value will be zero in any global optimum (a solution of the NCP). However, since the program is not convex, it could happen that the NLP algorithm stops earlier, at a point where the objective is still positive. In practice, this difficulty can often be tackled by some tricks, of which we mention the following:

1. *Change the norm in the objective.* It is often effective to change the choice of norm, say, every thousand iterations (e.g., switching between the 1-norm and the ∞-norm) and possibly also the scaling.

2. *Solve a simpler model first.* When the model is known to possess a subset of equations that is particularly hard to solve, say, some bound or terminal

condition, it is advisable to proceed in rounds, allowing for some violation of these constraints in the first round in order to obtain good starting values for the next round.

3. *Change format.* With a convex programming package it is relatively easy to solve a welfare program or a mathematical program as defined in chapter 3. Hence it is often practical, especially when the dimensions of the problem are large (as in the dynamic models of chapters 7 and 8), to initialize by solving such programs at fixed welfare weights and final demand, respectively.

4. *Normalize all prices.* Whereas in AGE models quantities tend to remain bounded via the specification of the production technology, marginal values (and hence prices) may become very large, especially when the model contains CES or Cobb-Douglas forms, which have infinite slopes along the axes (as discussed in section 4.4.1). Though prices may be finite in the equilibrium, the NLP algorithm may proceed along a path where prices tend to become very large. This will in turn have an effect on incomes, demand, and so on, and often it leads to the diagnostic that the problem is unbounded or infeasible. This result can be avoided by normalizing all prices on the simplex.

5. *Impose bounds.* Excessively large values can also be avoided by imposing bounds. However, artificial bounds should be used with great care because they can make the program infeasible. If they happen to be binding in the final solution, this will not solve the original NCP.

One further advantage of using a nonlinear programming algorithm is that it becomes relatively straightforward to calibrate and tune the model. Let $\beta \in B$ denote the vector of unknown parameters, such as the exponents and the constant of a Cobb-Douglas form, B a given compact set (defined on the basis of preliminary parameter estimations), and let $L(x - \hat{x}, y - \hat{y})$ be the density function that measures the likelihood of a deviation of the model value (x, y) from the vector of observations $(\hat{x}, \hat{y})$. If we attribute a sufficiently small relative weight γ to this density function, the other terms in the objective function will be zero in any global optimum, so this global optimum will be an equilibrium. The function F is then dependent on both x and β, and the program reads as

$$\min x \cdot y + \|s^u\| + \|s^l\| - \gamma L(x - \hat{x}, y - \hat{y}),$$

$$x, y, s^l, s^u \geqslant 0, \beta \in B,$$

subject to

$$y = F(x, \beta) + s^u - s^l.$$

This program of course is usually highly nonconvex. More important, experience suggests that unless the constraint set B is restricted sufficiently, the parameter values tend to drift away to implausible values. One way to avoid this problem is to include a priori bounds on elasticities. This amounts to extending the set of constraints (and once locally optimal estimates β^* have been obtained, these values can be used as given constants in simulations of policy alternatives). The procedure has the advantage that it enables the AGE modeler to perform parameter estimation, base year calibration, as well as dynamic tuning, all within a single program.

B A GAMS Application

The GAMS software package makes it relatively easy to transform mathematical specifications of AGE models into executable computer programs. This was already shown in section 3.6.4 for the basic CGE prototype. Here we perform a more elaborate exercise. Although the model that we will implement is also a prototype—in the sense that its parameters have arbitrary numerical values, the dimensions are small, and the functional forms kept as simple as possible—it combines the main characteristics of the models in chapters 4–7: international trade with the import price exceeding the export price, transport and processing margins, taxes and tariffs, and price rigidities supported through buffer stocks. The dynamics will be of the recursive type, but we will indicate how the program can be modified for simulation of a *T*-period equilibrium model. This appendix, however, is not intended as a tutorial to the GAMS language. For an introduction to GAMS, see Brooke et al. (1992). The application presented here is available for distribution via Internet. See Keyzer (1997).

B.1 Defining the Model in GAMS

A Sketch of the Recursively Dynamic Model

The model distinguishes four consumers, indexed I, and seven commodities, indexed K, of which there are three goods, indexed G, and four factors, indexed F. Of the four factors two can be traded internationally (and are indexed FT). Goods are produced under constant returns to scale with CES-type technology. Factors are not produced. Consumer demand is specified according to a linear expenditure system. Indirect taxes and tariffs can be levied at different rates on consumption, input, production, imports, and exports. Price wedges are also created by transport and processing margins. For the two tradable factors there are price rigidities due to the perfectly elastic import and export or to buffer stock adjustment. Finally, maximum quotas and fixed trade commitments can be imposed on imports and exports.

The regime in which the equilibrium is found will determine whether price bounds are effective.[1] A recursively dynamic simulation (covered in section 7.2.3) is conducted for three consecutive years, with an exogenous shift in factor endowments and stock updates as the only dynamic adjustments. For simplicity, we disregard savings and investments, though, as discussed in section 7.4.1, the direct taxes and the committed consumptions can be interpreted as savings and investments, respectively. The closure is assured via the adjustment of direct taxes, as in the CGE model of section 5.4.1.

We focus on recursive dynamics because, as was explained in section 7.2, the *T*-period competitive equilibrium merely amounts to a relabeling of

1. See the discussion in section 6.2.1 and figure 6.1.

commodities and equations in the static model. We will indicate how this relabeling can be implemented.

Index Sets

The application starts with a definition of the main index sets and subsets (I for consumers and K for commodities). Sets should be declared before their subsets. Every declaration consists of a logical name, a label field (optional), followed by a list of elements of the index set (optional). The logical name can be used in computation, and the label field in report writing. A subset, say, I of IT is declared as I(IT). Several sets may be defined within one SET statement, which, as all GAMS statements, ends with a semicolon.[2]

```
  *== Dollar controls to suppress some listings == *

$OFFSYMXREF
$OFFDOLLAR
$INLINECOM { }

*== Index sets ==*

* the set DEF defines Labels to initialize the ordering of various sets
* the meaning of these labels will be specified later on

    SET DEF /
        'CONSUMPT. ', 'CONS.SUB. ', INPUT, 'END STOCK',
        'INP.SUB. ' , EXPORT, 'EXP.SUB. ' ,
        PROCESSING, PROCESSD, AVAILABLE, SUPPLY, 'SUP.TAX' ,
        PROCESSM, IMPORT, 'IMP.TAX' , 'INIT.STOCK' , IMBALANCE,
        PRICE, 'EQU.PRICE' ,
        'EXP.PR.CL. ', 'IMP.PR.CL. ' , 'EXP.PR.BD. ' , 'IMP.PR.BD. ' ,
        'EXP.PR.TR. ' , 'IMP.PR.TR. ' , 'PUR.PR.ST. ' , 'SEL.PR.ST. ' ,
        ENDOWMENT, FOOD, TEXTILE, TRADE,
        FARMERS, WORKERS, EMPLOYERS, HOUSEHOLD, EXPORTER,
        EXPORTS, IMPORTS, PRIVATE, GOVERNMENT, NATIONAL,
        'GDP FC' , 'GDP MP' , 'TRADEDEF. ' , NETINDIRT, NETDIRT,
        NETSTOCK, TRANSFERS, 'FOR.PROFIT' , REVENUE,
        'UNC.INCOME' ,  ' ** GAP ** '
            /

    IT      consumer aggregates
            / FARMERS            " Farmers"
              WORKERS            " Workers"
              EMPLOYERS          " EMPLOYERS"
              PRIVATE           " Private"
              GOVERNMENT      " Government"
              NATIONAL        "National"
```

2. Comment lines are denoted by an * in the first column or kept between curled parentheses { }. A $ sign is used for the so-called $-control commands ($TITLE, etc.), here at the beginning of the program. A $-controlled command does not end with a semicolon.

```
        /
I(IT)   consumers
        / FARMERS, WORKERS, EMPLOYERS, GOVERNMENT/

IP(I)   private consumers
        / FARMERS, WORKERS, EMPLOYERS/

KT      all commodities
        / FOOD      " Food"
          TEXTILE   " Textile"
          TRADE     " Trade"
          LABOR     " Labor"
          EQUIP     " Equipment"
          FUEL      " Fuel"
          MACH      " Machinery"
          TOTAL     "Total"
        /

K(KT)   commodities
        / FOOD, TEXTILE, TRADE, LABOR, EQUIP, FUEL, MACH /

KN(K)   commodities excluding equipment
        / FOOD, TEXTILE, TRADE, LABOR, FUEL, MACH /

G(K)    goods
        / FOOD, TEXTILE, TRADE /

GPR(G) trade good
        / TRADE /

F(K)    factors
        / LABOR, EQUIP, FUEL, MACH /

FT(F)   tradable factors
        / FUEL, MACH /

FNT(F)  nontradable factors
        / LABOR, EQUIP /

L       levels in processing
        / CONSUMER, INPUT, PRODUCER, IMPORT, EXPORT/

LN(L)   consumer and input level in processing
        / CONSUMER, INPUT /

TALL    potential entries for dynamics
        / 1980*2020, GROWTH /

;
  ALIAS  (G,GP);
  ALIAS  (K,KP);
```

The ALIAS statement defines an alternative name for an index set (subscript). It is used, for example, in matrix operations when summation is to be performed over two subscripts of a single index set:

```
ALIAS (K,KP);
PARAMETERS Z, A0(K,KP), B(KP);
Z = SUM(K, SUM(KP, A0(K,KP)*B(KP)));
```

Parameter Declaration and Initialization

The example of the ALIAS-statement illustrates how parameters can be declared through a PARAMETERS statement. Parameters have a numerical, real value (as opposed to character, integer, or logical). Parameters are declared and initialized in the next segment of the application:

```
*== Parameters and Tables: declare and initialize ==*

PARAMETERS
        AO(K,G)        input output constants
        BETA(I,K)      budget shares (fixed: Cobb Douglas utility)
        COMMIT(I,K)    commited consumption
        ECOM(F)        committed exports (fixed quantity)
        EUP(FT)        quota on exports
        MCOM(F)        committed imports
        MUP(FT)        quota on imports
        OMEGA(I,F)     endowments
        OMEGTOT(F)     total endowments
        PE(F)          export price (f.o.b.)
        PL(FT)         purchasing price of buffer stock
        PM(F)          import price (c.i.f.)
        PNORM          price norm
        PROFW(I)       profits from abroad
        PROFWE         total profit outflow ( = trade surplus)
        PU(FT)         selling price of buffer stock
        QSL(FT)        upper bound on stock purchase
        QSO(F)         initial stocks
        SIGMA(G)       input output CES-substitution elasticities
        TAU(K,L)       rates of indirect tax
        TAUT(I)        shares to distribute direct taxes
        TRAD(K,L)      processing requirements (transport coefficients)
;

TABLE A0(K,G) input output constants

              FOOD       TEXTILE      TRADE
FOOD          0.0          0.0          0.0
TEXTILE       0.0          0.0          0.0
LABOR         0.2          0.4          0.1
EQUIP         0.1          0.1          0.1
FUEL          0.01         0.2          0.1
MACH          0.1          0.1          0.1
;
```

```
PARAMETER SIGMA(G) input output CES-substitution elasticities

                  / FOOD 2., TEXTILE .5, TRADE .2 /;

TABLE BETA(I,K) budget shares (fixed: Cobb Douglas utility)

                  FOOD   TEXTILE   LABOR   EQUIP   FUEL   MACH
FARMERS           .461      .239    .130    .170
WORKERS           .532      .128    .180    .160
EMPLOYERS         .263      .137    .080    .420   .100
GOVERNMENT                          .200    .800
;
* check summation to unity
PARAMETER BETATOT(I);
BETATOT(I) = SUM(K, BETA(I,K));
DISPLAY BETATOT;

TABLE COMMIT(I,K) committed consumption

                 FOOD     TEXTILE      MACH
FARMERS          0.00        0.00
WORKERS          0.00        0.00
EMPLOYERS        0.00        0.00
;

TABLE OMEGA(I,F) factor endowments

                 LABOUR     EQUIP     FUEL      MACH
FARMERS          110.00     10.00     0.00      0.00
WORKERS           20.00     10.00     0.00      0.00
EMPLOYERS         10.00    120.00   110.00    210.00
GOVERNMENT                 100.00   100.00    200.00
;
* total endowments

OMEGTOT(F) = SUM(I,OMEGA(I,F));

PARAMETERS
PROFW(I)  / FARMERS        12.0, WORKERS   3.0, EMPLOYERS 100.0,
            GOVERNMENT  110.0/
TAUT(I)   / FARMERS         0.0, WORKERS   0.0, EMPLOYERS   1.0,
            GOVERNMENT    0.0/
;
PROFWE  =  -SUM(I, PROFW(I));

PARAMETERS for import export quantities

ECOM(F) /LABOR   0., EQUIP   0., FUEL      0., MACH      0./
EUP(FT)                         /FUEL 2000., MACH  2000./
MCOM(F) /LABOR   0., EQUIP   0., FUEL      0., MACH      0./
MUP(FT)                         /FUEL 2000., MACH  2000./
;
```

```
PARAMETERS for prices

PE(F)   /LABOR     0., EQUIP     0., FUEL    50.,  MACH    80./
PL(FT)                               /FUEL    40.,  MACH    70./
PM(F)   /LABOR  1000., EQUIP  1000., FUEL   120.,  MACH   100./
PU(FT)                               /FUEL   100.,  MACH    80./
;

PARAMETERS for stocks

QSL(FT)                              /FUEL    0., MACH    0./
QSO(F)  /LABOR     0., EQUIP     0., FUEL   10., MACH   20./
;
```

TABLE TAU(K,L) rates of indirect tax

	CONSUMER	INPUT	PRODUCER	IMPORT	EXPORT
FOOD	0.00	0.00	0.00	0.00	0.00
TEXTILE	0.00	0.00	0.00	0.00	0.00
TRADE	0.00	0.00	0.00	0.00	0.00
LABOR	0.00	0.00	0.00	0.00	0.00
EQUIP	0.00	0.10	0.00	0.00	0.00
FUEL	0.10	0.00	0.00	0.10	0.00
MACH	0.00	0.00	0.00	0.00	0.10

;

TABLE TRAD(K,L) processing requirements (transport coefficients)

	CONSUMER	INPUT	PRODUCER	IMPORT	EXPORT
FOOD	0.10	0.00	0.00	0.00	0.00
TEXTILE	0.10	0.00	0.00	0.00	0.00
TRADE	0.00	0.00	0.00	0.00	0.00
LABOR	0.00	0.00	0.00	0.00	0.00
EQUIP	0.00	0.00	0.00	0.00	0.00
FUEL	0.00	0.00	0.00	0.10	0.00
MACH	0.00	0.00	0.00	0.10	0.00

;

Values of Derived Parameters

The GAMS language owes a great deal of its practical value to the fact that it does not distinguish between index sets and subscripts. This makes it possible to write statements in a very compact form, such as

```
C(I) = A(I) + B(I);
```

This means that the parameter C(I) is computed as the sum of parameters A(I) and B(I), for every consumer in the set I.

In the recursive dynamic formulation there is no need to distinguish symbols (parameters, variables, equations) by time period because the model generates a sequence of single-period equilibria. For a T-period model one should declare an index set for time periods and declare parameters, say, as

```
SET time periods /1982*1984/;
PARAMETER A0(K,KP,T);
```

There are two practical ways of initializing such parameter values if one does not want to enter these directly for every period:

1. *Calculate as time trends.* For this one can use the ORD function on (ordered) sets, which computes the position of member T in the set.[3]

```
A0(K,KP,T) = B0(K,KP)*ORD(T) + B1(K,KP);
```

2. *Use a loop statement.* The LOOP statement (a loop is opened and closed by parentheses) takes the form

```
LOOP(T, A0(K,KP,T + 1) = A0(K,KP,T) + B1(K,KP));
```

A loop-statement can also be used to program iterative, algorithmlike procedures.

Variables and Equations

All symbols belonging to the list of choice variables in the mathematical program should be declared as VARIABLES, not as PARAMETERS. Every declaration has, as for parameters, a logical name followed by a label field (optional). Equations are also denoted by symbols. Hence every equation can be referred to by its logical name. For example, for every good G, the balance of goods is

```
BALGOOD(G) .. SUM(I, X(I,G) + SUM(GP, A(G,GP)*Q(GP))
                              + TPROC$GPR(G) = E = Q(G);
```

where the "=E=" represents an equality sign; greater-or-equal is written as =G=, smaller-or-equal as =L=. The meaning of this equation is straightforward, except for the dollar expression $GPR(G). This expression indicates that the value of the variable TPROC should be added only if the expression that follows is true, that is, if the specific element of G belongs to GPR (members of index sets are treated as logical variables that are true if the set contains them). A grouping of equations that constitutes a mathematical program is called MODEL and is also given a name. The same equation may appear in different models.

```
*== Variables, equations and model definition ==*

POSITIVE VARIABLES
  A(K,G)          input output matrix
  E(F)            export quantity
  H(I)            uncommitted income
```

3. To compute the number of elements in the set T, one can use the CARD(T) function.

```
  M(F)             import quantity
  MUE(FT)          price margin on exports
  MUSL(FT)         price margin when buffer stock purchase is zero
  MUSU(FT)         price margin when buffer stock sale is zero
  MUM(FT)          price margins on imports
  MUEB(FT)         price margin when export quota is effective
  MUMB(FT)         price margin when import quota is effective
  MUSLB(FT)        price margin when buffer stock purchase at upper bound
  MUSUB(FT)        price margin when buffer stock purchase at lower bound
  P(K)             clearing price
  PT(K,L)          transaction prices (consumer input export import)
  Q(G)             production of goods quantity
  RHO              scaling factor to normalize prices
  SL(F)            buffer stock purchase
  SU(F)            buffer stock sale
  X(I,K)           consumption quantity
  ZAM(K,G)         input demand deficit (slack)
  ZAP(K,G)         excess on input demand (slack)
  ZPM(G)           price deficit (slack)
  ZPP(G)           excess on price (slack)
  ZXM(I,K)         consumption deficit (slack)
  ZXP(I,K)         excess consumption (slack)
;
VARIABLES
  TMAX             objective function (max-of-deviations)
  TMAX1            contribution to objective direct bounds on prices
  TMAX2            contribution to objective (due to bounds on quantities)
  TMAX3            contribution to objective (slack on pricing of goods)
  TMAX4            contribution to objective (slack on input output table)
  TMAX5            contribution to objective (slack on consumption)
                   processing requirement
  TSUM             objective function (sum-of-deviations)
  TX               tax adjustment
;

* Equations: declaration

EQUATIONS
  ACAL(K,G)        input output function
                   {  compute the input output coefficient using
                      Shephard's lemma for CES production function  }

  BALGOOD(G)       balance of goods

  BALFAC(F)        balance of factors

  DPROC            processing demand
                   {  intermediate demand for inputs into the trade
                        margins  }

  INC(I)           income equation of consumers
                   {  uncommitted income = value of endowments
                                          +  foreign profits
                                          +  transfers
                                          -  committed demand  }
```

PAYBAL	balance of payments { 0 = value of imports + value of exports − foreign outflow }
PECAL(FT)	export price { domestic price at clearing level = scaled tariff ridden export price − trade margin on export + premium if there is import − premium if quota is effective }
PLCAL(FT)	stock purchasing price { domestic price at clearing level = scaled stock purchasing price + premium if stock purchase at lower bound − premium if stock purchase at upper bound }
PMCAL(FT)	import price { domestic price at clearing level = scaled tariff ridden import price + trade margin on import − premium if there is export + premium if quota is effective }
PRIGOOD(G)	markup pricing of goods { price of a good = unit cost of inputs + producer tax }
PRITR(K,LN)	transaction prices { consumer price = clearing price + consumer processing + proportional consumer tax }
PRINORM	price normalization { value of total endowments is fixed }
PUCAL(FT)	stock selling price { domestic price at clearing level = scaled stock selling price + premium on price if stock sale at lower bound − premium on price if stock sale at upper bound }
TMAX1CAL	calculate TMAX1 { complementary slacks on lower bounds for trade and stock changes }
TMAX1OBJ	contribution of TMAX1
TMAX2CAL	calculate TMAX2 { complementary slacks on upper bounds for trade and stock changes }
TMAX2OBJ	contribution of TMAX2
TMAX3CAL	calculate TMAX3 { slacks on pricing of goods }

```
TMAX3OBJ        contribution of TMAX3

TMAX4CAL        calculate TMAX4
                {  slacks on input coefficients  }
TMAX4OBJ        contribution of TMAX4

TMAX5CAL        calculate TMAX5
                {  slacks on consumption  }
TMAX5OBJ        contribution of TMAX5

TSUMOBJ         calculate TSUM

XCAL(I,K)       consumption function
                {  uncommitted part of linear expenditure system  }
;

ACAL(K,G)$(A0(K,G) GT 0) ..
                A(K,G) + ZAP(K,G) =E= ZAM(K,G) + A0(K,G)
                 *((P(G)/(1 + TAU(G, 'PRODUCER')))
                 /PT(K, 'INPUT'))**SIGMA(G);

XCAL(I,K)     .. X(I,K) + ZXP(I,K) =E= COMMIT(I,K) + ZXM(I,K)
                                   +BETA(I,K)*H(I)/PT(K,'CONSUMER');


DPROC         ..  =E= SUM(K, SUM(I, TRAD(K,'CONSUMER')*X(I,K)))
                        + SUM(K, SUM(G, TRAD(K,'INPUT')*A(K,G)*Q(G)))
                        + SUM(F, TRAD(F,'IMPORT')*M(F))
                        + SUM(F, TRAD(F,'EXPORT')*E(F));

BALGOOD(G)    .. SUM(I, X(I,G)) + SUM(GP, A(G,GP)*Q(GP))
                 + TPROC$GPR(G) =E= Q(G);

BALFAC(F)     .. SUM(I, X(I,F)) + SUM(GP, A(F,GP)*Q(GP)) + E(F) + SL(F)
                                   =E= OMEGTOT(F) + M(F) + SU(F);

INC(I)        .. H(I) =E= SUM(F, P(F)*OMEGA(I,F)) + TAUT(I)*TX
                       + RHO*PROFW(I)
               - SUM(K, PT(K, 'CONSUMER')*COMMIT(I,K));

PAYBAL        .. 0. =E= SUM(FT, PM(FT)*M(FT)-PE(FT)*E(FT)) + PROFWE;

PRITR(K,LN)   .. PT(K,LN) =E= (P(K)+P('TRADE')*TRAD(K,LN))
                             *(1+TAU(K,LN));

PRIGOOD(G)    .. P(G) + ZPP(G) =E= ZPM(G) +
                 (1+TAU(G, 'PRODUCER'))*(SUM(GP,PT(GP, 'INPUT')*A(GP,G))
                                   + SUM(F, PT(F, 'INPUT')*A(F,G)));

PECAL(FT)     .. P(FT) =E= RHO*PE(FT)*(1-TAU(FT, 'EXPORT'))
                        - P('TRADE')*TRAD(FT, 'EXPORT')
                        + MUE(FT) - MUEB(FT);

PMCAL(FT)     .. P(FT) =E= RHO*PM(FT)*(1+TAU(FT, 'IMPORT'))
                        + P('TRADE')*TRAD(FT, 'IMPORT')
                        - MUM(FT) + MUMB(FT);
```

```
PLCAL(FT )    .. P(FT) =E= RHO*PL(FT) + MUSL(FT) - MUSLB(FT);

PUCAL(FT)     .. P(FT) =E= RHO*PU(FT) - MUSU(FT) + MUSUB(FT);

PRINORM       .. SUM(F, P(F)*OMEGTOT(F)) =E= PNORM;

TMAX1CAL      .. TMAX1 =E= SUM(FT, .99*MUM(FT)*(M(FT)-M.LO(FT))
                                 + MUE(FT)*(E(FT)-E.LO(FT))
                         + SUM(FT, .99*MUSU(FT)*(SU(FT)-SU.LO(FT))
                                 + MUSL(FT)*(SL(FT)-SL.LO(FT)));

TMAX2CAL      .. TMAX2 =E= SUM(FT, MUEB(FT)*(E.UP(FT)-E(FT)))
                         + SUM(FT, NUMB(FT)*(M.UP(FT)-M(FT)))
                         + SUM(FT, MUSLB(FT)*(SL.UP(FT)-SL(FT)))
                         + SUM(FT, MUSUB(FT)*(SU.UP(FT)-SU(FT)));

TMAX3CAL      .. TMAX3 =E= 1000*SUM(G, ZPP(G) + ZPM(G));

TMAX4CAL      .. TMAX4 =E= 1000*SUM(G, SUM(K, ZAP(K,G)+ZAM(K,G)));

TMAX5CAL      .. TMAX5 =E= 10*SUM(I, SUM(K, ZXP(I,K)+ZXM(I,K)));

TMAX1OBJ      .. TMAX =G= TMAX1;
TMAX2OBJ      .. TMAX =G= TMAX2;
TMAX3OBJ      .. TMAX =G= TMAX3;
TMAX4OBJ      .. TMAX =G= TMAX4;
TMAX5OBJ      .. TMAX =G= TMAX5;

TSUMOBJ       .. TSUM E= TMAX1 + TMAX2 + TMAX3 + TMAX4 + TMAX5;

* Model definition

MODEL AGESUM /ACAL, XCAL, DPROC, BALGOOD, BALFAC,
             INC, PAYBAL,
             PRITR, PRIGOOD, PECAL, PMCAL, PMCAL, PLCAL, PUCAL, PRINORM
             TMAX1CAL, TMAX2CAL, TMAX3CAL, TMAX4CAL, TMAX5CAL,
             TSUMOBJ/;

MODEL AGEMAX /ACAL, XCAL, DPROC, BALGOOD, BALFAC,
             INC, PAYBAL,
             PRITR, PRIGOOD, PECAL, PMCAL, PLCAL, PUCAL, PRINORM,
             TMAX1CAL, TMAX2CAL, TMAX3CAL, TMAX4CAL, TMAX5CAL,
             TMAX1OBJ, TMAX2OBJ, TMAX3OBJ, TMAX4OBJ, TMAX5OBJ/;
```

In a *T*-period formulation, all time-specific variables and equations have to carry a time subscript; consumption is written X(I,K,T), but the price normalization PRINORM and the income INC(I) remain as they are, since they apply to the full period. Initial conditions are introduced by treating the associated variables as fixed (via the .FX suffix described below), and terminal conditions through equations that only apply for the last period. They are declared as

```
XCAL(I,K,T)$(ORD(T) EQ CARD(T)) ..
```

Initialization and Bounds

A GAMS variable is also characterized by a suffix:

.L Current level of the variable
.M Shadow price on the bound (not used here)
.LO Lower bound
.UP Upper bound
.FX Fixed (lower bound = upper bound)

Before a model is solved, it is necessary to initialize all choice variables (.L suffix) and all relevant bounds. Bounds are treated in the same way as parameters. The variables (.L values) keep their value from one solution to the next assignment. Unassigned upper bounds are set at plus infinity, and noninitialized lower bounds at minus infinity. While EQUATIONS are only executed when the MODEL to which they belong is solved, direct assignments (with = signs) are executed as they appear. In direct assignments variables are referenced with their suffixes. Here the initialization is made at arbitrary values in order to test the computational procedure. In empirical applications it is advisable to initialize the variables at their SAM values.

```
*== Impose bounds ==*

 A.LO(K,G) = .0; A.LO(K,G) = .0001$(A0(K,G) GT 0);
 A.FX(K,G)$(A0(K,G) EQ 0) = 0;
 P.LO(K) = .001; PT.LO(K,L) = .001;
 RHO.LO = .0001;

*== Initialize variables ==*

 A.L(K,G) = A0(K,G); A.L(K,G) = 1.$(A0(K,G) GT 0) ;
 H.L(I) = 10.;
 E.L(F) = .1; M.L(F) = .1;
 MUE.L(FT) = PM(FT) - .5*PE(FT); MUEB.L(FT) = 0.;
 MUM.L(FT) = 0 ; MUMB.L(FT) = 0.; MUSL.L(FT) = 0.; MUSLB.L(FT) = 0.;
 MUSU.L(FT) = 0.; MUSUB.L(FT) = 0.;
 P.L(K) = 10.; P.L(FT) = .5*(PM(FT) +PE(FT)); PT.L(K,LN) = P.L(K);
 PNORM = SUM(FT, P.L(FT)*OMEGTOT(FT))/10;
 Q.L(G) = 10.;
 SL.L(FT) = 0; SU.L(FT) = 0;
 TMAX1.L = 100.; TMAX2.L = 100.; TMAX3.L = 100.; TMAX4.L = 100.;
 TMAX5.L = 100.; TMAX.L = 100.; TSUM.L = 0;
 TPROC.L = 0.;
 TX.L = 0.;
 X.L(I,K) = 10;
 ZAM.L(K,G) = 1.; ZAP.L(K,G) = 1.;
 ZPM.L(G) = 1.; ZPP.L(G) = .1;
 ZXM.L(I,K) = 1.; ZXP.L(I,K) = 1.;

* derived values

 H.LO(I) = -30*PNORM;
```

```
  TX.UP  =  60000.*PNORM;
  TX.LO  =  -60000.*PNORM;

* bounds from parameters

E.LO(FT)  =  ECOM(FT);
E.FX(FNT)  =  ECOM(FNT);
E.UP(FT)  =  EUP(FT);
M.LO(FT)  =  MCOM(FT);
M.FX(FNT)  =  MCOM(FNT);
M.UP(FT)  =  MUP(FT);
SL.FX(FNT)  =  QSO(FNT);
SL.UP(FT)  =  QSL(FT);
SU.FX(FNT)  =  0.;
SU.UP(FT)  =  QSO(FT);
```

Controlling the Period of Simulation

One should start a simulation run by specifying the name of the scenario (here, Baserun) and the years for which it applies (the sets T(TALL) and T(TT) below). In printing the report, any of these years can be selected. The set TSEL(TALL) below lists the potential candidates. In the set TCOR1(TALL) members are included if they belong to T; in the set TCOR2(TALL) members are included if they belong to the intersection of TCOR1(TALL) and TSEL(TALL) or to TGR(TALL). The intersection operator is denoted by "*" and the "or" by "+".

```
*== Define sets and parameters for reporting ==*

  PARAMETERS LRUN       LENGTH OF RUN
             RUNOPT     RUN OPTION
             STYR       STARTING YEAR
  ;

$TITLE Baserun
  LRUN=003       ;
  STYR=1982      ;
  RUNOPT=1       ;

  SET
      TSEL(TALL)     years to be selected if computed
                     /1980*1995, 2000, 2005, 2010, 2015, 2020, GROWTH/

      TGR(TALL)      percentage growth rate
                     /GROWTH/

      TCOR1(TALL)    intermediate correspondence 1
      TCOR2(TALL)    intermediate correspondence 2
      TS(TT)         selected years for report

      TFIRST(T)      first year
      TLAST(T)       last year
  ;

* initialize the selection of time periods

  TCOR1(TT)   =  YES$T(TT);
```

```
TCOR2(TALL) = YES$(TCOR1(TALL)*TSEL(TALL) + TGR(TALL));
TS(TT) = YES$TCOR2(TT);

TFIRST(T) = YES$(ORD(T) = 1);
TLAST(T) = YES$(ORD(T) = CARD(T));
```

Index Sets and Parameters for Accounting

The last step in preparing the model is to define the index sets and parameters that will be needed to perform the accounting and to store the results. For this we make use of two-dimensional index sets. For example, the expression:

```
A(J) = SUM(I$M(I,J), B(I));
```

means that $A_j = \Sigma_{i \in I_j} B_i$, where the set I_j is defined so that i belongs to it whenever the mapping between i and j is "true" (i.e., M(I,J) holds). Below we define two-dimensional index sets such as CI(I, IT) to specify the aggregation mapping from I to IT; the qualifier $CI(I,IT) will map any given member of I not only to its equivalent in IT but also to all relevant subtotals and totals in IT.

```
* Define other sets and mappings

SET   CI(I,IT)       Correspondence between I and IT

      GK(KT)         Subset of goods
                   / FOOD, TEXTILE, TRADE, TOTAL /

      CK(K,KT)       Correspondence between K and KT

      GT             Sectors
                   / FOOD          " Food"
                     TEXTILE       " Textile"
                     TRADE         " Trade"
                     HOUSEHOLD     " Self employed"
                     NATIONAL      National
                   /
      CG(G,GT)       Correspondence between G and GT
                   / FOOD.FOOD, TEXTILE.TEXTILE, TRADE.TRADE/

      NT1            Items in TABLE1
                   / 'CONSUMPT.'       Consumption
                     'INPUT'           Input
                     'EXPORT'          Export
                     'END STOCK'       Final stock
                     'PROCESSING'      Processing
                     'AVAILABLE'       Available
                     'SUPPLY'          Supply
                     'IMPORT'          Import
                     'INIT.STOCK'      Initial stock
                     'IMBALANCE'       Imbalance
```

```
                     'EQU.PRICE'        Clearing price
                     'EXP.PR.CL.'       Export lower price bound
                     'IMP.PR.CL.'       Import upper price bound
                     'EXP.PR.BD.'       Export price (tar.inc.)
                     'IMP.PR.BD.'       Import price (tar.inc.)
                     'EXP.PR.TR.'       Export price f.o.b.
                     'IMP.PR.TR.'       Import price c.i.f.
                     'PUR.PR.ST.'       Stock lower price bound
                     'SEL.PR.ST.'       Stock upper price bound
                   /
NT2                  Items in TABLE2
                   / 'CONSUMPT.'        Consumption
                     'CONS.SUB.'        Consumer subsidy
                     'INPUT'            Input
                     'INP.SUB.'         Input subsidy
                     'PROCESSD'         processing requirement
                     'EXPORT'           Exports
                     'EXP.SUB.'         Export subsidy
                     'END STOCK'        Final stock
                     'AVAILABLE'        Available
                     'PROCESSM'         Processing margins
                     'SUPPLY'           Supply
                     'SUP.TAX'          Producer tax
                     'IMPORT'           Imports
                     'IMP.TAX'          Import tax
                     'INIT.STOCK'       Initial stock
                   /
NT4                  Items in TABLE4
                   / 'CONSUMPT.'         " Consumption"
                     'ENDOWMENT'         " Endowment"
                     'TRANSFERS'         " Transfers"
                     'FOR.PROFIT'        " Foreign profits"
                     'REVENUE'           "Revenue"
                     'UNC.INCOME'        " Uncommitted income"
                   /
NT3(NT4)             Items in TABLE3
                   / 'CONSUMPT.'    Consumption
                     'ENDOWMENT'    Endowment
                   /
NT5                  Items in TABLE5
                   / 'EXPORTS'      Exports
                     'IMPORTS'      Imports
                   /
NT7                  Items in TABLE7
                   / 'GDP   FC'     GDP at factor cost
                     'GDP   MP'     GDP at market prices
                    'TRADEDEF.'     Trade deficit
                    'NETINDIRT'     Net indirect taxes
                    'NETDIRT'       Net direct taxes
                    'NETSTOCK'      Net stock sales
```

```
                    '** GAP **'          Accounting discrepancy
                   /
 ;
* Define tables
 PARAMETERS tables for report
                 TABLE1 (K,NT1,TT)
                 TABLE2 (KT,NT2,TT)
                 TABLE3 (NT3,IT,KT,TT)
                 TABLE4 (NT4,IT,TT)
                 TABLE5 (NT5,KT,TT)
                 TABLE6 (GT,TT)
                 TABLE7 (NT7,TT)
                 ;
* Initialize mappings for aggregation

 CI(I,I) = YES;
 CI(I, 'PRIVATE') = YES;
 CI('GOVERNMENT', 'PRIVATE') = NO;
 CI('GOVERNMENT', 'GOVERNMENT') = YES;
 CI(I, 'NATIONAL') = YES;

 CK(K, 'TOTAL') = YES;
 CK(K,K) = YES;
 CG(G, 'NATIONAL')= YES;

* define post-equilibrium variables for accounting

PARAMETERS
 DFT(F)          1 for FT and 0 otherwise

 GDP             total GDP
 GDPS(G)         sectoral GDP

 PCLEAR(K)       absolute level of clearing price
 PCONS(K)        consumer price
 PEXC(FT)        export price at clearing level
 PEXP(F)         tariff inclusive price exports
 PIMC(FT)        import price at clearing level
 PIMP(F)         tariff inclusive price imports
 PINT(K)         input price
 PPROD(K)        producer price
 PPROC           price of processing commodity
 PPURC(K)        lower bound purchasing price of stock
 PSELL(K)        upper bound selling price of stock

 QCONS(K)        consumer demand quantity
 QEXP(K)         exported quantity
 QIMP(K)         imported quantity
 QINT(K)         intermediate demand quantity
 QPROC(K)        processing requirement
 QQ(K)           domestic supply
 QQ1(K)          domestic supply from endowments
```

```
QQ2(K)           domestic supply from production
QSTOCK(K)        end stock level
QSTOCK0(K)       initial stock level

SCONS(K)         subsidy on consumer price (per unit)
SEXP(F)                    exports
SIMP(F)                    imports
SINT(K)                    inputs
SPROD(G)                   producer price

TCONS(K)         tax       on consumer price (per unit)
TEXP(F)                    exports
TIMP(F)                    imports
TINT(K)                    inputs
TPROD(G)                   producer price

VCONS(K)         processing value consumer price (per unit)
VEXP(K)                    exports
VIMP(K)                    imports
VINT(K)                    inputs

TAXIND           indirect tax receipts (total)
TRADEF           trade deficit
TRANSFER(I)      transfers
TSTOCK           total expenditures on stocks
TTRANS           total transfers
;
DFT(FT) = 1;
```

B.2 Computation of Equilibrium

The GAMS statement to solve the mathematical program defined by the model AGESUM with objective TSUM, using the MINOS5 nonlinear programming algorithm NLP, reads as

```
SOLVE AGESUM USING NLP MINIMIZING TSUM;
```

For a recursive dynamic simulation one could use the LOOP statement. It will execute the SOLVE command for every period in the index set T:

```
LOOP(T, SOLVE AGESUM USING NLP MINIMIZING TSUM;)
```

To prevent the program from stopping before it has found an equilibrium, it is advisable, as discussed in appendix A.9 to include additional solve statements that will be executed until an equilibrium is found or an iteration limit hit. Here this is implemented by alternating between iterations with an objective TSUM that calculates the absolute sum of deviations and TMAX that uses the maximum of five types of deviations.

```
*== Recursive dynamic simulation and tabulation ==*

   AGEMAX.ITERLIM=5000;
   AGEMAX.LIMROW=0;
   AGEMAX.LIMCOL=0;
   AGEMAX.OPTFILE=1;

   AGESUM.INTERLIM=5000;
   AGESUM.LIMROW=0;
   AGESUM.LIMCOL=0;
   AGESUM.OPTFILE=1;

   FILE OPT Minos option file / MINOS5.OPT /
      PUT OPT;
      PUT 'BEGIN' /
           ' Superbasics limit 600     ' /
           ' Minor iterations  400     ' /
           ' Penalty factor 1.e4       ' /
           ' END' ;
      PUTCLOSE OPT;

   SCALAR TAROBJ;
   SET ITEROUT Iteration set /1*2/ ;

* :::::::::::::::::::::::::::::::::::::::::::::::::::::::
   LOOP(T,
        TAROBJ = 100;
        LOOP(ITEROUT$(TAROBJ GT 10*EPS),

* X.a :::::::::::::::::: call NLP-algorithm ::::::::::::::

* Phase 1
              SOLVE AGESUM USING NLP MINIMIZING TSUM;
              TAROBJ = TSUM.L;
* Phase 2
              IF(TAROBJ GT EPS,
              SOLVE AGEMAX USING NLP MINIMIZING TMAX;
              TAROBJ = TMAX.L;
            );
        );
```

Phases can be defined in many other ways, including a rescaling of the terms in the objective. The application described here actually converges in phase 1 already if the MINOS/GAMS parameters are set in the way specified by the PUT statement (see also section B.4 for further details on the PUT statement). If default values are used for the MINOS options, several rounds are needed.

B.3 Accounting

When an equilibrium solution has been computed, the results have to be stored in the matrices TABLE1–TABLE7 for tabulation of the SAM accounts (commodity balances, prices, consumer budgets, etc.) in the report section. Note (by counting the number of closing brackets) that the T loop has not been closed as yet. Hence the results for tabulation will be stored for every year of simulation. Since this part of the program does not involve any simultaneous system of equations, it can be executed line by line through mere calculations on parameters.

```
*::::::::::::::::::::::::::::::::::    Accounting    ::::::::::::::::::::::

* Aggregate quantities of commodities

  QCONS(K) = SUM(I, X.L(I,K));
  QEXP(FT) = E.L(FT);
  QIMP(FT) = M.L(FT);
  QINT(K) = SUM(G, A.L(K,G)*Q.L(G));
  QPROC('TRADE') = TPROC.L;
  QQ(F) = OMEGTOT(F);
  QQ(G) = Q.L(G);
  QQ1(F) = OMEGTOT(F);
  QQ2(G) = Q.L(G);
  QSTOCK0(F) = QSO(F);
  QSTOCK(F) = QSO(F) + SL.L(F) - SU.L(F);

* Prices (division by rho to return to nominal prices)

  PPROC = P.L('TRADE')/RHO.L;
  PCLEAR(K)  = P.L(K)/RHO.L;

  PCONS(K) = PT.L(K, 'CONSUMER')/RHO.L;
  VCONS(K) = PPROC*TRAD(K, 'CONSUMER');
  SCONS(K) = -MIN((PCLEAR(K)+VCONS(K))*TAU(K, 'CONSUMER'), 0);
  TCONS(K) =  MAX((PCLEAR(K)+VCONS(K))*TAU(K, 'CONSUMER'), 0);

  VEXP(F) = PPROC*TRAD(F, 'EXPORT');
  PEXC(FT) = PE(FT)*(1-TAU(FT, 'EXPORT')) - VEXP(FT);
  PEXP(F) = (PCLEAR(F) + VEXP(F))$ (QEXP(F) GT E.LO(F))
          + PE(F)*(1-TAU(F, 'EXPORT'))$ ((QEXP(F) EQ E.LO(F))
                                        AND (DFT(F) EQ 1));

  SEXP(F) = MAX(PEXP(F) - PE(F), 0);
  TEXP(F) = MAX(PE(F) - PEXP(F), 0);

  VIMP(F) = PROC*TRAD(F, 'IMPORT');
  PIMC(FT) = PM(FT)*(1+TAU(FT, 'IMPORT')) + VIMP(FT);
  PIMP(F) = (PCLEAR(F) - VIMP(F))$ (QIMP(F) GT M.LO(F))
          + PM(F)*(1+TAU(F, 'IMPORT'))$ ((QIMP(F) EQ M.LO(F))
                                        AND (DFT(F) EQ 1));

  SIMP(F) = MAX(PM(F) - PIMP(F), 0);
  TIMP(F) = MAX(PIMP(F) - PM(F), 0);

  PINT(K) = PT.L(K, 'INPUT')/RHO.L;
  VINT(K) = PPROC*TRAD(K, 'INPUT');
  SINT(K) = -MIN((PCLEAR(K) + VINT(K))*TAU(K, 'INPUT'), 0);
  TINT(K) =  MAX((PCLEAR(K) + VINT(K))*TAU(K, 'INPUT'), 0);

  PPROD(G) = PCLEAR(G)/(1+TAU(G, 'PRODUCER'));
```

```
  SPROD(G) = MAX(PPROD(G) - PCLEAR(G), 0);
  TPROD(G) = MAX(PCLEAR(G) - PPROD(G), 0);

  PPURC(FT) = PL(FT)$ (SL.UP(FT) GT 0);
  PSELL(FT) = PU(FT)$ (SU.UP(FT) GT 0);

* GDP calculation
  GDP = SUM(F, PCLEAR(F)*OMEGTOT(F));
  GDPS(G) = PPROD(G)*QQ(G) - SUM(GP,PINT(GP)*A.L(GP,G))*QQ(G)
                           - SUM(FT,PINT(FT)*A.L(FT,G))*QQ(G);
* Transfer by consumer
  TRANSFER(I) = TAUT(I)*TX.L/RHO.L;

* :::::::::::::::::  Fill tables  ::::::::::::::::::::::::::::::::::::

* Commodity account in quantity terms

  TABLE1(K , ' CONSUMPT. ' ,T)   = QCONS(K);
  TABLE1(K , ' INPUT ' ,T)       = QINT(K);
  TABLE1(K , ' EXPORT ' ,T)      = QEXP(K);
  TABLE1(K , ' END STOCK ' ,T)   = QSTOCK(K);
  TABLE1(K , ' PROCESSING ' ,T)  = QPROC(K);
  TABLE1(K , ' AVAILABLE ' ,T)   = QQ(K)+QIMP(K)+QSTOCK0(K);
  TABLE1(K , ' SUPPLY ' ,T)      = QQ(K);
  TABLE1(K , ' IMPORT ' ,T)      = QIMP(K);
  TABLE1(K , ' INIT.STOCK ' ,T)  = QSTOCK0(K);

* imbalance should be zero in equilibrium
  TABLE1(K , ' IMBALANCE ' ,T)   = QCONS(K)+QINT(K)+QEXP(K)+QPROC(K)
                                 + QSTOCK(K)-QQ(K)-QIMP(K)-QSTOCK0(K);

* prices
  TABLE1(K , ' EQU.PRICE ' ,T)   = PCLEAR(K);
  TABLE1(FT , ' EXP.PR.CL. ' ,T) = PEXC(FT);
  TABLE1(FT , ' IMP.PR.CL. ' ,T) = PIMC(FT);
  TABLE1(F , ' EXP.PR.BD. ' ,T)  = PEXP(F);
  TABLE1(F , ' IMP.PR.BD. ' ,T)  = PIMP(F);
  TABLE1(F , ' EXP.PR.TR. ' ,T)  = PE(F)$ ((QEMP(F) GT E.LO(F))
                                        OR (DFT(F) EQ 1));
  TABLE1(F , ' IMP.PR.TR. ' ,T)  = PM(F)$ ((QIMP(F) GT M.LO(F))
                                        OR (DFT(F) EQ 1));
  TABLE1(FT, ' PUR.PR.ST. ' ,T)  = PPURC(FT);
  TABLE1(FT, ' SEL.PR.ST. ' ,T)  = PSELL(FT);

* Commodity account in value terms

  TABLE2(KT, ' CONSUMPT. ' ,T)   = SUM(K $CK(K,KT), PCONS(K)*QCONS(K));
  TABLE2(KT, ' CONS.SUB. ' ,T)   =
                 SUM(K $CK(K,KT), (SCONS(K)-TCONS(K))*QCONS(K));
  TABLE2(KT, ' INPUT ' ,T)       = SUM(K $CK(K,KT), PINT(K)*QINT(K));
  TABLE2(KT, ' INP.SUB. ' ,T)    =
                 SUM(K $CK(K,KT), (SINT(K)-TINT(K))*QINT(K));
  TABLE2(KT, ' PROCESSD ' ,T)    = SUM(K $CK(K,KT), PCLEAR(K)*QPROC(K));
  TABLE2(KT, ' EXPORT ' ,T)      = SUM(F $CK(F,KT), PE(F)*QEXP(F));
  TABLE2(KT, ' EXP.SUB. ' ,T)    =
                 SUM(F $CK(F,KT), (SEXP(F)-TEXP(F))*QEXP(F));
  TABLE2(KT, ' END STOCK ' ,T)   =
                 SUM(FT $CK(FT,KT), PCLEAR(FT)*QSTOCK(FT));
```

```
TABLE2(KT, ' AVAILABLE ' ,T)    =
              SUM(K $CK(K,KT), PCLEAR(K)*(QQ(K)+QIMP(K))
                             + VCONS(K)*QCONS(K)
                             + VINT(K)*QINT(K)
                             + VEXP(K)*QEXP(K)
                             + VIMP(K)*QIMP(K)
                             );
TABLE2(KT, ' PROCESSM ' ,T)     = SUM(K $CK(K,KT),
                               VCONS(K)*QCONS(K)
                             + VINT(K)*QINT(K)
                             + VEXP(K)*QEXP(K)
                             + VIMP(K)*QIMP(K)
                             );
TABLE2(KT, 'SUPPLY ' ,T)        = SUM(K $CK(K,KT), PCLEAR(K)*QQ1(K)
                                  + PROD(K)*QQ2(K));
TABLE2(KT, ' SUP.TAX ' ,T)      =
              SUM(G $CK(G,KT), (TPROD(G)-SPROD(G))*QQ2(G));
TABLE2(KT, ' IMPORT ' ,T)       = SUM(F $CK(F,KT), PM(F)*QIMP(F));
TABLE2(KT, ' IMP.TAX ' ,T)      =
              SUM(F $CK(F,KT), (TIMP(F)-SIMP(F))*QIMP(F));
TABLE2(KT, ' INIT.STOCK ' ,T)   =
              SUM(FT $CK(FT,KT), PCLEAR(FT)*QSTOCK0(FT));

* Account by commodity and class

TABLE3( ' CONSUMPT. ' ,IT,K,T)     = SUM(I $CI(I,IT), PCONS(K) *X.L(I,K));
TABLE3( ' ENDOWMENT ' ,IT,F,T)     = SUM(I $CI(I,IT), PCLEAR(F)*OMEGA(I,F));
TABLE3( ' CONSUMPT. ' ,IT, ' TOTAL ' ,T)
                                   = SUM(K, TABLE3( ' CONSUMPT. ' ,IT,K,T));
TABLE3( ' ENDOWMENT ' ,IT, ' TOTAL ' ,T)
                                   = SUM(K, TABLE3( ' ENDOWMENT ' ,IT,K,T));

TABLE4(NT3,IT,T)                   = TABLE3(NT3,IT, ' TOTAL ' ,T);
TABLE4( ' TRANSFERS ' ,IT,T)       = SUM(I $CI(I,IT), TRANSFER(I));
TABLE4( ' FOR.PROFIT ' ,IT,T)      = SUM(I $CI(I, IT), PROFW(I));
TABLE4( ' REVENUE ' ,IT,T)         = TABLE4( ' ENDOWMENT ' ,IT,T)
                                   + TABLE4( ' TRANSFERS ' ,IT,T)
                                   + TABLE4( ' FOR.PROFIT ' ,IT,T);
TABLE4( ' UNC.INCOME ' ,IT,T)      = SUM(I $CI(I,IT), H.L(I));

* Foreign account

TABLE5( ' EXPORTS ' ,F,T)          = PE(F)*QEXP(F);
TABLE5( ' IMPORTS ' ,F,T)          = PM(F)*QIMP(F);
TABLE5( ' EXPORTS ' , ' TOTAL ' ,T) = SUM(F, TABLE5( ' EXPORTS ' ,F,T));
TABLE5( ' IMPORTS ' , ' TOTAL ',T) = SUM(F, TABLE5( ' IMPORTS ' ,F,T));

* Value added at factor cost

TABLE6(GT,T)                       = SUM(G $ CG(G,GT), GDPS(G));
TABLE6( ' HOUSEHOLD ' ,T)          = GDP - SUM(G, GDPS(G));
TABLE6( ' NATIONAL ' ,T)           = GDP;

TAXIND                             = - TABLE3( ' TOTAL ' , ' CONS.SUB. ' ,T)
                                     - TABLE3( ' TOTAL ' , ' INP.SUB. ' ,T)
                                     - TABLE2( ' TOTAL ' , ' EXP.SUB. ' ,T)
                                     + TABLE2( ' TOTAL ' , ' SUP.TAX ' ,T)
```

```
                              + TABLE2( ' TOTAL ' , ' IMP.TAX ' ,T);
TRADEF                  =       TABLE5( ' IMPORTS ' , ' TOTAL ' ,T)
                              - TABLE5( ' EXPORTS ' , ' TOTAL ' ,T);
TTRANS                  =       TABLE4( ' TRANSFERS ' , ' NATIONAL ' ,T);
TSTOCK                  =       TABLE2( ' TOTAL ' , ' END STOCK ' ,T)
                              - TABLE2( ' TOTAL ' , ' INT.STOCK ' ,T);

* Macro aggregates and aggregate accounting discrepancy (GAP)

TABLE7( ' GDP FC ' ,T)          = GDP;
TABLE7( ' GDP MP ' ,T)          = GDP + TAXIND;
TABLE7( ' TRADEDEF. ' ,T)       = TRADEF;
TABLE7( ' NETINDIRT ' ,T)       = TAXIND;
TABLE7( ' NETDIRT ' ,T)         = -TTRANS;
TABLE7( ' NETSTOCK ' ,T)        = -TSTOCK;
TABLE7( ' ** GAP ** ' ,T)       = TTRANS + TSTOCK - TAXIND;
```

Parameter Updates

We are still within the recursive dynamic part of the program; the last step of the dynamic simulation is to update all parameters for the equilibrium model of the next period. In an actual application this section will usually be rather long. It will contain (1) an updating of all stock variables (population, capital stocks, natural resources, etc.), (2) adjustment of time-dependent coefficients, (3) reading of exogenous variables (e.g., world prices PE and PM), and (4) adjustments of policy coefficients (e.g., indirect taxes TAU).

Here we keep the adjustment as simple as possible and impose a 1 percent growth rate on factor endowments. Obviously in a *T*-period model there would be no T loop over SOLVE statements and no parameter update. The tabulation part could be exactly the same, within a T loop, except that the levels of all time-dependent variables would have to be indexed by T (e.g., P.L(I,K,T) for equilibrium prices).

```
* :::::::::::::  Update parameters for next period  ::::::::::::::::::::::::

* Factor endowments rise at rate of one percent annually
  OMEGA(I,F) = OMEGA(I,F)*1.01;
  OMEGTOT(F) = SUM(I, OMEGA(I,F));
* Stock update
  SU.UP(FT) = QSTOCK(FT);
  QSO(FT) = QSTOCK(FT);
  );
* ::::::::::::::::::  End of time loop  ::::::::::::::::::::::::::::::::::::
```

B.4 Report Writing

It is easy to retrieve all variables in a "rough" printing using the DISPLAY command as in section 3.6.4. However, to generate a report quality printing, GAMS requires the user to control many details, almost as many as in a programming language like FORTRAN or C. Yet this is a crucial step in AGE modeling. It controls the communication between the modeler and the policy analyst, who will be reluctant to spend effort on deciphering a rough printing. Therefore we discuss this printing aspect in some detail, but to avoid repetition, we only show the macro (file) PRT2 for the first table. The following notation is used:

Symbols

/	linefeed
@	skip to column
#	skip to line
:< >	centered justification
:12:2	print with 12 positions and 2 decimals

Commands

PUT	access a specified file (here LISTAB or RESULT) and write
PUTHD	write in page-header
PUTTL	write in page-title
PUTPAGE	new page

File suffices

.HDCC	column count in header
.HDLL	Last line count in header
.LL	Last line count
.LP	Last page count
.NZ	Number of significant digits
.PS	Page size
.TM	Top margin
.WS	number of free lines on page

System suffices

.DATE	calendar date
.TIME	time
.TITLE	title of model as specified in $TITLE statement

The report-writing program contains $BATINCLUDE statements that call five different macros: GRATE (compute growth rate), HDWP (print headers), PRT1 (print table with one index against time), PRT2 (print table with two indexes), PRT32 (print table with three index and one with two). The main program reads as

```
*==                  Report writing                  ==*
*== Calculate growth rates ==*

  PARAMETER RT, VFIRST, VLAST;
  RT = 1/ (CARD(T) - 1);

$BATINCLUDE GRATE TABLE1 K,NT1
$BATINCLUDE GRATE TABLE2 K,NT2
$BATINCLUDE GRATE TABLE3 NT3, IT,KT
$BATINCLUDE GRATE TABLE4 NT4,IT
$BATINCLUDE GRATE TABLE5 NT5,KT
$BATINCLUDE GRATE TABLE6 GT
$BATINCLUDE GRATE TABLE7 NT7

*== Preliminaries: definitions and initializations ==*

  Set CL columns /1*4/
      RW rows    /1*20/
  ;
  SCALARS INDENT0, INDENT1, INDENT1B, INDENT2, INDENT3, INDENT4, INDENT5,
        PGAR, SCHK, TCOL, TCONT, TEND, TK, TKLAST;

  INDENT0 = 15; INDENT1 = 30; INDENT2 = 33; INDENT3 = 60; INDENT4 = 60;
  INDENT5 = 20;

* calculate order in computed set TS(TT)

  PARAMETER CORD(TT);
  TKLAST = 0;
  LOOP(TT$TS(TT),
        CORD(TT) = 1$(ORD(TT) EQ1) + TKLAST$(ORD(TT) GT 1);
        TKLAST = CORD(TT) + 1;
        );
  TCOL = CARD(CL);
  TKLAST = CARD(TS);

*== Initialize creation of table of contents on file LISTAB ==*

  FILE LISTAB /LISTAB/ ; LISTAB.PC=3; LISTAB.PS=88; LISTAB.TM=2;
  LISTAB.HDCC= INDENT2;
  PUT LISTAB @INDENT3, SYSTEM.DATE, ' ', SYSTEM.TIME;
  PUT / #10     "      List of Tables "//;
  PUT LISTAB " === ", SYSTEM.TITLE, " === "//

*== Initialize file RESULT ==*

  FILE RESULT /RESULT/ ; RESULT.PC=3; RESULT.PS=88; RESULT.TM=0;
  RESULT.NZ = .006; RESULT.PW = 130; PGNR = 0;
  PUTTL RESULT @INDENT3, SYSTEM.DATE, ' page ', SYSTEM.PAGE //;

*== Print ==*

* Table 1
```

```
$BATINCLUDE HDWP  ' "Commodity account in quantity terms and prices" ' 0
$BATINCLUDE PRT2 TABLE1 K KT NT1 NT1

* Table 2
$BATINCLUDE HDWP ' "Commodity account in value terms" ' 0
$BATINCLUDE PRT2 TABLE2 KT KT NT2 NT2

* Table 3 & 4
$BATINCLUDE HDWP ' "Account by commodity and class" ' 0
$BATINCLUDE PRT32 TABLE3 NT3 NT3 IT IT KT KT TABLE4 NT4 NT4 IT IT

* Table 5
$BATINCLUDE HDWP ' "Foreign Account" ' 0
$BATINCLUDE PRT2 TABLE5 NT5 NT5 KT KT

* Table 6
$BATINCLUDE HDWP ' "Value added at factor cost" ' 10
$BATINCLUDE PRT1 TABLE6 GT GT

* Table 7
$BATINCLUDE HDWP ' "Macro aggregates" ' 10
$BATINCLUDE PRT1 TABLE7 NT7 NT7
```

The macro GRATE computes annual growth rates in percentages, based on beginning and end-year values. The value 999 will be assigned if one (and only one) of both is nonpositive.

```
* calculate growth rate
   LOOP((%2),
            VFIRST = SUM(TFIRST, %1(%2,TFIRST));
            VLAST = SUM(TLAST, %1(%2,TLAST));

            IF(VFIRST*VLAST GT 0,
              %1(%2, 'GROWTH' ) = 100*((VLAST/VFIRST)**RT-1.);
              ELSE
              IF(VFIRST+VLAST EQ 0,
                %1(%2, 'GROWTH' ) = 0.;
                ELSE
* one of the two is negative
                %1(%2, 'GROWTH' ) = 9999.;
                );
              );
            );
```

Here %1 and %2 denote the first and second parameter. The macro HDWP controls the printing of headers:

```
*== File HDWP, macro for header in A4-format print ==*
* parameter 1 header
* parameter 2 if 0 newpage, otherwise test for sufficient lines on page

  SCHK = %2;
  IF(SCHK EQ 0,
    IF (PGNR GT 0,
       PUTPAGE RESULT;
       );
    ELSE
```

```
    IF (RESULT.LL+SCHK GT RESULT.WS,
       PUTPAGE RESULT;
       );
    );
  IF (PGNR GT 0,
     PGNR = RESULT.LP + 1;
     ELSE
     PGNR = 1;
     );
  PUT LISTAB /"     "
  %1:<>,
  @INDENT4, " p. ", PGNR:3:0;
  RESULT.HDLL = 0;
  PUT RESULT /;
  PUT /@INDENT0"::::::::::::::::::::::::::::::::::::::::::::::::::::":<>;
  PUT /@INDENT0"::                                                ::":<>;
  PUT /@INDENT0"::", %1:<>48, "::";
  PUT /@INDENT0"::                                                ::":<>;
  PUT /@INDENT0"::::::::::::::::::::::::::::::::::::::::::::::::::::":<>;
  PUT //;
  PUTHD @INDENT5,
         %1 " (continued) "
         //;
```

Finally, the macro PRT2 manages the printing of tables, suppressing lines with zeros only (the C%1-variables), and controlling the looping when time-series results do not fit on a single line (the looping over RW and CL).

```
*== File PRT2, macro for table with two indexes ==*

* parameters
* 1    name of table
* 2    first index in table
* 3    index set containing the labels of first index
* 4    second index in table
* 5    index set containing the labels of second index

* The procedure suppresses lines with zeros only and prints up to a
* maximum of TCOL columns for time periods

  PARAMETER C%1(%2,%4);

  TCONT = 1;
  LOOP(RW$(TCONT = 1),
        IF(RESULT.LL+4 GT RESULT.WS,
          PUTPAGE;
          ELSE
          PUT //@INDENT2;
          );
        PUTHD #4@INDENT2;
        LOOP(%2,
              LOOP(%4,
                    C%1(%2,%4) = 0;
                    LOOP(CL,
                          TK = (ORD(RW)-1)*TCOL+ORD(CL);
                          IF(TK LE TKLAST, C%1(%2,%4) = C%1(%2,%4)
```

```
                        + SUM(TS$(CORD(TS) EQ TK), ABS(%1(%2,%4,TS)));
                        );
                      );
                  );
              );
    SCHK = SUM((%2,%4), C%1(%2,%4));
    IF(SCHK GT EPS,
      LOOP(CL,
            TK = (ORD(RW)-1)*TCOL+ORD(CL);
            IF(TK LE TKLAST,
              LOOP(TS$(CORD(TS) EQ TK), PUTHD TS.TL:<>12:0);
              LOOP(TS$(CORD(TS) EQ TK), PUT TS.TL:<>12:0);
              ELSE
              PUTHD "            ";
              PUT "              ";
              )
            );
      PUTHD //;
      LOOP(%2,
            IF(RESULT.LL+5 GT RESULT.WS,
              PUTPAGE;
              );
            PUT //' == ' %3.TE(%2) ' =='/;
            LOOP(%4,
                  IF(C%1(%2,%4) GT EPS,
                    PUT /%5.TE(%4)@INDENT1;
                    LOOP(CL,
                          TK = (ORD(RW)-1)*TCOL+ORD(CL);
                          IF(TK LE TKLAST,
                            LOOP(TS$(CORD(TS) EQ TK), PUT %1(%2,%4,TS):12:2)
                            );
                          );
                    );
                    TEND = CARD(%2) - ORD(%2) + TKLAST - TK;
                    TCONT = 0.$(TEND EQ 0) + 1.$(TEND GT 0);
                  );
              );
          );
        );
```

B.5 Running the GAMS Job

Suppose that the application listed in sections B.1 to B.4 were written on the ASCII file EXFULL. Then, on a personal computer in a DOS environment, a GAMS job can be run by entering:

```
GAMS EXFULL
```

However, it is often impractical to run the full application as a single job. A stepwise procedure is more convenient for testing, development, and policy exercises, also because the running of alternative scenarios does not require execution of the initial steps. GAMS allows one to proceed sequentially. For example, for the code of sections B.1 to B.4 written on the files EXDEF, EXSIM, and EXREP, respectively, for model definition, simulation, accounting, and report writing, the sequence of calls could be

```
GAMS EXDEF S = WKDEF
GAMS EXSIM R = WKDEF S = WKSIM
GAMS EXREP R = WKSIM
```

where S = WKDEF denotes the work files on which the results are saved, and R = WKDEF those from which the program will read its information for initialization.

Yet this is a rather rudimentary way of managing the jobs. In practice, it is more convenient to develop a batch control program (a .BAT-file which calls GAMS as well as other executable programs written, say, in FORTRAN) to manage the GAMS operations. Most GAMS users create such batch files. In Keyzer (1997) MS.DOS and WINDOWS.NT procedures are described that adjust GAMS parameters interactively and circumvent the drawback of the GAMS system that its work files keep on accumulating all parameters and variables that are defined, while the system is unable to write GAMS readable ASCII files. This is impractical in many applications because even small-scale models often require extensive manipulations to process the data as found in statistical publications into model parameters and initializations.

Storing the Reference Run

The listing of results to be presented below only refers to a single scenario "Baserun," but in applications one usually compares outcomes from several scenarios. Then, it is sometimes convenient to compute differences between them. For this one could proceed as follows: (1) include declarations for parameter matrices, say, REF1–REF7 jointly with TABLE1–TABLE7, as part of the model definition; (2) store the TABLE1–7 values into the matrices REF1–7 when, say, RUNOPT = 1 for the reference scenario; (3) run this reference scenario with S = WKDEF (as opposed to WKSIM); (4) finally, perform the calculations for comparison between runs in EXREP. Note, however, that in such a setup, it would be necessary to store base-year values of parameters separately and to use these for reinitializing the parameters changed during simulation.[4]

4. This is of course only relevant for a recursive dynamic model, since in a T-period simulation there is no parameter adjustment from one time period to the next.

B.6 Listing of Results

The listing that follows is printed as it comes out of the computer, without any editing. This is to illustrate that it is indeed possible to produce report quality output directly from GAMS. The printing shows the file LISTAB (the table of contents), followed by the file RESULT. The column "GROWTH" denotes the annual percentage growth rate.

Consistency Checks

The structure of the accounts allows to trace possible inconsistencies: The imbalance on the Commodity account in quantity terms (TABLE1) measures the disequilibrium on the commodity market; the imbalance appears to be zero (ie., below the specified tolerance level) for all markets. That a line with an imbalance is shown in the print is only because the growth rate of the infeasibility is not zero. The Account by commodity and class (TABLE3 and TABLE4) makes it possible to check the equality between revenue and consumption in the account of the private consumers and the government-as-consumer and to verify that the uncommitted income is nonnegative; finally, the accounting discrepancy in the Macro-aggregates (the variable GAP of TABLE7) basically checks the consistency of the government budget and the indirect tax accounts in the SAM of table 5.1 (the trade deficit plus the three net receipts items add up to the accounting discrepancy, which should be zero).

Market Regime

The prices in the first table describe, for every commodity, the market regime in equilibrium. For example, fuel is not traded because its clearing price remains between the import upper bound and the export lower bound. Some of the initial stocks of fuel are sold in 1982, but the clearing price does not rise above the "stock upper bound" level. In subsequent years there is no stock adjustment, and the price remains below this level. All stocks are also sold in that year for machinery, but there the equilibrium price is at the import upper bound, so imports can adjust. The large number of negative growth rates is due to the fact that while factor endowments rise, the stocks are only available in the first year. Comparison of the Account by commodity and class with the Foreign account shows that foreign profits add up to the trade deficit, which is only used to finance imports of fuel. In this solution there are no exports.

List of Tables

=== Baserun ===

Commodity account in quantity terms and prices	p.	1
Commodity account in value terms	p.	2
Account by commodity and class	p.	3
Foreign Account	p.	5
Value added at factor cost	p.	5
Macro aggregates	p.	5

Commodity account in quantity terms and prices

	1982	1983	1984	GROWTH
== Food ==				
Consumption	109.85	108.20	109.28	−0.26
Available	109.85	108.20	109.28	−0.26
Supply	109.85	108.20	109.28	−0.26
Clearing price	679.58	669.44	669.43	−0.75
== Textile ==				
Consumption	145.94	146.01	147.46	0.52
Available	145.94	146.01	147.46	0.52
Supply	145.94	146.01	147.46	0.52
Imbalance	0.00	0.00	0.00	9999.00
Clearing price	240.00	232.60	232.59	−1.56
== Trade ==				
Processing	25.80	25.65	25.90	0.18
Available	25.80	25.65	25.90	0.18
Supply	25.80	25.65	25.90	0.18
Clearing price	114.75	111.25	111.24	−1.54
== Labor ==				
Consumption	71.92	72.61	73.34	0.98
Input	68.08	68.79	69.48	1.02
Available	140.00	141.40	142.81	1.00
Supply	140.00	141.40	142.81	1.00
Imbalance	0.00	0.00	0.00	0.00
Clearing price	591.26	571.57	571.55	−1.68
== Equipment ==				
Consumption	220.47	222.62	224.84	0.99
Input	19.53	19.78	19.98	1.15

Available	240.00	242.40	244.82	1.00
Supply	240.00	242.40	244.82	1.00
Imbalance	0.00	0.00	0.00	0.00
Clearing price	667.60	644.43	644.40	−1.75

== Fuel ==

Consumption	114.07	112.94	114.06	0.00
Input	98.60	99.16	100.16	0.79
Final stock	7.33	7.33	7.33	0.00
Available	220.00	219.43	221.55	0.35
Supply	210.00	212.10	214.22	1.00
Initial stock	10.00	7.33	7.33	−14.39
Imbalance	0.00	0.00	0.00	9999.00
Clearing price	100.00	97.17	97.17	−1.43
Export lower price bound	50.00	50.00	50.00	0.00
Import upper price bound	143.47	143.12	143.12	−0.12
Export price (tar. inc.)	50.00	50.00	50.00	0.00
Import price (tar. inc.)	132.00	132.00	132.00	0.00
Export price f.o.b.	50.00	50.00	50.00	0.00
Import price c.i.f.	120.00	120.00	120.00	0.00
Stock upper price bound	100.00	100.00	100.00	0.00

== Machinery ==

Input	432.25	416.35	420.49	−1.37
Available	432.25	416.35	420.49	−1.37
Supply	410.00	414.10	418.24	1.00
Import	2.25	2.25	2.25	0.00
Initial stock	20.00	0.00	0.00	9999.00
Clearing price	111.47	111.12	111.12	−0.16
Export lower price bound	72.00	72.00	72.00	0.00
Import upper price bound	111.47	111.12	111.12	−0.16
Export price (tar.inc.)	72.00	72.00	72.00	0.00
Import price (tar.inc.)	100.00	100.00	100.00	0.00
Export price f.o.b.	80.00	80.00	80.00	0.00
Import price c.i.f.	100.00	100.00	100.00	0.00
Stock upper price bound	80.00	0.00	0.00	9999.00

Commodity account in value terms

== FOOD ==	1982	1983	1984	GROWTH
Consumption	75910.40	73634.88	74367.96	−1.02
Available	75910.40	73634.88	74367.96	−1.02
Processing margins	1260.49	1203.64	1215.59	−1.80
Supply	74649.92	72431.25	73152.37	−1.01

== Textile ==				
Consumption	36701.41	35584.64	35938.90	−1.04
Available	36701.41	35584.64	35938.90	−1.04
Processing margins	1674.67	1624.23	1640.40	−1.03
Supply	35026.74	33960.40	34298.50	−1.05
== Trade ==				
processing requirement	2960.98	2852.90	2881.03	−1.36
Available	2960.98	2852.90	2881.03	−1.36
Supply	2960.98	2852.90	2881.03	−1.36
== Labor ==				
Consumption	42526.37	41502.21	41915.35	−0.72
Input	40250.06	39317.75	39709.44	−0.67
Available	82776.42	80819.96	81624.79	−0.70
Supply	82776.42	80819.96	81624.79	−0.70
== Equipment ==				
Consumption	147186.36	143460.46	144888.50	−0.78
Input	14342.35	14024.06	14163.71	−0.62
Input subsidy	−1303.85	−1274.91	−1287.61	−0.62
Available	160224.85	156209.60	157764.61	−0.77
Supply	160224.85	156209.60	157764.61	−0.77
== Fuel ==				
Consumption	12547.59	12071.53	12191.70	−1.43
Consumer subsidy	−1140.69	−1097.41	−1108.34	−1.43
Input	9860.19	9636.05	9732.07	−0.65
Final stock	732.91	712.18	712.15	−1.43
Available	21000.00	20610.17	20815.44	−0.44
Supply	21000.00	20610.17	20815.44	−0.44
Initial stock	1000.00	712.18	712.15	−15.61
== Machinery ==				
Input	48185.03	46266.69	46726.67	−1.52
Available	45981.35	46291.72	46751.70	0.83
Processing margins	25.82	25.03	25.03	−1.54
Supply	45704.72	46016.66	46476.64	0.84
Imports	225.00	225.00	225.00	0.00
Import tax	0.00	0.00	0.00	−29.29
Initial stock	2229.50	0.00	0.00	9999.00
== Total ==				
Consumption	314872.13	306253.72	309302.42	0.00
Consumer subsidy	−1140.69	−1097.41	−1108.34	0.00
Input	112637.63	109244.55	110331.90	0.00
Final stock	732.91	712.18	712.15	0.00

Input subsidy	−1303.85	−1274.91	−1287.61	0.00
processing requirement	2960.98	2852.90	2881.03	0.00
Available	425555.42	416003.87	420144.42	0.00
Processing margins	2960.98	2852.90	2881.03	0.00
Supply	422343.63	412900.94	417013.37	0.00
Imports	225.00	225.00	225.00	0.00
Import tax	0.00	0.00	0.00	0.00
Initial stock	3229.50	712.18	712.15	0.00

Account by commodity and class

	1982	1983	1984	GROWTH
** Consumption **				
== Farmers ==				
Food	33065.99	32280.20	32601.59	−0.70
Textile	17142.67	16735.29	16901.91	−0.70
Labor	9324.46	9102.88	9193.51	−0.70
Equipment	12193.53	11903.76	12022.28	−0.70
Total	71726.65	70022.13	70719.29	−0.70
== Workers ==				
Food	9844.26	9606.56	9702.20	−0.72
Textile	2368.54	2311.35	2334.36	−0.72
Labor	3330.76	3250.34	3282.70	−0.72
Equipment	2960.68	2889.19	2917.95	−0.72
Total	18504.24	18057.44	18237.21	−0.72
== Employers ==				
Food	33000.16	31748.12	32064.18	−1.43
Textile	17190.20	16538.00	16702.63	−1.43
Labor	10038.07	9657.22	9753.36	−1.43
Equipment	52699.88	50700.42	51205.15	−1.43
Fuel	12547.59	12071.53	12191.70	−1.43
Total	125475.89	120715.29	121917.03	−1.43
== Private ==				
Food	75910.40	73634.88	74367.96	−1.02
Textile	36701.41	35584.64	35938.90	−1.04
Labor	22693.30	22010.44	22229.57	−1.03
Equipment	67854.08	65493.38	66145.38	−1.27
Fuel	12547.59	12071.53	12191.70	−1.43
Total	215706.79	208794.86	210873.52	−1.13

== Government ==				
Labor	19833.07	19491.77	19685.78	−0.37
Equipment	79332.27	77967.08	78743.12	−0.37
Total	99165.34	97458.85	98428.90	−0.37
== National ==				
Food	75910.40	73634.88	74367.96	−1.02
Textile	36701.41	35584.64	35938.90	−1.04
Labor	42526.37	41502.21	41915.35	−0.72
Equipment	147186.36	143460.46	144888.50	−0.78
Fuel	12547.59	12071.53	12191.70	−1.43
Total	314872.13	306253.72	309302.42	−0.89
** Endowment **				
== Farmers ==				
Labor	65038.62	63501.39	64133.76	−0.70
Equipment	6676.04	6508.73	6573.53	−0.77
Total	71714.65	70010.13	70707.29	−0.70
== Workers ==				
Labor	11825.20	11545.71	11660.68	−0.70
Equipment	6676.04	6508.73	6573.53	−0.77
Total	71714.65	70010.13	70707.29	−0.70
== Employers ==				
Labor	5912.60	5772.85	5830.34	−0.70
Equipment	80112.43	78104.80	78882.30	−0.77
Fuel	11000.00	10795.80	10903.32	−0.44
Machinery	23409.73	23569.51	23805.11	0.84
Total	120434.76	118242.97	119421.08	−0.42
== Private ==				
Labor	82776.42	80819.96	81624.79	−0.70
Equipment	93464.50	91122.27	92029.35	−0.77
Fuel	11000.00	10795.80	10903.32	−0.44
Machinery	23409.73	23569.51	23805.11	0.84
Total	210650.65	206307.54	208362.57	−0.54
== Government ==				
Equipment	66760.36	65087.33	65735.25	−0.77
Fuel	10000.00	9814.37	9912.11	−0.44
Machinery	22294.98	22447.15	22671.53	0.84
Total	99055.34	97348.85	98318.90	−0.37
== National ==				
Labor	82776.42	80819.96	81624.79	−0.70
Equipment	160224.85	156209.60	157764.61	−0.77

Fuel	21000.00	20610.17	20815.44	−0.44
Machinery	45704.72	46016.66	46476.64	0.84
Total	309705.99	303656.39	306681.47	−0.49
** Consumption **				
Farmers	71726.65	70022.13	70719.29	−0.70
Workers	18504.24	18057.44	18237.21	−0.72
Employers	125475.89	120715.29	121917.03	−1.43
Private	215706.79	208794.86	210873.52	−1.13
Government	99165.34	97458.85	98428.90	−0.37
National	314872.13	306253.72	309302.42	−0.89
** Endowment **				
Farmers	71714.65	70010.13	70707.29	−0.70
Workers	18501.24	18054.44	18234.21	−0.72
Employers	120434.76	118242.97	119421.08	−0.42
Private	210650.65	206307.54	208362.57	−0.54
Government	99055.34	97348.85	98318.90	−0.37
National	309705.99	303656.39	306681.47	−0.49
** Transfers **				
Employers	4941.13	2372.33	2395.95	−30.37
Private	4941.13	2372.33	2395.95	−30.37
National	4941.13	2372.33	2395.95	−30.37
** Foreign profits **				
Farmers	12.00	12.00	12.00	0.00
Workers	3.00	3.00	3.00	0.00
Employers	100.00	100.00	100.00	0.00
Private	115.00	115.00	115.00	0.00
Government	110.00	110.00	110.00	0.00
National	225.00	225.00	225.00	0.00
** Revenue **				
Farmers	71726.65	70022.13	70719.29	−0.70
Workers	18504.24	18057.44	18237.21	−0.72
Employers	125475.89	120715.29	121917.03	−1.43
Private	215706.79	208794.86	210873.52	−1.13
Government	99165.34	97458.85	98428.90	−0.37
National	314872.13	306253.72	309302.42	−0.89
** Uncommitted income **				
Farmers	1267.99	1262.52	1262.51	−0.22
Workers	327.12	325.58	325.58	−0.24
Employers	2218.17	2176.53	2176.51	−0.94
Private	3813.28	3764.62	3764.60	−0.64
Government	1753.05	1757.21	1757.19	0.12
National	5566.33	5521.83	5521.79	−0.40

Foreign Account

	1982	1983	1984	GROWTH
== Exports ==				
== Imports ==				
Machinery	225.00	225.00	225.00	0.00
Total	225.00	225.00	225.00	0.00

Value added at factor cost

	1982	1983	1984	GROWTH
Food	24068.28	23806.98	24044.56	−0.05
Textile	28117.70	27223.00	27493.99	−1.12
Trade	2406.42	2311.82	2334.61	−1.50
Self employed	255113.59	250314.59	252808.32	−0.45
National	309705.99	303656.39	306681.47	−0.49

Macro aggregates

	1982	1983	1984	GROWTH
GDP at factor cost	309705.99	303656.39	306681.47	−0.49
GDP at market prices	312150.53	306028.72	309077.42	−0.49
Trade deficit	225.00	225.00	225.00	0.00
Net indirect taxes	2444.54	2372.33	2395.95	−1.00
Net direct taxes	−4941.13	−2372.33	−2395.95	−30.37
Net stock sales	2496.59	0.00	0.00	9999.00
Accounting discrepancy	0.00	0.00	0.00	−70.00

C Assumptions

We begin with a list of assumptions that are used throughout the book. Then we list assumptions used in certain chapters that are seldom referred to elsewhere in the book.

C.1 General Assumptions

ASSUMPTION C1 (Utility functions) The utility function $u_i: R^r_+ \to R_+$, $u_i(x_i)$ of every consumer i is continuous, strictly concave, increasing, and, without loss of generality $u_i(0) = 0$.

ASSUMPTION C1′ (Utility functions) The utility function $u_i: R^r_+ \to R_+$, $u_i(x_i)$ is continuous, strictly concave, nonsatiated, and satisfies $u_i(0) = 0$ for all i; for $i = 1$ it is increasing with respect to all commodities.

ASSUMPTION C2 (Endowments) The endowment ω_i of every consumer i is strictly positive.

ASSUMPTION C2′ (Endowments) $\omega_i \geqslant 0$, $\omega_i \neq 0$ for all i and $\omega_i > 0$ for $i = 1$.

ASSUMPTION P1 (Production sets) The production set Y_j of every producer j has possibility of inaction ($0 \in Y_j$); it is compact and strictly convex.

ASSUMPTION P1′ (Production sets) The production set Y_j of every producer j has possibility of inaction ($0 \in Y_j$); it is compact and convex.

ASSUMPTION S (Standard assumptions) The standard assumptions are assumptions P1′, C1′, and C2′.

ASSUMPTION T (Transfer distribution) Lump-sum transfers (also called direct taxes) are distributed according to $T_i(p, T, h_1^0, h_2^0, \ldots, h_m^0)$ for all i. This function satisfies (1) continuity; (2) homogeneity of degree 1 in $(p, T, h_1^0, h_2^0, \ldots, h_m^0)$; (3) adding-up: $\Sigma_i T_i(.) \geqslant T$ with equality whenever $T + \Sigma_i h_i^0 > 0$; (4) positiveness $h_i^0 + T_i > 0$; and (5) there exists a constant $\bar{\eta} \in (0, 1)$ such that if $T + (1 - \bar{\eta})\Sigma_i h_i^0 \leqslant 0$, then all consumers are taxed at the same proportional rate so that for all i, $h_i = \eta h_i^0$ for $\eta \in [0, \bar{\eta}]$.

C.2 Competitive Equilibrium (Chapter 1)

ASSUMPTION Z1 (Single-valuedness and continuity) $z(p)$ is a single-valued and continuous function, defined for $p \geqslant 0$, $p \neq 0$.

ASSUMPTION Z2 (Continuous differentiability) $z(p)$ is continuously differentiable for $p > 0$.

ASSUMPTION Z3 (Homogeneity) $z(p)$ is homogeneous of degree zero in p.

ASSUMPTION **Z4** (Walras's law) $pz(p) = 0$.

ASSUMPTION **Z5** (Desirability) $p_k = 0$ implies that $z_k(p) > 0$ for $k = 1, 2, \ldots, r$.

C.3 Optimal Production and Consumption Allocations (Chapter 2)

ASSUMPTION **C1″** (Local strong concavity of $u_i(x_i)$ and infinite marginal utilities) The utility function $u_i: R^r_+ \to R_+$, $u_i(x_i)$ is continuous, strictly concave, nonsatiated, and satisfies $u_i(0) = 0$ for all i; for $i = 1$ it is increasing with respect to all commodities. Moreover it is locally strongly concave in the optimal points chosen by the consumer and satisfies the boundary condition $\lim_{x \to x^0} \| \partial u_i(x_i)/\partial x_i \| = \infty$ whenever $x_k^0 = 0$.

ASSUMPTION **GARP** If $x^t R x^s$, then it cannot be that $x^s P^0 x^t$.

ASSUMPTION **P1″** (Local strong convexity of the boundary of Y_j) The production set Y_j of every producer j has possibility of inaction ($0 \in Y_j$); it is compact and has a boundary that is locally strongly convex in the optimal points chosen by the producer.

C.4 Price Rigidities (Chapter 6)

ASSUMPTION **CR1** The utility function u_i: $R^r_+ \to R_+$, $u_i(x_i)$ is continuous, strictly concave, and satisfies $u_i(0) = 0$ for all i. Moreover it is continuously differentiable on R^r_{++}, satisfies the boundary condition $\lim_{x \to x^0} \| \partial u_i(x_i)/\partial x_i \| = \infty$ whenever $x_k^0 = 0$ and is uniformly bounded away from zero for the numéraire $k = 1$, that is, $\partial u_i(x_i)/\partial x_{i1} \geqslant \varepsilon > 0$.

ASSUMPTION **R1** (Positive price interval) $p^l < p^u$.

ASSUMPTION **R2** (No rationing for one commodity) For at least one commodity, say, the first, $p_1^l < \bar{p}_1^e$, $p_1^u > \bar{p}_1^m$, and $\bar{p}_1^m = \bar{p}_1^e > 0$ so that the price rigidities are always ineffective.

C.5 Finite-Horizon Dynamics (Chapter 7)

ASSUMPTION **CT1** (The recursive utility function) (1) $W_i: R^r_+ \times R \to R$, $W_i(x_i, u_i)$ is continuous, strictly concave, and continuously differentiable on $R^r_{++} \times R$; (2) $W_i(x_i, u_i)$ is nonsatiated in x_i and increasing in x_i for $i = 1$; (3) $W_i(0, 0) = 0$.

ASSUMPTION **E1** (Expectations) Expectations $\phi_j(p_1)$, $\phi_i(p_1)$, $\Pi_{i2}(p_1)$ are nonnegative, continuous, and uniformly bounded above. Moreover, for $i = 1$, $\phi_i(p_1) \neq 0$ whenever $p_1 \neq 0$.

ASSUMPTION E2 (Nonnegativity of initial net worth) $V_{i1} \geqslant 0$ for all i, and $V_{i1} > 0$ for $i = 1$; $V_{j1} \geqslant 0$ for all j.

ASSUMPTION E2′ (Positive aggregate net worth) $\Sigma_i V_{i1} + \Sigma_j V_{j1} > 0$ and $V_{i1} > 0$ for $i = 1$.

ASSUMPTION PT1 (The transformation function) The transformation function $F_{jt}: R^a \times R^a_+ \times R^a_+ \to R$ is continuous, convex, nondecreasing in $(y_{jt}, \tilde{k}_{j,t+1}, -k_{jt})$. The constraint $F_{jt}(y_{jt}, \tilde{k}_{j,t+1}, k_{jt}) \leqslant 0$ satisfies (1) possibility of inaction: $F_{jt}(0, 0, k_{jt}) \leqslant 0$ for all $k_{jt} \geqslant 0$; (2) boundedness: for any given $k_{jt} \geqslant 0$, $p_t \geqslant 0$, $\psi_{jt} \geqslant 0$, the set $\{y_{jt}, \tilde{k}_{j,t+1} | p_t y_{jt} + \psi_{j,t+1}\tilde{k}_{j,t+1} \geqslant 0, F_{jt}(y_{jt}, \tilde{k}_{j,t+1}, k_{jt}) \leqslant 0\}$ is bounded; and (3) $F_{jt}(y_{jt}, \tilde{k}_{j,t+1}, k_{jt})$ is continuously differentiable.

ASSUMPTION TD (Wealth distribution) Wealth is distributed according to $T_i(p_1, 0, h^0)$ for consumer i and to $T_j(p_1, p, h^0)$ for producer j, where h^0 is the vector with the expected wealth of every agent as elements. This function satisfies (1) continuity; (2) homogeneity of degree 1 in $(p_1, 0, h^0)$; (3) adding-up: $\Sigma_i T_i(.) + \Sigma_j T_j(.) = 0$; and (4) positiveness: $h_i^0 + T_i > 0$ for consumer i and $h_j^0 + T_j > 0$ for producer j.

C.6 Infinite-Horizon Dynamics (Chapter 8)

ASSUMPTION CD1 (The dynastic utility function) (1) W_i: $R^r_+ \times R \to R$, $W_i(x_i, u_i)$ is continuous, strictly concave, and continuously differentiable on $R^r_{++} \times R$; (2) $W_i(x_i, u_i)$ is nonsatiated in x_i and increasing in x_i for $i = 1$; (3) $W_i(0, 0) = 0$; (4) discounting: $0 < \partial W_i(.)/\partial u_i \leqslant \gamma$ for some given $\gamma \in (0, 1)$ common to all i, and (5) $W_i(x_i, u_i)$ uniformly bounded above with upper bound $\bar{v}$.

ASSUMPTION CD1′ (The dynastic utility function with fixed discount factor) $W_i(x_i, u_i)$ satisfies parts 1 to 5 in assumption CD1 with the additional assumption that the discount factor is constant and common to all i: $\partial W_i/\partial u_i = \beta$ so that $W_i(x_{it}, u_{i,t+1})$ is linear in $u_{i,t+1}$ and can be written as $u_i(x_{it}) + \beta u_{i,t+1}$; then $u_{i1} = \Sigma_{t=1}^{\infty}(\beta)^{t-1} u_i(x_{it})$.

ASSUMPTION CO1 The consumer born in t has utility function $u^t: R^r_+ \times R^r_+ \to R_+$; $u(x_t^t, x_{t+1}^t)$ is strictly concave, uniformly bounded above, nondecreasing for all t; it is continuously differentiable for positive consumption. Consumption is necessary in both periods: $\lim_{x \to 0} \| \partial u(x, y)/\partial x \| = \infty$ for all $y \geqslant 0$ and $\lim_{y \to 0} \| \partial u(x, y)/\partial y \| = \infty$ for all $x \geqslant 0$. Finally, the utility function is increasing for at least one commodity k in both in periods t and $t + 1$, and we denote by K the set of such commodities.

ASSUMPTION CO2 (Nonzero endowments) Endowments are nonnegative and stationary (and thus bounded) so $\omega_t^{t-1} = \omega_1^0$ and $\omega_t^t = \omega_1^1$, and

$\omega_1^0 + \omega_1^1 > 0$. Each consumer has a positive endowment of at least one commodity k in which his utility is increasing in t or in $t + 1$.

ASSUMPTION **CO3** (Feasibility of the claim) $\omega_{k,1}^0 + m_k^0 \geqslant 0$ and strictly positive for some commodity $k \in K$ and $\omega_{k,1}^1 - m_k^0 \geqslant 0$ for all k.

ASSUMPTION **CO4** (Slopes of the indifference curves) There exists a positive constant δ such that $(\partial u(\,.\,)/\partial x_{h,t+1}^t)/(\partial u(\,.\,)/\partial x_{k,t}^t) \leqslant \delta$, for all commodities h and k and all nonzero consumptions.

ASSUMPTION **K** (Initial stocks) In period 1 each consumer i owns nonnegative and nonzero initial stocks ($\bar{k}_{i1} \geqslant 0, \bar{k}_{i1} \neq 0$) and a positive stock of the first element of the vector ($\bar{k}_{i1,1} > 0$), say, labor. We denote $\Sigma_i \bar{k}_{i1}$ by $\bar{k}_1$.

ASSUMPTION **PD1** (The transformation function) (1) $F(y_t, \bar{k}_{t+1}, k_t)$ is convex, nondecreasing in $(y_t, \bar{k}_{t+1}, -k_t)$, and increasing in $-k_{1t}$, where k_{1t} denotes the aggregate stock of capital good 1 in period t. The constraint $F(y_t, \bar{k}_{t+1}, k_t) \leqslant 0$ satisfies (2) homogeneity of degree one: if $F(y_t, \bar{k}_{t+1}, k_t) \leqslant 0$, then $F(\lambda y_t, \lambda\bar{k}_{t+1}, \lambda k_t) \leqslant 0$ also for any $\lambda > 0$; (3) boundedness: y_t and $\bar{k}_{t+1}$ are bounded above for any given bounded k_t; (4) the set $M = \{k > 0 \,|\, F(y, k, k) < 0, \quad y > 0\}$ is nonempty; (5) there exists $k \in M \cap (0, \bar{k}_1)$, where $\bar{k}_1$ is the initial stock.

ASSUMPTION **PD1′** The number of capital stocks is equal to the number of commodities. The transformation function satisfies assumption PD1 and can be written $F(y_t, k_{t+1}, k_t) = \max_h(y_{ht} + G_h(k_{t+1}) - k_{ht})$.

C.7 Externalities (Chapter 9)

ASSUMPTION **CG1** (Utility functions with nonrival commodities) The utility function $u_i\colon R_+^r \times R_+^r \to R_+$, $u_i(x_i, x_g)$ is continuous, concave, and satisfies $u_i(0, x_g) = 0$ for all i; for $i = 1$ it is strictly concave, increasing in (x_i, x_g), and for $i = 2, 3, \ldots, m$, it is strictly concave, nonsatiated in x_i, and nondecreasing in x_g.

ASSUMPTION **CW1** (The utility function with empathy)

1. $W_i\colon R_+^r \times R_+^m \to R_+$, $W_i(x_i, v_i)$ is continuously differentiable, concave, strictly concave in x_i, nondecreasing in v_i, and satisfies $W_i(0, v_i) = 0$ for all i; for $i = 1$ it is increasing in x_i; for $i = 2, 3, \ldots, m$, it is nonsatiated in x_i.
2. Limited altruism: $\lim_{\|u\| \to \infty} \max_i(u_i - W_i(x_i, u)) > 0$.

ASSUMPTION **CX1** (Externalities in the utility function) $W_i\colon R_+^r \times R_+^{rm} \to R_+$, $W_i(x_i, \xi_i)$ is continuously differentiable, concave, strictly concave in x_i, nondecreasing in ξ_i, and satisfies $W_i(0, \xi_i) = 0$ for all i; for $i = 1$ it is increasing in x_i; for $i = 2, 3, \ldots, m$, it is nonsatiated in x_i.

ASSUMPTION CX2 (Consumption dependent endowment) The function $\omega_i(x_i)$ is continuous, concave, nondecreasing, and uniformly bounded from above; $\omega_i(x_i) \geqslant 0$, $\omega_i(x_i) \neq 0$; for $i = 1$, $\omega_i(0) > 0$.

ASSUMPTION PG1 (Transformation function with nonrival inputs) The transformation function $F_j: R^r \times R^r_- \to R$ is continuous, strictly quasi-convex, nonsatiated in inputs, and has possibility of inaction ($F_j(0, 0) = 0$); the associated profit function is compact valued.

C.8 Nonconvexities (Chapter 10)

ASSUMPTION CX2′ (Nonconcave efficiency-wage relation) The function $\omega_i(x_i)$ is continuous, nondecreasing, and uniformly bounded from above; $\omega_i(x_i) \geqslant 0$, $\omega_i(x_i) \neq 0$. Moreover $\omega_1(0) > 0$.

ASSUMPTION PI1 (Production sets under increasing returns)

1. Firms $j \in J_1$ (e.g., the first n_1) with increasing returns produce a single output, with a production function $f_j(v_j)$ that is strictly quasi-concave, nondecreasing, and such that $f_j(0) = 0$; their production set is given by $Y_j = \{y_j | y_j \leqslant b_j q_j - v_j, q_j \leqslant f_j(v_j), q_j \geqslant 0, v_j \geqslant 0\}$.
2. Firms $j \in J_2$ ($n_1 + 1$ to n) satisfy assumption P1′.
3. The set $X = \{x \geqslant 0 | x \geqslant \Sigma_j y_j + \Sigma_i \omega_i, y_j \in Y_j\}$ is bounded above.

ASSUMPTION PN1 The production set $\tilde{Y}$ is compact and nonempty.

ASSUMPTION PS1 (Production set under setup costs) The production set of producer j is $Y_j = \{y_j | y_j = \tilde{y}_j \delta_j, \tilde{y}_j \in \tilde{Y}_j, \delta_j \in \{0, 1\}\}$ where $\tilde{Y}_j$ is nonempty, compact, and convex.

C.9 Imperfect Competition (Chapter 11)

ASSUMPTION PM (Markup functions) Markup functions $\mu_j^+(.)$, $\mu_j^-(.)$ are continuous, nonnegative, and homogeneous of degree one in prices.

C.10 Money and Incomplete Asset Markets (Chapter 12)

ASSUMPTION CA1 (Utility function) The utility function $u_i: R_+^{r(1+S)} \to R_+$, $u_i(x_{i0}, x_{i1}, \ldots, x_{iS})$ is continuously differentiable and strictly concave for all i; for $i = 1$ it is increasing in all commodities in all states including $s = 0$ and $\partial u_1/\partial x_{1ks} \geqslant \delta_0$ for some positive constant δ_0 and $x_{1ks} \leqslant \Sigma_h \omega_{hks}$. For $i = 2, 3, \ldots, m$, it is nonsatiated in state $s = 0$. Moreover $u_i(0) = 0$, and $\lim_{x \to x_0, n \downarrow 0} n u_i(x_i/n) = 0$ for every $x_0 \geqslant 0$ and $n \geqslant 0$.

ASSUMPTION **CA1′** The utility function $u_i\colon R_+^{r(1+S)} \to R_+$, $u_i(x_{i0}, x_{i1}, \ldots, x_{iS})$ satisfies CA1 with the additional requirement that for consumers $i = 2, 3, \ldots, m$, it is nonsatiated in all states.

ASSUMPTION **CA2** The endowment ω_{is} of consumer i in state s is nonnegative, nonzero for all i and all s; $\Sigma_i \omega_{is} > 0$ for all s.

ASSUMPTION **CA2′** $\omega_{is} > 0$ for all i and s.

ASSUMPTION **M** (Money supply) The money supply function $M(\bar{p})$ is positive, continuous, and homogeneous of degree one in $\bar{p}$ for all nonzero, nonnegative $\bar{p}$.

ASSUMPTION **R** The return functions $D_{sc}(\bar{p}, \bar{\phi})$ are nonnegative, continuous, and homogeneous of degree zero in $(\bar{p}, \bar{\phi})$. For given c, $D_{sc}(\bar{p}, \bar{\phi})$ is different from zero in some state s.

ASSUMPTION **R′** · Returns in state s depend on $\bar{p}_s$ only; $D_{sc}(\bar{p}_s)$ is nonnegative, continuous, and homogeneous of degree zero in $\bar{p}_s$. For given c, $D_{sc}(\bar{p}_s)$ should not be zero in all states.

References

Abel, A. B., and O. J. Blanchard. 1986. The present value of profits and cyclical movements in investment. *Econometrica* 54: 249–73.

Adelman, I., and S. Robinson. 1978. *Income Distribution Policy in Developing Countries: A Case Study of Korea.* Oxford: Oxford University Press.

Afriat, S. 1967. The construction of a utility function from expenditure data. *International Economic Review* 8: 67–77.

Afriat, S. 1973. On a system of inequalities in demand analysis: An extension of the classical method. *International Economic Review* 14: 460–72.

Allgower, E., and K. Georg. 1980. Simplicial and continuation methods for approximating fixed points and solutions to systems of equations. *SIAM Review* 22: 28–85.

Amemiya, T. 1985. *Advanced Econometrics.* Cambridge: Harvard University Press.

Archibald, G. C., and D. Donaldson. 1976. Non-paternalism and the basic theorems of welfare economics. *Canadian Journal of Economics* 9: 492–507.

Arellano, M., and C. Meghir. 1992. Female labour supply and on-the-job search: An empirical model estimated using complementary data sets. *Review of Economic Studies* 59: 537–77.

Armington, P. 1969. A theory of demand for products distinguished by place of production. *IMF Staff Papers* 16: 159–78.

Arrow, K. J. 1954. Import substitution in Leontief models. *Econometrica* 22: 481–92.

Arrow, K. J. 1970. *Essays in the Theory of Risk Bearing.* Amsterdam: North-Holland.

Arrow, K. J. 1981. Optimal and voluntary income distribution. Reprinted in *Collected Papers.* Oxford: Basil Blackwell, 1984.

Arrow, K. J., and F. Hahn. 1971. *General Competitive Analysis.* San Francisco: Holden-Day.

Arrow, K. J., and M. Kurz. 1980. *Public Investment, the Rate of Return and Optimal Fiscal Policy.* Baltimore: Johns Hopkins University Press.

d'Aspremont, C., R. Dos Santos, and L.-A. Gérard-Varet. 1992. General equilibrium concepts under imperfect competition: A Cournotian approach. CORE Discussion Paper 9257. Université Catholique de Louvain, Louvain-la-Neuve.

Atkinson, A. B., and J. E. Stiglitz. 1980. *Lectures on Public Economics.* New York: McGraw-Hill.

Auerbach, A., and L. Kotlikoff. 1987. *Dynamic Fiscal Policy.* Cambridge: Cambridge University Press.

Aumann, R. 1964. Markets with a continuum of traders. *Econometrica* 32: 39–50.

Aumann, R. 1966. Existence of competitive equilibrium with a continuum of traders. *Econometrica* 34: 1–17.

Avriel, M. 1976. *Nonlinear Programming.* Englewood Cliffs, NJ: Prentice Hall.

Avriel, M., W. E. Diewert, S. Schaible, and I. Zang. 1988. *Generalized Concavity.* New York: Plenum.

Backus, D., and P. Kehoe. 1989. International evidence on business cycles. Working Paper 425. Federal Reserve Bank of Minneapolis Research Department.

Balasko, Y., D. Cass, and K. Shell. 1980. Existence of a competitive equilibrium in a general overlapping generations model. *Journal of Economic Theory* 23: 307–22.

Barnett, W. A., J. Geweke, and M. Wolfe. 1991. Semi-nonparametric Bayesian estimation of consumer and factor demand models. In W. A. Barnett, B. Cornet, C. d'Aspremont, J. J. Gabszewicz, and A. Mas-Colell, eds., *Equilibrium Theory and Applications.* Cambridge: Cambridge University Press.

Barten, A. P., and V. Böhm. 1982. Consumer theory. In K. J. Arrow and M. D. Intrilligator, eds., *Handbook of Mathematical Economics*, vol. 2. Amsterdam: North-Holland.

Barten, A. P., and E. Geyskens. 1975. The negativity condition in consumer demand. *European Economic Review* 6: 227–60.

Bartik, T. J. 1987. The estimation of demand parameters in hedonic price models. *Journal of Political Economy* 95: 81–88.

Becker, G. 1993. *A Treatise on the Family.* Cambridge: Harvard University Press.

Benassy, J. P. 1975. Neo-keynesian disequilibrium in a monetary economy. *Review of Economic Studies* 42: 503–23.

Benassy, J. P. 1982. *The Economics of Market Disequilibrium.* New York: Academic Press.

Benassy, J. P. 1986. *Macroeconomics: An Introduction to the Non-Walrasian Approach.* Orlando: Academic Press.

Benassy, J. P. 1988. The objective demand curve in general equilibrium with price makers. *Economic Journal* 98 (suppl.): 37–49.

Benassy, J. P. 1991. Monopolistic Competition. In W. Hildenbrand and H. Sonnenschein, eds., *Handbook of Mathematical Economics*, vol. 4. Amsterdam: North-Holland.

Benjamin, N. 1994. Investment, expectations and Dutch disease. A comparative study: Bolivia, Cameroon, Indonesia. In J. Mercenier and T. N. Srinivasan, eds., *Applied General Equilibrium and Economic Development.* Ann Arbor: University of Michigan Press.

Bergman, L., D. Jorgenson, and E. Zalai, eds. 1990. *General Equilibrium Modeling and Economic Policy Analysis.* Cambridge, MA: Basil Blackwell.

Bergman, L., and S. Lundgren. 1990. General equilibrium approaches to energy policy analysis in Sweden. In L. Bergman, D. Jorgenson, and E. Zalai, eds., *General Equilibrium Modeling and Economic Policy Analysis.* Cambridge, MA: Basil Blackwell.

Berthelemy, J. C., and F. Bourguignon. 1994. North–South–OPEC trade relations. In J. Mercenier and T. N. Srinivasan, eds., *Applied General Equilibrium Analysis and Economic Development*, Ann Arbor: University of Michigan Press.

Bevan, D. L., P. Collier, and J. W. Gunning. 1990. *Controlled Open Economies.* Oxford: Oxford University Press.

Bhagwati, J. N., R. A. Brecher, and T. Hatta. 1983. The generalized theory of transfers and welfare: Bilateral transfers in a multilateral world. *American Economic Review* 73: 606–18.

Blackorby, C., D. Primont, and R. R. Russell. 1978. *Duality, Separability and Functional Structure: Theory and Economic Applications.* New York: Elsevier.

Blanchard, O., and S. Fischer. 1989. *Lectures on Macroeconomics.* Cambridge: MIT Press.

Blitzer, C. R., P. B. Clark, and L. Taylor, eds. 1976. *Economy-wide Models and Development Planning.* Oxford: Oxford University Press.

Blitzer, C. R., and R. Eckaus. 1986. Energy-economy interactions in Mexico: A multiperiod general equilibrium model. *Journal of Development Economics* 21: 259–82.

Blitzer, C. R., R. S. Eckaus, S. Lahiri, and A. Meeraus. 1994. A general equilibrium analysis of the effects of carbon emissions restrictions on economic growth in a developing country. In J. Mercenier and T. N. Srinivasan, eds., *Applied General Equilibrium and Economic Development.* Ann Arbor: University of Michigan Press.

Böhm, V. 1978. Disequilibrium dynamics in a simple macroeconomic model. *Journal of Economic Theory* 17: 179–99.

Böhm, V. 1990. General equilibrium with profit maximizing oligopolists. Discussion Paper 414/90, Universität Mannheim.

Böhm, V., and P. Levine. 1979. Temporary equilibria with quantity rationing. *Review of Economic Studies* 46: 361–77.

Boldrin, M., and L. Montrucchio. 1986. On the indeterminacy of capital accumulation paths. *Journal of Economic Theory* 40: 26–39.

Bonanno, G. 1990. General equilibrium theory with imperfect competition. *Journal of Economic Surveys* 4: 297–328.

Bonnisseau, J. M., and B. Cornet. 1991. General equilibrium theory with increasing returns: The existence problem. In W. Barnett, B. Cornet, C. d'Aspremont, J. Gabszewicz, and A. Mas-Colell, eds., *Equilibrium Theory and Applications.* Cambridge: Cambridge University Press.

Bourguignon, F., W. Branson, and J. de Melo. 1992. Adjustment and income distribution: A micro-macro model for counterfactual analysis. *Journal of Development Economics* 38: 17–39.

Bourguignon, F., and C. Morrison. 1992. *Adjustment and Equity in Developing Countries: A New Approach.* Paris: OECD.

Bovenberg, A. L., and L. Goulder. 1989. Promoting international capital mobility: An international general equilibrium analysis. NBER Working Paper 3139.

Bovenberg, A. L., and L. Goulder. 1991. Introducing intertemporal and open economy features in applied general equilibrium models. In H. Don, Th. van de Klundert, and J. van Sinderen, eds., *Applied General Equilibrium Modelling*. Dordrecht: Kluwer.

Boyce, R. 1975. Estimation of dynamic Gorman polar form utility functions. *Annals of Economic and Social Measurement* 4: 103–16.

Brooke, A., D. Kendrick, and A. Meeraus. 1992. *GAMS: A User's Guide. Release 2.25*. San Francisco: Scientific Press.

Brown, D. J. 1991. Equilibrium analysis with non-convex technologies. In W. Hildenbrand and H. Sonnenschein, eds., *Handbook of Mathematical Economics*, vol. 4. Amsterdam: North-Holland.

Brown, D. J., and J. D. Geanakoplos. 1982. Understanding overlapping generations economies as lack of market clearing at infinity. Cowles Foundation Discussion Paper. Yale University.

Bruno, M. 1966. A programming model for Israel. In I. Adelman and E. Thorbecke, eds., *The Theory and Design of Economic Development*. Baltimore: Johns Hopkins University Press.

Burniaux, J.-M., J. Martin, and G. Nicoletti. 1991. GREEN: A multi-sector, multi-region general equilibrium model for quantifying the cost of curbing CO2 emissions. A technical manual. Economics and Statistics Department. Paris: OECD.

Burniaux, J.-M., and D. van der Mensbrugghe. 1990. The RUNS model, a rural–urban North–South general equilibrium model for agricultural policy analysis. OECD Development Center Technical Paper 33. Paris: OECD.

Burniaux, J.-M., and J. Waelbroeck. 1992. Preliminary results from two experimental models of general equilibrium with imperfect equilibrium. *Journal of Policy Modelling* 14: 65–92.

Burniaux, J. M., and J. Waelbroeck. 1994. Market reform in a heterogeneous world: A general equilibrium assessment of the single market program. In M. Dewatripont and V. Ginsburgh, eds., *European Economic Integration: A Challenge in a Changing World*. Amsterdam: North-Holland.

Carlson, D. A., and A. Haurie. 1987. *Infinite Horizon Optimal Control*. Berlin: Springer-Verlag.

Carson, R. T., and R. C. Mitchell. 1993. The issue of scope in contingent valuation studies. *American Journal of Agricultural Economics* 75: 1263–67.

Chae, S. 1988. Existence of competitive equilibrium with incomplete markets. *Journal of Economic Theory* 44: 179–88.

Chander, P., and H. Tulkens. 1994. A core-theoretic solution for the design of cooperative agreements on transfrontier pollution. CORE Discussion Paper 9448. Université Catholique de Louvain, Louvain-la-Neuve.

Chang, L. J. 1995. Business cycles with distorting taxes and disaggregated capital markets. *Journal of Economic Dynamics and Control* 19: 985–1009.

Chenery, H. B., and A. M. Strout. 1966. Foreign assistance and economic development. *American Economic Review* 56: 680–733.

Christensen, L. R., D. W. Jorgenson, and L. J. Lau. 1971. Conjugate duality and the transcendental logarithmic production function. *Econometrica* 39: 255–56.

Clarete, R. L., and J. A. Roumasset. 1990. The relative welfare cost of industrial and agricultural policy distortions: A Philippine illustration. *Oxford Economic Papers* 42: 462–72.

Clower, R. W. 1960. Keynes and the classics: A dynamical perspective. *Quarterly Journal of Economics* 74: 318–23.

Codognato, G. 1994. *Cournot-Walras Equilibrium in Exchange Economies*. Ph.D. dissertation. Université Catholique de Louvain. Louvain-la-Neuve: CIACO.

Codognato, G., and J. J. Gabzsewicz. 1991. Cournot-Walras equilibria in pure exchange economies. CORE Discussion Paper 9110. Université Catholique de Louvain, Louvain-la-Neuve.

Codsi, G., K. R. Pearson, and P. J. Wilcoxen. 1992. *General Purpose Software for Intertemporal Economic Models.* Amsterdam: Kluwer.

Collier, P., and J. W. Gunning. 1992. Aid and exchange rate adjustments in African trade liberalisations. *Economic Journal* 102: 925–39.

Condon, T., V. Corbo, and J. de Melo. 1990. Exchange rate based disinflation, wage rigidity and capital inflows. *Journal of Development Economics* 32: 113–31.

Cooley, T. F., ed. 1995. *Frontiers of Business Cycle Research.* Princeton: Princeton University Press.

Cornet, B. 1988. General equilibrium theory and increasing returns: Presentation. *Journal of Mathematical Economics* 17: 103–18.

Cornielje, O. 1990. *Rationing and Capital Mobility in Applied General Equilibrium Models.* Amsterdam: VU University Press.

Cornielje, O., and G. van der Laan. 1986. The computation of quantity constrained equilibria by virtual taxes. *Economics Letters* 22: 1–6.

Cottle, R., and G. B. Dantzig. 1968. Complementary pivot theory of mathematical programming. *Linear Algebra and Its Applications* 1: 103–25.

Cox, D., and R. Harris. 1985. Trade liberalization and industrial organization: Some estimates for Canada. *Journal of Political Economy* 93: 115–45.

Cripps, M. W., and G. D. Myles. 1988. General equilibrium and imperfect competition: Profit feedback effects and price normalisations. Discussion Paper 295. University of Warwick.

Dahl, H., and P. Mitra. 1990. Applying tax policy models in country economic work: Bangladesh, China and India. Washington: World Bank.

Danthine, J.-P., and J. B. Donaldson. 1989. Risk sharing, the minimum wage, and the business cycle. Mimeo. Columbia University.

Danthine, J.-P., and J. B. Donaldson. 1992. Risk sharing in the business cycle. *European Economic Review* 36: 468–75.

Danthine, J.-P., and J. B. Donaldson. 1993. Methodological and empirical issues in real business cycle theory. *European Economic Review* 37: 1–35.

Danthine, J.-P., and J. B. Donaldson. 1995. Computing equilibria of nonoptimal economies. In T. F. Cooley, ed., *Frontiers of Business Cycle Research.* Princeton: Princeton University Press.

Dasgupta, P. 1993. The economics of destitution. In A. de Zeeuw, ed., *Advanced Lectures in Quantitative Economics.* San Diego: Academic Press.

Dasgupta, P., and D. Ray. 1986. Inequality as determinant of malnutrition and unemployment: Theory. *Economic Journal* 96: 1011–34.

Davidson, R., and J. G. MacKinnon. 1993. *Estimation and Inference in Econometrics.* Oxford: Oxford University Press.

Davies, J., J. Whalley, and B. Hamilton. 1989. Capital income taxation in a two commodities lifecycle model: The role of factor intensities and asset capitalization effects. *Journal of Public Economics* 39: 109–26.

Deardorff, A. V., and R. M. Stern. 1986. The structure and sample results of the Michigan computational model of world production and trade. In T. N. Srinivasan and J. Whalley, eds., *General Equilibrium Trade Policy Modeling.* Cambridge: MIT Press.

Deaton, A. 1985. Life cycle models of consumption: Is the evidence consistent with the theory? In T. F. Bewley, ed., *Advances in Econometrics.* New York: Cambridge University Press.

Deaton, A. 1986. Demand analysis. In Z. Griliches and M. D. Intrilligator, eds., *Handbook of Econometrics.* Amsterdam: North-Holland.

Deaton, A. 1991. Saving and liquidity constraints. *Econometrica* 59: 1221–48.

Deaton, A. 1995. Data and econometric tools for development analysis. In J. Behrman and T. N. Srinivasan, eds., *Handbook of Development Economics*, vol. 3. Amsterdam: North-Holland.

Deaton, A., and J. Muellbauer. 1980. *Economics and Consumer Behavior.* Cambridge: Cambridge University Press.

Debreu, G. 1954. Valuation equilibrium and Pareto optimum. *Proceedings of the National Academy of Sciences* 40: 588–92.

Debreu, G. 1959. *Theory of Value*. New-York: Wiley.

Debreu, G. 1974. Excess demand functions. *Journal of Mathematical Economics* 1: 15–23.

Debreu, G. 1983. *Mathematical Economics: Twenty Papers of Gérard Debreu*. Cambridge: Cambridge University Press.

Decaluwé, B., and A. Martens. 1988. CGE modeling and developing economies: A concise empirical survey of 73 applications to 26 countries. *Journal of Policy Modeling* 10: 520–68.

Decaluwé, B., and A. Martens. 1989. Developing countries and GE models: A review of the empirical literature. IDRC Report No MR 155e. International Development Research Center, Ottawa.

Delorme, F., and D. van der Mensbrugghe. 1990. Assessing the role of scale economies and imperfect competition in the context of agricultural trade liberalization: A Canadian case study. OECD Economic Studies No 13. OECD Development Center, Paris.

Dervis, K., J. de Melo, and S. Robinson. 1982. *General Equilibrium Models for Development Policy*. New York: Cambridge University Press.

Devarajan, S., C. Jones, and M. Roemer. 1989. Markets under price controls in partial and general equilibrium. *World Development* 17: 1881–93.

Devarajan, S., J. D. Lewis, and S. Robinson. 1986. A bibliography of computable general equilibrium (CGE) models applied to developing countries. Mimeo. Department of Agricultural and Resource Economics, University of California at Berkeley.

Devarajan, S., and D. Rodrick. 1991. Pro-competitive effects of trade reform: Results from a CGE-model of Cameroon. *European Economic Review* 35: 1157–84.

Dewatripont, M., and G. Michel. 1987. On closure rules, homogeneity and dynamics in applied general equilibrium models. *Journal of Development Economics* 26: 65–76.

Dierker, E. 1972. Two remarks on the number of equilibria of an economy. *Econometrica* 40: 951–53.

Dierker, E., and B. Grodal. 1994. Profit maximization, relative prices, and the maximization of shareholders real wealth. Department of Economics, University of Copenhagen.

Dierker, H., and B. Grodal. 1986. Nonexistence of Cournot-Walras equilibrium in a general equilibrium model with two oligopolists. In W. Hildenbrand and A. Mas-Colell, eds., *Contributions to Mathematical Economics. Essays in Honour of Gérard Debreu*. Amsterdam: North-Holland.

Diewert, W. E. 1971. An application of the Shephard duality theorem: A generalized Leontief production function. *Journal of Political Economy* 79: 481–507.

Diewert, W. E. 1974. Applications of duality theory. In M. D. Intrilligator and D. A. Kendrick, eds., *Frontiers of Quantitative Economics*. Amsterdam: North-Holland.

Diewert, W. E. 1977. Generalized Slutsky conditions for aggregate consumer demand functions. *Journal of Economic Theory* 15: 336–53.

Diewert, W. E. 1982. Duality approaches to microeconomic theory. In K. J. Arrow and M. D. Intrilligator, eds., *Handbook of Mathematical Economics*, vol. 2. Amsterdam: North-Holland.

Diewert, W. E., and T. J. Wales. 1987. Flexible functional forms and global curvature conditions. *Econometrica* 55: 43–68.

Dixit, A., and J. Stiglitz. 1977. Monopolistic competition and optimum product diversity. *American Economic Review* 67: 297–308.

Dixon, P. 1975. *The Theory of Joint Maximization*. Amsterdam: North-Holland.

Dixon, P. 1983. Notes and problems in applied general equilibrium economics. Chapter 3: The Johansen approach to general equilibrium modelling. Mimeo.

Dixon, P. 1990. A general equilibrium approach to public utility pricing: Determining prices for a water authority. *Journal of Policy Modelling* 12: 745–67.

Dixon, P., B. Parmenter, A. Powell, and P. J. Wilcoxen. 1992. *Notes and Problems in Applied General Equilibrium Economics*. Amsterdam: North-Holland.

Dixon, P., B. Parmenter, G. Ryland, and J. Sutton. 1977. *Orani: A General Equilibrium Model for the Australian Economy*. Canberra: Government Printing Office.

Dixon, P., B. Parmenter, J. Sutton, and D. Vincent. 1982. *Orani: A Multisectoral Model of the Australian Economy*. Amsterdam: North-Holland.

Don, H., T. van de Klundert, and J. van Sinderen. 1991. *Applied General Equilibrium Modelling*. Dordrecht: Kluwer.

Drèze, J. 1974. Investment under private ownership: Optimality, equilibrium and stability. In J. Drèze, ed., *Allocation under Uncertainty: Equilibrium and Optimality*. London: Macmillan.

Drèze, J. 1975. Existence of an exchange equilibrium under price rigidities, *International Economic Review* 16: 301–20.

Drèze, J. 1982. Decision criteria for business firms. In M. Hazewinkel and A. H. G. Rinnooy Kan, eds., *Current Developments in the Interface: Economics, Econometrics, Mathematics*. Amsterdam: Reidel.

Drèze, J., and K. Hagen. 1978. Choice of product quality: Equilibrium and efficiency. *Econometrica* 46: 493–513.

Drèze, J., and D. de la Vallée Poussin. 1971. A tatonnement process for public goods. *Review of Economic Studies* 38: 133–50.

Duffie, D. 1991. The theory of value in security markets. In W. Hildenbrand and H. Sonnenschein, eds., *Handbook of Mathematical Economics*, vol. 4. Amsterdam: North-Holland.

Easterly, W. 1990. Portfolio effects in a CGE model: Devaluation in a dollarized economy. In L. Taylor, ed., *Socially Relevant Policy Analysis*. Cambridge: MIT Press.

Elbers, C. 1992. *Spatial Disaggregation in General Equilibrium Models, with an Application to the Nepalese Economy*. Amsterdam: VU University Press.

Epple, D. 1987. Hedonic prices and implicit markets: Estimating demand and supply functions for differentiated products. *Journal of Political Economy* 95: 59–79.

Epstein, L. G. 1987. The global stability of efficient intertemporal allocations. *Econometrica* 55: 329–55.

Epstein, L., and M. Denny. 1983. The multivariate flexible accelerator model: Its empirical restrictions and an application to US manufacturing. *Econometrica* 51: 647–74.

Erlich, S., V. Ginsburgh, and L. Van der Heyden. 1987. Where do real wage policies lead Belgium: A general equilibrium analysis. *European Economic Review* 31: 1369–83.

Ermoliev, Yu. M., and R. J. B. Wets, eds. 1988. *Numerical Techniques for Stochastic Optimization*. Berlin: Springer-Verlag.

Fair, R. C., and J. B. Taylor. 1983. Solution and maximum likelihood estimates of dynamic nonlinear rational expectations models, *Econometrica* 51: 1169–85.

Farmer, R. E. 1993. *The Macroeconomics of Self-fulfilling Expectations*. Cambridge: MIT Press.

Feltenstein, A. 1986. An intertemporal general equilibrium analysis of financial crowding out: A policy model and an application to Australia. *Journal of Public Economics* 31: 79–104.

Feltenstein, A., D. Lebow, and A. Sibert. 1988. An analysis of the welfare implications of alternative exchange rate regimes: An intertemporal model with applications to Australia. *Journal of Policy Modelling* 10: 611–29.

Fischer, G., K. Frohberg, M. A. Keyzer, and K. S. Parikh. 1988. *Linked National Models: A Tool for International Food Policy Analysis*. Dordrecht: Kluwer.

Fischer, G., K. Frohberg, M. A. Keyzer, K. S. Parikh, and W. Tims. 1991. Hunger beyond the reach of the invisible hand. IIASA Research Report 91-15. Laxenburg, Austria.

Flam, S. D. and R. J. B. Wets. 1987. Existence results and finite horizon approximation of infinite horizon optimization problems. *Econometrica* 55: 1187–1209.

Flores, R. 1993. Assessing the gains from Mercosul: A general equilibrium imperfect competition evaluation. Paper presented at the Latin American Meeting of the Econometric Society, Tucumán, Argentina.

Fullerton, D., and R. H. Gordon. 1983. A reexamination of tax distortions in general equilibrium models. In M. Feldstein, ed., *Behavioral Simulation Methods in Tax Policy Analysis*. Chicago: University of Chicago Press.

Fuss, M., and D. McFadden, eds. 1978. *Production Economics: A Dual Approach to Theory and Application.* Amsterdam: North-Holland.

Fuss, M., D. McFadden, and Y. Mundlak. 1978. A survey of functional forms in the economic analysis of production. In M. Fuss and D. McFadden, eds., *Production Economics: A Dual Approach to Theory and Application.* Amsterdam: North-Holland.

Gabszewicz, J. J., and P. Michel. 1992. Oligopoly equilibria in exchange economies. CORE Discussion Paper 9247. Université Catholique de Louvain, Louvain-la-Neuve.

Gabszewicz, J. J., and J. P. Vial. 1972. Oligopoly à la Cournot in general equilibrium analysis. *Journal of Economic Theory* 4: 381–400.

Garcia, C. B., and W. I. Zangwill. 1981. *Pathways to Solutions, Fixed Points, and Equilibria.* Englewood Cliffs, N.J.: Prentice Hall.

Galor, O. 1992. A two sector overlapping generations model: A global characterization of the dynamical system. *Econometrica* 60: 1351–86.

Garcia-Mila, T., A. Marcet, and E. Ventura. 1995. Supply-side interventions and redistribution. Economics Working Paper 115. Universitat Pompeu Fabra, Barcelona.

Gary-Bobo, R. 1989. Cournot-Walras and locally consistent equilibria. *Journal of Economic Theory* 49: 10–32.

Gasiorek, M., A. Smith, and A. J. Venables. 1992a. Completing the internal market in the EC: Factor demands and comparative advantage. In L. A. Winters and A. J. Venables, eds., *European Integration: Trade and Industry.* Cambridge: Cambridge University Press.

Gasiorek, M., A. Smith, and A. J. Venables. 1992b. 1992, trade and welfare: A general equilibrium model. In L. A. Winters, ed., *Trade Flows and Trade Policy after 1992.* Cambridge: Cambridge University Press.

Geanakoplos, J. D. 1990. An introduction to general equilibrium with incomplete asset markets. *Journal of Mathematical Economics* 19: 1–38.

Geanakoplos, J. D., and H. M. Polemarchakis. 1986. Existence, regularity and constrained suboptimality of competitive allocations when the asset market is incomplete. In W. Heller, R. Starr, and D. Starrett, eds., *Uncertainty, Information and Communication. Essays in Honour of Kenneth Arrow.* Cambridge: Cambridge University Press.

Geanakoplos, J. D., and H. M. Polemarchakis. 1991. Overlapping generations. In W. Hildenbrand and H. Sonnenschein, eds., *Handbook of Mathematical Economics,* vol. 4. Amsterdam: North-Holland.

Gertler, P., and J. van der Gaag. 1988. Measuring the willingness-to-pay for social services in developing countries. Living Standards Measurement Study, Working Paper 45. Washington: World Bank.

Ginsburgh, V. 1994. In the Cournot-Walras general equilibrium model, there may be "more to gain" by changing the numéraire than by eliminating imperfections: A two-good economy example. In J. Mercenier and T. N. Srinivasan, eds., *Applied General Equilibrium Analysis and Economic Development.* Ann Arbor: University of Michigan Press.

Ginsburgh, V., and M. A. Keyzer. 1994. Existence and efficiency properties of an approximate equilibrium when asset markets are incomplete: A welfare approach. Discussion Paper TI 94-83. Tinbergen Institute, Amsterdam.

Ginsburgh, V., and H. Sneessens. 1992. Structural shocks and investment subsidies in an overlapping generations model with perfect foresight. In J. J. Gabszewicz, J.-F. Richard, and L. A. Wolsey, eds., *Economic Decision Making: Games, Econometrics and Optimization.* Amsterdam: North-Holland.

Ginsburgh, V., and L. Van der Heyden. 1985. General equilibrium with wage rigidities: An application to Belgium. *Mathematical Programming Study* 23: 23–39.

Ginsburgh, V., and L. Van der Heyden. 1988. On extending the Negishi approach to computing equilibria. *Journal of Economic Theory* 44: 168–78.

Ginsburgh, V., and J. Waelbroeck. 1976. Computational experience with a large general equilibrium model of the world economy. In J. and M. Los, eds., *Computing Equilibria: How and Why.* Amsterdam: North-Holland.

Ginsburgh, V., and J. Waelbroeck. 1981. *Activity Analysis and General Equilibrium Modelling.* Amsterdam: North Holland.

Goldin, I., and O. Knudsen, eds. 1990. *Agricultural Trade Liberalization.* Paris: OECD Development Center.

Goldin, I., and A. Winters, eds. 1992. *Open Economies: Adjustment and Agriculture.* Cambridge: Cambridge University Press.

Gorman, W. M. 1953. Community preference fields. *Econometrica* 21: 63–80.

Goulder, L. 1991. Effect of carbon taxes in an economy with prior tax distortions: An intertemporal general equilibrium analysis for the U.S. Mimeo.

Gourieroux, Ch., J.-J. Laffont, and A. Montfort. 1980. Coherency condition in simultaneous linear equations models with endogenous switching regimes. *Econometrica* 48: 675–96.

Grandmont, J. M. 1974. On the short run equilibrium in a monetary economy. In J. Drèze, ed., *Allocation under Uncertainty, Equilibrium and Optimality.* New York: Macmillan.

Grandmont, J. M. 1977a. The logic of the fix-price method. *Scandinavian Journal of Economics* 79: 169–86.

Grandmont, J. M. 1977b. Temporary general equilibrium theory. *Econometrica* 45: 535–72.

Grandmont, J. M. 1982. Temporary general equilibrium theory. In K. J. Arrow and M. D. Intrilligator, eds., *Handbook of Mathematical Economics,* vol. 2. Amsterdam: North-Holland.

Grandmont, J. M., and G. Laroque. 1975. On money and banking. *Review of Economic Studies* 42: 207–36.

Greene, W. H. 1991. *Econometric Analysis.* New York: Macmillan.

Gregory, R., S. Lichtenstein, and P. Slovic. 1993. Valuing environmental resources: A constructive approach. *Journal of Risk and Uncertainty* 7: 177–97.

Griliches, Z., ed. 1971. *Price Indexes and Quality Change.* Cambridge: Harvard University Press.

Grodal, B. 1992. Profit maximization and imperfect competition. Manuscript. University of Copenhagen.

Grodal, B. 1993. Strategic behaviour in imperfectly competitive markets. Manuscript. University of Copenhagen.

Guesnerie, R. 1977. On the direction of tax reform. *Journal of Public Economics* 7: 179–202.

Gunning, J. W. 1983. Rationing in an open economy. *European Economic Review* 23: 71–98.

Gunning, J. W., G. Carrin, J. Waelbroeck, J. M. Burniaux, and J. Mercenier. 1982. Growth and trade of developing countries: A general equilibrium analysis. Discussion Paper 8210. Centre d'Economie Mathématique et d'Econométrie. Université Libre de Bruxelles.

Gunning, J. W., and M. A. Keyzer. 1995. Applied general equilibrium models for policy analysis. In J. Behrman and T. N. Srinivasan, eds., *Handbook of Development Economics,* vol. 3. Amsterdam: North-Holland.

Hahn, F. 1984. *Equilibrium and Macroeconomics.* Oxford: Basil Blackwell.

Hamilton, B., and J. Whalley. 1984. Efficiency and distributional implications of global restrictions on labour mobility: Calculations and policy implications. *Journal of Development Economics* 14: 61–75.

Hansen, G. D., and E. C. Prescott. 1995. Recursive methods for computing equilibria of business cycle models. In T. F. Cooley, ed., *Frontiers of Business Cycle Research,* Princeton: Princeton University Press.

Härdle, W., W. Hildenbrand, and M. Jerison. 1991. Empirical evidence on the law of demand. *Econometrica* 59: 1525–49.

Harris, R. 1984. Applied general equilibrium analysis of small open economies with scale economies and imperfect competition. *American Economic Review* 74: 1016–32.

Harris, J. R., and M. P. Todaro. 1970. Migration, unemployment and development: A two-sector analysis. *American Economic Review* 60: 126–42.

Hart, O. 1975. On the optimality of equilibrium when the market structure is incomplete. *Journal of Economic Theory* 11: 418–33.

Hart, O. 1985. Imperfect competition in general equilibrium: An overview of recent work. In K. J. Arrow and S. Honkapohja, eds., *Frontiers of Economics*. Oxford: Basil Blackwell.

Harvey, A. C., and A. Jaeger. 1993. Detrending stylized facts and the business cycle. *Journal of Applied Econometrics* 8: 231–47.

Hausman, D. M., and M. S. McPherson. 1993. Taking ethics seriously: Economics and contemporary moral philosophy. *Journal of Economic Literature* 31: 671–731.

Hayashi, F. 1982. Tobin's marginal *q* and average *q*: A neo-classical interpretation. *Econometrica* 50: 213–24.

Helpman, E., and P. R. Krugman. 1989. *Trade Policy and Market Structure*. Cambridge: MIT Press.

Helpman, E., and P. R. Krugman. 1990. *Market Structure and Foreign Trade*. Cambridge: MIT Press.

Hildenbrand, K., and W. Hildenbrand. 1986. On the mean income effect: A data analysis of the U.K. family expenditure survey. In W. Hildenbrand and A. Mas-Colell, eds., *Contributions to Mathematical Economics in Honour of Gérard Debreu*. Amsterdam: North-Holland.

Hildenbrand, W., and H. Sonnenschein, eds. 1991. *Handbook of Mathematical Economics*, vol. 4. Amsterdam: North-Holland.

Hirsch, M. W., and S. Smale. 1973. *Differential Equations, Dynamical Systems and Linear Algebra*. New York: Academic Press.

Hirschleifer, J. 1985. The expanding domain of economics. *American Economic Review* (December suppl.): 53–68.

Hodrick, R. J., and E. C. Prescott. 1980. Post-war U.S. business cycles: An empirical investigation. Mimeo. Carnegie Mellon University, Pittsburgh.

Hotelling, H. 1929. Stability in competition. *Economic Journal* 39: 41–57.

Imam, H., and J. Whalley. 1982. General equilibrium with price intervention policies: A computational approach. *Journal of Public Economics* 18: 105–19.

Ingram, B. F., N. R. Kocherlakota, and N. E. Savin. (1994). Explaining business cycles: A multiple-shock approach. *Journal of Monetary Economics* 34: 415–28.

Ito, T. 1980. Disequilibrium growth theory. *Journal of Economic Theory* 23: 380–409.

de Janvry, A., and E. Sadoulet. 1990. Efficiency, welfare effects and political feasibility of alternative anti-poverty and adjustement programs. Paris: OECD.

de Janvry, A., E. Sadoulet, and A. Fargeix. 1991. Politically feasible and equitable adjustment: Some alternatives for Ecuador. *World Development* 19: 1577–99.

Johansen, L. 1960. *A Multisectoral Study of Economic Growth*. Amsterdam: North-Holland.

Johansen, L. 1972. *Production Functions: An Integraton of Micro and Macro Short Run and Long Run Aspects*. Amsterdam: North-Holland.

Jorgenson, D. W. 1984. Econometric methods for applied general equilibrium analysis. In H. Scarf and J. Shoven, eds., *Applied General Equilibrium Analysis*. Cambridge: Cambridge University Press.

Jorgenson, D. W. 1986. Econometric methods for modelling producer behavior. In Z. Griliches and M. D. Intrilligator, eds., *Handbook of Econometrics*. Amsterdam: North-Holland.

Jorgenson, D. W., and D. T. Slesnick. 1985. General equilibrium analysis of economic policy. In J. Pigott and J. Whalley, eds., *New Developments in Applied General Equilibrium Analysis*. Cambridge: Cambridge University Press.

Jorgenson, D. W., D. T. Slesnick, and P. J. Wilcoxen. 1992. Carbon taxes and economic welfare. *Brookings Papers: Microeconomics 1992*. Washington: Brookings Institution.

Jorgenson, D. W., and P. J. Wilcoxen. 1990. Environmental regulation and U.S. economic growth. *Rand Journal of Economics* 21: 314–40.

Jorgenson, D. W., and P. J. Wilcoxen. 1993. Energy, the environment and economic growth. In A. V. Kneese and J. L. Sweeney, eds., *Handbook of Natural Resource and Energy Economics*, vol. 3. Amsterdam: North-Holland.

Kaitala, V., K.-G. Mäler, and H. Tulkens. 1992. The acid rain game as a resource allocation process with an application to the cooperation among Finland, Russia and Estonia. CORE Discussion Paper 9242. Université Catholique de Louvain, Louvain-la-Neuve.

Kalecki, M. 1943. Costs and prices. Reprinted in M. Kalecki. 1972. *Selected Essays on the Dynamics of the Capitalist Economy*. Cambridge: Cambridge University Press.

Karlin, S. 1959. *Mathematical Methods and Theory in Games, Programming and Economics*. London: Pergamon.

Kehoe, T. J. 1991. Computation and multiplicity of equilibria. In W. Hildenbrand and H. Sonnenschein, eds., *Handbook of Mathematical Economics*, vol. 4. Amsterdam: North-Holland.

Kehoe, T. J., and D. K. Levine. 1985. Comparative statics and perfect foresight in infinite-horizon economies. *Econometrica* 53: 433–53.

Kehoe, T. J., D. K. Levine, and P. M. Romer. 1989. Characterizing equilibria of models with externalities and taxes as solutions to optimization problems. Working Paper. Federal Reserve Bank of Minneapolis.

Kehoe, T. J., D. K. Levine, and P. M. Romer. 1990. Determinacy of equilibria in dynamic models with finitely many consumers. *Journal of Economic Theory* 50: 1–21.

Kehoe, T. J., D. K. Levine, A. Mas-Colell, and W. R. Zame. 1989. Determinacy of equilibrium in large-scale economies. *Journal of Mathematical Economics* 18: 231–62.

Kehoe, T. J., and J. Whalley. 1985. Uniqueness of equilibrium in large scale numerical general equilibrium models. *Journal of Public Economics* 28: 247–54.

Keller, W. J. 1980. *Tax Incidence: A General Equilibrium Approach*. Amsterdam: North-Holland.

Kelley, A. C., and J. G. Williamson. 1984. *What Drives Third World City Growth? A Dynamic General Equilibrium Approach*. Princeton: Princeton University Press.

Kemp, M. G. 1968. *The Pure Theory of International Trade and Investment*. Englewood Cliffs, NJ: Prentice Hall.

Keynes, J. M. 1936. *The General Theory of Employment, Interest and Money*. New York: Harcourt Brace.

Keyzer, M. A. 1977. Analysis of a national model with domestic price policies and quota on international trade. IIASA Research Memorandum RM-77-19. Laxenburg, Austria.

Keyzer, M. A. 1981. *The International Linkage of Open Exchange Economies*. Amsterdam: Vrije Univesiteit Amsterdam.

Keyzer, M. A. 1989. Application of a continuation method to the solution of nonlinear complementarity problems. Research Memorandum 89-02. Centre for World Food Studies, Amsterdam.

Keyzer, M. A. 1991. On the approximation of infinite horizon allocations. In H. Don, Th. van de Klundert, and J. van Sinderen, eds., *Applied General Equilibrium Modelling*. Dordrecht: Kluwer.

Keyzer, M. A. 1995. Using dual variables to compute markup rates in AGE models of imperfect competition. Working Paper 95-09. Centre for World Food Studies, Amsterdam.

Keyzer, M. A., W. van Veen, and W. Tims. 1992. The SOW applied general equilibrium model. In E. Thorbecke, ed., *Adjustment and Equity in Indonesia*. Paris: OECD Development Center.

Keyzer, M. A. 1997. Building applied general equilibrium models in GAMS: A user's guide and a library of models. MIT Press Internet site (http://www.mitpress.mit.edu).

Koopmans, T. C. 1957. *Three Essays on the State of Economic Science*. New York: McGraw-Hill.

Kuhn, H. W., and J. G. MacKinnon. 1975. Sandwich methods for finding a fixed point. *Journal of Optimization Theory and Applications* 17: 82–100.

Kurz, M. 1982. Unemployment equilibrium in an economy with linked prices. *Journal of Economic Theory* 26: 100–23.

Kydland, F. 1995. Business cycles and aggregate labor market fluctuations. In T. F. Cooley, ed., *Frontiers of Business Cycle Research.* Princeton: Princeton University Press.

Kydland, F., and E. C. Prescott. 1982. Time to build and aggregate fluctuations. *Econometrica* 50: 1345–70.

Laffont, J. J. 1987. Incentives and the allocation of public goods. In A. J. Auerbach and M. Feldstein, eds., *Handbook of Public Economics*, vol. 2. Amsterdam: North-Holland.

Lancaster, K. 1966. A new approach to consumer theory. *Journal of Political Economy* 74: 132–57.

Lancaster, K. 1968. *Mathematical Economics.* New York: Macmillan.

Lancaster, K. 1971. *Consumer Demand: A New Approach.* New York: Columbia University Press.

Laroque, G. 1981. On the local uniqueness of fixed price equilibria. *Review of Economic Studies* 48: 113–29.

Lau, L. J. 1980. Applications of profit functions. In M. Fuss and D. McFadden, eds., *Production Economics: A Dual Approach to Theory and Application.* Amsterdam: North-Holland.

Lau, L. J. 1982. A note on the fundamental theorem of exact aggregation. *Economics Letters* 9: 119–26.

Lau, L. J. 1986. Functional forms in econometric model building. In Z. Griliches and M. D. Intrilligator, eds., *Handbook of Econometrics.* Amsterdam: North-Holland.

Leibenstein, H. 1959. *Economic Backwardness and Economic Growth.* New York: Wiley.

Leijonhufvud, A. 1968. *On Keynesian Economics and the Economics of Keynes.* New York: Oxford University Press.

Levy, S., and S. van Wijnbergen. 1991. Transition problems in economic reform: Agriculture in the Mexico–U.S. free trade agreement. Development Center Research Paper. OECD Development Center, Paris.

Lewbel, A. 1987. Characterizing some Gorman Engel curves. *Econometrica* 55: 1451–59.

Lewis, J. 1994. Macroeconomic stabilization and adjustment policies in a general equilibrium model with financial markets. In J. Mercenier and T. N. Srinivasan, eds., *Applied General Equilibrium and Economic Development.* Ann Arbor: University of Michigan Press.

Lluch, C., A. A. Powell, and R. A. Williams. 1977. *Patterns in Household Demand and Saving.* Oxford: Oxford University Press.

Long, J. B., and C. I. Plosser. 1983. Real business cycles. *Journal of Political Economy* 91: 39–69.

Lucas, R. E., and N. L. Stokey. 1984. Optimal growth with many consumers. *Journal of Economic Theory* 32: 139–71.

Lucas, R. E., and N. L. Stokey. 1987. Money and interest in a cash-in-advance economy. *Econometrica* 55: 491–513.

Maasland, A. 1990. Continuing the tradition of equity in Sri Lanka: Policy options in a CGE model. In L. Taylor, ed., *Socially Relevant Policy Analysis.* Cambridge: MIT Press.

Magill, M., and M. Quinzii. 1991. The nonneutrality of money in a production economy with nominal assets. In W. A. Barnett, ed., *Equilibrium Theory and Applications.* New York: Cambridge University Press.

Magill, M., and M. Quinzii. 1992. Theory of incomplete markets. Manuscript. University of Southern California, Los Angeles.

Magill, M., and W. Shafer. 1991. Incomplete markets. In W. Hildenbrand and H. Sonnenschein, eds., *Handbook of Mathematical Economics*, vol. 4. Amsterdam: North-Holland.

Makarov, V., and A. Rubinov. 1977. *Mathematical Theory of Economic Dynamics and Equilibria.* New York: Springer-Verlag.

Malinvaud, E. 1971. A planning approach to the public good production. *Swedish Journal of Economics* 73: 96–112.

Malinvaud, E. 1977. *The Theory of Unemployment Reconsidered.* Oxford: Basil Blackwell.

Manne, A. S. 1985. On the formulation and solution of economic equilibrium models. *Mathematical Programming Study* 23: 1–22.

Manne, A. S., H. P. Chao, and R. Wilson. 1980. Computation of competitive equilibria by a sequence of linear programs. *Econometrica* 48: 1595–1615.

Manne, A. S., and R. G. Richels. 1990. CO2 emission limits: An economic analysis for the U.S.A. *Energy Journal* 11: 51–85.

Manne, A. S., and R. G. Richels. 1992. *Buying Greenhouse Insurance: The Economic Consequences of Carbon Dioxide Emission Limits.* Cambridge: MIT Press.

Mansur, A., and J. Whalley. 1984. Numerical specification of applied general equilibrium models: Estimation, calibration and data. In H. Scarf and J. Shoven, eds., *Applied General Equilibrium Analysis.* Cambridge: Cambridge University Press.

Mantel, R. R. 1974. On the characterization of aggregate excess demand. *Journal of Economic Theory* 7: 348–53.

Marcet, A., and D. A. Marshall. 1994. Solving nonlinear rational expectations models by parametrized expectations: Convergence to stationary solutions. Working Paper, Department of Finance. Northwestern University, Evanston, and Universitat Pompeu Fabra, Barcelona.

Marschak, T., and R. Selten. 1974. *General Equilibrium with Price Making Firms.* Berlin: Springer-Verlag.

Mas-Colell, A. 1982. The Cournotial foundations of Walrasian equilibrium theory: An exposition and recent theory. In W. Hildenbrand, ed., *Advances in Economic Theory.* Cambridge: Cambrige University Press.

Mas-Colell, A. 1985. *The Theory of General Economic Equilibrium: A Differentiable Approach.* Cambridge: Cambridge University Press.

Mas-Colell, A., and W. D. Zame. 1991. Equilibrium theory in infinite dimensional spaces. In W. Hildenbrand and H. Sonnenschein, eds., *Handbook of Mathematical Economics,* vol. 4. Amsterdam: North-Holland.

Matthiesen, L. 1985. Computation of economic equilibria by a sequence of linear complementarity problems. *Mathematical Programming Study* 23: 144–62.

McFadden, D. 1978a. Cost, revenue and profit functions. In M. Fuss and D. McFadden, eds., *Production Economics: A Dual Approach to Theory and Application.* Amsterdam: North-Holland, pp. 3–109.

McFadden, D. 1978b. The general linear profit function. In M. Fuss and D. McFadden, eds., *Production Economics: A Dual Approach to Theory and Application.* Amsterdam: North-Holland, pp. 269–86.

McMahon, G. 1990. Financial computable general equilibrium models of developing countries: A critical assessment. Mimeo. International Development Research Centre, Ottawa.

de Melo, J. 1988. Computable general equilibrium models for trade policy analysis in developing countries: A survey. *Journal of Policy Modeling* 10: 469–503.

de Melo, J., and D. Roland-Holst. 1994. Tariffs and export subsidies when domestic markets are oligopolistic. In J. Mercenier and T. N. Srinivasan, eds., *Applied General Equilibrium Analysis and Economic Development.* Ann Arbor: University of Michigan Press.

de Melo, J., and D. Tarr. 1992. *A General Equilibrium Analysis of United States Foreign Trade Policy.* Cambridge: MIT Press.

Mercenier, J. 1994. Completing the European internal market: A general equilibrium evaluation under alternative market structure assumptions. In M. Dewatripont and V. Ginsburgh, eds., *European Economic Integration: A Challenge in a Changing World.* Amsterdam: North-Holland.

Mercenier, J., and P. Michel. 1994a. Discrete-time finite horizon approximation of infinite horizon optimization problems with steady-state invariance. *Econometrica* 62: 635–56.

Mercenier, J., and P. Michel. 1994b. A criterion for time aggregation in an intertemporal dynamic model. *Mathematical Programming* 64: 179–97.

Mercenier, J., and M. Sampaio de Souza. 1994. Structural adjustment and growth in a highly indebted market economy: Brazil. In J. Mercenier and T. N. Srinivasan, eds., *Applied General Equilibrium and Economic Development.* Ann Arbor: University of Michigan Press.

Mercenier, J., and T. N. Srinivasan, eds. 1994. *Applied General Equilibrium and Economic Development.* Ann Arbor: University of Michigan Press.

Mintz, J., and H. Tulkens. 1986. Commodity tax competition between member states of a federation: Equilibrium and efficiency. *Journal of Public Economics* 29: 133–72.

Mintz, J., and H. Tulkens. 1990. Strategic use of tax rates and credits in a model of international corporate income tax competition. CORE Discussion Paper 9073. Université Catholique de Louvain, Louvain-la-Neuve.

Mirrlees, J. A. 1975. A pure theory of underdeveloped economies. In L. Reynolds, ed., *Agriculture in Development Theory.* New Haven: Yale University Press.

Mitchell, R. C., and R. T. Carson. 1989. Using surveys to value public goods: The contingent valuation method. Washington: Resources for the Future.

Mitra, P. K. 1992. Tariff design and reform in revenue-constrained economy: Theory and an illustration from India. *Journal of Public Economics* 47: 227–51.

Modigliani, F., and M. H. Miller. 1958. The cost of capital, corporation finance and the theory of investment. *American Economic Review* 48: 261–97.

Morgenstern, O., and G. L. Thompson. 1976. *Mathematical Theory of Expanding and Contracting Economies.* Lexington, MA: Lexington Books.

Morrison, C. 1991. *Adjustment and Equity in Morocco.* Paris: OECD Development Center.

Moulin, H. 1985. Egalitarianism and utilitarianism in quasi-linear bargaining. *Econometrica* 53: 49–67.

Muellbauer, J. 1975. Aggregation, income distribution and consumer demand. *Review of Economic Studies* 62: 269–83.

Muellbauer, J. 1976. Community preferences and the representative consumer. *Econometrica* 44: 525–43.

Munkres, J. R. 1975. *Topology: A First Course.* Englewood Cliffs, NJ: Prentice Hall.

Nadiri, M. I. 1982. Producers theory. In K. J. Arrow and M. D. Intrilligator, eds., *Handbook of Mathematical Economics*, vol. 2. Amsterdam: North-Holland.

Narayana, N. S. S., K. S. Parikh, and T. N. Srinivasan. 1990. *Agriculture, Growth and Redistribution of Income: Policy Analysis with a General Equilibrium Model of India.* Amsterdam: North-Holland.

Neary, J. P., and K. Roberts. 1980. The theory of household behaviour under rationing. *European Economic Review* 13: 25–42.

Negishi, T. 1960. Welfare economics and existence of an equilibrium for a competitive economy. *Metroeconomica* 12: 92–97.

Negishi, T. 1961. Monopolistic competition and general equilibrium. *Review of Economic Studies* 28: 196–201.

Negishi, T. 1972. *General Equilibrium Theory and International Trade.* Amsterdam: North-Holland.

Nguyen, T., and J. Whalley. 1985. Equilibrium with price controls under endogenous transaction costs. *Journal of Economic Theory* 29: 290–300.

Nikaido, H. 1975. *Monopolistic Competition and Effective Demand.* Princeton: Princeton University Press.

Nsengiyumva, F., and B. Decaluwé. 1992. Impact of economic policies in a credit rationed economy: Illustration with a real and financial CGE model applied to Rwanda. Department of Economics, Laval University, Canada.

Ortega, J. M., and W. C. Rheinboldt. 1970. *Iterative Solution of Nonlinear Equations in Several Variables.* New York: Academic Press.

Pardalos, P. M., and J. B. Rosen. 1987. *Constrained Global Optimization: Algorithms and Applications.* Berlin: Springer-Verlag.

Parikh, K. S., G. K. Fischer, K. Froherg, and O. Gulbrandsen. 1987. *Towards Free Trade in Agriculture.* Dordrecht: Martinus Nijhoff.

Pereira, A., and J. Shoven. 1988. Survey of dynamic computational general equilibrium models for tax policy evaluation. *Journal of Policy Modeling* 10: 401–36.

Perroni, C., and T. H. F. Rutherford. 1995. Regular flexibility of nested CES functions. *European Economic Review* 39: 335–43.

Phimister, E. 1993. *Savings and Investment in Farm Households.* Aldershot: Avebury.

Piggott, J., and J. Whalley. 1980. Economic effects of U.K. tax subsidy policies: A general equilibrium appraisal. Mimeo.

Polemarchakis, H. M. 1990a. Competitive allocations when the asset market is incomplete. *The Geneva Papers on Risk and Insurance* 15: 5–16.

Polemarchakis, H. M. 1990b. The economic implications of an incomplete asset market. *American Economic Review Papers and Proceedings* 80: 280–283.

Powell, A. A., and F. H. Gruen. 1968. The constant elasticity of transformation production frontier and linear supply system. *International Economic Review* 9: 315–28.

Powell, A. A., and T. Lawson. 1990. A decade of applied general equilibrium modelling for policy work. In L. Bergman, D. Jorgenson, and E. Zalai, eds., *General Equilibrium Modeling and Economic Policy Analysis.* Cambridge, MA: Basil Blackwell.

Prescott, E. C., and R. E. Lucas, Jr. 1972. A note on price systems in infinite dimensional space. *International Economic Review* 13: 416–22.

Pyatt, G., and J. I. Round. 1979. Accounting and fixed price multipliers in a social accounting matrix framework. *Economic Journal* 89: 850–73.

Pyatt, G., and J. I. Round, eds. 1985. *Social Accounting Matrices. A Basis for Planning.* Washington: World Bank.

Radner, R. 1972. Existence of equilibrium of plans, prices and price expectations. *Econometrica* 36: 31–58.

Radner, R. 1982. Equilibrium under uncertainty. In K. J. Arrow and M. D. Intrilligator, eds., *Handbook of Mathematical Economics*, vol. 2. Amsterdam: North-Holland.

Rios-Rull, J. V. 1995. Models with heterogeneous agents. In T. F. Cooley, ed., *Frontiers of Business Cycle Research.* Princeton: Princeton University Press.

Roberts, K. 1980. Interpersonal comparability and social choice theory. *Review of Economic Studies* 47: 421–39.

Robinson, J., and J. Eatwell. 1973. *An Introduction to Modern Economics.* New York: McGraw-Hill.

Robinson, S. 1989. Multisectoral models. In H. Chenery and T. N. Srinivasan, eds., *Handbook of Development Economics*, vol. 2. Amsterdam: North-Holland.

Rodgers, G. 1975. Nutritionally based wage determination in the low income labour market. *Oxford Economic Papers* 28: 185–207.

Romer, P. 1991. Increasing returns and new developments in the theory of growth. In W. Barnett, B. Cornet, C. d'Aspremont, J. Gabszewicz, and A. Mas-Colell, eds., *Equilibrium Theory and Applications.* Cambridge: Cambridge University Press.

Rosen, S. 1974. Hedonic prices and implicit markets: Product differentiation in pure competition. *Journal of Political Economy* 82: 34–55.

Rosenzweig, J. A., and L. Taylor. 1990. Devaluation, capital flows, and crowding out: A CGE model with portfolio choice for Thailand. In L. Taylor, ed., *Socially Relevant Policy Analysis.* Cambridge: MIT Press.

Rotemberg, J., and M. Woodford. 1989. Dynamic general equilibrium models with imperfect product markets. In T. F. Cooley, ed., *Frontiers of Business Cycle Research.* Princeton: Princeton University Press.

Rutherford, T. 1982. Computing general equilibrium in a complementarity format. NORGE: A general equilibrium model for the Norwegian economy. Department of Operations Research, Stanford University.

Rutherford, T. 1994. Applied general equilibrium modeling with MPSGE as a GAMS-subsystem. In T. Rutherford, ed., *The GAMS/MPSGE and GAMS/MILES User Notes.* Washington: GAMS Development Corporation.

Ryan, S. M., and J. C. Bean. 1989. Degeneracy in infinite horizon optimization. *Mathematical Programming* 43: 305–15.

Salauddin Ahmad, W. van der Geest, M. A. Keyzer, and M. K. Mujeri. 1986. The applied general equilibrium model for the third five-year plan of Bangladesh. *Bangladesh Journal of Political Economy* 7: 35–99.

Salvatore, D., ed. 1993. *Protectionism and World Welfare*. Cambridge: Cambridge University Press.

Samuelson, P. A. 1954. The pure theory of public expenditures. *Review of Economics and Statistics* 36: 387–89.

Sandee, J. 1960. *A Demonstration Planning Model for India*. Bombay: Asia Publishing House.

Sato, R. 1967. A two-level constant elasticity of substitution production function. *Review of Economic Studies* 34: 201–18.

Sato, R. 1975. *Production Functions and Aggregation*. Amsterdam: North-Holland.

Sato, R. 1976. Self-dual preferences. *Econometrica* 44: 1017–32.

Scarf, H. 1967. The approximation of fixed points of a continuous mapping. *SIAM Journal of Applied Mathematics* 15: 1328–43.

Scarf, H., with T. Hansen. 1973. *The Computation of Economic Equilibria*. New Haven: Yale University Press.

Scarf, H., and J. Shoven, eds. 1984. *Applied General Equilibrium Analysis*. Cambridge: Cambridge University Press.

Sen, A. K. 1963. Neo-classical and neo-Keynesian theories of distribution. *Economic Record* 39: 53–64.

Shafer, W., and H. Sonnenschein. 1982. Market demand and excess demand functions. In K. J. Arrow and M. D. Intrilligator, eds., *Handbook of Mathematical Economics*, vol. 2. Amsterdam: North-Holland.

Shoven, J. B. 1972. A proof of existence of a general equilibrium with ad valorem commodity taxes. *Journal of Economic Theory* 8: 1–25.

Shoven, J. B., and J. Whalley. 1972. A general equilibrium calculation of the effects of differential taxation of income from capital in the U.S. *Journal of Public Economics* 1: 281–322.

Shoven, J. B., and J. Whalley. 1974a. General equilibrium with taxes: A computational procedure and existence proof. *Review of Economic Studies* 40: 475–89.

Shoven, J. B., and J. Whalley. 1974b. On the computation of competitive equilibrium on international markets with tariffs. *Journal of International Economics* 4: 341–54.

Shoven, J. B., and J. Whalley. 1984. Applied general equilibrium models of taxation and international trade. *Journal of Economic Literature* 22: 1007–51.

Shoven, J. B., and J. Whalley. 1992. *Applying General Equilibrium*. New York: Cambridge University Press.

Sierra, H. F. 1982. Price rigidities in a general equilibrium model for the Mexican economy. Ph. D. dissertation. Stanford University.

Smale, S. 1976. A convergent process of price adjustment and global Newton methods. *Journal of Mathematical Economics* 3: 1–14.

Smith, A., and A. Venables. 1988. Completing the internal market in the European Community. *European Economic Review* 32: 1501–25.

Sonnenschein, H. 1972. Market excess demand functions. *Econometrica* 40: 549–63.

Sonnenschein, H. 1973. Do Walras' identity and continuity characterize the class of community excess demand functions? *Journal of Economic Theory* 6: 345–54.

Srinivasan, T. N. 1989. Food aid: A cause of development failure or instrument for success? *World Bank Economic Review* 3: 39–65.

Srinivasan, T. N., and J. Whalley, eds. 1986. *General Equilibrium Trade Policy Modeling*. Cambridge: MIT Press.

Stadler, G. W. 1994. Real business cycles. *Journal of Economic Literature* 32: 1750–83.

Stiglitz, J. 1976. The efficiency wage hypothesis, surplus labour and the distribution of income in LDC's. *Oxford Economic Papers* 28: 185–207.

Stoker, T. M. 1993. Empirical approaches to the problem of aggregation over individuals. *Journal of Economic Literature* 31: 1827–74.

Stokey, N. L., and R. E. Lucas with E. C. Prescott. 1989. *Recursive Methods in Economic Dynamics*. Cambridge: Harvard University Press.

Strauss, S., and D. Thomas. 1995. Human resources: Household decisions and markets. In J. Behrman and T. N. Srinivasan, eds., *Handbook of Development Economics*, vol. 3. Amsterdam: North Holland.

Svejnar, J. 1986. Bargaining power, fear of disagreement, and wage settlements: Theory and evidence from U.S. industry. *Econometrica* 54: 1055–78.

Tabor, S. R. 1989. *Price and Quantity of Rice in Java*. Amsterdam: Free University Press.

Takayama, A. 1985. *Mathematical Economics*, 2d ed. Cambridge: Cambridge University Press.

Taylor, J. B., and H. Uhlig. 1990. Solving nonlinear stochastic growth models: A comparison of alternative solution methods. *Journal of Business and Economic Statistics* 8: 1–17.

Taylor, L. 1979. *Macro Models for Developing Countries*. New York: McGraw-Hill.

Taylor, L. 1983. *Structuralist Macroeconomics*. New York: Basic Books.

Taylor, L. 1990. Structuralist CGE models. In L. Taylor, ed., *Socially Relevant Policy Analysis*. Cambridge: MIT Press.

Taylor, L., and F. J. Lysy. 1979. Vanishing income redistributions: Keynesian clues about model surprises in the short-run. *Journal of Development Economics* 6: 11–30.

Thorbecke, E., in collaboration with B. Kim, D. Roland-Holst, and D. Berrian. 1992. A computable general equilibrium model integrating real and financial transactions. In E. Thorbecke, ed., *Adjustment and Equity in Indonesia*. Paris: OECD Development Center.

Tirole, J. 1988. *The Theory of Industrial Organization*. Cambridge: MIT Press.

Tobin, J. 1969. A general equilibrium approach to monetary theory. *Journal of Money, Credit and Banking* 1: 15–29.

Tobin, J., and D. D. Hester. 1967. *Studies of Portfolio Behaviour*. New York: Wiley.

Trzeciakowski, W. 1971. Systems of indirect management in a planned economy. Cowles Foundation Discussion Paper 312. Yale University, New Haven.

Uhlig, H. 1995. A toolkit for analyzing nonlinear dynamic stochastic models easily. CentER Discussion Paper 9597. Tilburg University, Tilburg.

van der Laan, G., and A. J. J. Talman. 1980. An improvement of fixed point algorithms by using a good triangulation. *Mathematical Programming* 18: 274–85.

van der Laan, G., and A. J. J. Talman. 1985. Adjustment processes for finding economic equilibria. In G. van der Laan and A. J. J. Talman, eds., *The Computation and Modelling of Economic Equilibria*. Amsterdam: North-Holland.

van Wijnbergen, S. 1982. Stagflationary effects of monetary stabilization policies: A quantitative analysis of South Korea. *Journal of Development Economics* 10: 133–169.

Varian, H. 1982a. The nonparametric approach to demand analysis. *Econometrica* 50: 945–73.

Varian, H. 1982b. The nonparametric approach to production analysis. *Econometrica* 52: 579–97.

Varian, H. 1992. *Microeconomic Analysis*. New York: Norton.

Whalley, J. 1985. *Trade Liberalization among Major World Trading Areas*. Cambridge: MIT Press.

Whalley, J., and R. Wigle. 1991. Cutting CO2 emissions: The effects of alternative policy approaches. *Energy Journal* 12: 109–24.

Wonham, M. 1979. *Linear Multivariate Control: A Geometric Approach*. New York: Springer-Verlag.

Index

Absolute price level, 459–60
AGE. *See* Applied general equilibrium (AGE)
Agents
 classifying, 117
 types, 12
Aggregation
 over consumers, 79–81, 117
 over commodities, 78
Anticipation stage
 alternative specifications, 397–98, 404–406
 market-clearing stage and, 391–93, 425
 representations of, 408–16
Applied general equilibrium (AGE), 92
 calibration, 120–22
 classifying agents for, 117
 classifying commodities for, 116–17
 construction steps, 115–16
 data organization, 118–19
 format selection, 119–20
 functional forms selection, 119–20
 parameter estimation, 120–22
 policy analysis with, 128–29
 simulations for, 128
Applied modeling, 115–29
 externalities and, 345–47
 under finite-horizon dynamics, 383, 309, 318, 320, 324
 under imperfect competition, 417–18
 under infinite-horizon dynamics, 288, 324
 international trade in, 145–46
 money and financial assets with, 460, 461–63
 nonconvexities with, 386
 price rigidities with, 218–25
 taxes, tariffs, and quotas in, 183–88
Average cost
 pricing, 368–69
 schedules, 53–55

Benassy equilibrium, 208–11
Benassy rationing, 194, 198–99, 208–11, 423
Bertrand-Nash behavior, 391
Bertrand paradox, 405
Bertrand price setting, 405
Brouwer's fixed point theorem, 6–8, 477
Buffer stocks
 equilibrium with, 196
 operations, 199
 SAM structure with, 226, 227
Boundary point, 465
Bounded set, 465
Brouwer fixed point theorem, 477

Calibration, AGE model, 120–22
Capital stocks, non-firm, 260
Cash-in-advance model, 427–36
Centrally held stocks, 195–98, 215–16
CGE. *See* Computable general equilibrium (CGE)
Chamberlin price setting, 405406
Classifying
 agents, 117
 commodities, 419–20
Closed economies, 102, 357–58
Closed set, 465
Closure rules, 111–12
Collusion, in imperfect competition, 407
Commodities
 classifying, 116–22
 demands and supplies, 1–2
 incomplete asset markets and, 455–56
 indivisible (*see* Indivisible commodities)
 missing, 122–23
 non-rival, in consumption, 325–36
 not traded, 125
 traded, not fully represented, 125–26
Commodity markets, restricted access and, 437–38
Compact set, 465
Competitive equilibrium, 1. *See also* General competitive equilibrium
 assumptions, 1518
 open economy format, 104
 under small-country assumption, 143–44
 with transfers, 22
Composition, 468, 477
Computable general equilibrium (CGE), 92
 assumptions, 114
 dynamics in, 255–57
 format, 107–12
 GAMS application, 129–38
 international trade in, 145–47
 markups and bounds, 126–27
 nonconvexities, 384–86
 price rigidities in, 226
 taxes and tariffs, 183ff
Computation of equilibrium, 14–15, 479–92
Concave production function, 351, 353
Conjectural variations, in imperfect competition, 407–408
Consumer behavior
 in applied models, 7481
 assumptions, 6165
 in competitive equilibrium, 59
 decision dynamics, 117, 235–39, 261
 in perfect competition, 23
Consumer rationing, 198–202, 216–18, 226
Continuation path, 484
Continuous, 469
Convex program, 470–72
Convex set, 465
Cooperation, in imperfect competition, 407
Cost function
 concave-convex, 370–71
 input demand and, 48–51
Cost minimization, by firms, 366–67
Cournot duopoly, 388–91, 393, 404, 424
Cournot-Walras model, 388

Demand schedule, of customers, 387
Differential equations, solving, 483–88
Direct taxes. *See* Lump-sum transfers

Drèze equilibrium, 201–208
Drèze rationing, 194, 198–99, 201–208
Dual approach, vs. primal approach, 55, 74–75
Dynamic implications, of market equilibrium conditions, 239–40
Dynamics, elements of, 229–30
Dynastic models, 267
competitive equilibrium, 276–78
Negishi program, 267–70
vs. OLG model, 319–21

Economies of scope, 374–76
Efficiency
of equilibrium, 452–55
equity and, 18–29
Efficiency wages, 341–45
nonconvexity and, 379–84
Empathy, 332–36
Endowments
assumptions, 16
CGE model, 108
consumer behavior and, 61
consumption dependent, 341–45
increasing returns of, 379–81
Envelope theorem, 475
Equity, efficiency and, 1829
Exact penalty function, 469
Excess demand equilibrium
assumptions, 45, 243–47
computation, 13–14
definition of, 2
existence proofs, 5–10
general competitive equilibrium and, 91
general properties, lack of, 12–13
multiplicity, 10–12
vs. other formats, 112–13
Excess demand function, 10, 11, 13
Exchange rates, 434, 460
Expenditure function, 67–68
Expenditure minimization, 67
Externalities, 325–49
Extreme point, 465

Fiat money, 434–35
Financial assets, 461
Finite-horizon approximation
in applied modeling, 321–22
dynamics, 229–66
dynastic models, 283–85
OLG models, 309–310
OLG models with production, 318–19
Firms
technology of, 36–41
variable number of, 419–20
First welfare theorem, 19–20
Fixed-point mapping, 94, 113, 114, 204, 476–78
Fixed prices, in CGE model, 218–23
Free entry, 403
Full-format, 97–101

Functional forms, 55–59, 75, 119–20
Functional separability, 59, 75–76

General algebraic modeling system (GAMS), 127
accounting, 510–14
CGE model, basic application of, 129–38
computation of equilibrium, 509–10
listing of results, 521–28
model definition, 493–509
nonlinear programming in, 488–92
report writing, 514–19
Government
general equilibrium and, 173
taxation and, 157–62
Gross substitutability, 481

Heterogeneous commodities, 151–52
Hicksian demand, 67–68
Homogeneity, 16
Homotopy path, 486

Imperfect competition, 387–425, 417
Import quotas, 159
Inaction, possibility of, 37
Incomplete markets
asset, 436–58
with firms, 455–57
Increasing, 465
Indeterminacy, of equilibrium, 278–79, 307, 317
Indirect taxes
on consumption, 159–60, 163–68, 174–78
on intermediate demand, 168–70, 178–79
on production, 178–79
in small open economy, 179–80
vs. direct taxes, 157–58
Indirect utility function, 68–70
Infinite-horizon models. *See* Dynastic models; Overlapping generations models (OLG)
Infinitesimally small agents, 382–84
Input demand, 48–50, 51
Institutions, nonconvexities and, 362–63
Integer-valued jobs, 381
Integrability
of demand functions, 72
recovering consumer behavior via, 71–72
Interdependence, in production, 339–41
Interdependent consumptions, 336–39
Interdependent utilities, 332–33
International linkage, 141–42
International trade
in CGE model, 145–47
competitive equilibrium model and, 137–38
country model with, 153–55
dynamics in, 249–50
in general equilibrium model, 138–43
model specification issues, 147–53
quotas, 178–79
tariffs, 178–79
welfare implications, 143–44

Inverse demand functions, 396–97
Investment, 257–59

Kakutani fixed point theorem, 477

Lagrange inequalities, 448
Lagrange multipliers, 24
Leader-follower equilibrium, 407
Learning-by-doing, 374–76
Lindahl equilibrium, 329–32
Linear activity analysis, 39–41
Locally nonsatiated, 471
Lump-sum transfers
 allocations and, 162
 distribution, 160–61
 proportional taxes and, 343–44
 sidies vs. taxes, 25
 vs. indirect taxes, 157–58

Market-clearing stage, anticipation stage and, 391–93, 425
Markets
 for characteristics, 124–25
 clearing stage, 391–93, 404, 406–08, 425
 commodity, restricted access and, 437–38
 incomplete (*see* Incomplete markets)
 missing, 122–23, 344–45
Markup
 advantages, 411
 bounds and, 126–27
 CGE model, 112
 disadvantages, 421
 market clearing of arbitrary allocations, 411–12
 objective, 414–16
 pricing, equilibrium in, 413–14
Marshallian demand, 6870, 82
Maximum theorem, 468
Measure zero, 482
Money
 cash in advance, 427–36
 fiat, 434–35
 supply, flexible, 428–30
 transaction demand for, 460–61, 462–63
Monotone, 466, 480
Monotonic, 466
Multiplicity, of equilibrium, 10–12, 96–97, 278–79, 479–83
Multistage games, 422–23

Nash equilibrium, 406
Negative externalities, pollution and, 345–47, 348
Negative returns, 443–44
Negishi equilibrium
 fixed discount factor β, 270–71
 format, 92–97, 113, 119, 424
 with recursive utility function, 272–76
 subjective demand approach, 408–11
Negishi theorem, 22, 92–96
Nonarbitrage, 452–55
Nonconcave utility functions, 377–78
Nonconvexity
 at above-firm level, 372–76
 CGE applications, 384–86
 in consumption, 376–84
 firm-level, 355–72
 in production, 351–52
 within-firm, 352–55
Nonconvex program, 478–79
Nondecreasing, 466
Nonhomogeneities, 459–60, 462–63
Nonlinear complementarity problems, 127, 488–92
Non-rival commodities, in consumption, 325–36
Nonsatiated, 466
Normalization, 402–03
Number of agents, 261–63

Objective markups, 414–16
OLG models. *See* Overlapping generations models (OLG)
Open economy
 competitive equilibrium, 104
 extensions of, 106–107
 format, 101–107, 114, 358
Overlapping generations models (OLG), 324
 infinite horizon dynamics in, 300
 with production, 310–19
 of pure exchange, 291–310
 vs. dynastic models, 319–21

Parameter estimation, 120–22, 226
Parametric convex program, 472–76
Pareto-efficiency, 29
 definition of, 18
 first welfare theorem and, 19–20
 second welfare theorem and, 24–25
 of welfare optimum, 21–22
Partial equilibrium, 388–93
Perfect competition, 2
Perturbation theorem, 475
P-matrices, 481
Policy analysis, 182–83
Pollution, negative externalities and, 345–47, 348
Positive externalities, 325
Price setting
 Bertrand, 405
 Chamberlin, 405–406
 Cournot, 404
Primal approach, vs. dual approach, 55, 74–75
Producer behavior
 in applied models, 35–36, 55–59
 definition of, 2–3
 testing, 73–74
Producer decisions
 dynamics in, 232–35
 investment in, 257–59
 through profit functions, 259

Producer rationing, 213–15, 216–18
Product differentiation, 417–19
Production, nonconvexity, 351–52
Production function, 44–45
Production sets, 16, 37, 38–39, 108
Profit function, 45–48, 51, 259
Profit maximization, 41–43, 360, 387

Quotas, 159, 162–73

Real business cycle (RBC) model, 285–91
Recursive utility function, 109–10, 272–78
Reform
 with compensation, 27
 national taxes, 190–91
 without compensation, 27–29
Research, development and, 372–74
Restricted profit function, 48
Revealed preference, 71–72
Rough markets, 123–24

SAM. *See* Social accounting matrix (SAM)
Sard theorem, 482
Scale, returns to
 constant, 51–52
 decreasing, 52–53, 110–11
 increasing, 53, 363–65
Second welfare theorem, 24–25
Sets, and functions, 465–70
Setup costs, 356–57, 361–63
Shephard's lemma, 50
Simplex, 477
Simulations
 AGE model, 116, 127–28
 dynamics, 266, 322–24
 international trade and, 155
 price rigidities, 226, 228
 tariffs/taxes and, 188–92
Single-consumer case, 25–26, 219–23
Single-period equilibrium, 247–49, 259–60
Small-country assumption, 142–44
Social accounting matrix (SAM)
 construction of, 225–27
 country model with international trade, 153–55
 data organization, 118–19
 investment with, 263–64, 265
 parameter estimation, 121–22
 savings with, 263–64, 265
Soft rationing, 198–200, 212–13
Standard assumptions, supply and demand under, 81–84
Stationary, 466
Steady state
 infinite-horizon dynamics and, 279–82
 OLG model and, 307–309, 317–18
Stocks
 buffer, 196, 199, 226, 227
 centrally held, 195–98, 215–16
 vs. rationing, 223–25
Strategic consumers, 393–96
Strict convexity, 38
Strictly convex set, 467
Strict (quasi)concave, 466
Strict (quasi)convex, 465, 467
Strict diagonal dominance, 481
Strong (quasi)concave, 466
Strong (quasi)convex, 465, 468
Subdifferential, 467
Subgradients, 467
Subjective demand approach, 408–11
Subsidies, 158
Supply and demand, under standard assumptions, 81–84

Tariffs
 in CGE model, 183–85
 international reform of, 191–92
 international trade, 170–73, 178–79
 reform at national level, 190–91
 SAM with, 187–89
 in small country model, 170–71
 in small open economy, 179–80
 welfare program, 180–81
Tatonnement, 15
Taxes
 in CGE model, 183–85
 direct (*see* Lump-sum transfers)
 dynamics in, 250–51
 government and, 157–62
 indirect (*see* Indirect taxes)
 quota and, in general equilibrium, 159–60, 162–73
 reforms, national, 190–91
 SAM with, 187–89
 virtual, 198–200, 211–12
Technology, of firm, 36–41
Temporary equilibrium
 definition, 240–43
 expectations in, 259
 international trade and, 248–49
 price rigidities and, 250
 proof, 243–47
 single-period, 247–48
 taxes and, 249–50
T-period competitive equilibrium, 254, 297–98
 calibration of, 264
 consumer decision dynamics, 235–39
 definition of, 230–31, 231–32
 estimation of, 264
 producer decision dynamics, 233–35
Trade
 international (*see* International trade)
 with transportation costs, 148–50
 without transportation costs, 147–48
Trade pool approach, 150–51
Trade welfare function, in international trade, 152–53
Transactions demand for money, 460–61, 462–63
Transformation function, 44–45

Two-country case, 142–43
Tychonoff theorem, 479

Uniqueness of equilibrium, 120
Uppersemicontinuity, 472
Utility function
 CGE model, 108–10
 in competitive equilibrium, 16
 consumer behavior and, 62–65
 dynastic, 268–69
 indirect, 68–70
 nonconvexities in, 376–77
 recursive, 109–10, 272–78
 separability of, 76–79
Utility maximization, 65–67

Variable number of firms, 419–20
Virtual taxes, 198–200, 211–12
Vouchers, 225

Walrasian concept, for anticipation stage, 423–24
Weierstrass approximation, 480
Welfare
 in absence of compensation, 181–82, 183
 basic program, 445–47
 decentralization, 357–61, 363–65
 finite horizon dynamics and, 252–53
 gains from reduced taxes, 176–77
 international trade and, 143–44
 loss from consumption tax, 175
 measures, 87–90
 optimum, 20–24, 84–87
 price rigidities and, 215–18
 reform consequences, 26–29
 taxes and, 173–74
Within-firm nonconvexities, 352–55